Distillation Engineering

REINHARD BILLET

translated by

M. WULFINGHOFF

1979
Chemical Publishing Co.
New York, N.Y.

Distillation Engineering

ISBN: 978-0-8206-0215-8

Chemical Publishing Company:
www.chemical-publishing.com
www.chemicalpublishing.net

First edition:

© Chemical Publishing Company, Inc. – New York, 1979

Second Impression:

Chemical Publishing Company, Inc. - 2011

Printed in the United States of America

Preface to German Edition

Distillation and its related separation methods are undoubtedly the most predominant of the basic thermal process engineering operations, and the amount of literature on this subject has greatly increased recently. The need for a uniform summary of the various fields of distillation is apparent. It would not be expedient to extensively cover the whole field of distillation in one book, because of the great number of special problems which exist. A book is therefore needed which is both technical in its approach and orientated towards application, whereby focal points can be seen, in theory, for example, in particular consideration of phase equilibrium, as well as in the construction and planning of distillation installations for industrial application. This book is intended to give an introduction to the latter, while the theory is dealt with only from the viewpoint of practice-related tasks and problems.

The various problems in industrial distillation mostly require individual treatment of the questions connected with calculation and construction. For example, optimization of the process and the installation is now made absolutely necessary by the application of more expensive materials, or the strict observance of particular operating conditions, whether they be the need for minimal product temperature in the distillation equipment of a separation column for the least possible residence time of the product, or a limited heating medium temperature. Particular attention is therefore directed to considering the choice of the optimum separation apparatus for a particular separation problem, and to investigating the optimum operating conditions. Also taken into account is that the chemical decomposition and polymerization of materials require a transfer of the chemical compound separation through distillation in the field of lower pressures, frequently under application of special apparatus.

Where separation column internals are concerned, only those which have prominently stood the test in recent years in the most varied separation processes of the chemical and allied industries and which are also therefore more frequently used, are dealt with in greater detail. In order to make some judgment on them from both cost and process engineering points of view, the result of an extensive study is presented here. This study was practicable only on the basis of numerous experimental investigations. Consequently, an evaluation of separation installations from the points of view of operation, process engineering and cost, became possible.

In addition to the findings obtained both through theory and practice, easily-applied calculation methods are suggested. Where possible, representation and utilization have been simplified if the effect of such simplification on the results obtained is unimportant when practically applied.

Where it is of importance for the construction of industrial distillation installations, methods for transferring results received either directly, or through simulation, to full-scale installations, are detailed and shown during the course of laboratory distillation.

In the section dealing with accessories for distillation installations, measurement and control equipment is also detailed. In order to give some examples of application, the suitability of various types of control for a separation column is explained through the discussion of superior aspects.

The use of computers for industrial application is discussed only in so far as it is relevant to the object of this book.

The numerical examples of particular chapters have been selected so that the material presented can be better understood, and to illustrate the application of this material in the construction and operation of distillation installations.

Ludwigshafen, July 1972 R. Billet

Preface to American Edition

In the past, the theory of distillation and rectification has appeared in great detail in the form of both single publications and monographies. However, until now, this particular field has lacked a book on thermal separation processes which concentrates on the practical application, design and planning of total plants from an industrial point of view. It was therefore the author's aim in writing "Distillation Engineering" to cover the industrial side of distillation and rectification. Following publication of the German edition in 1973, the great demand for this book confirmed that this intention was right.

The increasing use of distillation apparatus and plants in the field of technology and environmental protection, particularly in the chemical and allied industries, justified the inclusion in the American edition of a complete chapter concerning distillation technology in environmental protection. Furthermore, the author also felt it necessary to discuss Vacuum Distillation in greater detail. In addition to this, certain chapters have been extended in this edition in order to include new industrially significant findings which have been revealed in the light of further developments during the last five years.

Bochum, June 1978 R. Billet

Table of Contents

I The Thermal Separation of Liquids

The field of distillation engineering, as presented here, is comprised of the currently prevalent practices for thermal splitting as based on known facts of simple distillation theory. Many chemical reactions supply liquid or gas mixtures that have to be decomposed by heat. Some mixtures occurring in nature must be broken down to recover specific constituents, such as aromatics, petroleum distillates serving as fuels, air liquefied to produce nitrogen, oxygen and rare gases, or water distilled for use in nuclear installations.

Obviously, the many uses of distillation justify a broad coverage of its fundamentals.

The primary operations in the thermal separation of liquids are distillation and rectification[3]. While simple distillation will generally achieve a rough separation of constituents, because the vapors arising from a mixture will have a composition different from that of the mother liquid, rectification provides a better separation and a recovery of virtually pure components.

Essentially, in rectification the vapor mixture generated by distilling a liquid mixture is passed counter to a stream of condensed vapor in such a way that an intimate contact of the liquid and vapor phases occurs. The exchange of materials thus brought about will effectively achieve a decomposition of the mixture.

In accordance with this, there will have to be an evaporator to supply the vapor mixture. The mixture will go to a rectifier where its components will separate, after

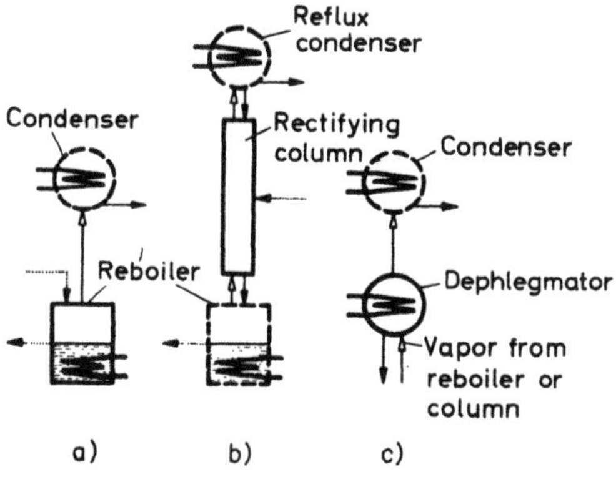

Fig 1.1 is a schematic presentation of separations by distillation; a) Partial distillation, b) Rectification, c) Partial condensation.

Fig. 1.2 shows the corresponding equipment as a part of an up-to-date chemical production plant.

which a condenser will liquefy the low-boiling fractions from the rectifier. Part of the condensate formed is recycled as reflux. Part is recovered as product, and part is reused to serve as coolant in the final cooling stage. The residue from rectification is the high-boiling fraction; this may undergo reworking for secondary products if desired.

Splitting up of mixed vapors by means of partial condensation is of secondary significance. It may be regarded as partial distillation in reverse. It may be combined with the aforementioned separation method by allowing only that part of the vapor needed for refluxing to be liquefied, while the balance of vapor is recovered in a secondary, subordinate condenser.

Such separations may take place on a continuous or discontinuous basis. Both methods will be discussed after the physical fundamentals have been reviewed. Furthermore, there is a semi-continuous mode of operation which results from the origin of the mixture on the one hand, and its integration with the separation scheme on the other.

While rectifiers of the kind under discussion work adiabatically, a nonadiabatic mode of operation is available for specific separations, characterized by a lateral vaporization and condensation superimposed on a counterflow scheme for the liquid and vapor phases.

Finally, a direct, partial-condensation process may serve to achieve an effectual separation.

II Thermodynamics of Mixtures

The devices of thermodynamics and physical chemistry essential for an understanding of distillation problems and for finding a solution by calculation comprise: (a) Dalton's Law, in combination with the equation of state for perfect gases; (b) Raoult's Law, as derived from the Gibbs-Duhem-Margules equation; and (c) the notion of the relative volatility of a mixture. Departures of the actual performance of mixtures from the ideal should be noted; they require the use of activities and activity coefficients in the representation of evaporation processes.

In the past, many investigations have been undertaken into the complex interrelations of phase equilibria for real binary and multicomponent systems. They frequently permit the fixing of the boiling performances of mixtures by calculations or by a combination of calculations and tests with a minimum of laboratory equipment. Nevertheless, industrial practice still prefers straight experimentation, especially when the mixture contains novel components whose properties are not well known.

1. DEFINITIONS AND RELATIONSHIPS

A. Separability of a Liquid Mixture

The *relative volatility* is a direct index of the tendency of a liquid mixture to separate upon distillation or rectification. In general, the low boiler will evaporate more readily than the high boiler, and the vapor mixture in equilibrium with a boiling liquid mixture will have a higher percentage of the low boiler than the liquid itself. The degree of enrichment of the low boiler in the vapor depends on the relative volatility α. In a binary mixture

$$\alpha_{12} = \frac{\dfrac{y_1}{x_1}}{\dfrac{y_2}{x_2}} = \frac{y_1 x_2}{y_2 x_1} \tag{2-1}$$

where y_1 (mol-%) equals concentration of low boiler, and y_2 (mol-%) equals concentration of high boiler in the vapor mixture at phase equilibrium with the liquid having a composition of x_1 (mol-%) of the low boiler and x_2 (mol-%) of the high boiler.

A high relative volatility indicates easy separation. In general, α_{12} varies with the boiling pressure and hence with the boiling temperature. Consequently, varying the pressure will affect the boiling equilibrium of a multicomponent mixture.

For binary mixtures,

$$x_2 = 100 - x_1 \tag{2-2}$$

$$y_2 = 100 - y_1 \tag{2-3}$$

and eq. (2-1) will supply

$$y_1 = \frac{\alpha_{12} x_1}{100 + x_1(\alpha_{12} - 1)} 100 \tag{2-4}$$

If the relative volatility of a binary mixture is known, the vapor concentrations y_1 corresponding to various liquid concentrations can be stated. The relative volatility can be determined readily for ideal mixtures only and the vapor curves of the single components must be known. In general, the relative volatility is ascertained by test.

For the virtually ideal mixture of benzene and toluene, with benzene as the low boiler, Fig. 2.1 represents the relative volatility as functions of the liquid concentrations for different pressures. Note that pressure changes affect the value greatly,

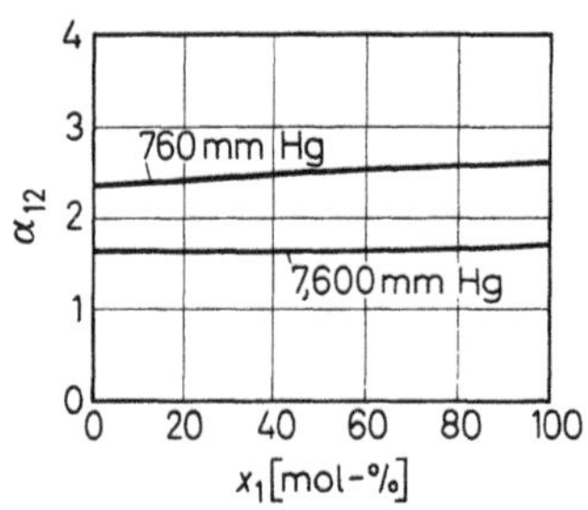

Fig. 2.1 Relative volatility α_{12} of the benzene-toluene mixture for two different distillation pressures.

while the concentration effect is much less. However, this applies only for ideal mixtures with boiling temperatures not too far apart. Still, there are liquid mixtures with two concentration ranges characterized by relative volatilities in one range exceeding unity, while in some other range remaining below unity. Liquid mixtures characterized by their relative volatilities being below unity will yield, when distilled, vapors with a low-boiler concentration below that of the original mixture. For a relative volatility exceeding unity, the concentration of low boiler in the vapor will exceed that in the liquid. If, however, the relative volatility becomes unity, it will be seen from eq. (2-4) that the composition of the vapor will be the same as that of the liquid—simple distillation or rectification cannot produce separation. The unique concentration yielding a relative volatility equal to unity is termed *azeotropic*. Azeotropic mixtures cannot, in general, be separated beyond the azeotropic point that appears at a specific concentration regardless of pressure (usually).

Example 1 How much benzene is contained in a benzene-toluene vapor mixture at phase equilibrium with a benzene mixture where $x_1 = 30$ mol-% benzene? From Fig. 2.1, the relative volatility of the virtually ideal mixture at constant boiling pressure is almost independent of the composition of the mixture; at atmospheric pressure $\alpha_{12} = 2.4$.

Answer Eq. (2-4) yields, for the above data, a concentration of benzene in the vapor of

$$y_1 = \frac{2.4 \cdot 30}{100 + 30(2.4 - 1)} \cdot 100 = 50.7 \text{ mol-\%}$$

B. Partial Pressures in Vapor Mixtures

In a multicomponent vapor mixture, let $p_1, p_2, p_3, \ldots p_n$ mm Hg = partial pressures of components; $y_1, y_2, y_3, \ldots y_n$ mol-% = corresponding concentrations in vapor. Then, according to Dalton, the total pressure p (mm Hg) of the mixture equals the sum of the partial pressures p_c ($c = 1, 2, 3, \ldots n$) of the individual components

$$p_1 + p_2 + p_3 + \ldots p_n = \sum p_c = p \qquad (2\text{-}5)$$

Using the equation of state for ideal gases, the partial pressure for component c, p_c, and the total pressure, p, would be:

$$p_c = y_c \frac{KT}{v} \qquad (2\text{-}6)$$

$$p = 100 \frac{KT}{v} \qquad (2\text{-}7)$$

where:

K = the gas constant
T = the absolute temperature (K)
v = the volume of 100 mol of the mixed vapor, (m^3).

Writing the ratios of partial pressure p_c and total pressure p will yield Dalton's partial pressure relations[1].

$$y_1 = \frac{p_1}{p} 100$$

$$y_c = \frac{p_c}{p} 100 \qquad (2\text{-}8)$$

$$y_n = \frac{p_n}{p} 100$$

This means that the ratio of the individual partial pressures and the total pressure is the same as that of the fractions and the entire number of moles. The mole fraction $y_c/100$ equals the ratio p_c/p (partial pressure over total pressure).

In calculations of this type it is advantageous to describe the quantity of a mixture in terms of kilomoles, and its composition in terms of mole percents. The kilomole figures are obtained by dividing the mass given in kilograms by the molecular weight.

C. Evaporation of Liquid Mixtures

The Gibbs-Duhem-Margules equation establishes a relation between the composition x (in mole-%) of a multicomponent liquid mixture and the partial pressures p (in mm Hg) of the mixed vapor in phase equilibrium with it. For a binary mixture[2-5],

$$x_1 \left[\frac{\partial}{\partial x_1} (\ln p_1) \right]_{p,T} + x_2 \left[\frac{\partial}{\partial x_1} (\ln p_2) \right]_{p,T} = 0 \tag{2-9}$$

Transformation based on eq. (2-2) will yield, letting $\partial x_1 = -\partial(100 - x_1)$:

$$x_1 \left[\frac{\partial}{\partial x_1} (\ln p_1) \right]_{p,T} - \left\{ (100 - x_1) \left[\frac{\partial (\ln p_2)}{\partial (100 - x_1)} \right]_{p,T} \right\} = 0 \tag{2-10}$$

In eq. (2-10) let both terms equal unity, then integrate[3]. By Raoult's Law of partial pressures for ideal mixtures

$$p_1 = \frac{x_1}{100} P_1 \quad \text{and} \quad p_2 = \frac{x_2}{100} P_2 \tag{2-11}$$

P_1 and P_2 (in mm Hg) are the boiling pressures of the pure components at the temperature considered.[3] For ideal mixtures, according to Raoult, the partial vapor pressure $p_1, p_2, \ldots$ of a component over the ideal mixture is less than that over the pure phase by the mole fraction ratio $x_1/100, x_2/100, \ldots$

D. Real Systems

When the simple Raoult's Law is inapplicable to the boiling performance of a mixture, eq. (2-11) must be extended by the "activity coefficients" γ_1 and γ_2

$$p_1 = \frac{\gamma_1 x_1}{100} P_1 \quad \text{and} \quad p_2 = \frac{\gamma_2 x_2}{100} P_2 \tag{2-12}$$

Then, the Gibbs-Duhem-Margules equation will become

$$x_1 \left\{ \frac{\partial}{\partial x_1} [\ln (\gamma_1 x_1)] \right\}_{p,T} + x_2 \left\{ \frac{\partial}{\partial x_2} [\ln (\gamma_2 x_2)] \right\}_{p,T} = 0 \tag{2-13}$$

Van Laar[6] derives

$$\ln \gamma_1 = \frac{K_1}{(1 + K_1 x_1/K_2 x_2)^2} \quad \text{and} \quad \ln \gamma_2 = \frac{K_2}{(1 + K_2 x_2/K_1 x_1)^2} \tag{2-14}$$

The integration constants following from

$$K_1 = \ln\gamma_1 (1 + x_2 \ln\gamma_2 / x_1 \ln\gamma_1)^2 \quad \text{and} \quad K_2 = \ln\gamma_2 \; (1 + x_1 \ln\gamma_1 / x_2 \ln\gamma_2)^2 \tag{2-15}$$

Once the activity coefficients γ_1 and γ_2 are known for one pair of concentrations x_1, y_1 and x_2, y_2, constants K_1 and K_2 can be computed from eq. (2-15) and activity coefficients $\gamma_1 = f(x_1)$ and $\gamma_2 = f(x_2)$ can be found from eq. (2-14).

The boiling performances of real systems may now be characterized by use of the activity coefficients γ, replacing the concentrations x with the activities, that is, the products of activity coefficients and concentrations.

$$a_1 = \gamma_1 x_1 \quad \text{and} \quad a_2 = \gamma_2 x_2 \tag{2-16}$$

2. TYPES OF MIXTURES

A. Ideal Binary Mixtures

The ideal binary mixture obeys Raoult's Law. A component's vaporizing tendency depends on its concentration in the liquid and its vapor pressure at the temperature in question. Fig. 2.2 is a graph of an ideal binary mixture described in eqs. (2-5) and (2-11).

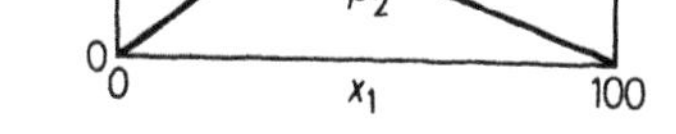

Fig. 2.2 Partial pressures p_1, p_2, and total pressure p for an ideal binary mixture as functions of liquid composition x_1 at constant boiling point t.

Combining Dalton's Law of partial pressures, eq. (2-8), and Raoult's Law, eq. (2-11) yields

$$\frac{y_1}{x_1} = \frac{P_1}{p} \quad \text{and} \quad \frac{y_2}{x_2} = \frac{P_2}{p} \tag{2-17}$$

By eq. (2-1) the relative volatility is equal to the ratio of the boiling pressures of the pure components

$$\alpha_{12} = \frac{P_1}{P_2} \tag{2-18}$$

When the vapor pressure curves P_1 and $P_2 = f(t)$ are known for the pure substances, the relative volatility α_{12} for a constant total pressure p can be determined as a function of the boiling temperature t; note the scheme in Fig. 2.3. From the vapor pressure curves, locate P_1 and P_2 (1 and 2 in the graph) for two or more different boiling temperatures located between t_1 and t_2, and find the corresponding α_{12} values from eq. (2-18). At the extremes of the pure components the boiling pressures equal the total pressure.

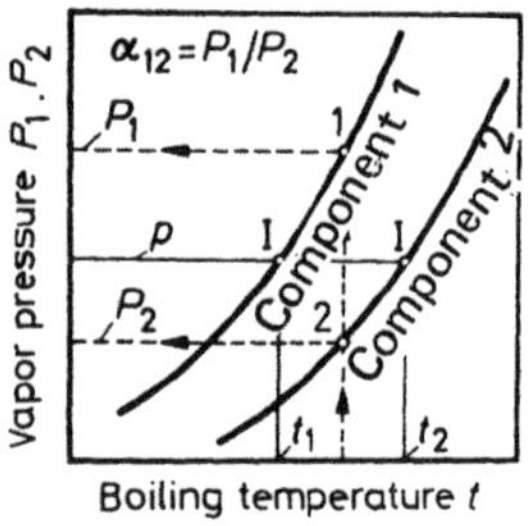

Fig. 2.3 Finding the relative volatility $\alpha_{1,2}$ of an ideal binary mixture from the vapor pressure curves $P_1 = f(t)$ and $P_2 = f(t)$.

For x_1, relations (2-17) and (2-4) yield

$$x_1 = \frac{\frac{p}{P_1}\alpha_{12} - 1}{\alpha_{12} - 1} \cdot 100 \tag{2-19}$$

from which the liquid compositions corresponding to certain boiling temperatures can be found. Eq.(2-4) will disclose the corresponding equilibrium vapor data. Plotting the y_1 values thus determined over the corresponding x_1 values gives the equilibrium

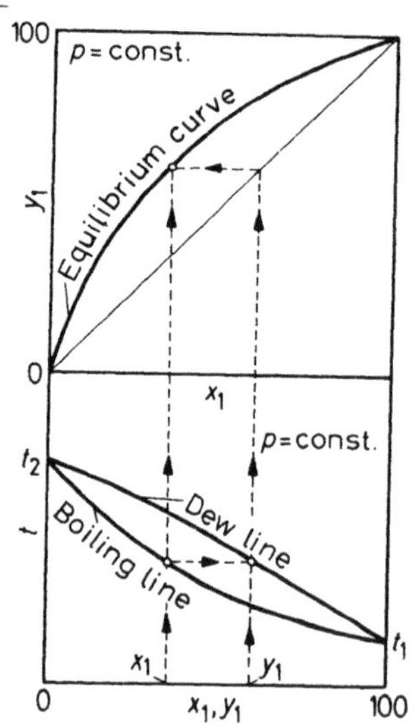

Fig. 2.4 Equilibrium curve $y_1 = f(x_1)$, boiling line $t = f(x_1)$, and dew line $t = f(x_1)$ for an ideal binary mixture.

curve as outlined in the upper part of Fig. 2.4. The vapor concentration y_1 (mol-%), in phase equilibrium with the boiling liquid of composition x_1 (mol-%) at constant p (mm Hg), can be read off readily.

The lower portion of Fig. 2.4 shows the boiling and dew point curves. The former correlates the liquid concentration x_1 and the boiling point, while the latter correlates the vapor concentration y_1 and the condensation point of the mixed vapor. Interrelated concentrations x_1 and y_1 of liquid and vapor for a given boiling or condensing temperature cut off the boiling or dew point curve by a parallel to the abscissa, and will yield a point on the equilibrium curve.

On the whole, the vapor pressure ratio is a function of temperature and therefore variable. However, for ideal mixtures it may be virtually constant across the entire boiling point range and hence over the concentration range, especially when the vapor pressure lines of the individual components are close to each other.

B. Nonideal Binary Mixtures

Raoult's Law for ideal mixtures is actually a special variant of the Gibbs-Duhem-Margules relation. Unfortunately most mixtures do not behave ideally. Because one cannot predict how a given mixture will boil, experimentation is unavoidable. The molecules of the individual components do not perform independently of each other, and their tendency to exchange the liquid phase for the vapor is not always in proportion to the vapor pressure. Although in some cases the boiling equilibrium of nonideal mixtures can be predicted, by way of Van Laar's activity coefficient[6] or by Othmer's method[7], partial pressure and equilibrium curves, and boiling and dewpoint lines are commonly determined by tests.

Of the nonideal mixtures, the ones characterized by an azeotropic point, and hence a reversion of the relative volatility at the azeotropic concentration, will be discussed first. They are the mixtures with a maximum and a minimum vapor pressure. Within a given concentration range, apart from the azeotropic, the one component is the more volatile, while in the opposite range, for the other component of the binary mixtures, the situation is reversed. Also, insolubility and partial solubility will be discussed.

Figs. 2.5 and 2.6 show characteristics of an azeotropic mixture with a vapor pressure maximum or boiling point minimum. Compared with Fig. 2.2, in Fig. 2.5 the lines of the partial pressures lie above the straight partial pressure lines of ideal mixtures. Only in the range of high concentrations of low or high boiling substances do the partial pressure curves enter tangentially into the P_1 or P_2 lines and Raoult's relation becomes applicable. At a certain concentration x_a the total pressure line has a maximum at constant boiling temperature. In Fig. 2.6 that concentration corresponds to the azeotropic point P, at which the vapor has the same composition as the liquid, and at which the relative volatility becomes unity. Simple distillation or rectification cannot effect a separation beyond the azeotropic concentration, and evaporation takes place like that of a pure liquid at a temperature which corresponds to a boiling point minimum in the lower part of Fig. 2.6. A boiling point minimum in a temperature-concentration plot for $p(=\text{constant})$ and is the equivalent of a vapor pressure maximum in a pressure versus concentration plot for $t(=$ constant). To the left of the boiling point minimum, where the relative volatility exceeds unity, the vapor is richer in component 1 than the liquid at phase equilibrium; to the right, component 2 prevails. Mixtures of ethanol and water, propanol and water, benzene and cyclohexane, acetone and water, and carbon disulfide and acetone, for example exhibit minimum boiling points and vapor pressure maxima. For the azeotropic composition, their boiling points are below those of the pure components.

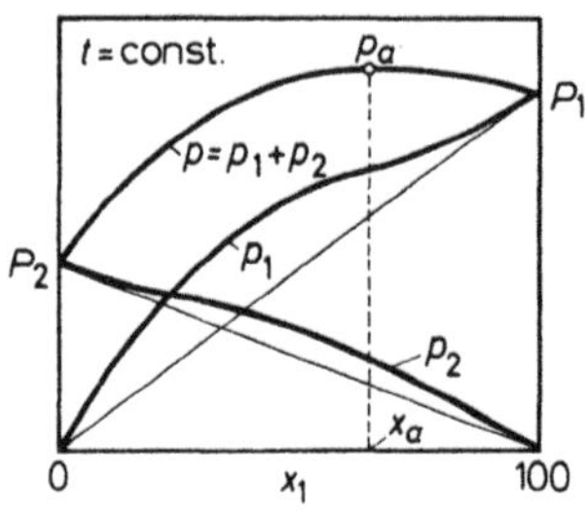

Fig. 2.5 Partial pressures p_1, p_2, and total pressure p of a binary mixture with vapor pressure maximum as a function of liquid composition x_1 at constant boiling point t.

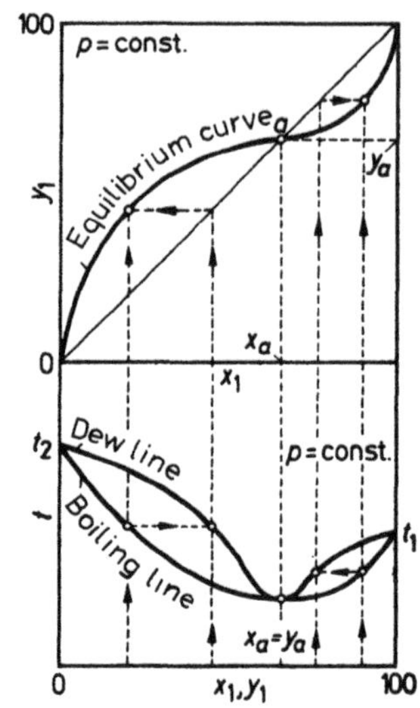

Fig. 2.6 Equilibrium curve $y_1 = f(x_1)$, boiling line $t = f(x_1)$, and dew line $t = f(y_1)$ of a binary mixture with vapor pressure maximum (boiling point minimum).

By contrast, minimum vapor pressures of azeotropic binary mixtures are due to the fact that the partial pressure curves $p_1 = f(x_1)$ and $p_2 = f(x_1)$, at constant boiling temperature, are bent downward as in Fig. 2.7. This results in a boiling point maximum, and the boiling point of the mixture is located above the boiling points of the two pure components (note Fig. 2.8). To the left of azeotropic point P the relative volatility is below unity, and the vapor phase contains less of component 1 than does the liquid phase. To the right of the azeotropic point the relative volatility exceeds unity, and the vapor is richer in the more volatile component than the liquid. Mixtures of water and sulfuric acid, nitric acid and water, hydrochloric acid and water, and other acid-water systems exhibit minimum vapor pressures and maximum boiling points.
In analogy with relations (2-17), for a nonideal mixture

$$\frac{y_1}{x_1} = \gamma_1 \frac{P_1}{p} \quad \text{and} \quad \frac{y_2}{x_2} = \gamma_2 \frac{P_2}{p} \tag{2-20}$$

With this, Eq.(2-1) will yield the relation for the relative volatility

$$\alpha_{12} = \frac{\gamma_1 P_1}{\gamma_2 P_2} \tag{2-21}$$

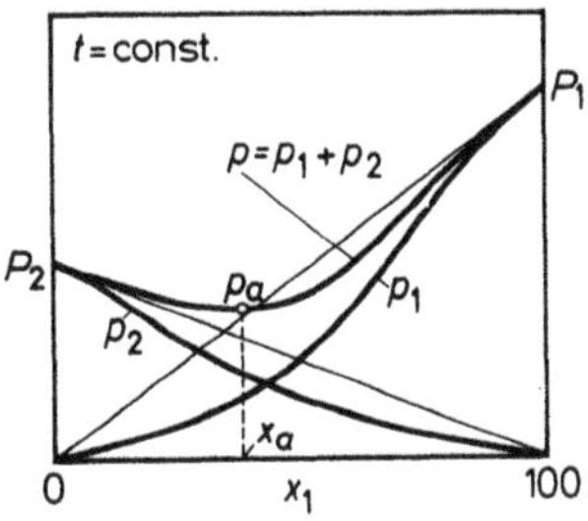

Fig. 2.7 Partial pressures p_1, p_2, and total pressure p of a binary mixture with vapor pressure minimum as a function of liquid composition x_1 at a constant boiling temperature t.

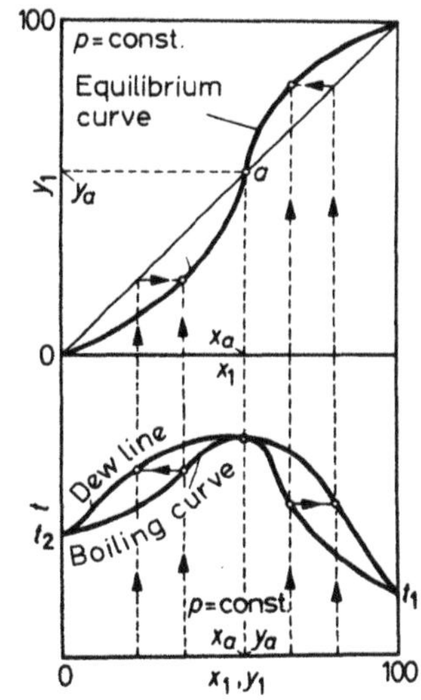

Fig. 2.8 Equilibrium curve $y_1 = f(x_1)$, boiling line $t = f(x)$, and dewline $t = f(y_1)$ of a binary mixture with vapor pressure minimum (boiling point maximum).

and Eq.(2-4) will supply:

$$x_1 = \frac{\dfrac{p\alpha_{12}}{P_1\gamma_1} - 1}{\alpha_{12} - 1} 100 \qquad (2\text{-}22)$$

When, for a nonideal binary mixture, one equilibrium point is defined by x_1, y_1 or x_2, y_2, and the boiling temperature t, as for the azeotropic composition, the boiling pressures P_1 and P_2 of the pure components can be read off the boiling pressure lines. Then the corresponding activities can be eliminated and calculated by means of Eq. (2-20), thus

$$\gamma_1 = \frac{y_1 p}{x_1 P_1} \quad \text{and} \quad \gamma_2 = \frac{y_2 p}{x_2 P_2} \qquad (2\text{-}23)$$

In Section 1.B. of this chapter a method was described for finding the activity coefficients $\gamma_1 = f(x_1)$ and $\gamma_2 = f(x_2)$ as functions of the liquid concentration. Now, the equilibrium curve can be established from Eq.(2-4), (2-21), and (2-22). Select a boiling point located between t_1 and t_2 at p = constant. The boiling pressures P_1 and P_2 can be determined from the vapor pressure curves; after choosing a liquid concentration for x_1 read the corresponding γ_1 and γ_2 from the lines $\gamma_1 = f(x_1)$ and $\gamma_2 = f(x_2)$. Note the upper portion of Fig. 2.9. Eq.(2-21) will then supply a provisional α_1 and Eq.(2-22) will supply a provisional x_1; note point I in the lower part of Fig. 2.9. Repeat this operation for additional x_1 values at the same temperature t. Plot the x_1 values thus computed over the assumed ones; note points I, II, and III. Draw a line through these points. Intersection i of that line with the diagonal gives the x_1 value meeting the desired relationship. From Eq.(2-4) this x_1 will yield the corresponding equilibrium concentration y_1 for the vapor. In this manner, any number of points of the equilibrium curve can be found.

Since, in view of relations (2-20),

$$y_1 + y_2 = \frac{P_1}{p}\gamma_1 x_1 + \frac{P_2}{p}\gamma_2 x_2 \tag{2-24}$$

the equilibrium curve $y_1 = f(x_1)$ may also be determined by assuming a pair of values, (x_1, x_2) corresponding to (γ_1, γ_2) for various boiling points, and reading off the boiling pressures P_1 and P_2 from the vapor pressure lines. Of the temperatures tried, the correct one will be the one yielding

$$y_1 + y_2 = 100 \text{ mol-\%} \tag{2-25}$$

by application of Eq.(2-24). A graphic solution consists in plotting the sums resulting from Eq.(2-24) against the corresponding selected boiling temperatures. The line joining the points thus computed will intersect the parallel to the abscissa for $y_1 + y_2 = 100$ at the boiling temperature belonging to the equilibrium point.

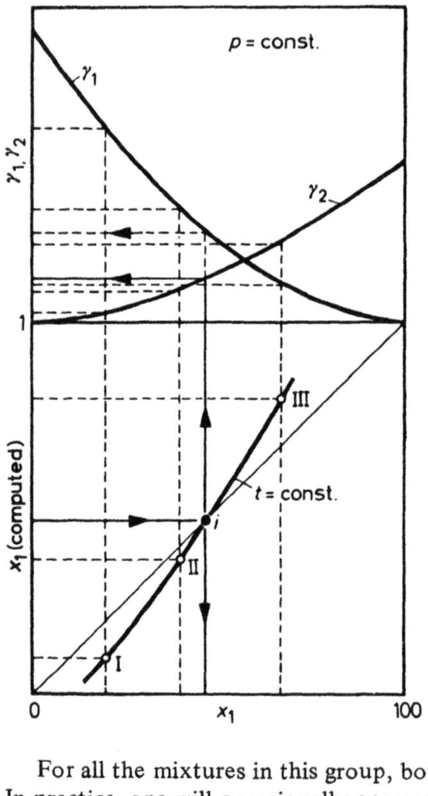

Fig. 2.9 Finding the phase equilibrium of non-ideal binary mixtures

For all the mixtures in this group, both components must be perfectly intersoluble. In practice, one will occasionally encounter almost insoluble mixtures such as benzene and water. When such mixtures are distilled each component will vaporize at its own vapor pressure independently of the other, and the total pressure p becomes the sum of the boiling pressures P_1 and P_2 at the temperature prevailing

$$p = P_1 + P_2 \tag{2-26}$$

In Fig. 2.10 the boiling point t_b for the insoluble binary mixture is always below those of the individual components, and if the vapor pressure curves are known it can

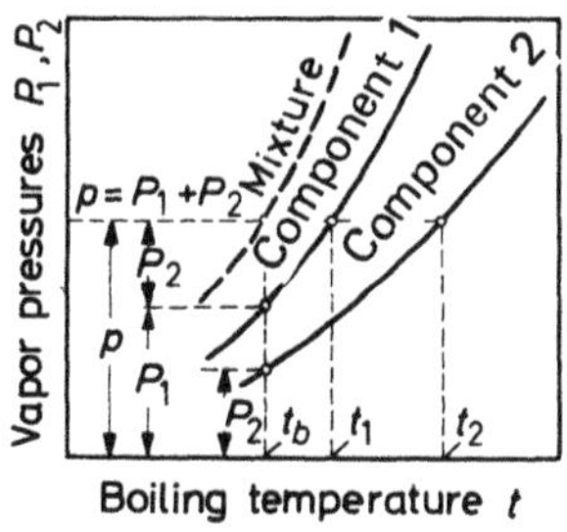

Fig. 2.10 Finding the boiling point t_b for an insoluble binary mixture.

be found readily. Since $p_1 = P_1$ and $p_2 = P_2$ the vapors will always have a constant composition. This is independent of the liquid composition, and can be determined from Eq.(2-8).

$$y_1 = \frac{P_1}{p} \cdot 100 \quad \text{and} \quad y_2 = \frac{P_2}{p} \cdot 100 \tag{2-27}$$

This boiling performance of insoluble mixtures is taken advantage of in the so-called steam distillation process[2,8] to recover a liquid of elevated boiling point, for example, a temperature-sensitive organic insoluble in water.

Certain mixtures, such as *i*-propanol and water, and water and *n*-butanol[9], exhibit partial insolubility. Fig. 2.11 renders a generalized equilibrium plot. Compositions and

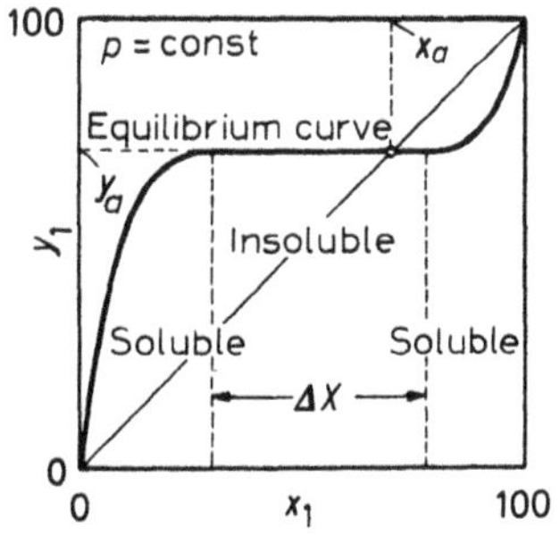

Fig. 2.11 Equilibrium curve $y_1 = f(x_1)$ of a partially insoluble mixture.

boiling points are constant for certain ranges of Δx. The boiling point corresponds to a minimum. Some solubility exists for mixtures with compositions outside the miscibility gap Δx. A partially insoluble mixture is a specialized instance of mixtures of minimum boiling point.

A nonideal mixture need not necessarily be azeotropic. Many systems have equilibrium curves, boiling and dewpoint lines similar to those of ideal mixtures and still do not follow Raoult's Law; note the bibliography (1–14).

All concentration plots shown above refer to the lower boiling component to avoid errors. For the same reason, for mixtures the lower boiler is always named first.

C. Ideal Multicomponent Mixtures

As in the case of ideal binary mixtures, the phase equilibrium can be predicted since Raoult's Law, Eq.(2-11), will hold for each component c ($c = 1, 2, 3, \ldots n$) of the mixture with n components. Referring to Dalton's Law, Eq.(2-5)

$$p = \frac{x_1}{100}P_1 + \frac{x_2}{100}P_2 + \frac{x_3}{100}P_3 + \cdots \frac{x_n}{100}P_n \tag{2-28}$$

With the composition $x_1, x_2, x_3, \ldots x$ (mol-%) of the multicomponent mixture boiling at constant pressure p (mm Hg) given, the boiling temperature t_b can be found from the vapor pressure curves $P_1 = f(t)$, $P_2 = f(t)$, $P_3 = f(t)$, $\ldots P_n = f(t)$ (Fig. 2.12). Within the temperature interval $t_1 < t < t_n$, one selects various boiling points t_{I}, t_{II}, etc., then reads the boiling pressures for the pure components and computes the corres-

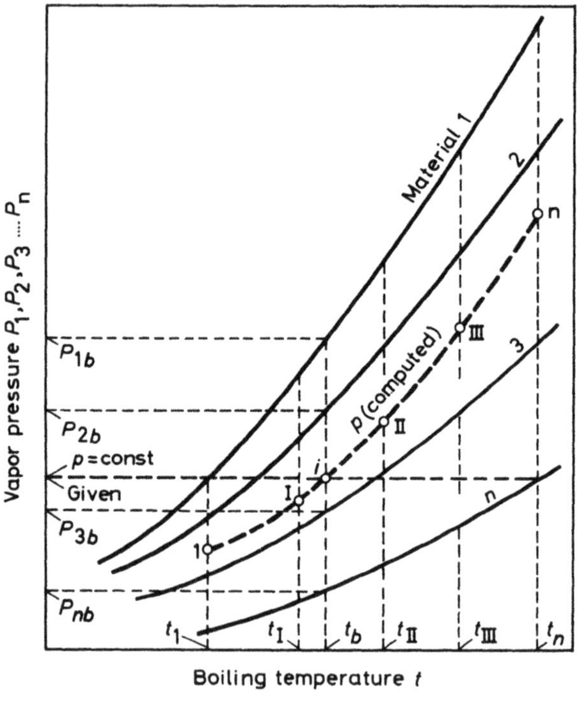

Fig. 2.12 Finding the boiling point t_b of an ideal multicomponent mixture of given composition from the vapor pressure curves $P_1 = f(t)$, $P_2 = f(t)$, $P_3 = \ldots, P_n = f(t)$ of the individual components.

ponding p values from Eq.(2-28). Point I applies to t_{I}, point II applies to t_{II}, etc., and for t_1 and t_n points 1 and n will apply. Join points 1, I, II, III, . . . n by a line, and make this intersect with the line for p = constant. Intersection i will be at the wanted boiling temperature t_b, which is required to be found.

The boiling point t_b has the corresponding boiling pressures $P_1 = P_{1b}, P_2 = P_{2b}, P_3 = P_{3b}, \ldots P_n = P_{nb}$ of the pure components, which must be known in order to find the composition $y_1, y_2, y_3, \ldots y_n$ of the mixed vapors at phase equilibrium. In analogy with Eq. (2-17)

$$y_1 = \frac{P_1}{p} x_1$$

$$y_c = \frac{P_c}{p} x_c \tag{2-29}$$

$$y_n = \frac{P_n}{p} x_n$$

Example 2 What is the boiling temperature, t_b, of an ideal ternary mixture benzene-toluene-xylene, composed of x_1 = 30 mol-% benzene, x_2 = 60 mol-% toluene, and x_3 = 30 mol-% xylene? The total pressure p = 758 mm Hg. What is the composition of the vapor in equilibrium with the liquid?

Answer By the method outlined above, the vapor pressure curves for benzene, toluene, and xylene will indicate, for p = 758 mm Hg, a boiling point of t_b = 100 C. From this, the boiling pressure curves will yield P_1 = 1350 mm Hg for benzene, P_2 = 550 mm Hg for toluene, and P_3 = 230 mm Hg for xylene. Eq. (2-29) will furnish the vapor compositions of y_1 = 53.5 mol-% benzene, y_2 = 43.5 mol-% toluene, y_3 = 3 mol-% xylene, a total of 100 mol-%. The boiling temperature of 100 C may be verified by Eq.(2-28)

$$p = \frac{30}{100}(1350) + \frac{60}{100}(550) + \frac{10}{100}(230) = 758 \text{ mm Hg}$$

Another method of predicting the boiling performance of an ideal multicomponent mixture is offered by Eq.(2-29). The denominator is replaced by Eq.(2-28). Dividing the numerator and denominator by the boiling pressure P_n of the pure high boiler and introducing the relative volatilities will yield

$$\alpha_{1n} = \frac{P_1}{P_n} \text{ for the first component}$$

$$\alpha_{cn} = \frac{P_c}{P_n} \text{ for the } c\text{th component} \tag{2-30}$$

$$\alpha_{nn} = 1 \quad \text{for the } n\text{th component}$$

Relations for the vapor concentrations:

$$y_1 = \frac{x_1}{\frac{1}{\alpha_{1n}}(\alpha_{1n}x_1 + \alpha_{2n}x_2 + \alpha_{3n}x_3 + \cdots x_n)}(100) \text{ for the first component}$$

$$y_c = \frac{x_c}{\frac{1}{\alpha_{cn}}(\alpha_{1n}x_1 + \alpha_{2n}x_2 + \alpha_{3n}x_3 + \cdots x_n)}(100) \text{ for the } c\text{th component}$$

$$y_n = \frac{x_n}{\alpha_{1n}x_1 + \alpha_{2n}x_2 + \alpha_{3n}x_3 + \cdots x_n}(100) \qquad (2\text{-}31)$$

for the nth component of the n-component mixture

For a precise determination of the vapor compositions $y_1, y_2, y_3, \ldots y_n$ (mol-%) at constant boiling point p, assume again various boiling temperatures t_{I}, t_{II}, and t_{III}, within the range $t_1 < t < t_n$. Also, from the corresponding vapor pressure curves, read the boiling points for the pure components. By means of Eqs.(2-30) find the relative volatilities, and use these in Eqs. (2-31) to obtain the vapor concentrations. For the correct boiling points, the sum

$$\sum y_c = y_1 + y_2 + y_3 \cdots y_n = 100 \qquad (2\text{-}32)$$

A graphic method consists in plotting the summands $(y_1 + y_2 + \ldots y_n)$ against the selected temperatures; note Fig. 2.13, points I, II, III. The resulting curve intersects the horizontal Σy_c = 100 mol-%. The abscissa of i gives the boiling temperature t_b.

If at constant pressure $t_1 < t < t_n$ give fairly constant α_{cn} values, the vapor compositions may be obtained directly from Eq.(2-31)[15].

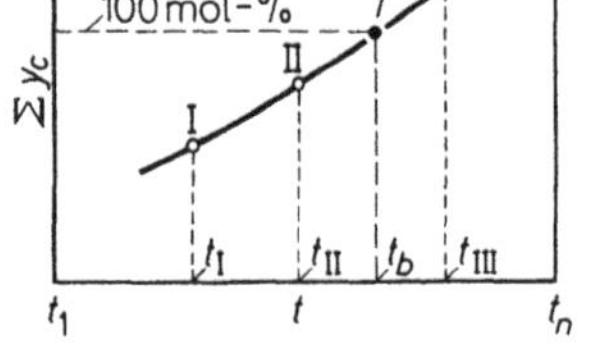

Fig. 2.13 Boiling point t_b of an ideal multicomponent mixture.

Example 3 For the data from Example 2, find the vapor compositions by means of Eqs.(2-31)

Answer First find the relative volatilities by means of Eq.(2-30):

$$\alpha_{13} = \frac{P_1}{P_3} = \frac{1350}{230} = 5.87 \qquad \alpha_{13} = \frac{P_2}{P_3} = \frac{550}{230} = 2.39 \qquad \alpha_{13} = \frac{P_3}{P_3} = \frac{230}{230} = 1$$

Then Eq.(2-31) will supply the vapor concentrations:

$$y_1 = \frac{30}{\frac{1}{5.87}[(5.87\cdot 30) + (2.39\cdot 60) + 10]} \cdot 100 \quad = 53.5 \text{ mol-\% benzene}$$

$$y_2 = \frac{60}{\frac{1}{2.39}[(5.87\cdot 30) + (2.39\cdot 60) + 10]} \cdot 100 = 43.5 \text{ mol-\% toluene}$$

$$y_3 = \frac{10}{(5.87\cdot 30) + (2.39\cdot 60) + 10} \cdot 100 \quad = 3 \text{ mol-\% xylene}$$

as before.

D. Nonideal Multicomponent Mixtures

Here the methods described in the previous paragraph will serve if the liquid concentrations are replaced by the activities, that is multiplied by the activity coefficients. Eq.(2-28) will then supply

$$p = \frac{\gamma_1 x_1}{100} P_1 + \frac{\gamma_2 x_2}{100} P_2 + \frac{\gamma_3 x_3}{100} P_3 + \ldots \frac{\gamma_n x_n}{100} P_n \tag{2-33}$$

and by Eq.(2-29)

$$y_1 = \frac{P_1}{p}\gamma_1 x_1 \qquad y_c = \frac{P_c}{p}\gamma_c x_c \qquad y_n = \frac{P_n}{p} x_n \tag{2-34}$$

To use these systems of equations, the functions

$$\begin{aligned} \gamma_1 &= f(x_1) \\ \gamma_c &= f(x_c) \\ \gamma_n &= f(x_n) \end{aligned} \tag{2-35}$$

must be known for the multicomponent mixture. Then one proceeds in accordance with Section 2.C. and Fig. 2.12.

Also, one may apply the alternate method of the previous paragraph, and divide Eqs.(2-34) by Eq.(2-33)

$$y_1 = \frac{x_1}{x_1 + \frac{\gamma_2 P_2}{\gamma_1 P_1} x_2 + \frac{\gamma_3 P_3}{\gamma_1 P_1} x_3 + \ldots \frac{\gamma_n P_n}{\gamma_1 P_1} x_n} (100)$$

$$y_c = \frac{x_c}{\frac{\gamma_1 P_1}{\gamma_c P_c} x_1 + \frac{\gamma_2 P_2}{\gamma_c P_c} x_2 + \frac{\gamma_3 P_3}{\gamma_c P_c} x_3 + \ldots \frac{\gamma_n P_n}{\gamma_c P_c} x_n} (100) \qquad (2\text{-}36)$$

$$y_n = \frac{x_n}{\frac{\gamma_1 P_1}{\gamma_n P_n} x_1 + \frac{\gamma_2 P_2}{\gamma_n P_n} x_2 + \frac{\gamma_3 P_3}{\gamma_n P_n} x_3 + \ldots x_n} (100)$$

Again, the activity coefficients must be known as functions of the concentrations in order to determine vapor compositions $x_1 \ldots x_c \ldots x_n$ by the alternate method for a given liquid composition (Section 2.2.3, note Fig. 2.13).

The earlier bibliography[16-21] discusses methods for finding the boiling equilibria of nonideal multicomponent mixtures. Recent work[22,23] is based upon the data by Redlich and Kister[24], Prausnitz and collaborators,[25] and Wilson[27] and permits the determination of vapor equilibria for many multicomponent systems in a satisfactory manner. A shortcut procedure[22] yields activity coefficients based on the binary equilibrium data of various pairs of components.

E. Presentation of Ternary Mixtures

For mixtures of up to three components, a pictorial presentation of the phase equilibrium at constant boiling pressure is shown in Fig. 2.14. On the sides of an equilateral

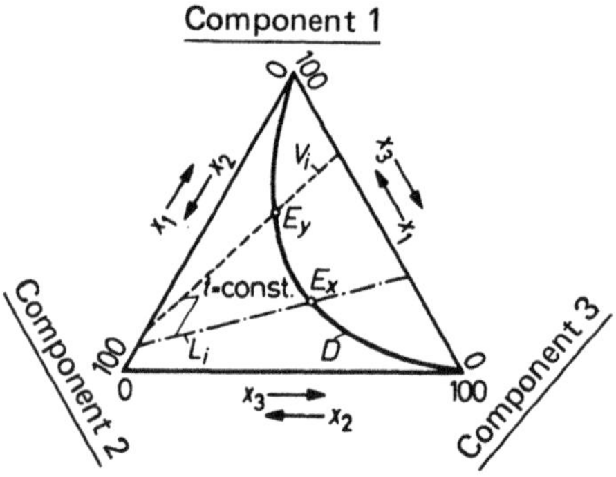

Fig. 2.14 Presentation of a ternary mixture and generalized location of the liquid isothermal, vapor isothermal, and distillation line.

triangle the percentages for one pair of materials have been plotted so that the three corners of the triangle each indicate 100% of the component considered. A parallel to one side drawn through a given point of one triangular surface will cut off, on the two other sides of the triangle, the concentration of the material whose corner is opposite to that parallel. The field of presentation will accept lines of equal boiling point, liquid isothermals, lines of equal dewpoints, vapor isothermals, and distillation lines. These indicate the direction in which the vapor concentrations will vary at boiling as compared to those of the liquid.

For an ideal ternary mixture the significant lines may be computed. With a constant boiling pressure p and a constant boiling temperature t, combining Raoult's and Dalton's

Laws with the concentration balance of the liquid phase gives

$$x_1 + x_2 + x_3 = 100 \tag{2-37}$$

and the equation of the liquid isothermals becomes

$$x_1 = \frac{p-P_3}{P_1-P_3} \cdot 100 - \frac{P_2-P_3}{P_1-P_3} x_2 \tag{2-38}$$

which in turn will yield the vapor isothermal by substituting for x_1 and x_2 the proper terms of Eq.(2-29)

$$y_1 = \frac{(p-P_3)P_1}{(P_1-P_3)p}(100) - \frac{(P_2-P_3)P_1}{(P_1-P_3)P_2} y_2 \tag{2-39}$$

To evaluate Eqs.(2-38) and (2-39) and represent them in the distillation triangle the vapor pressure curves of the pure components must be known. For an ideal binary mixture at p = constant and for a given boiling point t there is only one possible composition whereas an infinity of compositions is feasible for ideal ternary mixtures. In Eqs.(2-38)and (2-39) at p = constant for a given temperature t, only x_1 and x_2 and y_1 and y_2 are variables; they represent straight lines that form the locus for all ternary mixtures of equal boiling point.

Consider one point E_x on the liquid isothermal L_i (Fig. 2.14). The correlated point E_y of the vapor at equilibrium when boiling and of the corresponding liquid must be located on the proper isothermal, V_1, for the same temperature t, with the point fixed by Eqs.(2-29). However, through point E_y there will also pass a liquid isothermal for a temperature below t, which has a vapor isothermal for which the new equilibrium point will be found from Eqs.(2-29). The connecting line through the equilibrium points thus determined is the "distillation curve" D.[2] Many such lines can be obtained in the manner described. Together with the liquid and vapor isothermals they define the distillation path for a ternary mixture.

III Continuous Rectifiers

Thermal splitting by rectification is a countercurrent procedure in which material exchange takes place between the vapor and the liquid phase of the mixture to be decomposed.

1. MODE OF OPERATION

The continuous rectifier is more amenable to calculations than the batch type and will be treated first. Fig. 3.1 explains its construction and function. The stripping

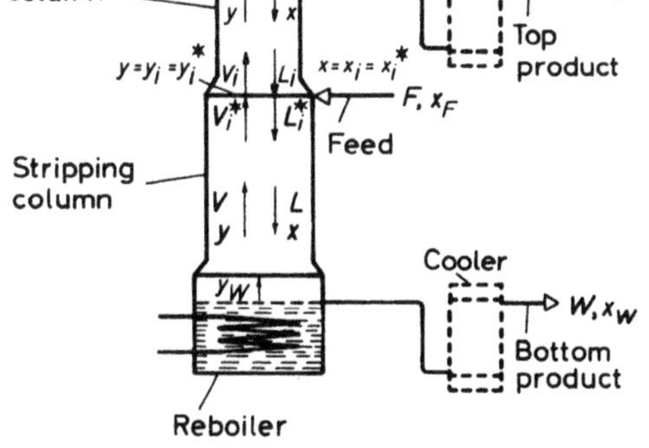

Fig. 3.1 Continuous rectifier, diagrammatic. In the rectifying column $V = V_e$ and $L = L_e$, and in the stripping column $V = V_s$ and $L = L_s$.

column is mounted on top of the reboiler, and on top of the stripping column is the rectifier followed by a reflux condenser and a product cooler. The stripper is supplied with the mixture to be worked, F(kmol/h) continuously, either as a liquid, or as a vapor, or part liquid part vapor.

The reboiler continuously generates V_s(kmol/h) which passes, counter to reflux liquid L_s (kmol/h), to the stripper. By the resulting material interchange, the more volatile components are driven off. The mixed vapors eventually enter the rectifier. Its quantity is V_e (kmol/h) and the reflux quantity running counter to it equals L_v (kmol/h). The volatile vapors become enriched in the rectifier top and usually condense totally in the reflux condenser. Part of the condensate returns to the rectifier top, and part is drawn off, after passing through a product cooler yielding the top product D (kmol/h). Runoff W may also be drawn from the reboiler at a constant rate (if necessary). When charges of more than two components are worked top product and runoff may be fed to subsequent columns without further cooling.

Of course, one may condense only as much vapor in the reflux condenser as is required to provide reflux. In that event, the reflux condenser, now termed "dephlegmator" will supply a residual vapor of composition different from that entering. That residual vapor condenses in a "condenser-cooler" and is recovered as a top product.

Ordinarily, the molar quantities V_s, V_e, L_s, and L_e vary along the flow path and may be practically constant only for equal molar heats of vaporization for individual components. It will be seen that at any given column level the vapor will be hotter than the liquid. The contact interface produces a concentration differential corresponding to the temperature differential between vapor and liquid; this achieves the material exchange inasmuch as the more volatile components will vaporize, and the less volatile will tend to condense. When the molar heats of evaporation are equal, the molar quantities of vapor and liquid will remain virtually constant as such, in both the stripper and the rectifier.

2. OPERATING LINES

To ascertain the dimensions of a rectifier, the relative liquid and vapor compositions at any given column level must be known. If the molar heats of vaporization are of the same order, the exchange line[1] will reveal that interrelation in a concentration plot in which the ordinate is the vapor concentration and the abscissa is the liquid concentration. The calculation of the exchange line rests on the basic rectification concept[2] by which at the phase interface high-boiling material condenses from the vapor, and the heat given off serves to vaporize the low boiler from the liquid. Normally, no consideration need be given to the fact that heats of solution may be present, or that the vapor from each component is superheated; since the partial vapor pressure of a material is below the saturation pressure of the pure material at the same temperature. That lenience in the determination is usually admissible in practice, and for design purposes the operating lines may be established from the molar heats of evaporation[1] and the basic concepts of rectification[3]. In what follows, no special separating action of the reflux condenser will be assumed.

The following conclusions are for a binary mixture for which the composition is known from its content of the low boiling component. At a given level of the separating column, the mixed vapor will contain y (mol-%) and the liquid will contain x (mol-%) of low boiler. Quantities and concentrations at the feed point will show the subscript i (note Fig. 3.1).

Consider an infinitesimal path of contact of vapor and liquid inside the column, along which the mass of mixed vapor V (kmol/h) is varied by the infinitesimal amount dV (kmol/h), and its concentration y (mol-%) is changed because in the material exchange by rectification an infintesimal amount dh (kmol/h) of high boiler condenses into the liquid, and at the same time an infinitesimal amount dl (kmol/h) of low boiler from the liquid will evaporate into the vapor. With the heats q_h and q_l (kcal/mol) released by 1 kmol of high boiler, or absorbed by 1 kmol of low boiler, and with

$$dh \cdot q_h = dl \cdot q_l \quad \text{and} \quad dl - dh = dV$$

based on a mass balance for the low boiler, following the rectification theory there results the differential equation

$$\frac{dV}{V} = \frac{d_y}{\dfrac{q_h}{q_h - q_l}(100 - y)} \tag{3-1}$$

Integration between the limits $V = V_a$ (kmol/h), the mass of mixed vapor at a given level a of the column having the vapor concentration y_a (mol-%), and $V = V_b$ (kmol/h) of the mass of mixed vapor in a section b located a, with the vapor concentration y_b (mol-%) yields

$$\ln \frac{V_a}{V_b} = \int_{y = y_b}^{y = y_a} \frac{dy}{q - y} \tag{3-2}$$

with

$$q = \frac{q_h}{q_h - q_l}(100) \tag{3-3}$$

At those two levels a and b through which pass the masses of vapor V_a and V_b (kmol/h) and the masses of liquid L_a and L_b (mol/h) with x_a and x_b (mol-%) of the low boiler, countercurrently, the mass balances will be

$$V_b + L_a = V_a + L_b \tag{3-4}$$

and

$$V_b y_b + L_a x_a = V_a y_a + L_b x_b \tag{3-5}$$

Solving Eq.(3-4) for L_b, and inserting the result in Eq.(3-5) will give

$$V_b y_b + L_a x_a = V_a y_a + V_b x_b + L_a x_b - V_a x_b \tag{3-6}$$

A. Enrichment Line

A mass balance for the rectifying section of the column will yield

$$L_a = \frac{r_a}{r_a + 1} V_a \tag{3-7}$$

where

r_a = reflux ratio at a given level a of the rectifier, equals L_a/D

D = mass of product at the column top

By substituting in Eq.(3-6) the term L_a with Eq.(3-7), and introducing the result into Eq.(3-2), with V_a/V_b[1]

$$\ln\left[\frac{y_b - x_b}{y_a - \left[\frac{r_a}{r_a+1}(x_a - x_b)\right] - x_b}\right] = \int_{y_b}^{y_a} \frac{dy}{q-y} \tag{3-8}$$

When no enriching effect due to partial condensation is undergone by the flowing vapor, the concentrations of vapor, reflux, and product are identical. If a is the cross section at the column top, and b some other cross section of the rectifier, then

$y_a = x_a = x_D$ = concentration of top product

$y_b = y$

$r_a = r_D$ = reflux ratio at column top,

and from Eq.(3-8)

$$\ln\,(r_D + 1)\frac{y-x}{x_D - x} = \int_{y}^{x_D} \frac{dy}{q-y} \tag{3-9}$$

Eq.(3-9) is the equation of the enrichment line in a y/x plot; it gives the relation between vapor and liquid concentration at any given column level. When q_h, q_i, and hence q depend on concentrations, the evaluation is part graphic and part calculation; the integral on the right-hand side of Eq.(3-9) must be done graphically. First the q values must be stated in terms of vapor concentration y. Then plot curve $1/(q-y)$ versus y. Graphic integration beginning at $y = x_D$, the product concentration at the top of the column, gives the integral curve. This will furnish the $\int dy/(q\text{-}y)$ values for any given y and the corresponding x[Eq.(3-9)]. Plotting the y values against x gives the enrichment line.

The effects of superheat and of heats of solution are mostly negligible, and q becomes equal to a constant K; q_h becomes equal to the molar heat of vaporization λ_h (kcal/kmol) of the high boiler, and q_1 becomes equal to the molar heat of evaporation λ_1 (kcal/kmol) of the low boiler. The right-hand side of Eq.(3-9) can be integrated.

Ultimately, the equation of the enrichment line becomes

$$y = \frac{\left(\frac{r_D}{r_D+1} - \frac{x_D}{K}\right)x + \frac{x_D}{r_D+1}}{1 - \frac{1}{r_D+1}\frac{x}{K} - \frac{r_D}{r_D+1}\frac{x_D}{K}} \tag{3-10}$$

where

$$K = \frac{\lambda_h}{\lambda_h - \lambda_l} 100 = q \tag{3-11}$$

Eq.(3-10) will furnish for $y = y_o$ at point $x = 0$:

$$y_o = \frac{x_D}{r_D \left(1 - \frac{x_D}{K}\right) + 1} \tag{3-12}$$

By definition, reflux ratio r_D is the ratio of the reflux quantity $L = R$ at the column top and the quantity of top product D

$$r_D = \frac{R}{D} \tag{3-13}$$

Point D is the intersection of the diagonal and the enrichment line; point D is located at $y = x = x_D$ when $y = x$ in Eq.(3-10) (Fig. 3.2).

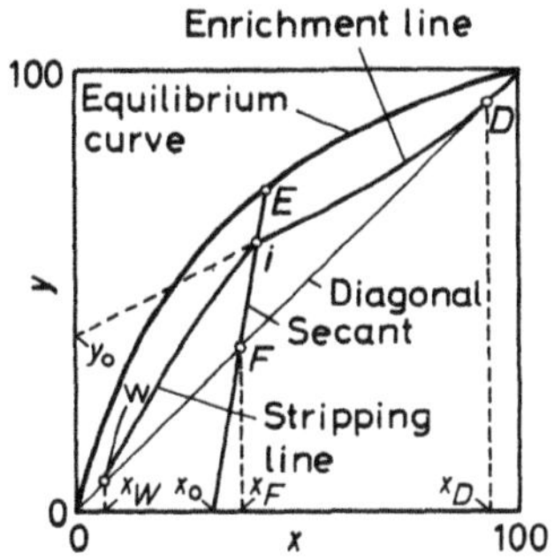

Fig. 3.2 Location of operating lines in y/x plot.

An evaluation of Eq.(3.9) will demonstrate that the simplification of Eq.(3-10) is for the most part justified. Even if for certain mixtures the heats of solution are appreciable, and the heats of condensation and vaporization are affected, Eq.(3-10) will still be usable, although corrected λ_h and λ_l data must be applied.

According to the rule by Trouton and Pictet, the molar heats of evaporation are virtually constant for the individual components of many mixtures[4]. Then, with $q_h = q_l$, or $\lambda_h = \lambda_l$, from Eq.(3-11) K becomes infinite, and from Eq.(3-10) there evolves a relation for a line[5] – the so-called enrichment line

$$y = \frac{r_D}{r_D + 1} x + \frac{x_D}{r_D + 1} \tag{3-14}$$

In an y/x plot, that line will cut off an ordinate value of

$$y_o = \frac{x_D}{r_D + 1} \tag{3-15}$$

Eq.(3-12) will give the same result for $r_D = \infty$. Since, with $\lambda_h = \lambda_l$ the molar liquid and vapor quantities remain virtually constant, in Eqs.(3-14) and (3-15), one may write for r_D

$$r_D = \frac{R}{D} = \frac{L}{D} = r = \text{constant} \tag{3-16}$$

In the y/x plot the diagonal again intersects the enrichment line at the point $y = x = x_D$.

B. The Stripping Line

Let: F (kmol/h) = mass of mixture entering rectifier, f_i = factor allowing for its entrance enthalpy,[6,7] and L_i (kmol/h) = mass of reflux from rectifier entering. Then, according to Fig. 3.1

$$L_i^* = L_i + f_i F \tag{3-17}$$

In that relation, f_i equals unity only if the mixture enters at boiling temperature. L_i^* is a multiple m^* of V_i^*, the mass of vapor entering from the stripper[1]

$$L_i^* = m^* V_i^* \tag{3-18}$$

From Eqs.(3-17) and (3-18)

$$m^* = \frac{L_i + f_i F}{V_i^*} \tag{3-19}$$

In Eq.(3-19) insert for V_i^* the following term, resulting from a mass balance for the entrance

$$V_i^* = V_i + F(f_i - 1) \tag{3-20}$$

where V_i is the mass of vapor entering from the rectifier. Furthermore, for

$$L_i = D r_i \tag{3-21}$$

with D = product quantity at the column top, reflux ratio, r_i applied immediately above the feed level, and equating

$$V_i = D(r_i + 1) \tag{3-22}$$

D becomes, from a mass balance for the entire installation,

$$D = F\,\frac{x_F - x_w}{x_D - x_w} \tag{3-23}$$

where

x_F = concentration of feed

x_w = concentration at runoff of the column base

x_D = concentration of distillate at the column top,

so that in conclusion[1]

$$m^* = \frac{\frac{x_F - x_W}{x_D - x_W} r_i + f_i}{\frac{x_F - x_W}{x_D - x_W}(r_i + 1) + (f_i - 1)} \tag{3-24}$$

In the simple mass balance for r_i

$$r_i = \frac{x_D - y_i}{y_i - x_i} \tag{3-25}$$

$y_i = y_i^*$ and $x_i = x_i^*$ are the coordinates of the intersection of enrichment line and the stripping line. In analogy with Eq.(3-8) there is the relation

$$\ln \frac{y - x}{y_i^* - m^*(x_i^* - x) - x} = \int_y^{y_i^*} \frac{dy}{q - y} \tag{3-26}$$

which again must be resolved by a combined graphic and mathematical approach. When $q = K$ [note Eq.(3-11)] the relation for the stripping line follows[1]

$$y = \frac{\left(m^* - \frac{y_i^*}{K}\right) x + (y_i^* - m^* \cdot x_i^*)}{1 - (1 - m^*)\frac{x}{K} - m^* \frac{x_i^*}{K}} \tag{3-27}$$

Eq.(3-27) combined with Eqs.(3-24) and (3-25) permits determination of the stripping line, if the intersection coordinates $y_i^* = y_i$, $x_i^* = x$ are known. The diagonal of the y/x plot is intersected by the stripping line at W for $y = x = x_W$ (Fig. 3.2).

Again, if the molar heats of evaporation of the components are of the same order K in Eq. (3-11) becomes infinity, and Eq. (3-27) becomes the stripping line relation

$$y = \frac{w}{w - 1} x - \frac{x_W}{w - 1} \tag{3-28}$$

where:

$$w = \frac{L_s}{W} \tag{3-29}$$

is the runoff ratio—the quotient of the mass of liquid $L = L_s$ in the stripping section and the mass of bottom product W. The stripping line will intersect the diagonal of the y/x plot at $y = x = x_W$.

The relations obtained from the data for the operating lines on the enrichment and stripping lines with $K = \infty$ may also be derived from simple mass balances.

The locus of intersections i on the enrichment line and the stripping line is a straight line

$$y_i = \frac{f_i}{f_i - 1} x_i - \frac{x_F}{f_i - 1} \tag{3-30}$$

termed the secant[6,7] and is obtained from a mass balance taken at the feed level. It will intersect the abscissa of the y/x plot at

$$x_o = \frac{x_F}{f_i} \tag{3-31}$$

and the diagonal of the plot—Fig. 3.2 at point F located at

$$y = x = x_F \tag{3-32}$$

Eventually, a heat balance around the feed point will yield the factor f_i characterizing the thermal condition of the charge,

$$f_i = 1 + \frac{h'_F - h_F}{\lambda} \tag{3-33}$$

where: h'_F = molar enthalpy of the mixture at boiling, (kcal/kmol)
h_F = same at charging temperature
λ = molar heat of condensation at charging point, (kcal/kmol)

According to Eq. (3-33), f_i will exceed unity when the charge is cold ($h_F < h'_F$), because in that event some vapor condenses in the feed area to heat the mixture to the boiling point (according to Eq. (3-20) $V_i < V_i^*$). For a charge at boiling temperature ($h_F = h'_F$), f_i becomes unity; according to Eq. (3-17), $L_i^* = L_i + F$, and from Eq. (3-20) $V_i = V_i^*$. When the feed mixture is a vapor ($h_F = h'_F + \lambda$), $f = 0$; in that case, from Eq. (3-17), $L_i = L_i^*$, and from Eq. (3-20), $V_i = V_i^* + F$. Furthermore, the column charge may be a wet vapor, then f_i is less than unity, or it may be a superheated vapor which makes f_i less than zero (Fig. 3.3). For an investigation of conditions in rectifiers,

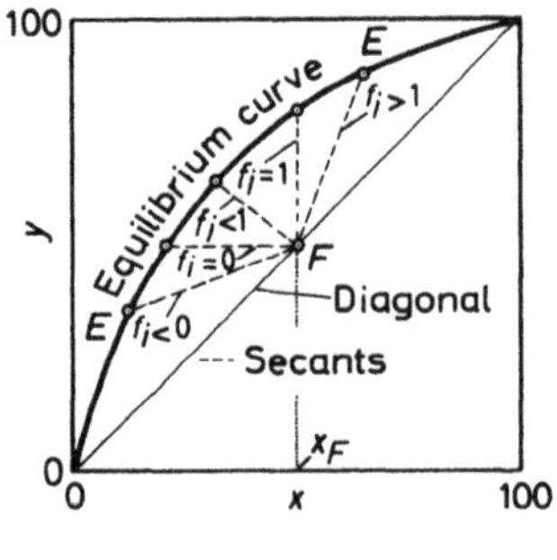

Fig. 3.3 Location of secant.

only those intersections i that are located between stripping and rectifying lines, and between points E and F cut off by the secant on the equilibrium curve and on the diagonal are significant (Fig. 3.2).

Example 1 The molar heat of evaporation of the high boiler is $\lambda_h = 10{,}000$ kcal/kmol, and that of the low boiler $\lambda_l = 6{,}000$ kcal/kmol. The reflux ratio at the column head is $r_D = 4$, and the product concentration is $x_D =$ 100 mol-%. The feed is at boiling point, $f_i = 1$, and has $x_F = 50$ mol-% of low boiler, while the runoff from the stripper shall have $x_W = 0$ mol-% concentration.

Answer From Eq. (3-30), for $f_i = 1$, $x_i = x_F = 50$ mol-%. From Eq. (3-11), K becomes 250. Eq. (3-10) will yield y_i for $x = x_i = 50$ mol-%. Then $y =$

$$y_i = \frac{\left(\frac{4}{5} - \frac{100}{250}\right)(50) + \frac{100}{5}}{1 - \frac{1}{5}\frac{50}{250} - \frac{4}{5}\frac{100}{250}} = 62.5 \text{ mol-\%}$$

From Eq. (3-25),

$$r_i = \frac{100 - 62.5}{62.5 - 50} = 3$$

and from Eq. (3-24),

$$m^* = \frac{\frac{50 - 0}{100 - 0}(3) + 1}{\frac{50 - 0}{100 - 0}(4)} = 1.25$$

In this instance, when $x_W = 0$, m^* may be found more readily. When $x = x_W = 0$, y must equal 0. From Eq. (3-27) this is feasible only when the second term in the numerator of the equation becomes zero. Hence,

$$(m^*)_{x_W=0} = \frac{y_i^*}{x_i^*}$$

and in this case

$$m^* = \frac{62.5}{50} = 1.25$$

The coordinates of the intersection $y_i^* = y_i$ and $x_i^* = x_i$ on the enriching and stripping lines are known, and points of the stripping line can be computed by Eq. (3-27). For instance, for $x = 20$ mol-%

$$y = \frac{\left(1.25 - \frac{62.5}{250}\right)(20) + \left[62.5 - (1.25)(50)\right]}{1 - (1 - 1.25)\left(\frac{20}{250}\right) - (1.25)\frac{50}{250}} = 26.4 \text{ mol-\%}$$

which establishes one point of the stripping line.

Fig. 3.4 represents the enriching and stripping lines computed from Eqs. (3-10) and (3-27). The dotted lines apply for $\lambda_h = \lambda_l$, other conditions being equal.

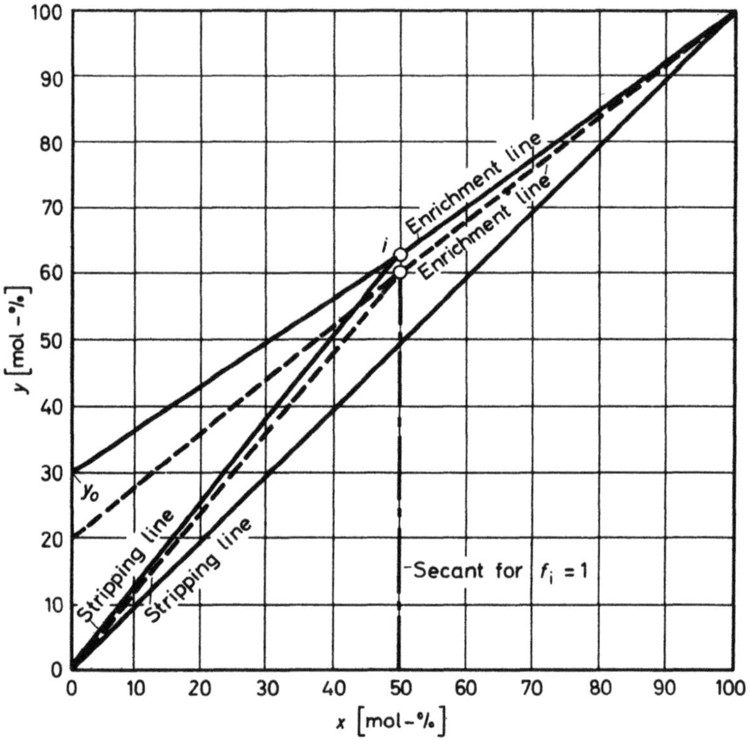

Fig. 3.4 Interchange lines for a binary mixture with λ_h = 10,000 kcal/mol and λ_l = 6000 kcal/kmol. Reflux ratio at column top r_D = 4. Interchange lines at r_D = 4.

From Eq. (3-12), the enriching line intersects the ordinate at

$$y_o = \frac{100}{4\left(1 - \frac{100}{250}\right) + 1} = 29.4 \text{ mol-\%}$$

while from Eq. (3-15) the enriching line intersects the ordinate at

$$y_o = \frac{100}{4+1} = 20 \text{ mol-\% for } x = 0.$$

When the molar heats of vaporization of the components are close use the enriching lines.

Generally, the enriching line will indicate, for the area between top product concentration x_D and the concentration x_i of the intersection with the intersection line, the trend of changes in vapor and liquid concentrations in the enrichment column. Changes of vapor and liquid concentrations in the stripper are represented by the stripping line between x and x_W in the y/x plot.

Fig. 3.5 explains qualitatively for a given point how the concentration difference $y - x$, and the corresponding temperature difference θ between vapor and liquid can

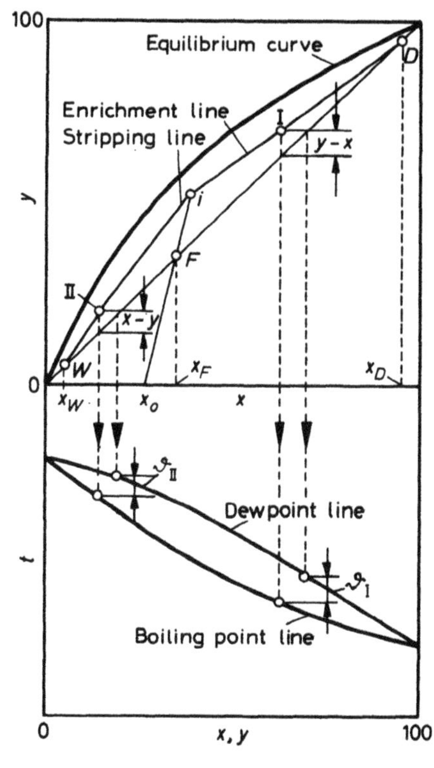

Fig. 3.5 Finding the concentration and temperature differences between vapor and liquid at a given location.

be determined for a given column level, for point I of the rectifier and point II of the stripper. That concentration difference offers a yardstick for the discrepancy in the equilibrium of vapor and liquid at that level, and it is required for a condensation of the low boiler from the liquid phase. The significance of the operating lines in the layout of a rectifier will be explained further.

3. STEPWISE SEPARATION IN RECTIFIERS

To intensify separation in a rectifier, the vapor-liquid interface must be enlarged. This is achieved through suitable internals. The efficacy of such internals is commonly judged on the basis of the performance of a "theoretical plate." Hence, that concept is important in column design, and will be dealt with at greater length.

A. Theoretical plates for separations of binaries

To determine the number of theoretical plates in a rectifying column, it is assumed that the vapor rising from one plate is in equilibrium with the liquid leaving the same plate. The correlated concentrations of vapor and liquid leaving these theoretical steps must yield points on the equilibrium curve in a y/x plot. McCabe and Thiele[8] have established a method to determine the number of theoretical plates for separating a binary mixture from a known equilibrium curve shown in a y/x plot. After fixing a reflux ratio one plots the interchange lines. Beginning with the concentration of the top product, x_D, a sequence of steps is drawn between the interchange lines and the equilibrium curve, until the runoff concentration x_W is reached (Fig. 3.6). The

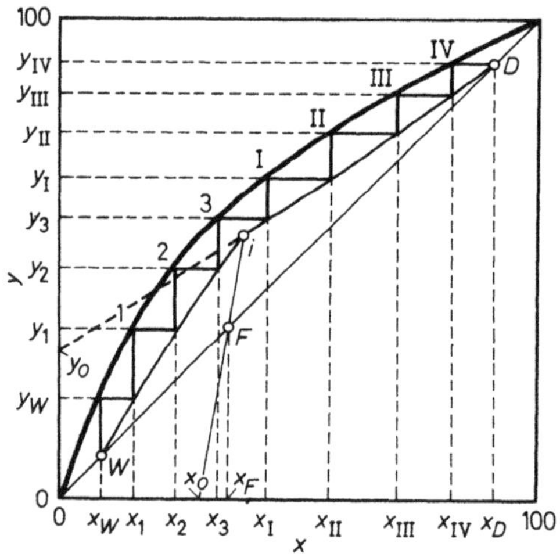

Fig. 3.6 Graphic determination of the number of theoretical steps.

theoretical number of plates, n_{ts} of the stripper equals the number of steps on the stripping line, that for the rectifier, n_{te}, the steps on the rectifying line. The total number of steps for the assembly then becomes $n_{ts} + n_{te} = n_t$. The corners located on the equilibrium, (1, 2, . . .) for the stripper, and (I, II, . . .) for the rectifier, indicate the sequence of concentrations for vapor and liquid issuing from one step, while the corners located on the operating lines illustrate the succession of vapor and liquid concentrations between steps.

In Fig. 3.6 there are four theoretical rectifying steps n_{te}, and three theoretical stripping steps n_{ts}. In this qualitative presentation, no separation through partial condensation was assumed. Therefore, the concentrations of vapor and reflux at the column top are identical with those in top product x_D. Also, it was assumed that the vapor entering the lowest stripping plate is at equilibrium with concentration x_W.

Fig. 3.7 is a sketch of the rectifier; at the levels shown as dotted lines, the concentrations are given for vapors and liquids corresponding to the plot in Fig. 3.6.

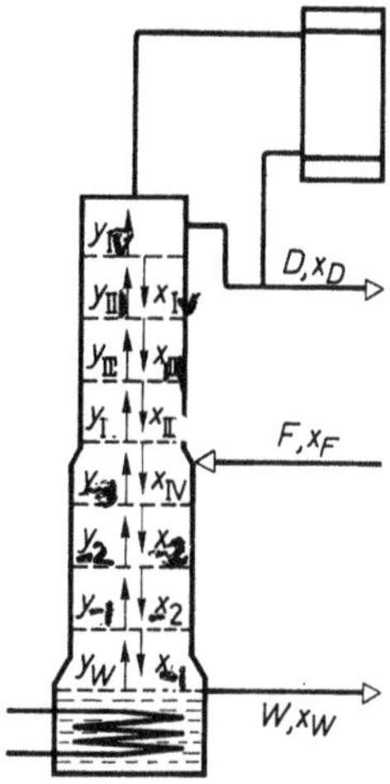

Fig. 3.7 Diagram of a continuous rectifier, indicating the vapor and liquid conditions on the steps from Fig. 3.6.

From the above, the mechanics of separation may be interpreted in that the vapor entering a given stage partially condenses the colder reflux liquid encountered. The heat of condensation liberated produces fresh vapor of volatile material. In the stripper the high boilers in the liquid are enriched up to the desired concentration x_W of the bottom product if the number of steps is adequate, while in the rectifier the concentration of low boiler increases to t_D—that of the top product.

B. The reflux ratio in the separation of binaries

With x_F, x_W, x_D, and f_i the number of steps depends upon the inclinations of the operating lines, and hence especially upon the reflux ratio r_D [Eq. (3-13) and Eq. (3-16)]. In any event, the reflux ratio must be large enough so that intersection i of the operating lines stays clear of the equilibrium curve. If it is still located on the equilibrium curve, $i = E$, the separation of the mixture would require an infinity of steps in both the rectifier and the stripper; note the sequence of steps in Fig. 3.8. There is a minimum reflux ratio r_m.

Another limiting condition results when no product D is drawn off at the column head; the reflux ratio would become infinite—$r_D = r = \infty$. The interchange lines then

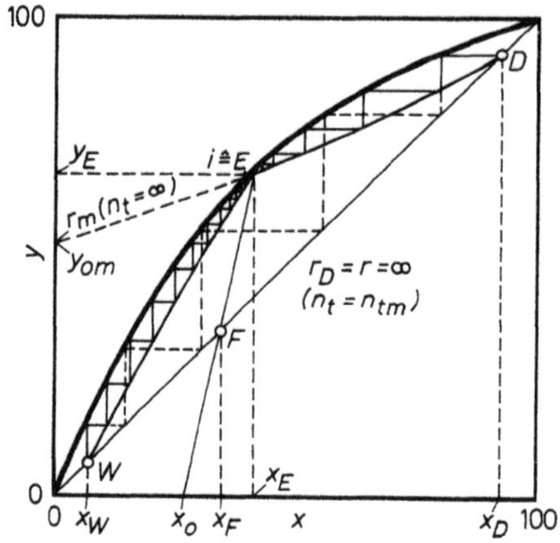

Fig. 3.8 Limiting conditions $n_t = \infty$ and $r = \infty$.

would merge with the diagonal, and the theoretical steps would become a minimum, $n = n_{tm}$. Note the sequence of dotted steps in Fig. 3.8.

For infinite reflux, the number of theoretical steps n_{tm} can also be found from the Fenske equation[9]

$$n_t = n_{tm} = \frac{1}{\log \alpha} \log \left(\frac{x_D}{100 - x_p} \frac{100 - x_W}{x_W} \right) - 1 \tag{3.34}$$

where $\alpha = \alpha_{1,2}$ is the relative volatility of the binary mixture.

In practice, reflux ratios lie between r_m and $r = \infty$. For the former, Eq. (3-10) offers a solution; the intersection coordinates y_E and x_E replace the current coordinates y and x, and r_m results from

$$r_D = r_m = \left(\frac{K - y_E}{K - x_D} \frac{x_D - x_E}{y_E - x_E} \right) - 1 \tag{3.35}$$

For constant molar heats of vaporization, $K = \infty$. That equation then becomes

$$r_D = r_m = \frac{x_D - y_E}{y_E - x_E} \tag{3.36}$$

Eqs. (3-35) and (3-36) may be inapplicable for nonideal mixtures whose equilibrium curves have a point of inflection in the range in question. In that event, a tangent rectifying line is applied to the equilibrium curve (Fig. 3.9). That line starts from point D and intersects the ordinate at y_{om}. Eq. (3.12) will supply r_m

$$r_D = r_m = \frac{\dfrac{x_D}{y_{om}} - 1}{1 - \dfrac{x_D}{K}} \tag{3.37}$$

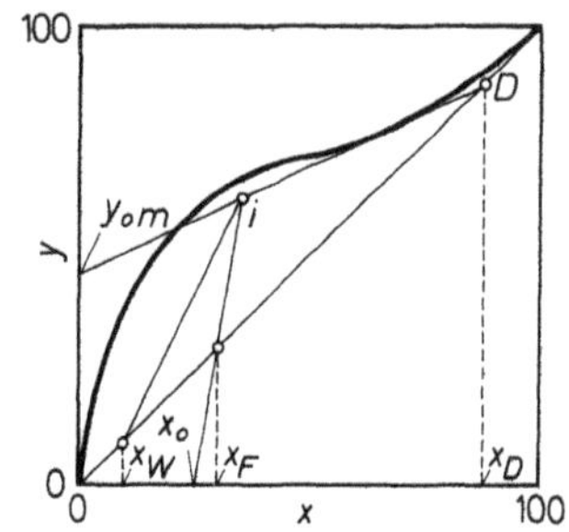

Fig. 3.9 Finding the minimum reflux ratio r_m for equilibrium curves with inflection point.

For equal molar heats of vaporization $K = \infty$, simply

$$r_D = r_m = \frac{x_D}{y_{om}} - 1 \tag{3.38}$$

To obtain a finite number of steps, the reflux ratio employed must be larger than r_m. By theory, the heat requirement is least for the minimum reflux ratio, for which the number of steps and hence the column height becomes infinite. For larger reflux ratios the number of steps is reduced and the heat requirement increases. The optimum reflux ratio can be determined only with foreknowledge of total investment C_T for a rectifier installation, that is, the sum of first cost and operating cost, as qualitatively explained in Fig. 3.10. In practice, reflux ratios are between 1.05 and 2, and higher.

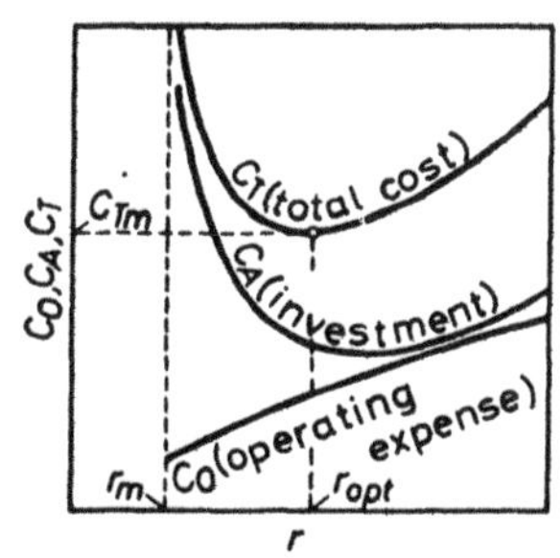

Fig. 3.10 Finding the optimum reflux ratio.

Example 2. A methanol-water mixture containing $x_F = 5.88$ mol-% methanol is to be split up by continuous rectification to supply a top product x_D with 99.2 mol-% methanol and a bottom product x_W with 0.28 mol-% methanol. The operating pressure is 1 atm. The mixture enters a boiling temperature. How many theoretical stages are required (n_{te} in the rectifying section, and n_{ts} in the stripping section), and what reflux ratio r_D is desired for the column?

If x_F becomes 21.3 mol-% instead of 5.88, what will be the values for n_{te}, n_{ts}, and r_D?

Answer. At atmospheric pressure, the molar heats of evaporation are, for methanol λ_t = 8420 kcal/kmol and for water λ_h = 9170 kcal/kmol. From Eq. (3-11) we find:

$$K = 753$$

For an inlet concentration of mol-%		5.88	21.3
from Eq (3.35)	r_m	3.37	1.19
hence	r_D	4	1.5
From Fig. 3.11	n_{te}	8	11
and	n_{ts}	7	6

Fig. 3.11 shows the characteristic concentrations, the interchange lines, and the separation stages.

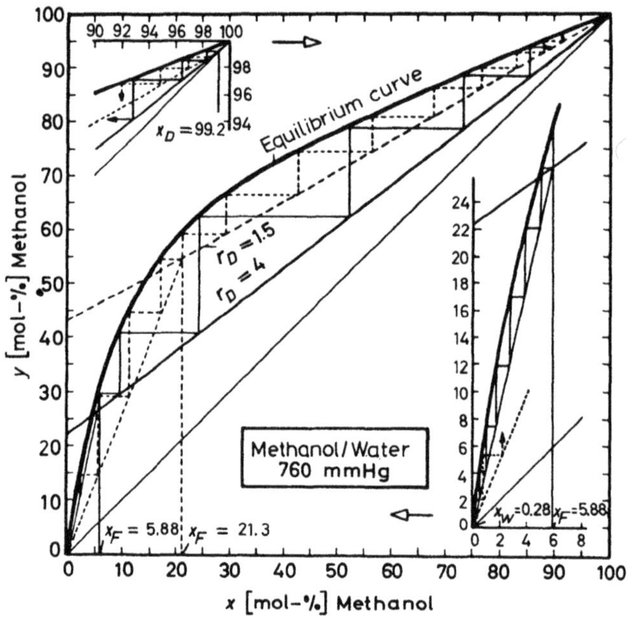

Fig. 3.11 Interchange lines and theoretical stages for the rectification of a methanol- water mixture at atmospheric pressure with x_F = 5.88 and 21.3 mol-% of methanol in the feed.

C. Multicomponent Mixtures

Essentially, the thinking devoted to a rectification of binary mixtures may also be applied to the decomposition of mixtures with three or more components. Equations for interchange lines can be written for each component. If the phase equilibria for a multicomponent mixture are known, the theoretical steps can be determined as for binaries. The assumption that the vapor leaving a stage is at phase equilibrium with the liquid establishes a relationship between the concentrations, of two adjacent steps. Since, however, the number of variables in a multicomponent system may be high, calculation of the number of steps becomes complex and use of computers may pay off (note chapter 14).

For ideal multicomponent mixtures, a determination of the theoretical stages of the vapor pressure curves are available, even though there may be no tests on the vapor equilibria. The relative volatilities α_{cn} follow from the vapor pressure curves using Eqs. (2-30), for each component c of the multicomponent mixture of materials. These relative volatilities may be constant within wide limits for ideal mixtures. In a y_c/x_c plot the interchange lines for each component c can be drawn. They run parallel with each other. For a given composition $(x_1)_W \ldots (x_c)_W \ldots (x_n)_W$ of the runoff, the composition $(y_1)_W \ldots (y_c)_W \ldots (y_n)_W$ of the vapor entering the lowest stripping step is found from Eqs. (2-31), allowing for the proper relative volatilities. The vapor concentrations thus computed, $(y_c)_W$ (for $c = 1, 2, 3 \ldots n$) are plotted over the corresponding $(x_c)_W$ values. For the cth component, one obtains point 0 in Fig. 3.12

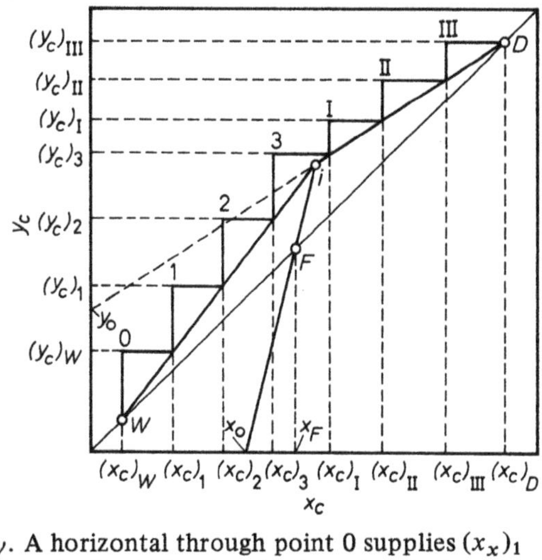

Fig. 3.12 Finding the number of theoretical steps for an ideal multicomponent mixture.

located above $(x_c)_W$ at a distance $(y_c)_W$. A horizontal through point 0 supplies $(x_x)_1$ which is the share of the cth component in the liquid leaving the lowest stripping step. From the individual y_c/x_c plots, one can find the composition $(x_1)_1 \ldots (x_c)_1 \ldots (y_n)_1$ of the vapor at equilibrium with it, rising from the lowest stripping step and entering the second lowest stripping step, again using Eqs. (2-31). For the cth component, one finds a $(y_c)_1$ to be plotted against $(x_c)_1$ in Fig. 3.12, point 1. The horizontal through point 1 cuts off abscissa $(x_c)_2$
cth component in the liquid leaving the second lowest stripping stage. Following Fig. 3.12, this routine can be continued until share $(x_c)_D$ of the cth component in the top product is reached. Then, points (1, 2 . . .) give the number of the theoretical stripping steps, and points (I, II . . .) give those of the rectifying steps.

A trial-and-error procedure yields the reflux conditions; eventually the minimum reflux r_m will be found for which the number of steps become infinite. Note (9-17) in the bibliography.

When the relative volatilities for the temperature range considered are not constant, the above procedure must be modified. For an ideal multicomponent mixture, first the boiling points for the individual steps must be located. Try several boiling points for the step in question, and determine the values for the relative volatilities. According to Eqs. (2-31) the concentrations can be computed for each of these tentative boiling points for the liquid runoffs at phase equilibrium with the vapors leaving the same step. An y_c/x_c plot must be drawn for each component. That stage temperature for which the sum of the concentrations in the vapor of that stage equals 100 is the correct one [Eqs. (2-32)].

For the total separation of an ideal multicomponent mixture, $(n-1)$ columns are required. In general, the components are withdrawn in the sequence of their boiling points as top products (Fig. 3.13).

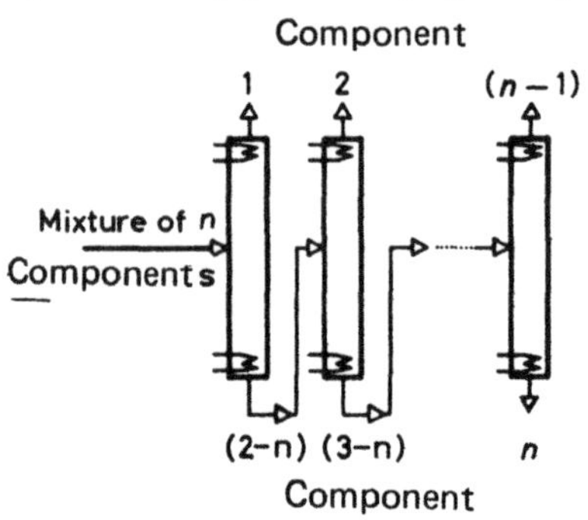

Fig. 3.13 Hookup of an installation for separating a multicomponent mixture.

For each column, the number of theoretical steps can be found as shown; according to Fig. 3.13 the number of components in question diminishes by one as compared to the prior one, in any given column. The theoretical stages of the last column are determined by the method given for binary mixtures. A combination of one principle column with several lateral columns will be discussed in Chapter 6, Section 4.A.

The bibliography (6, 7, 18–20) discloses still other methods for finding the number of theoretical stages for ideal multicomponent mixtures. The method described may be used for nonideal multicomponent mixtures, provided Eqs. (2-31) are replaced by the relations (2-36).

The height of a column is ultimately dependent on the equivalence of the number of theoretical separation stages and the effectiveness of the internals employed. This question is treated in Chapter 6.

4. COLUMN DIAMETER AND COLUMN THROUGHPUT

When the quantity of the vapor V (kmol/h) is known, diameter d_s (m) of the column follows from

$$d_s = \left(\frac{4}{\pi} V \frac{22.4}{3600} \frac{T}{273} \frac{760}{p} \frac{1}{u_V}\right)^{1/2} = 0.149\left(V \frac{T}{p} \frac{1}{u_V}\right)^{1/2} \tag{3-39}$$

where

T = absolute temperature, (K), at column level considered,
p = absolute pressure (mm Hg) at same level,
u_V = vapor velocity in free cross section of column, (m/s)

Since the vapor velocity must be maintained within certain limits and V, T, p, and u_v vary along the column, it is recommended that the calculation of d_s by Eq. (3-39) be made, for installations having a rectifier and a stripper, on a total of four sections which are at the column head, above and below the feed point, and at the bottom of the column.

Again neglecting superheats and heats of mixing, in differential Eq. (3-1) $q_h = \lambda_h$, and $q_1 = \lambda_1$; in Eq. (3-3) q = a constant K [Eq. (3-11)]. By integration of Eq. (3-1) within the limits $V = V_e$, the mass of mixed vapors at any given level of the rectifier with concentration y, V equalling V_D, and the vapor leaving the column top with concentration y_D, Eq. (3-23) combined with Eq. (3-22) will result—from a mass balance around the column top

$$V_D = D\,(r_D + 1) \tag{3-40}$$

and for the mass of vapor in the rectifier

$$V = V_e = F\,\frac{1 - \dfrac{y_D}{K}}{1 - \dfrac{y}{K}}\,(r_D + 1)\,\frac{x_F - x_W}{x_D - x_W} \tag{3-41}$$

Eq. (3-41) will give, for the mass of vapor at the rectifier top, with $y = y_D$

$$V_D = F(r_D + 1)\frac{x_F - x_W}{x_D - x_W} \tag{3-42}$$

and for the mass of mixed vapor flowing at the level of the lower rectifier section, above the feedpoint, and for $y = y_i$

$$V_i = V_D\,\frac{1 - \dfrac{y_D}{K}}{1 - \dfrac{y_i}{K}} \tag{3-43}$$

For a total condensation of the mixed vapors in the reflux condenser, $y_D = x_D$ (Fig. 3.1).

Analagously, integration of differential equation (3-1) within the limits of $V = V_s$, the vapor at some level of the stripper, and V_i, the vapor immediately above the feed-point, and the correlated vapor concentrations y to y_i^* and combination of Eq. (3-20) with (3-43) will yield

$$V_i^* = V_D \frac{1 - \frac{y_D}{K}}{1 - \frac{y_i}{K}} + F(f_i - 1) \tag{3-44}$$

and for the mass of mixed vapor flowing at some level of the stripper

$$V = V_s = \frac{1 - \frac{y_i}{K}}{1 - \frac{y}{K}} [V_i + F(f_i - 1)] \tag{3-45}$$

From this, the mass of vapor flowing at the base of the stripper

$$V_W = \frac{1 - \frac{y_i}{K}}{1 - \frac{y_W}{K}} [V_i + F(f_i - 1)] \tag{3-46}$$

where y_W is the phase equilibrium concentration for runoff concentration x_W (Fig. 3.1). Introducing the mass of reflux $L = L_e$ in the rectifier, and the reflux ratio

$$r = \frac{L_e}{D} \tag{3-47}$$

the vapor flow at that section becomes

$$V = V_e = F(r + 1) \frac{x_F - x_W}{x_D - x_W} \tag{3-48}$$

Equating this relation with Eq. (3-41) yields

$$r = (r_D + 1) \frac{1 - \frac{y_D}{K}}{1 - \frac{y}{K}} - 1 \tag{3-49}$$

by which the reflux ratio r_D at the head of the column can be converted for any given level of the rectifier. Combining Eq. (3-23) and (3-47) gives for the stream of liquid corresponding to the vapor mass V_e [Eq. (3-41)]

$$L = L_e = Fr \frac{x_F - x_W}{x_D - x_W} \tag{3-50}$$

with r given by Eq. (3-49). Thus, the molar reflux mass may be found for any section of the rectifier within the range $y_i < y < y_D$.

Finally, the mass of liquid correlated to the molar vapor quantity $V = V_s$ [Eq. (3-45)] at some level of the stripper resulting from a mass and concentration balance is found from

$$L = L_s = V_s + F\left(1 - \frac{x_F - x_W}{x_D - x_W}\right) \tag{3-51}$$

If the higher boiler has the greater molar heat of evaporation, $\lambda_h > \lambda_1$, according to Eq. (3-11) K will exceed zero, and the molar quantities must increase in the columns going upward because of Eqs. (3-40) to (3-51).

However, for equal molar heats of evaporation the molarities in the rectifier and the stripper will be virtually constant.

Example 3 Determine, for $F = 40$ kmol/h and for the data of Example 1, the molar vapor and liquid quantities in the stripper and the rectifier as functions of the concentrations involved.

Answer Eq. (3-42) will supply the mass of vapor at the column top which enters the reflux condenser

$$V = V_D = 40(4 + 1)\,\frac{50 - 0}{100 - 0} = 100 \text{ kmol/h}$$

From Eq. (3-43) the mass of vapor above the feedpoint becomes

$$V = V_i = 100\,\frac{1 - \dfrac{100}{250}}{1 - \dfrac{62.5}{250}} = 80 \text{ kmol/h}$$

From Eq. (3-44) the mass of vapor below the feedpoint becomes, with $f_i = 1$

$$V = V_i^* = V_i = 80 \text{ kmol/h}$$

From Eq. (3-46) the mass of vapor rising from the reboiler is

$$V_W = \frac{1 - \dfrac{62.5}{250}}{1 - \dfrac{0}{250}}\,[80 + 40(1 - 1)] = 60 \text{ kmol/h}$$

From Eq. (3-50) and $r = r_D = 4$, the reflux quantity at the column top becomes

$$L_e = L_D = 40.3\frac{50 - 0}{100 - 0} = 80 \text{ kmol/h}$$

The reflux ratio $r = r_i$ based on the feedpoint section equals [from Eq. (3-49)]

$$r = r_i = (4 + 1)\frac{1 - \dfrac{100}{250}}{1 - \dfrac{62.5}{250}} - 1 = 3$$

From Eq. (3-50) the mass of reflux at the feedpoint becomes

$$L_e = L_i = 4 \cdot 3\ \frac{50 - 0}{100 - 0} = 60 \text{ kmol/h}$$

Eq. (3-51) yields the quantity of reflux flowing from the feedpoint level to the stripper

$$L_s = L_i^* = 80 + 40\left(1 - \frac{50 - 0}{100 - 0}\right) = 100 \text{ kmol/h}$$

Finally, with $V = V_W = 60$ kmol/h, the reflux returning to the reboiler becomes

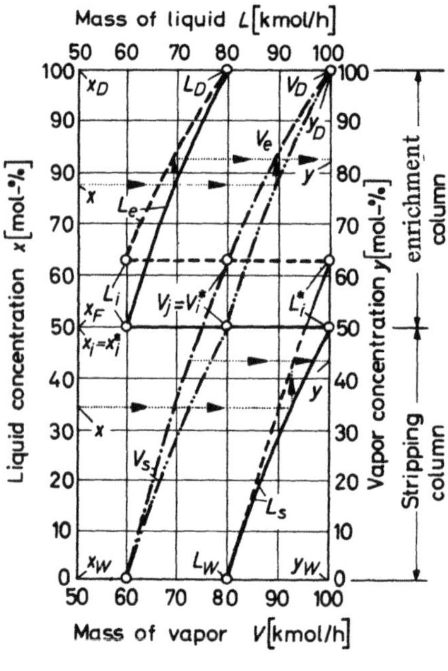

Fig. 3.14 Changes of molar liquid and vapor quantities in a rectifier working a binary mixture in which the ratio of molar heats of evaporation is $\lambda_l/\lambda_h = 6000/10{,}000$. $F = 40$ mol/h, $x_F = 50$ mol-%, $x_D = 100$ mol-%, $x_W = 0$ mol-%, $r_D = 4, f_i = 1$.

$$L_s = L_W = 60 + 40\left(1 - \frac{50 - 0}{100 - 0}\right) = 80 \text{ kmol/h}$$

Fig. 3.14 depicts the interrelation between vapor and liquid concentrations and masses in the rectifier and the stripper. Correlated concentrations are points of interchange lines according to Fig. 3.4 calculated by Eqs. (3-10) and (3-27).

The diameter of the column depends, according to Eq. (3-39), on the vapor flow and the pressure to temperature ratio p/T, and further on the admissible vapor velocity u_v–dependent upon the liquid to vapor ratio L/V. For the latter, calculation methods for various internals will be given in Chapter 8. In general, the diameters of the rectifier and the stripper will be those which turn out to be larger in the calculations.

5. HEAT REQUIREMENTS

The heat requirement Q (kcal/h) of a rectifier is readily found from a heat balance along the column (Fig. 3.15). The quantity of heat entering must equal that leaving. The heat supplied to the reboiler is Q (kcal/h), while the charge supplies Q_F(kcal/h). If total condensation takes place in the reflux condenser, for a vapor quantity V_D the heat Q_V (kcal/h) must be removed. In the dephlegmator hookup, the distillate vapor D (kmol/h) is condensed in the condenser-cooler, and part of Q_V (kcal/h) must be removed (Fig. 3.16). In either case, along with the liquefied top product D (mol/h) the heat Q_D will leave the equipment, with the runoff W supplying the heat Q_W (kcal/h).

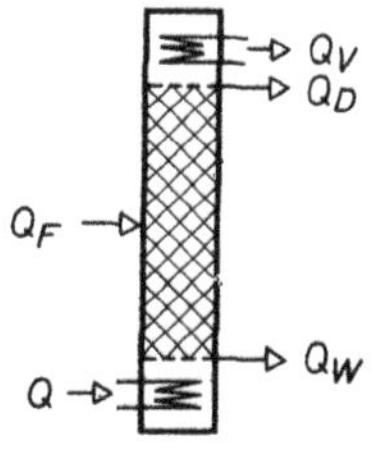

Fig. 3.15 Heat quantities with total condensation in the reflux condenser.

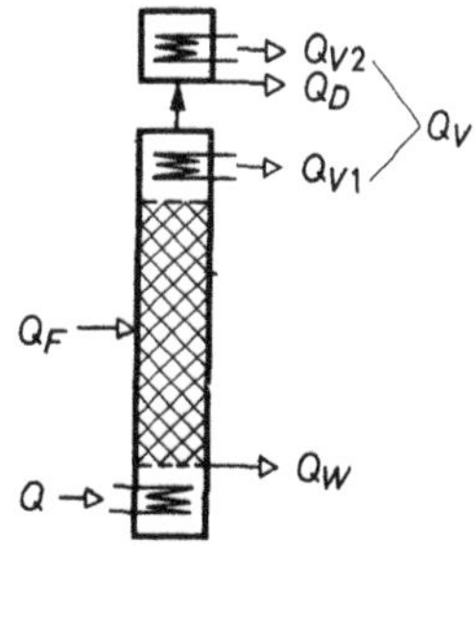

Fig. 3.16 Heat quantities with the dephlegmator hookup.

These heat quantities can be determined separately. For totally condensing vapor V_D (kmol/h), in the reflux condenser, the molar heat of condensation is $\lambda = \lambda_D$ (kcal/kmol). Eq. (3-42) then yields

$$Q_V = F(r_D + 1)\frac{x_F - x_W}{x_D - x_W}\lambda_D \tag{3-52}$$

In the dephlegmator hookup, the reflux condenser removes, with λ_R (kcal/mol) for molar heat of condensation of the reflux R (kmol/h), the heat

$$Q_{V1} = R\lambda_R = r_D D\lambda_R \tag{3-53}$$

and the condenser-cooler removes as pure condensation heat

$$Q_{V2} = D\lambda_D \tag{3-54}$$

Note Fig. 3.16. The total heat of condensation to be removed then equals

$$Q_V = Q_{V1} + Q_{V2} = D(r_D\lambda_R + \lambda_D) \tag{3-55}$$

which, from Eq. (3-23), becomes

$$Q_V = F(r_D\lambda_R + \lambda_D)\frac{x_F - x_W}{x_D - x_W} \tag{3-56}$$

Eq. (3-56) can be replaced with Eq. (3-52) if $\lambda_R = \lambda_D$ holds for equal molar heats of evaporation. For different molar heats of evaporation, for the individual components, this is still admissible provided the enrichment achieved in the dephlegmator is slight.[21,22,23] Hence, Eq. (3-52) is acceptable in practice.

Together with the liquefied top product D (kmol/h) (note Eq. 3-23), the heat

$$Q_D = Dh'_D = F\frac{x_F - x_W}{x_D - x_W}h'_D \tag{3-57}$$

will leave the equipment. Also, the bottom product

$$W = F - D \tag{3-58}$$

will remove the heat

$$Q_W = Wh'_W = F\left(1 - \frac{x_F - x_W}{x_D - x_W}\right)h'_W \tag{3-59}$$

h'_D and h'_W (kcal/kmol) are the molar enthalpies of the top and the bottom product at boiling temperature.

The enthalpy in the feed is h_F (kcal/kmol). It imports into the equipment the heat

$$Q_F = Fh_F \tag{3-60}$$

Now the heat balance becomes

$$Q = Q_V + Q_D + Q_W - Q_F \tag{3-61}$$

Summing up these heats, and allowing for the heat losses Q_l (kcal/h) yields a total heat requirement

$$Q = F\left\{\frac{x_F - x_W}{x_D - x_W}[(r_D + 1)\lambda_D + h'_D - h'_W] + h'_W - h_F\right\} + Q_l \qquad (3\text{-}62)$$

This relation reveals immediatley how the variables in rectification affect the heat distribution. Thus, when the mixture is heated by the runoff heat in a runoff cooler, or by the heat of condensation in the reflux condenser, the heat requirement is reduced as h_F increases. A reduction of reflux ratio, r_D, also reduces the heat requirement. Special means for economizing heat are discussed in Chapter 7.

Example 4 Calculate a continuous rectifier for making ethanol, using 324 kmol/h of an ethanol-water mash with 1.9 mol-% ethanol to produce a top product with 70.6 mol-% ethanol. The runoff should not contain more than 0.01 mol-% ethanol. The mash has an original temperature of 25 C; it is to be heated in the reflux condenser to 70 C prior to being charged into the rectifier. The plant operates at atmospheric pressure. No side products are intended.

Answer The specifications state:

$p = 760$ mm Hg $\qquad F = 324$ kmol/h $\qquad x_F = 1.9$ mol-%
$x_D = 70.6$ mol-% $\qquad x_W = 0.01$ mol-%

From Eq. (3-23) the quantity of top product becomes:

$$D = (324)\,\frac{1.9 - 0.01}{70.6 - 0.01} = 8.67 \text{ kmol/h}$$

From Eq. (3-58) the runoff quantity becomes:

$$W = 324 - 8.67 = 315.33 \text{ kmol/h}$$

Assume the mash to behave like a binary mixture of ethanol and water; disregard amyl alcohols and such additives. Physical Tables give, for the above mixture, for $x = 1.9$ mol-% an $h_F = 1260$ kcal/kmol, for 70 C, an $h'_F =$ 1730 kcal/kmol for 95 C and an $\lambda = 9600$ kcal/kmol, which is virtually independent of concentration, since the pure components ethanol and water have almost identical molar evaporation heats. For these enthalpies, Eq. (3–33) yields

$$f_i = 1 \cdot \frac{1730 - 1260}{9600} = 1.049$$

Then the secant can be plugged in the y/x diagram for ethanol/water.

$$x_o = \frac{1.9}{1.049} = 1.81 \text{ mol-\%}$$

and the equilibrium curve with the coordinates

$$y_E = 26 \text{ mol-\%, and } x_E = 3.1 \text{ mol-\%}$$

for which Eq. (3-36) gives a minimum reflux ratio of

$$r_m = \frac{70.6 - 26}{26 - 3.1} = 1.95$$

ecause of the location of x_D, x_F, and $x_o = x_F/f_i$, Eq. (3-36) cannot in this s case be used, although there is an inflection point in the equilibrium curve. An actual reflux ratio of

$$r = r_D = 1.8\, r_m = 3.5$$

is selected, for which the y/x plot allows between rectifying line and equilibrium curve

$$n_t = n_{te} = 4$$

theoretical rectifying steps, and, between the stripping line and the equilibrium curve

$$n_t = n_{ts} = 12$$

theoretical stripping steps.
The Physical Tables give, for x_D = 70.6 mol-% and 79 C boiling point, and for x_W = 0.01 mol-% and 100 C boiling point, the molar enthalpies h'_D = 2120 kcal/kmol, and h'_W = 1800 kcal/kmol for atmospheric pressure. With this, Eq. (3-62) yields a heat requirement

$$Q = 324 \left\{ \frac{1.9 - 0.01}{70.6 - 0.01} \left[(3.5 + 1)9600 \quad 2120 - 1800 \right] + 1800 - 1260 \right\}$$

$$= 554{,}000 \text{ kcal/h}$$

disregarding heat losses and the pressure rise at the column bottom caused by the pressure loss of the vapor flowing through the column.

Example 5 Raw methanol containing 1.4 mol-% of dimethyl ether is to be refined by continuous rectification so that the methanol can be withdrawn from the bottom of the column with a content of only 0.14 mol-% of dimethyl ether, while the latter is obtained at the column top, enriched to 95.7 mol-%. The operating pressure is to be 9 kg/cm^2.
The molar heats of evaporation of dimethyl ether and methanol are in the ratio of 4060 and 7250 kcal/kmol. The influence of the feed temperature is to be investigated between 100 and 132 C, while the boiling point of the original mixture is investigated at 9 kg/cm^2.

Answer Fig. 3.17 demonstrates how the intersections (y_E, x_E) of the secant and the equilibrium curve will migrate if the inlet temperature, t_F = 100 C, is increased up to the boiling temperature, $t_F = t_b$ = 132 C. In view of the steep equilibrium curve, a marked dependence of the minimum reflux ratio on the feed temperature must be expected. In Fig. 3.18, the minimum reflux ratio r_m is plotted against the vapor concentrations of the secant y_E. The upper r_m curve refers to the reflux quantity at the column top, and the lower curve r_E refers to the reflux quantity above the feedpoint. In accordance with Eq. (3-49) both are correlated

$$r_E = (r_m + 1) \frac{1 - \dfrac{x_D}{K}}{1 - \dfrac{y_E}{K}} \tag{3-63}$$

By Eq. (3-11) factor K was computed, with $\lambda_h = 7250$ kcal/kmol
$\lambda_l = 4060$ kcal/kmol

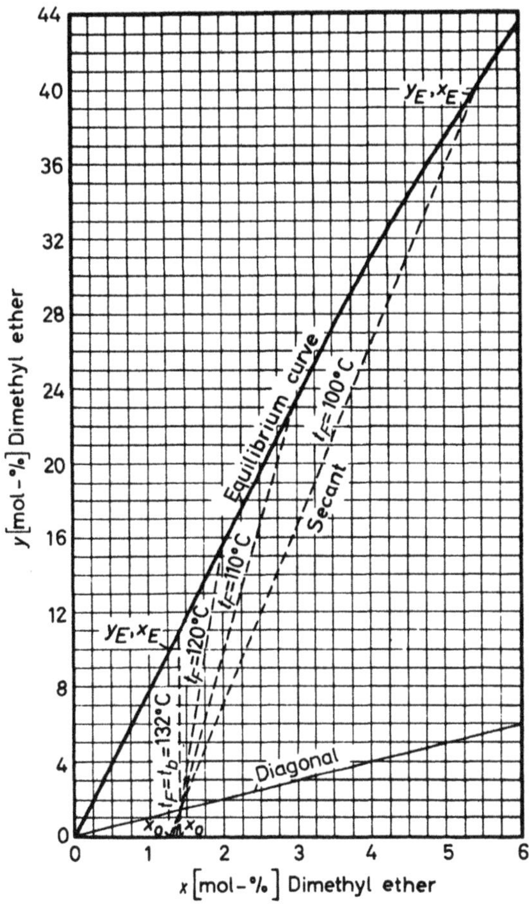

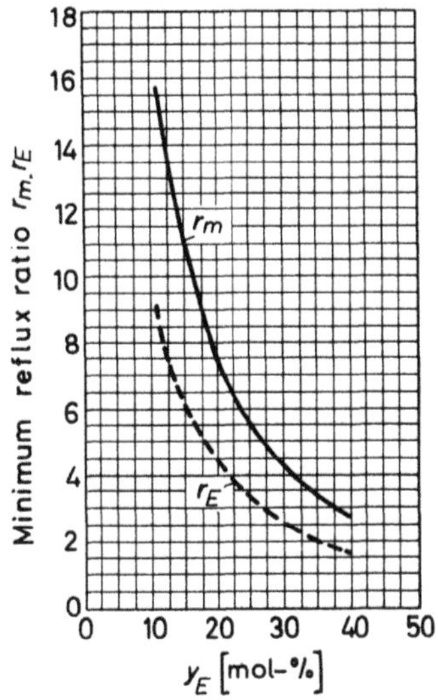

Fig. 3.18 Minimum reflux ratio as a function of vapor concentrations at intersection for the conditions of Example 5.

Fig. 3.17 Location of secant for different feedstock temperatures for dimethyl ether/methanol.

$$K = \frac{7250}{7250 - 4060}(100) = 227$$

If equal molar heats of vaporization were assumed, K would equal infinity, and Eq. (3-36) would yield with $t_F = 132$ C, $y_E = 10.8$ mol-%, and $x_E = x_F = 1.4$ mol-%

$$r_m = \frac{95.7 - 1.4}{10.8 - 1.4} - 1 = 9$$

However, by taking into account the different molar heats of evaporation, Eq. (3-35) will yield a minimum reflux ratio at the column top of

$$r_m = \frac{227 - 10.8}{227 - 95.7}\,\frac{95.7 - 1.4}{10.8 - 1.4} - 1 = 15.5$$

to which is correlated, by Eq. (3-63), a reflux ratio above feedpoint of

$$r_E = (15.5 + 1)\frac{1 - \dfrac{95.7}{227}}{1 - \dfrac{10.8}{227}} - 1 = 9$$

that is numerically equivalent with the reflux ratio based on equal molar heats of vaporization, and constant along the rectifier height. Obviously, the inequality of the molar heat of vaporization must be considered to avoid giving an inadequate minimum reflux ratio for the column top. Molar quantities in the rectifier will increase going up because more molecules of low boiler will evaporate from the liquid phase than molecules of high boiler will condense from the vapor phase, in view of the greater molar heat of evaporation of the higher boiler.

Figs. 3.17 and 3.18 permit evaluation of Fig. 3.19 which illustrates the dependence of the minimum reflux, r_m and r_E, on the feed temperatures, and that of the intersection coordinates y_E, x_E. Note that the validity of the curves is based on perfect mixing of liquid and vapor at the feedpoint, and hence the heating of the feed to the boiling point at that location.

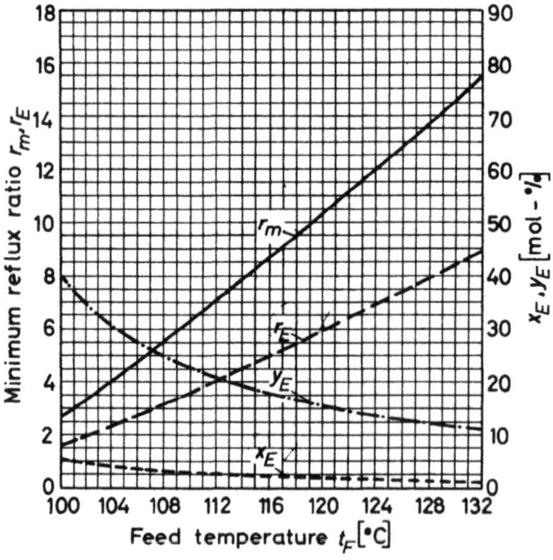

Fig. 3.19 Minimum reflux ratio and intersection coordinates (or secant and equilibrium curve) versus feed temperature for the conditions of Example 5.

It is assumed that in actual operation at the column top a reflux ratio of $r_D = 1.4 r_m$ is sufficient. With this, the rectifying lines are computed for different feed temperatures in the $100\ C < t_F < 132\ C$ range by means of Eq. (3-10). Then Eq. (3-12) determines the intersection points in the y/x plot.

The usual steps are drawn between the intersection points of the rectifying line and the diagonal, $y = x = x_D$, and the intersection of the rectifying line with the secant $y = y_i$, $x = x_i$, which yields the number of theoretical steps n_{te} for the rectifying column. In Fig. 3.20 these have been plotted over the feed temperatures t_F.

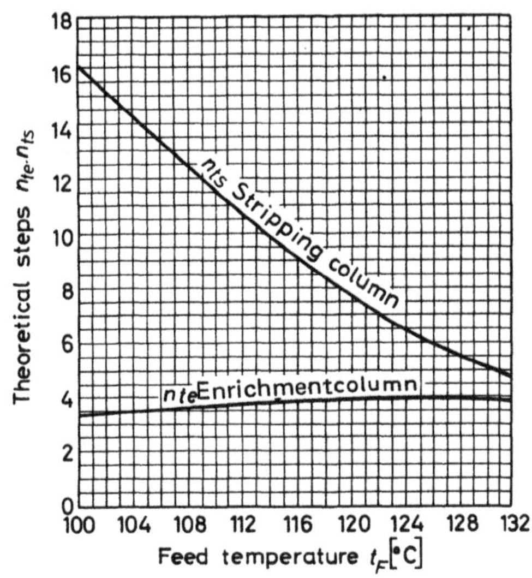

Fig. 3.20 Theoretical steps as function of feed temperature, for the conditions of Example 5.

Analogously, one finds the number of theoretical steps in the stripper from the sequence between the stripping line, Eq. (3-27), and the diagonal in the range $y = y_i$, $x = x_i$, and $y = x = x_W$. Fig. 3.20 also shows the stripping steps n_{ts} as a function of the feed temperature t_F. Eqs. (3-41) to (3-51) serve to determine the molar quantities of vapor and liquid of the charge in the stripper and the rectifier. Fig. 3.21 shows these quantities and the ensuing vapor and liquid ratios versus the feed temperature and versus, the reflux ratios for four column levels:

1. Top of column, suffix "D",
2. Above feedpoint, suffix "i",
3. Below feedpoint, suffix "i*",
4. Bottom of column, suffix "W".

The greater the feed temperature, and hence the reflux ratio, the higher the expected vapor load of the column. Characteristically, the molar liquid to vapor ratio of the stripper is about seven to ten times larger than that of the rectifier.

Referring to Fig. 3.5, Fig. 3.22 shows in two instances how the vapor and liquid temperature can be determined for a given level of a rectifier, with the aid of the equilibrium curve, the boiling, and the dewpoint lines. First, let $r_D = 22$ describe a

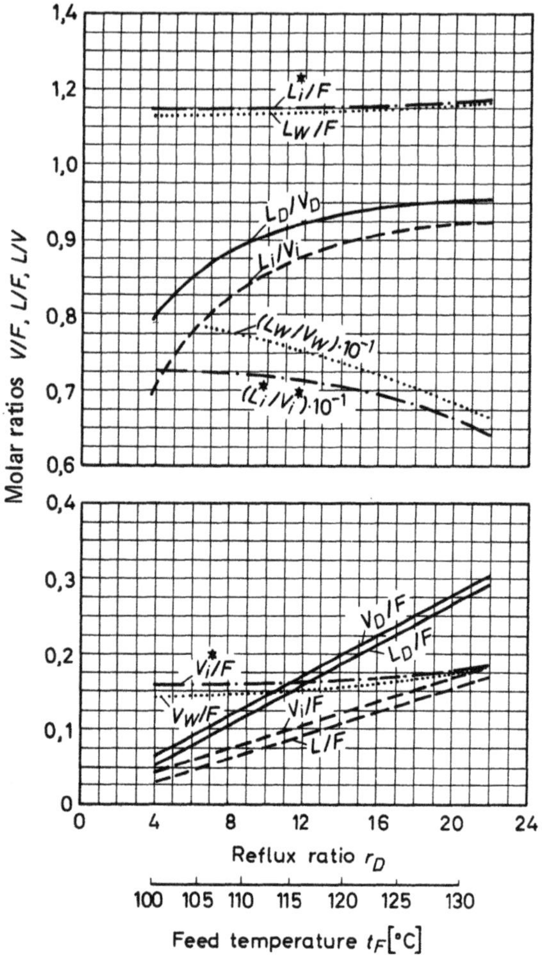

Fig. 3.21 Molar quantities as functions of the reflux ratio and the feed temperature for conditions of Example 5.

condition under which the top product is obtained with the desired top concentration of x_D = 95.7 mol-% of dimethyl ether. Second, let r_D = 10 describe a condition where the obtained top concentration is less because of too low a reflux rate, (for example, x_D = 82 mol-%).

The lines defined by circles indicate the procedure for finding the temperature difference ϑ between vapor and liquid at a given level of the rectifier; ϑ_D applies to column top. The vapor temperatures t_V and liquid temperatures t_L thus determined have been plotted in Fig. 3.23, and the corresponding temperature differences are in Fig. 3.24, correlated to the theoretical steps, counting down from the column top. The curves apply for the reflux at the boiling point. According to Fig. 3.24 the temperature differences between the vapor and the liquid are relatively large; they are up to 50 C. The heat consumption, related to the feed quantity, is determined from

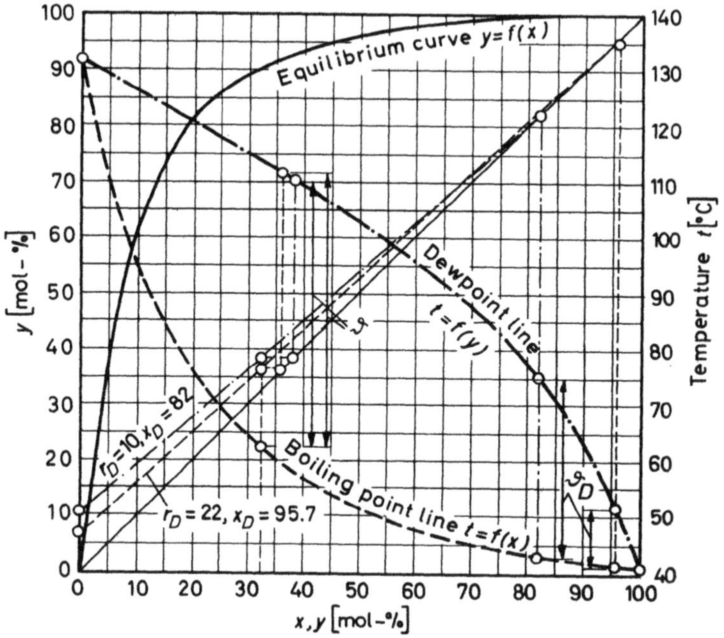

Fig. 3.22 Determination of the temperatures and temperature differences between vapor and liquid at given levels of the rectifier, for the conditions of Example 5—system dimethyl ether/methanol, at 9 kg/cm², $t = f(y) = t_V$, $t = f(x) = t_L$.

Eq. (3-62). In Fig. 3.25 the heavy curve gives the relation per one kilomole of feed for the temperature, or for the respective reflux ratio $r_D = 1.4\, r_m$. From this, the heat consumption increases rather slowly with the reflux ratio, if the feed is properly preheated.

By taking advantage of the runoff heat, the feed can be preheated up to a maximum of

$$t_F = t_f + \frac{x_D - x_F}{x_D - x_W} \frac{c\Big|_{t_w}^{t_W}}{c\Big|_{t_f}^{t_F}} (t_W - t_w) \tag{3-64}$$

if the heat exchanger is amply dimensioned.

In Eq. (3-64):

t_f = runoff temperature of the feed,

$c\Big|_{t_f}^{t_F}$ = mean specific heat of the feed between t_f and t_F

t_W = boiling point of runoff

t_w = the runoff temperature after cooling by the fresh feed

$c\Big|_{t_w}^{t_W}$ = specific heat of runoff between t_w and t_W

According to Eq. (3-64) the runoff heat will not suffice to preheat the feed to its boiling point even if the exchanger is ample. The dotted lines in Fig. 3.25 indicate how heat consumption increases if the reflux ratio is enlarged with constant feed temperature.

The heavy curve in Fig. 3.25 reveals that the heat consumption of 1082 kcal/kmol feed increases by only 8.5 % to 1175 kcal/kmol feed if the reflux ratio of 10 is increased to 18, or the feed is preheated to 126.2 C. If preheating to 118 C is feasible, heat consumption will increase by almost 28 % to 1383 kcal/kmol feed; note dotted curve $f(r_D)$, t_F = 118 C.

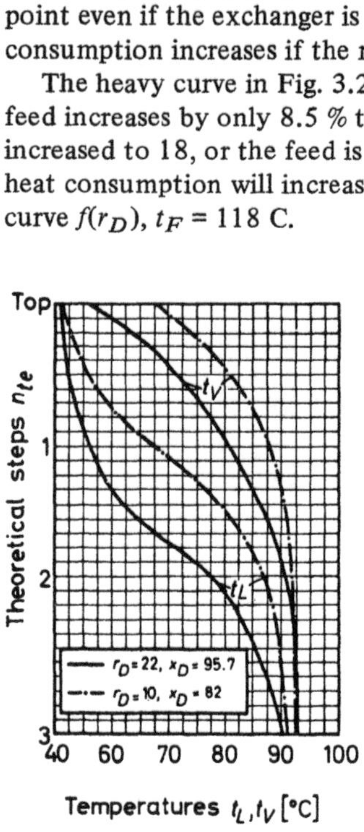

Fig 3.23 Number of theoretical steps versus the temperatures of vapor and liquid in correlated locations in the rectifier for the conditions of Example 5.

Fig 3.24 Number of theoretical steps versus temperature differences between vapor and liquid at corresponding levels of the rectifier for the conditions of Example 5.

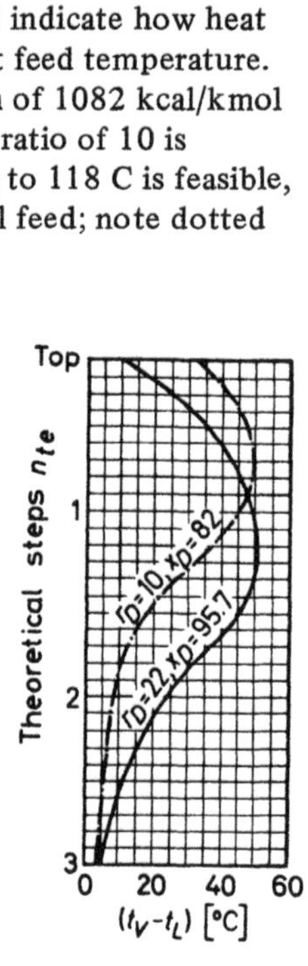

The optimum feed temperature, giving a minimum of heat requirement, would demand precise knowledge of the equilibrium curve and the enthalpies. That a minimum is achieved is demonstrated in that, with growing feed temperature, the subtrahend h_F becomes larger, and so does the reflux ratio r_D.

There remains the question of numerically representing the theoretically feasible mixing process in the feeding zone. For example, let a coincidence of t_F = 110 C feed temperature and a reflux ratio of r_D = 8.8 be assumed. Fig. 3.26, case a), would yield the following mass balances for the feedpoint level

$$x_F + \frac{L_i}{F} x_i + \frac{V_i^*}{F} y_i^* = \frac{V_i}{F} y_i + \frac{L_i^*}{F} x_i^* \tag{3-65}$$

Furthermore, Fig. 3.26 explains how, for the above example, the y_i and x_i values are determined. In Fig. 3.27 they are plotted against the feed inlet temperature, for

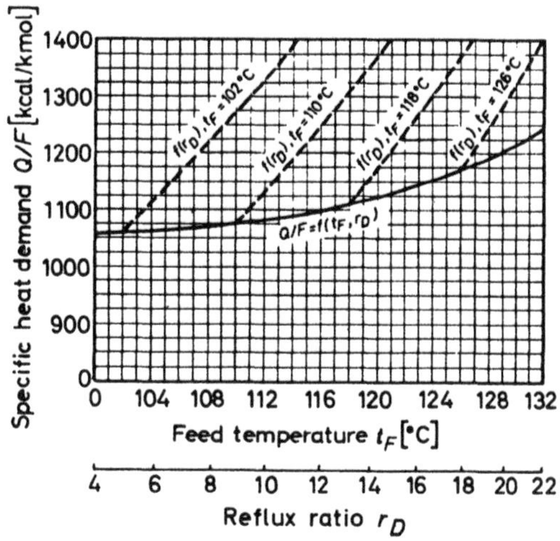

Fig. 3.25 Heat requirement as a function of the feed temperature and the reflux ratio for the conditions of Example 5.

a reflux ratio around $r_D = 1.4\, r_m$.

In theory, as the mixing at the feedpoint level corresponds to the intersection of the stripping and the enriching lines (Fig. 3.26) case a) yields

$$y_i^* = y_i = 17.6 \text{ mol-\%} \qquad x_i^* = x_i = 2.57 \text{ mol-\%}$$

From this, and with the data calculable by Eqs. (3-41) to (3-51) and obtainable from Fig. 3.21

$$\frac{L_i}{F} = 0.0676 \quad \frac{V_i}{F} = 0.0808 \quad \frac{L_i^*}{F} = 1.146 \quad \text{and} \quad \frac{V_i^*}{F} = 0.1592$$

the mass balance of Eq. (3-65) will yield

$$1.4 + (0.0676)(2.57) + (0.1592)(17.6) \cong (0.0808)(17.6) + (1.146)(2.57)$$
$$4.374 \cong 4.364$$

which is a satisfactory agreement.

However, the mass balance may be obtained by considering the feed level as one theoretical working stage, meaning that the vapor rising from the feed level is at phase equilibrium with the liquid leaving. Then, from Fig. 3.26, case b) yields

$$x_F + \frac{L_i}{F} x_i + \frac{V_i^*}{F} y_i^* = \frac{V_i}{F} y_i + \frac{L_i^*}{F} x_i^* \qquad (3\text{-}66)$$

with $x = 2.28$ mol-% and $y_i^* = 15.5$ mol-%

$$1.4 + (0.0676)(2.57) + (0.1592)(15.5) \cong (0.0808)(17.6) + (1.146)(2.28)$$
$$4.044 \cong 4.034$$

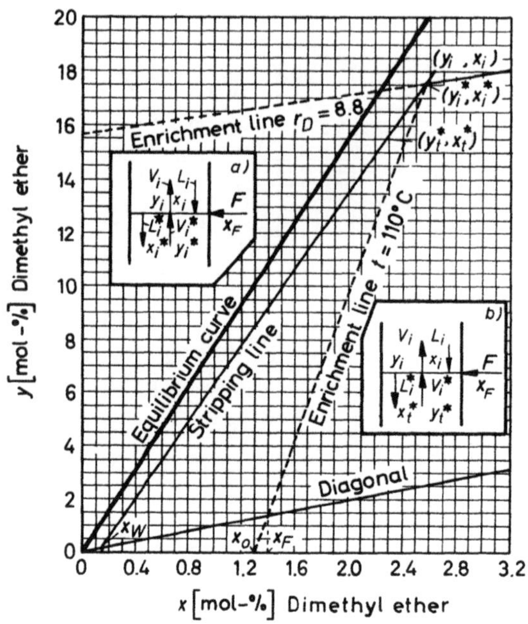

Fig. 3.26 Vapor and liquid concentrations at the feed-point for the conditions of Example 5.

It is seen that, for perfect mixing at the feed level, for cases a) and b) theoretically low reflux demands exist—if the feed is not supplied at its boiling temperature but instead at a lower temperature.

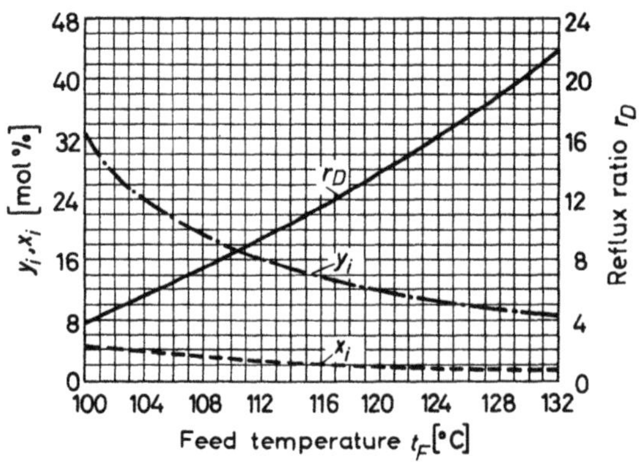

Fig. 3.27 Coordinates of intersection of the enrichment line and the stripping line, and the reflux ratio as a function of the feed temperature, for $r_D = 1.4\ r_m$, and for the conditions of Example 5.

IV The Batch Still

1. OPERATION

Discontinuous operation is preferred to steady operation when small quantities of raw materials arrive at irregular intervals, or if their compositions vary greatly. The necessary equipment then comprises a reboiler (kettle) of ample volume to receive the charge, a separating column, the customary heat exchangers to produce the reflux and to cool the product, and storage tanks.

When operating a batch still, the reflux ratio r_D at the column head may be held steady, or a constant product of composition x_D at the column top may be desired. The former procedure may prevail for batchwise separation of multicomponent mixtures, and the latter is mostly applied to separate batches of binaries. In many industries, however, both methods are combined to split up multicomponent mixtures.

A large number of variables renders the design rather complex for a continuously operating still. Variation of working conditions in time makes calculations for a batch still even more complex. In what follows, the two operating procedures mentioned above are discussed for the case of a binary mixture.

2. OPERATING LINE AND SEPARATION STEPS

Fig. 4.1 is a diagram of a batch still installation fitted with a still kettle, a column mounted on top, a reflux condenser and product cooler. It is assumed that the reflux condenser will have no enriching effect upon the stream of mixed vapors. Hence, the concentration of the column top product will be the same as that of the vapor entering the reflux condenser and that of the reflux from the condenser.

In batch rectification the column is the separator proper. Inside, the vapor from the kettle and the reflux from the condenser flow countercurrently. The correlated concentrations of the vapor and liquid are located on the rectifying line of the y/x plot and are computed from Eq. (3-10).

Qualitatively, the location of the rectifying line in the x/y plot is given in Fig. 4.2. The ordinate intersection is at y_o which is found from Eq. (3-12). Again, the reflux ratio r_D at the column top must exceed the minimum r_m that is obtained for a number n_t of theoretical steps between the equilibrium curve and the enrichment line,

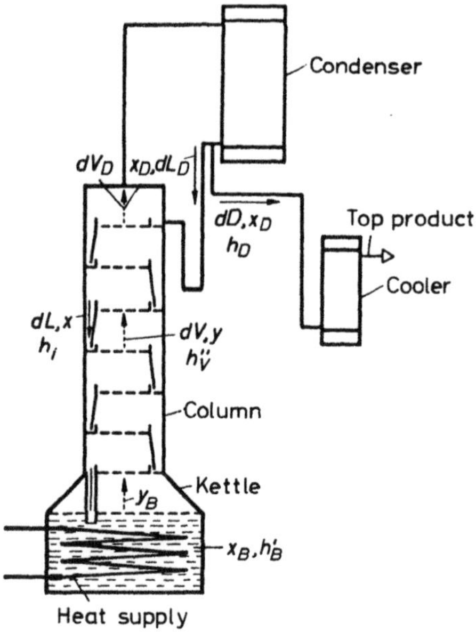

Fig. 4.1 Diagram of a batch still column mounted on a still kettle, reflux condenser, and product cooler.

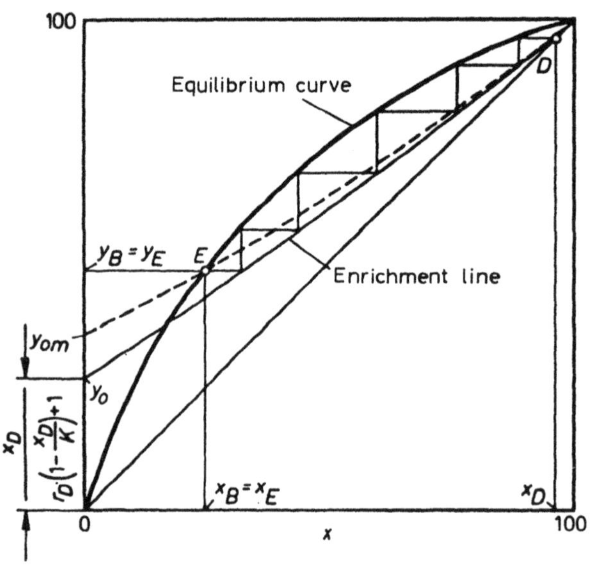

Fig. 4.2 Location of enriching line, and number of theoretical steps.

from the top concentration x_D to the still concentration x_B that becomes infinite (r_m becoming finite if enlarged by a small amount n_1).

For equilibrium curves without an inflection point, r_m may again be computed from Eq. (3-35), replacing y_E with y_B, and x_E with x_B. y_B is the phase-equilibrium concentration correlated to x_B for the mixed vapors supplied by the kettle. The enrichment line drawn in Fig. 4.2 corresponds to the minimum reflux ratio. For equal molar heats of evaporation of the individual components, Eqs. (3-10), (3-12), and (3-35) merge with (3-14), (3-15), and (3-36) which establish the enrichment line and the minimum reflux ratio.

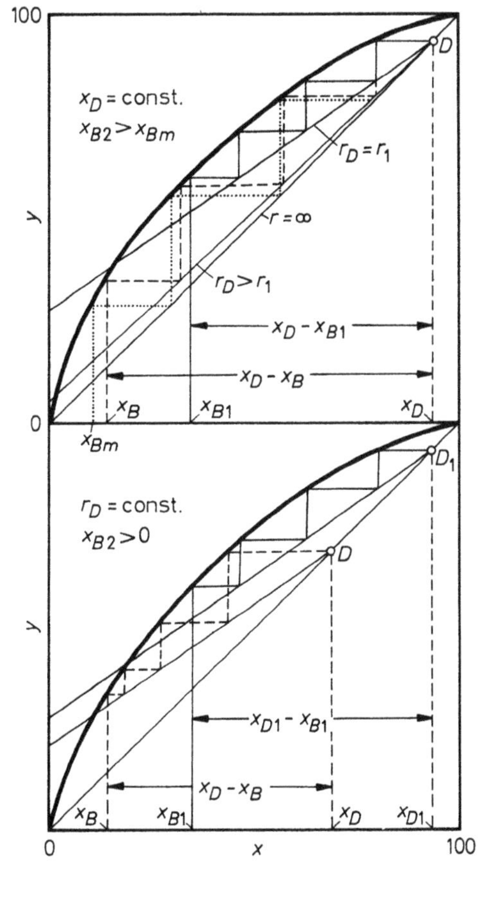

Fig. 4.3 Conditions x_D = constant and r_D = constant in batch runs.

Fig 4.3 is a juxtaposition of the two cases x_D = constant and r_D = constant for the binary mixture, referring to the connection between concentrations in the top product and the kettle charge on the one hand, and the number of theoretical steps on the other.

Simplifying assumptions include equal molar heats of evaporation and the validity of the enriching line. In either case, the initial still concentration x_B is the same, and so are the top product concentration $x_D = x_{D1}$, and the initial reflux ratio $r_D = r_1$. Thus, the enrichment lines for the start of the rectification will be parallel, and for this presentation, three theoretical steps will result: n_1, for x_D = constant, and for the case of r_D = constant. If the reflux ratio r_D is held constant, the product concentration will drop from its initial value x_{D1} to x_D when the concentration in the kettle has dropped from x_{B1} to x_B. If, however, a constant product concentration x_D is desired, the reflux ratio must be increased from its initial value of $r_D = r_1$ to $r_D > r_1$, if the kettle concentration drops from an initial x_{B1} to x_B.

Basically, the calculation of the number of steps is the same for batch operations and for continuous working. The reflux ratio must always exceed the minimum that would require an infinity of steps. Beyond that, in the case of x_D = constant the number of steps specified must exceed their minimum. The minimum number of steps is the one required to achieve the desired runoff concentration at an infinite reflux ratio. From Fig. 4.3, at constant x_D, 3 theoretical stages will secure the product concentration x_D only as long as concentration x_B (and x_{B1} at the end of the rectification) exceed x_{Bm}. In fact, for $x_B = x_{Bm}$, $n_t = 3$ is the minimum number of steps. For binary mixtures of constant relative volatility, it may also be computed from Eq. (1-34) by writing $x_W = x_{B2}$.

3. COLUMN DIAMETER, COLUMN THROUGHPUT AND HEAT REQUIREMENTS

The diameter of the column of a batch still, as outlined in Fig. 4.1, may again be found from Eq. (3-39). In this:

$V = V'$ (kmol/h) = mass of mixed vapors flowing per unit time through the column,
T (K) = the temperature of that vapor
p (mm Hg) pressure of that vapor,
u_v (m/s) = flow velocity through column section.

The variables of Eq. (3-39) will have to be so determined that during rectification the proper quantities of the charge material can be obtained, and that at the end of that period the residual material in the still retains a given concentration. Specifically, the charge is B_1 (kmol) with x_{B1} (mol-%) of low boiler. This is to be separated, within time Z_T (h), so that eventually the residue B_2 (mol) still contains x_{B2} (mol-%) of the low boiler, either with a constant concentration of the top product, x_D (mol-%) of the low boiler, or at a constant reflux ratio r_D at the column top. All quantities are, hereafter, expressed in terms of kmol and all compositions given in mol-% of the low boiler.

In the infinitesimal period dZ(h) the reflux condenser will condense the infinitesimal vapor quantity dV_D (kmol) (Fig. 4.1). The condensate formed is divided into the reflux dL_D (kmol) and the top product dD (kmol) that goes to the product cooler.

Now, let:

γ_D (kcal/kmol) = molar heat of condensation of mixed vapor,
x_D (mol-%) = concentration of mixed vapor, at condenser pressure,
h'_D (kcal/kmol) = molar enthalpy of liquid top product at boiling point.

Then, during rectification the heat balance is[1]

$$dQ = \lambda_D dV_D + h'_D dD, \tag{4-1}$$

disregarding heat losses. Heat dQ(kcal) is composed of the part dQ_H (kcal) due to the heat source, and the part dQ_B (kcal) corresponding to the reduction of the charge dB (kmol) during operation. The latter part equals the product $h'_B \cdot dB$ (kcal), with h'_B (kcal/kmol) = molar boiling entropy of the charge B (kmol) of composition x_B (mol-%) of the low boiler. Thus,

$$dQ = dQ_H + dQ_B \tag{4-2}$$

When an infinitesimal amount of vapor dV(kmol) passes through a certain level of the column in time dZ(h) countercurrently to an infinitesimal amount of reflux dL (kmol), Fig. 4.1 will give a mass balance with the differential equation

$$dV = dD + dL = (r + 1)dD \tag{4-3}$$

where r is the reflux ratio

$$r = \frac{dL}{dD} = \frac{dL/d\overline{Z}}{dD/d\overline{Z}} \tag{4-4}$$

referred to the reflux quantity flowing through the section considered. Analogously, the mixed vapor dV_D (kmol) liquefied in the reflux condenser during time dZ (h) is

$$dV_D = (r_D + 1)\, dD \tag{4-5}$$

where

r_D = reflux ratio at the top of the column.

With Eq. (4-5), Eq. (4-1) becomes the differential equation

$$dQ = \lambda_D (r_D + 1) dD + h'_D dD \tag{4-6}$$

Integration of (4-6) within the limits $D = 0$ and $D = D_T$ (kmol) will yield the equation for the heat requirement for the entire rectification period Z_T (h), for the constant x_D and for the constant r_D.

For a constant composition x_D (mol-%) of the top product, one obtains

$$Q_T = \lambda_D \int_0^{D_T} (r_D + 1) dD + h'_D D_T \tag{4-7}$$

The integral in Eq. (4-7) is solved graphically[1], the dependence of the reflux ratio r_D of the quantity of the top product D(kmol) being known. The latter is found from the relation

$$D = B_1 \frac{x_{B1} - x_B}{x_D - x_B} \tag{4-8}$$

which disregards the holdup of the column internals and the reflux condenser and makes use of the equilibrium curve and the enriching line, and where x_B (mol-%) is the variable still concentration. The latter (note Fig. 4.3, on top) will drop from x_{B1} (mol-%) at the start to x_{B2} at the end upon expiration of the total rectification time $Z = Z_T$ (h), in which the entire mass of top product $D = D_T$ is recovered. Since to each x_B and D there belongs a specific r_D, a curve $D = f(r_D + 1)$ can be drawn, and its graphic integration will yield the solution of Eq. (4-6).

For a constant r_D at the top of the column, the integration of differential Eq. (4-6) will give:

$$Q_r = (r_D + 1) \int_0^{D_T} \lambda_D \, dD + \int_0^{D_T} h'_D \, dD \tag{4-9}$$

Again, the two integrals in this equation are solved graphically, after the curves $\lambda_D = f(D)$ and $h'_D = f(D)$ have been evaluated. Since at r_D = constant the concentration of the product x_D varies with the rectification time, in a relation analogous to Eq. (4.8),

$$D = B_1 \frac{x_{B1} - x_B}{x_{Dm} - x_B} \tag{4-10}$$

for the mean product concentration x_{Dm} one must write:

$$x_{Dm} = \frac{B_1 x_{B1} - B x_B}{B_1 - B} \tag{4-11}$$

For the kettle content B (kmol) corresponding to a given still concentration x_B there exists an integral relationship derived from a differential equation for the low boiler

$$B = B_1 \exp - \int_{x_B}^{x_{B1}} \frac{dx_B}{x_D - x_B} \tag{4-12}$$

This can be solved graphically, from the equilibrium curve of the mixture and the enrichment line (Fig. 4.3). At r_D = constant, there is a specific h'_D for any x_D and D value, and if the molar heats of evaporation of the components vary there is also a specific λ_D Thus, by graphic integration of the curves $\lambda_D = f(D)$ and $h'_D = f(D)$, the two integrals in Eq. (4-9) are obtained. For mixtures of equal molar heats of evaporation, for the components, γ_D may be placed ahead of the integral which then becomes identical with D_T. It can be solved by Eqs. (4-10), (4-11), and (4-12) by writing x_{B2} for x_B.

If the Q_T values have thus been found, either for x_D = constant or for r_D = constant, integrating Eq. (4-2) will yield, with $dQ_B = h'_B \, dB$, the term

$$Q_H = Q_T - \int_0^{(B_1 - B_2)} h'_B \, dB \tag{4-13}$$

for the heat supplied during Z_T (h) for rectification.

In Eqs. (4-8), (4-10), (4-11), and (4-12), the influence of liquid holdup by the internals of the reflux condenser and column was disregarded because of the complexity of its determination. In estimates, it may be assumed that $Q_H \sim Q_T$ [from Eq. (4.13)]. Q_T is relatively small, anyway, and the liquid content of the kettle is not changed appreciably.

Heat Q_P because preheating the charge prior to the start of rectification equals the product of the initial charge, B_1 (kmol) and the enthalpy difference Δh (kcal/kmol) between charging temperature and boiling temperature. Hence

$$Q_p = B_1 \, \Delta h \tag{4-14}$$

A heat balance at a given level of the column, Fig. 4.1, will yield for a given instant

$$\frac{dQ}{dZ} = \frac{dV}{dZ} \cdot h''_V - \frac{dL}{dZ} h'_L \tag{4-15}$$

where dQ follows from Eq. (4-2). From this, with

$$dL = dV - dD, \tag{4-16}$$

and by combination with Eqs. (4-3) and (4-4), the following relations are obtained for a quantity V' (kmol/h) passing through a column section

$$V' = \frac{dV}{dZ} = \frac{dQ}{dZ} \frac{r + 1}{(r + 1)(h''_V - h'_L) + h'_L} \tag{4-17}$$

where h''_V and h'_L (kcal/kmol) are the molar enthalpies of mixed vapors and reflux liquid, and r is the corresponding reflux ratio in a given column section and for a given instant. h''_V is composed of the molar heat of evaporation λ (kcal/kmol) and the molar enthalpies h'_V (kcal/kmol) at boiling. h''_V and h'_L are correlated to concentrations y and x in the column section considered.

Enthalpies h'_V and h'_L always contain the heats of mixing. These, however, are mostly negligible.

For the conditions at the column top, Eq. (4-17) becomes more simple:

$$V'_D = \frac{dV_D}{dZ} = \frac{dQ}{dZ} \frac{r_D + 1}{(r_D + 1)\lambda_D + h'_D} \tag{4-18}$$

If the heat supplied, dQ_H/dZ (kcal/h) is held constant, quotient dQ/dZ (kcal/h) may be regarded as virtually constant, and

$$Q' = \frac{dQ}{dZ} \cong \frac{Q_T}{Z_T} \tag{4-19}$$

where Q_T must be determined depending on the mode of rectification, either by

Eq. (4-7) for x_D = constant, or by Eq. (4-9) for r_D = constant. Z_T is the entire duration of rectification.

Where the molar heats of evaporation of the components differ, usually the high boiler has the larger heat of evaporation. In that event, the molar quantities in the column become larger going up; when Eq. (3-39) is used to find the column diameter, V will equal the greater V'_D value from Eq. (4-18).

When x_D becomes constant, V'_D becomes a maximum for the largest r_D at the end of the run, and for the case of r_D = constant, V'_D will be largest at the start of the run, provided that the lower boiler has the smaller molar enthalpies.

Eq. (4-17) permits determination of value V' at any moment of the run for a given column section. The required reflux ratio r, corresponding to the mass of reflux in the section considered, will differ from r_D when the components of the mixture have different molar heats of evaporation. In that event, r is computed from the relation

$$r = \frac{x_D - y}{y - x} \tag{4-20}$$

which is analogous to Eq. (3-25)[1] and follows from a mass balance. y and x are correlated points of the enrichment line, Eq. (3-10), relating the concentration of mixed vapor y and that of the reflux liquid for a given column section. For a given r_D, the corresponding y can be computed from Eq. (3-10) if x is located between x_D and x_B, and r can be obtained by Eq. (4-20). The corresponding enthalpies h''_V of vapor and h'_L of liquid follow from these y and x by Eq. (4-17), and V'' is obtained. If the molar heats of evaporation are the same, virtually $r = r_D$.

However, r may also be found from Eq. (3-49), where with total condensation $y_D = x_D$. If the effect of liquid in both the column and reflux condenser is disregarded, and if $dB = dD$, Eq. (4-2) will supply the mass of vapor in a given column section by means of Eq. (4-7)

$$V' = \frac{dV}{dZ} = \frac{dQ_H}{dZ} \frac{r + 1}{(r + 1)(h''_V - h'_L) + h'_L - h'_B} \tag{4-21}$$

and the vapor flowing at the top of the column becomes

$$V'_D = \frac{dV_D}{dZ} = \frac{dQ_H}{dZ} \frac{r_D + 1}{(r_D + 1)\lambda_D + h'_D - h'_B} \tag{4-22}$$

By writing

$$\frac{dQ_H}{dZ} = \frac{Q_H}{Z_T} = \text{constant} \tag{4-23}$$

where Q_H is found from Eq. (4-13) through graphic integration, it will be seen that Eq. (4-22) will actually supply results which are only one percent smaller than Eq. (4-18)[2]. Hence, Eq. (4-18) is sufficient and more readily applied.

4. TIME FOR SEPARATION AND RELATED VARIABLES AT CONSTANT PRODUCT CONCENTRATION

For most binary mixtures, the phase equilibrium of vapor and liquid cannot be rendered by any simple analytical expression; the equations found must be evaluated graphically. Equilibrium curves and operating, lines are fundamental.

At this juncture, there should follow some annotations on batch rectifier control. Fundamentally, there are two approaches. One consists in keeping the still pressure constant and regulating the heat accordingly, the other in maintaining the heat supply at the same level. As far as load on the column is concerned, for continuous rectifiers both methods are equivalent since the vaporized quantities are alike per unit time. In batch stills, however, and especially for binary mixtures, widely different molar heats of evaporation for different components may cause variable results. In any event, the assumption will be made that the best use is made of the available exchanging surface.

When the production is controlled by maintaining constant still pressure, the average molar vapor quantity across the run will be virtually constant if the mean specific weight in the column and the mean molecular weight of the mixed vapor undergo little change in operation, as can be shown readily[1]; however, if the molar heats of the individual components are identical, the mass of mixed vapor flowing through the column per unit time (kmol) will also remain unchanged.

Therefore, two different conditions may arise in the evaluation of the fundamental equations

a) $V' = \dfrac{dV}{dZ} = \text{constant},$

b) $Q' = \dfrac{dQ}{dZ} = \text{constant}.$

In view of Eq. (4-17), these two conditions cannot be identical, since r, h''_V, and h'_L vary during the run.

A. Molar vapor load constant in time

For any column section

$$V' = \frac{dV}{dZ} = \text{constant} \tag{4-24}$$

or

$$dZ = \frac{1}{V'}\,dV \tag{4-25}$$

Substituting the term on the right hand side of Eq. (4-3) for the differential dV in Eq. (4-25) yields

$$dZ = \frac{1}{V'}(r + 1)\,dD \tag{4-26}$$

By integration of the left-hand side of differential Eq. (4-26) within the limits of $Z = 0$ and any given time $Z < Z_T$, and of the right-hand side within the limits of $D = 0$ and the mass of the top product D for the time Z, there follows for any given length of run

$$Z = \frac{1}{V'}\int_0^D (r + 1)\,dD = \frac{V}{V'} \tag{4-27}$$

Equation (4-27) states that, for a given time $Z < Z_T$ the vapor passing through the column V (kmol) plotted against Z in a two-dimensional graph must yield a straight line for V' = constant. With virtually equal molar heats of vaporization of the components, and negligible heats of solution, the vapor mass V' (kmol/h) is the same for the entire column height. It is related to the mass of vapor V_T, condensed in the reflux condenser through the entire run Z_T (h), by

$$V' = \frac{V_T}{Z_T} \tag{4-28}$$

in which V_T (kmol) must be found from Eq. (4-3) by integration within limits $0 < D < D_T$, the last term meaning the mass of top product obtained during period Z_T(h). Hence

$$V_T = \int_0^{D_T} (r + 1)\,dD \tag{4-29}$$

where

$$D_T = B_1 \frac{x_{B1} - x_{B2}}{x_D - x_{B2}} \tag{4-30}$$

For an actual evaluation of Eq. (4-27), first the mass of the top product must be known as a function of the reflux ratio r. That is achieved by combined analytic and graphic procedure using Eq. (4-8) and the equilibrium curve. Graphic integration of the curve $D = f(r + 1)$ will yield a curve

$$V = f'(r + 1) = \int_0^D (r + 1)\,dD \tag{4-31}$$

From this and Eq. (4-27) the time for the run, Z, follows as a function of the mass of mixed vapor V. Since to any V there belongs a specific D, and to the latter belong specific x_B and r values, for a known function $V = f(Z)$ the curves $D = f(Z)$, $x_B = f(Z)$, and $r = f(Z)$ can be drawn. The quantity of the top product produced per unit time $D' = dD = dZ$ and its dependence upon the run time Z can then be found, either by graphic differentiation of curve $D = f(Z)$, or analytically from the relation derived from Eq. (4-26)

$$D' = \frac{dD}{dZ} = V' \frac{1}{r+1} \tag{4-32}$$

jointly with function $r = f(Z)$. Eq. (4-27) can give the rectification time even when the column diameter is fixed. From Eq. (3.39) the molar vapor quantity per hour is found first for the admissible vapor velocity. For a batch still, this is identical with the vapor quantity V' (kmol) of Eq. (4-27). The equation is solved by graphic integration as before, based on the interrelation between x_D, x_B, and the number of steps n_t. The less the number of steps, the larger the average reflux ratio, and the longer the run. Differentiation of Eq. (4-8) with respect to x will yield[4]

$$dD = -B_1 \frac{(x_D - x_{B1})}{(x_D - x_B)^2} dx_B \tag{4-33}$$

and Eq. (4-27) can be transformed into

$$Z = \frac{B_1}{V'}(x_D - x_{B1}) \int_{x_{B1}}^{x_B} (r+1) \frac{dx_B}{-(x_D - x_B)^2} \tag{4-34}$$

The effects of the miscellaneous variables upon batch separation can be recognized, and a solution is feasible by graphic integration in the y/x plot, using the equilibrium curve. The total time follows by integration within the limits $x_B = x_{B1}$ and $x_B = x_{B2}$.

Example 1 A batch rectifier is to separate an ethanol-water mixture of $B = 17.7$ kmol with an ethanol content of $x_{B1} = 28.3$ mol-% in $Z = 8$h. During the run, a product concentration at the column top of 83.8 mol-% of ethanol is to be maintained. At the end of the run the residual still concentration is not to exceed $x_{B2} = 0.5$ mol-% of ethanol. The operating condenser pressure is one atmosphere. Total condensation of the mixed vapors is expected in the reflux condenser.

Answer Assuming equal molar heats of vaporization, the equilibrium curve for the ethanol-water mixture will yield, for an initial reflux ratio of $r_1 = 4$, $n_t = 11$ theoretical steps. At the end of the run, with $x_{B2} = 0.5$ mol-% of ethanol, the reflux ratio must equal $r_2 = 16$. Eq. (4-8) gives, for any reflux ratio r located between r_1 and r_2, the top product quantity D by graphic determination of the still concentrations x_B correlated to r; note Fig. 4.3, top. Plotting the D values thus found against the corresponding $(r + 1)$ values, one obtains, according to Eq. (4-31), the integral curve V by graphic integration of curve $D = f(r + 1)$. Table 1 presents, for various reflux ratios r, the graphically determined values of the still concentrations x_B, the D values found by Eq. (4-8) at the column top, and the V values obtained by graphic integration of curve $D = f(r + 1)$ plotted in Fig. 4.4 over $(r + 1)$, and also the corresponding D' and Z values. Part of the results are shown in Fig. 4.5. Note that the reflux ratio r and the amount of the top product D' (kmol/h) per unit time change at the end of the run in an accelerated manner, with the bulk of the ethanol distilled off; this can be explained from the characteristic shape of the equilibrium curve.

Table 4.1

Data for Example 1

Mixture: Ethanol and Water	V' = constant
Initial Batch B_1	B_1 = 17.7 mol
Initial Batch Concentration x_{B_1}	x_{B_1} = 28.3 mol-% Ethanol
Concentration of Top Product	x_D = 83.8 mol-% Ethanol
Theoretical Steps	n_t = 11
Final Still Concentration	x_{B_2} = 0.5 mol-% Ethanol
Run Time	Z_T = 8 h
Condenser Pressure	p = 1 atm.

r_D	4	4.5	5	7	9	16
x_B (mol-%)	28.3	3.9	2.2	1.1	0.7	0.5
D (kmol)	0	5.41	5.66	5.82	5.88	5.92
V (kmol)	0	27.0	9.8	30.7	31.1	31.9
D' (kmol/h)	0.798	0.725	0.665	0.499	0.399	0.235
Z (h)	0	6.76	7.46	7.70	7.80	8.00

Knowledge of curve $r = f(Z)$ is important in concentration control, when due to the behavior of the mixture a constant top product concentration can no longer be achieved through temperature regulation at top pressure. This takes place when the top product concentration falls in a range where a large change in concentration occurs for a slight temperature change, because of the boiling line configuration.

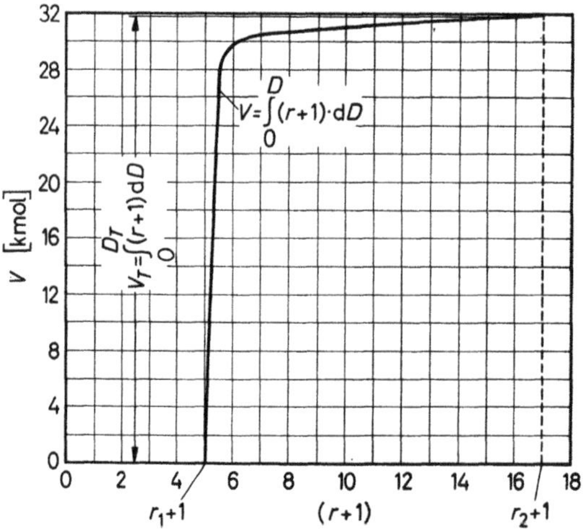

Fig 4.4 Quantity of mixed vapor as a function of the reflux ratio in Example 1.

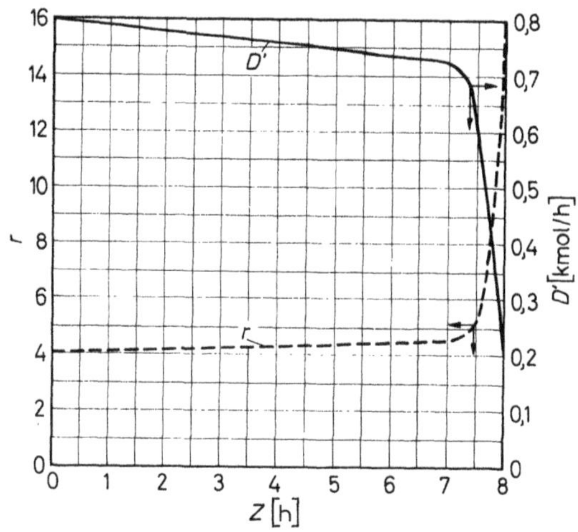

Fig. 4.5 Quantity of the top product produced per unit time and of the reflux ratio as functions of run time.

Some investigations have been undertaken[5] to determine the effect of the liquid content on measurements. For instance, for a liquid content equalling five percent of the kettle volume, the mean deviation of the D values would be less than five percent, whether the liquid volume is considered or not.

B. Heat requirement constant in time

Reference is made to Eqs. (4-17) and (4-18), and the heat requirement per unit time is to be held constant. Again, it is assumed that the molar heat of vaporization λ(kcal/kmol) is a function of the concentration. To apply Eq. (4-17) to a given column section, first the proper reflux ratio at the column top, r_D, must be considered with respect to the reflux volume passing through the column section by means of relations (3-10) and (4-20) or by Eq. (3-49). The concentrations y and x of the vapor and liquid for that r value correspond to enthalpies h''_V and h'_L. If Q' is obtained from Eq. (4-19) and (4-7), differential equation (4-17) will furnish the dV/dZ value for the reflux ratio given.

In Fig. 4.6 the differential quotients dV/dZ and their reciprocals, dZ/dV, are plotted against the corresponding vapor quantities which can be found for the given reflux ratios by graphic integration of Eq. (4-31). Since for the duration Z one must have

$$Z = \int_0^V \left(\frac{dZ}{dV}\right) dV \tag{4-35}$$

graphic integration of the curve $dZ/dV = f(V)$ can yield the vapor quantity (kmol) as a function of time Z (h) (Fig. 4.6). If $V = f(Z)$ is known, V', r or r_D, x_B, D; and D' can be stated as functions of Z. Analogously to Fig. 4.6, Fig. 4.7 demonstrates that Q' is

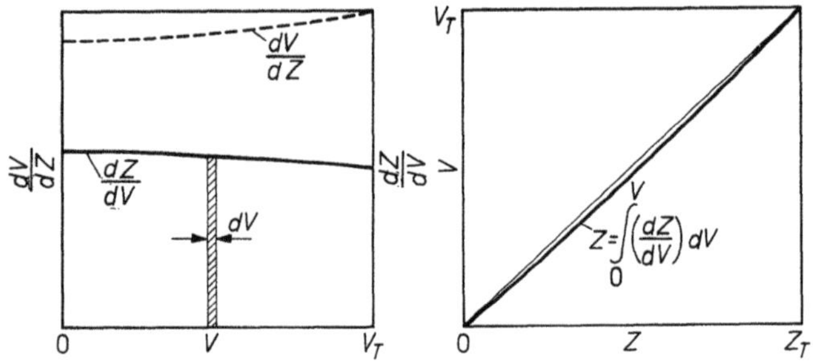

Fig. 4.6 Run time in batchwise operation; heat requirement constant in time.

not required in order to represent curve $V = f(Z)$. It is based on the relationship

$$Q'Z = \int_0^V Q'\left(\frac{dZ}{dV}\right) dV \tag{4-36}$$

and shows how graphic integration of curve $Q'(dZ/dV)$ against V yields the $Q'Z$ values as functions of V, and hence $V = f(Z)$. Since Q' = constant, the $(Q'Z)$ axis need only be subdivided into Z_T scale parts. For each $(Q'Z)$ value the corresponding Z value can be stated.

Figures 4.6 and 4.7 qualitatively give the graphic solution of differential equation (4-17). In Fig. 4.6 the assumption is made that dV/dZ increases with time, while in Fig. 4.7, dV/dZ diminishes in time. Either condition is possible in practice.

The rectification time may be found in other ways. By writing for dV the term of Eq. (4-3) in Eq. (4-17) and solving for dZ, there will result with relation (4-19) the differential equation

$$dZ = \frac{1}{Q'}[(r+1)(h''_V - h'_L) + h'_L]\, dD \tag{4-37}$$

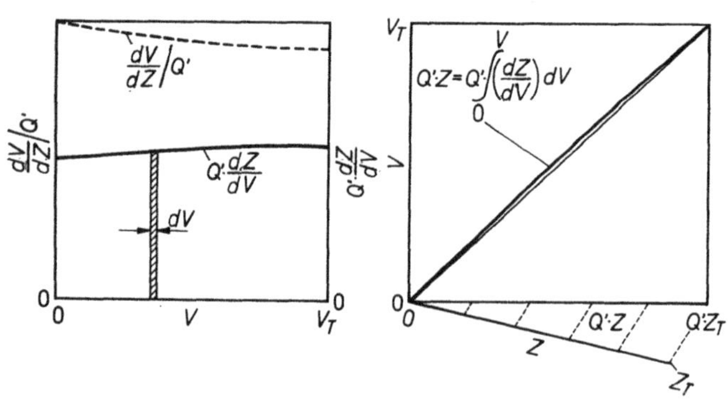

Fig. 4.7 Run time in batchwise operation; heat requirement constant in time.

which upon integration becomes

$$Z = \frac{1}{Q'} \int_0^D [(r+1)(h''_V - h'_L) + h'_L]\, dD \tag{4-38}$$

Graphic evaluation of the integral in Eq. (4-38) is achieved by plotting in a two-dimensional presentation the top product quantity D against the term in brackets from that equation, and subsequent graphic integration of the curve is obtained.

For the column top conditions, however, Eq. (4-38) and its solution may be simplified. In that case, $r = r_D$, $h''_V - h'_L = \lambda_D$ = constant, and $h'_L = h_D$ = constant, and Eq. (4-38) will yield

$$Z = \frac{1}{Q'} \left[\lambda_D \int_0^D (r_D + 1)\, dD + h'_D\, D \right] \tag{4-39}$$

and by substitution of

$$V_D = \int_0^D (r_D + 1)\, dD \tag{4-40}$$

which is the mass of mixed vapor passing through the column top up to time Z (h) and allowing for Eq. (4-19)

$$Z = \frac{Z_T}{Q_T} (\lambda_D V_D + h'_D D) \tag{4-41}$$

Equation (4-41) may also be found directly from Eq. (4-6) by using Eqs. (4-19) and (4-5):

$$dZ\, Q' = \lambda_D dV_D + h'_D dD \tag{4-42}$$

Upon integration, this will give Eq. (4-41)

Introducing the differential term of Eq. (4-33) into differential Eq. (4-37), and integrating the latter between the limits $Z = 0$ and $Z = Z$(h) and $x_B = x_{B1}$ (mol-%) at the start of the run up to the residual still concentration x_B (mol-%) after Z (h), the rectification time becomes

$$Z = \frac{B_1}{Q'} (x_D - x_{B1}) \int_{x_{B1}}^{x_B} [(r+1)(h''_V - h'_L) + h'_L] \frac{dx_B}{-(x_D - x_B)^2} \tag{4-43}$$

This relation contains all the variables characterizing rectification. Its analytical and graphic evaluation will also give the rectification time. The total time $Z = Z_T$ (h) is obtained by integration between $x_B = x_{B1}$ and $x_B = x_{B2}$.

Example 2 10 kmol of a mixture of acetic acid and acetic anhydride are to be separated in Z_T = 10 h by rectification. The concentration x_{B1} is 27.7 mol-% of acetic acid. The column top concentration, which is to be constant, is x_D = 98 mol-% of acetic acid. The still concentration at the end of a run shall not exceed x_{B2} = 4.9 mol-% of acetic acid. The heat requirement Q' (kcal/h) is to be constant. The gage pressure in the reflux condenser is to be 1 atmosphere, and there is no additional separation by that condenser.

Answer First the minimum reflux ratio r_m is established. The molar heats of evaporation are, for acetic acid λ_l = 5820 kcal/kmol, and for anhydride λ_h = 9420 kcal/kmol[6]. Equation (3-11) gives:

$$K = \frac{9420}{9420 - 5820}(100) = 262.$$

The equilibrium curve for the mixture, acetic acid–acetic anhydride, (Fig. 4.8) gives a concentration y_{B1} = 52.9 mol-% for an x_{B1} = 27.7 mol-%. By writing in Eq. (3-35) $y_E = y_{B1}$ and $x_E = x_{B1}$,

$$r_m = \frac{262 - 52.9}{262 - 98}\ \frac{98 - 27.7}{52.9 - 27.7} - 1 = 2.55$$

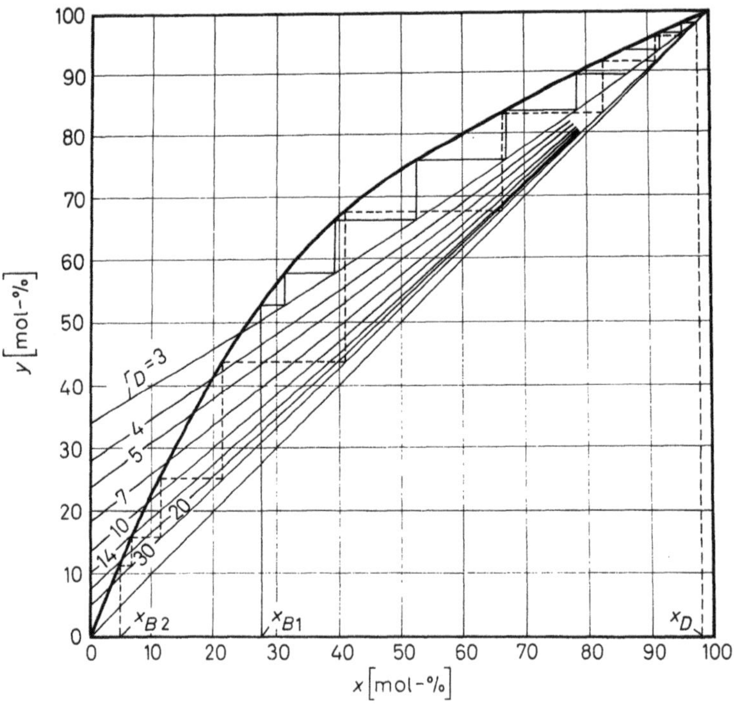

Fig. 4.8 Equilibrium curve of acetic acid and acetic anhydride at normal pressure, and enrichment lines for various reflux conditions at constant product concentration at the column top, and theoretical steps. Conditions of Example 2.

To find the number of theoretical steps, an initial reflux ratio of $r = r_1 = 3$ is assumed. Equation (3-10) gives the enrichment line shown in Fig. 4.8. The number of theoretical steps is $n_t = 8$. At the end of the run, the reflux ratio must equal $r = r_2 = 30$. Table 4.2 is based on the same reflux conditions as the enriching lines computed by Eq. (3-10) and plotted in Fig. 4.8. The graphically determined still concentrations x_B and the top product quantities D are plotted in Fig. 4.9 against $(r_D + 1)$. From this, integral curve V_D can be

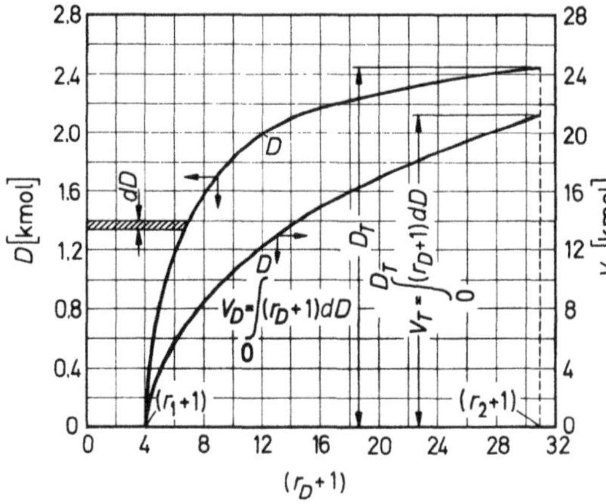

Fig 4.9 Quantity of the top product and of the reflux condensate as functions of the reflux ratio at the column top. Conditions of Example 2.

found which is related to r_D and D by Eq. (4-40). Eqs. (4-7) and (4-19) yield $Q' = 13,410$ kcal/h. For the reflux conditions of Table 4.2, Eq. (4-18) supplies the values $V'_D = dV_D/dZ$, and from this Eq. (4-32) gives the $D' = (dD/dZ)$ figures with $V' = V'_D$ and $r = r_D$. Fig. 4.10 shows the graphic deter-

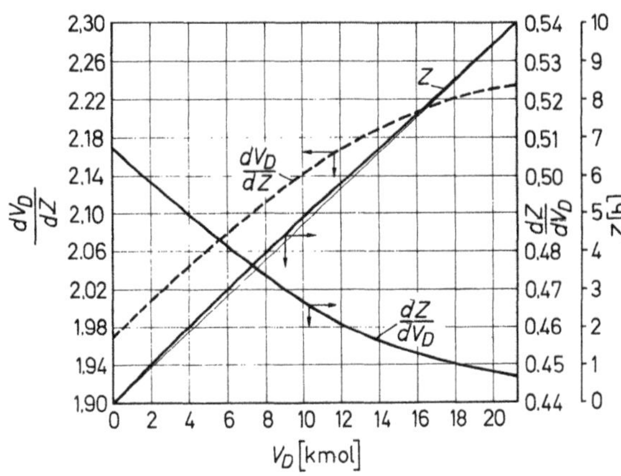

Fig. 4.10 Rectification time at constant heat consumption. Conditions of Example 2.

mination of rectification time Z, based on Eq. (4-35). In Fig. 4.11 the r_D, x_B, D, D', V_D, and V'_D values are plotted. Table 4.2 also presents for given reflux ratios the corresponding figures for V_D, V'_D, D', and Z.
The use of Eq. (4-41) will be demonstrated. Find time Z for the conditions of Table 2, for a reflux ratio of $r_D = 7$ and $Q' =$ constant. The table gives, for $r_D = 7$, $V = 8.45$ kmol and $D = 1.60$ kmol. The heat supplied hourly $Q' = Q_T/Z_T$ had been found to equal 13.410 kcal/h. To $x_D = 98$ mol-% there belongs a $\lambda_D = 5888$ kcal/kmol, and a $h'_D = 3630$ kcal/kmol. With this, and Eq. (4-41), the rectification time becomes

$$Z = \frac{1}{13{,}410}\,[(5888)(8.45) + (3630)(1.60)] = 4.14\ h$$

The same result was obtained graphically from Fig. 4.10

Table 4.2

Data for Example 2

Mixture: Acetic Acid and Anhydride	Q = constant
Initial Batch	B_1 = 10. mol
Initial Batch Concentration	x_{B1} = 27.7 mol-% Acetic Acid
Concentration of Top Product	x_D = 98.0 mol-% Acetic Acid
Theoretical Steps	n_t = 8
Final Still Concentration	x_{B2} = 4.9 mol-% Acetic Acid
Run Time	Z_T = 10h
Condenser Pressure	p = 1 atm.

r_D	3	4	5	7	10	14	20	30
x_B (mol-%)	27.7	21.7	18.6	14.3	11.2	8.6	6.8	4.9
D (kmol)	0	0.786	1.148	1.600	1.900	2.140	2.290	2.450
V_D (kmol)	0	3.51	5.60	8.45	11.41	14.35	17.17	21.28
V_D (kmol/h)	1.970	2.030	2.070	2.120	2.160	2.190	2.215	2.235
D (kmol/h)	0.492	0.406	0.345	0.265	0.196	0.146	0.105	0.072
Z (h)	0	1.75	2.79	4.14	5.51	6.85	8.13	10.00

For Examples 1 and 2 note that the molar heat of evaporation for a mixture, $\bar{\lambda}$, is determined from the molar heats of evaporation, λ_1 and λ_h, of the lower and the higher boiler, for the existing total pressure and the concentration y (mol-%) of the lower boiler, using equation

$$\lambda = \frac{y}{100}\lambda_l + \frac{100 - y}{100}\lambda_h \tag{4-44}$$

The differences in the results of the two methods of evaluation for $V' =$ constant and for $Q' =$ constant are not appreciable. Evaluating the figures of Table 1, for $Q' =$ constant one finds an increase in the $V' = V'_D$ values by a maximum of 3.3%, and if the figures of Table 2 are evaluated for $V' = V'_D$ = constant the Z values in the last line of the table show a maximum deviation of about 5%.

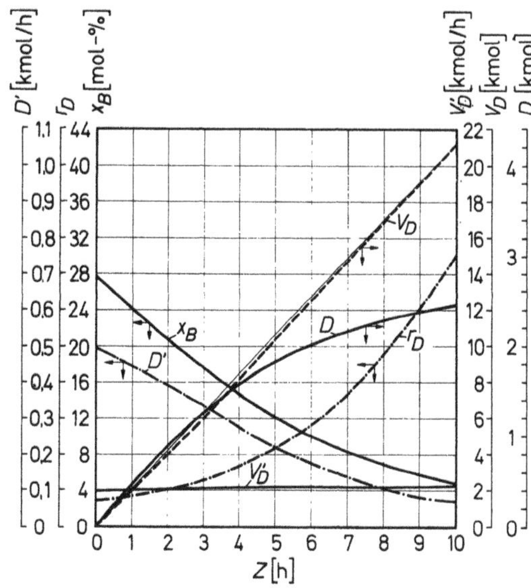

Fig. 4.11 Dependence of the top produce quantity, the mass of the reflux condensate, of the still concentration, and of the reflux ratio at the column top, on the duration of rectification. Conditions of Example 2.

For the two examples, Fig. 4.12 gives the heat requirement per unit of product as a function of the run time. The curves demonstrate emphatically that for batch rectification long runs must be avoided, and a residual amount of low boiler must

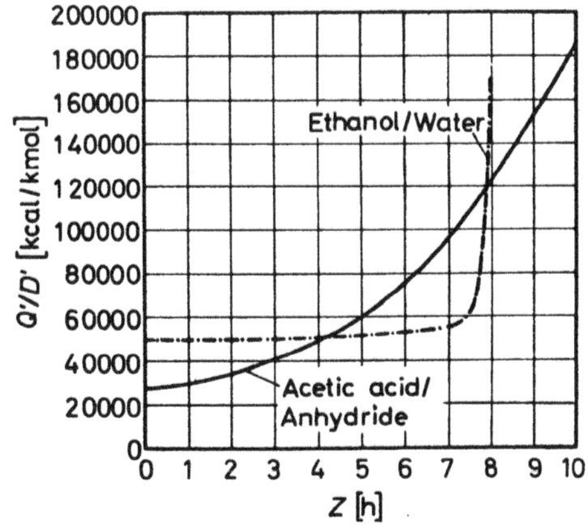

Fig. 4.12 Heat requirement referred to unit production of the top product as a function of the run time, for ethanol-water and acetic acid-anhydride mixtures. Conditions of Examples 1 and 2.

remain in the kettle, as otherwise the heat consumption due to the large refluxing demand would become excessive. In theory, a pure high boiler could be obtained at the end of the run only by means of an infinite reflux ratio.

Example 3 From B_1 = 169 kmol of crude phthalic anhydride, containing x_{B1} = 0.85 mol percent benzoic acid, that impurity is to be removed up to a residue in the still of x_{B2} = 0.3 mol percent or x_{B2} = 0.05 mol-% by batch rectification at constant reflux. The column top pressure is p = 260 mm Hg. What will be the quantity of top product and its share of benzoic acid? Find the quantity of top product D (kmol) and its mean composition x_{Dm} (mol-%) as a function of bottoms composition.

Answer Referring to Fig. 4.3, below, find, from the equilibrium curve of the mixture benzoic acid–phthalic anhydride, a theoretical number of steps n_t = 10 for a reflux ratio of r = 56. Herefor, a number of parallel stripping lines can be drawn, the differences corresponding to $(x_D - x_B)$ for n_t = 10 can be found, and Eq. (4-12) can be evaluated by graphic integration. For the still residue $B = B_2$ Eq. (4-11) can yield x_{Dm}, and from this Eq. (4-10) will give D as a function of $x_B = x_{B2}$. The result is presented in Fig. 4.13. It is seen that at

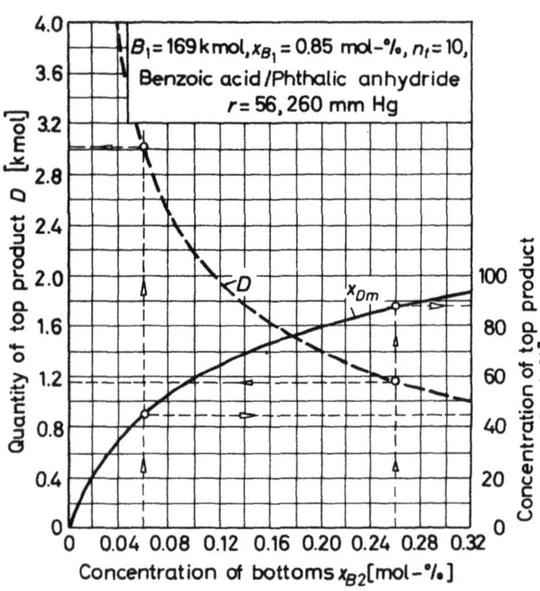

Fig. 4.13 Dependence of the top product quantity and concentration on the bottom concentration. Conditions of Example 3.

x_{B2} = 0.05 mol-% benzoic acid in the still residue, about D = 3.02 kmol of distillate with about x_{Dm} mol-%, corresponding to 100 – x_{Dm} = 55 mol-% phthalic anhydride will be recovered. With the given mode of operation, less phthalic anhydride is obtained in the top product (i.e., 100 – x_{Dm} = 13 mol-% if in the still residue more benzoic acid (x_{B2} = 0.3 mol-%) is allowed. In that event, the quantity of top product will be only about D = 1.16 kmol.

5. SEPARATION TIME FOR VARIABLE HEATING AREA

In the above mentioned circumstances, the effective heat exchange surfaces were not affected by the length of the run. That situation always exists if the heat transfer surface is fully wetted on the side of the heating medium and on that of the evaporate.

When, however, the kettle is of jacketed design, the heat transfer surface will be reduced as the still content becomes less, and the time of run Z (h), during which the batch has been diminished by $(B_1 - B)$, obeys the equation

$$Z = \frac{Q}{Q'_m} \tag{4-45}$$

where Q (kcal) is the heat producing the top product quantity

$$D = B_1 - B \tag{4-46}$$

That heat may be determined from differential Eq. (4-6), either for a constant top concentration x_D, by integration within the limits $D = 0$, and the product quantity D (kmol) given by Eq. (4-8)

$$Q = \lambda_D \int_0^D (r_D + 1)\, dD + h'_D D, \tag{4-47}$$

or for a constant reflux ratio r_D at the column top, through integration within $D = 0$ and product quantity D (kmol) defined by Eqs. (4-10), (4-11), and (4-12)

$$Q = (r_D + 1) \int_0^D \lambda_D \, dD + \int_0^D h'_D \, dD \tag{4-48}$$

Since the heat transferred per unit time Q' (kcal/h) diminishes with the still content B (kmol), in Eq. (4-45) an integral mean value[7] Q'_m (kcal/h) must be applied that can be found for the equipment considered from

$$Q' = f(A) \tag{4-49}$$

where A (m^2) is the effective heating surface of the still according to

$$Q'_m = \frac{1}{A_1 - A} \int_{A_1}^{A} Q' dA \tag{4-50}$$

where A_1 (m^2) = effective transfer area at the start of the run, when the still content is B_1 (kmol), and A (m^2) = remaining effective transfer area after the still contents have diminished to B (kmol).

The heat transferred per unit time Q' (kcal/h) is generally found from the equation

$$Q' = kA(t_H - t_B) \tag{4-51}$$

where k = heat transfer coefficient (kcal/m^2 h C), t_H = temperature of heating medium (C), and t_B = boiling temperature in still (C).

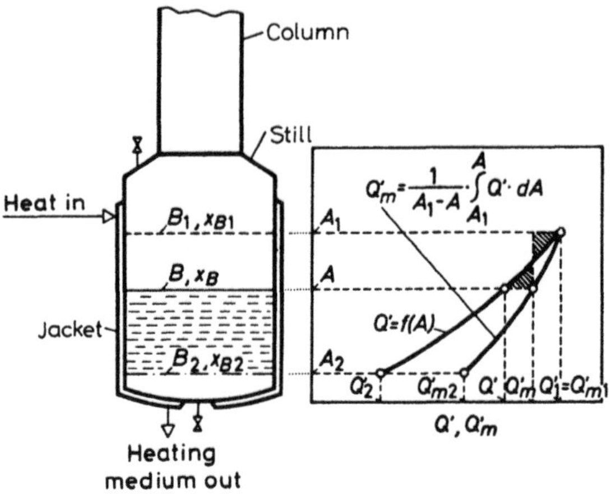

Fig. 4.14 Mean integral value per unit time for heat supplied to the still, with a diminishing area due to the decline of the still contents during the run.

Fig. 4.14 shows qualitatively the steps in evaluation of Eqs. (4-50) and (4-51).

For any moment of rectification time, the characteristic conditions can be determined as outlined above.

In Fig. 4.14, Q'_{m2} (kcal/h) is the average integral value of the heat transferred per unit time during the entire run Z_T with the still contents reduced from B_1 to B_2. Thus, Z_T can be found from Eq. (4-45) by writing Q'_{m2} for Q'_m, and Q_T for Q [Eqs. (4-7) and (4-9)].

V The Semicontinuous Still

1. OPERATION

When a mixture of substances is split up, sometimes the operational features of a batch still are combined with equipment common in continuous work, that is a column embodying a rectifier and a stripper section. This "semicontinuous" rectification is useful when, for instance, a chemical reaction takes place intermittently, producing a vapor mixture as a partial product that is decomposed on the spot, the low-boiling components being recovered for better production control, while the high-boiling bottoms are recycled. If in the course of the reaction the lower boiler continuously drops in concentration, the reflux ratio at the column top must be enhanced steadily if a top product and a bottom product are to be obtained at constant composition or purity.

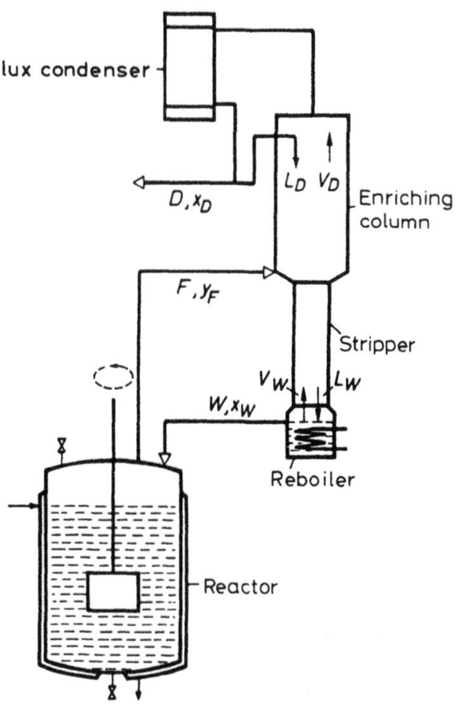

Fig. 5.1 Diagram of batch esterification combined with semicontinuous rectification.

One example of an application of semicontinuous rectification is an intermittent reverse esterification (Fig. 5.1). The equipment comprises a jacketed reactor and a rectifying column having a stripper section and enriching section with a reboiler and reflux condenser. The alcoholic vapors liberated during the reaction are passed on to the column where the lower boiling alcohol is removed and yields the top product. A rapid evolution of the reaction may be expected when the bottom product of the column is free of low-boiling alcohol. If along with the bottom appreciable amounts of low-boiling alcohol were recycled to the reactor, the reaction equilibrium would be displaced, favoring the original materials, and would claim more heat for more vaporization.

2. FINDING THE OPERATING LINES, THE SEPARATION STEPS, THE COLUMN LOAD, THE COLUMN SIZE, AND THE HEAT DEMAND

According to Fig. 5.2, during the run the reflux ratio r_1 must be increased to r_2 when the low-boiler concentration of the mixed vapor declines from x_{F1} to x_{F2}, and the top concentration x_D must be held constant. The limit values r_1 and r_2 can be

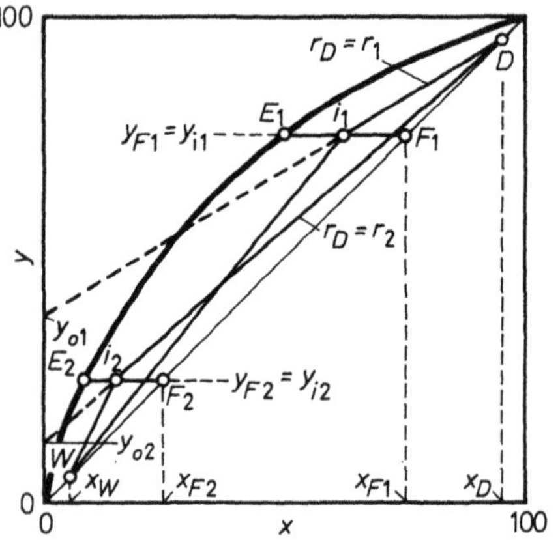

Fig. 5.2 Operating lines in y/x plot for the starting and ending conditions of a semicontinuous rectification with vapor feed.

determined from the minimum reflux ratios, Eqs. (3-35) and (3-36), and (3-37) and (3-38). As the reflux ratios change, even for constant column feed, the vapor load in the stripper must be modified in order to maintain a constant bottom composition. When the column is supplied with vapor in its saturated condition, $f_i = 0$; substituting Eq. (3-43) into Eq. (3-46) for the vapor quantity at the column base becomes

$$V_W = V_D \frac{1 - \dfrac{y_D}{K}}{1 - \dfrac{y_W}{K}} - F \frac{1 - \dfrac{y_F}{K}}{1 - \dfrac{y_W}{K}} \tag{5-1}$$

noting that $x_F = y_F = y_i$. Also, substitution of Eqs. (3-42) and (3-43) into Eq. (3-45) for the vapor quantity in a given section of the stripper

$$V_s = F\left[(r_D + 1)\frac{y_F - x_W}{x_D - x_W}\,\frac{1 - \frac{y_D}{K}}{1 - \frac{y}{K}} - \frac{1 - \frac{y_F}{K}}{1 - \frac{y}{K}}\right] \tag{5-2}$$

where $y_W < y < y_F$. Even with F = constant, V_s will vary because r_D and y_F vary.

In actual operation, it would be desirable to maintain an even heat supply at the still base to match the largest calculated V_s value for continuity of control. Then, the reflux ratio of Eq. (5-2), with $y = y_W$ requiring a vapor of $V_s = V_W$, and with $x_F = y_F = y_i$ equals

$$r_D = \left[\frac{V_W}{F}\,\frac{1 - \frac{y_W}{K}}{1 - \frac{y_D}{K}} + \frac{1 - \frac{y_F}{K}}{1 - \frac{y_D}{K}}\right]\frac{x_D - x_W}{y_F - x_W} - 1 \tag{5-3}$$

This permits determination of the reflux conditions with V_W = constant and variable F

Obviously, both the enriching and the stripping section of the column will be fitted with the most theoretical steps indicated. Note the method described in Chapter 3 for a binary mixture. In a y/x plot the interchange lines for various reflux conditions $r_1 < r < r_2$ are established, and the sequence of steps between the former and the equilibrium curve is drawn between x_W and x_D. From the r_D value of Eq. (5-3) by Eq. (3-42), with $x_F = y_F = y_i$

$$V_D = F\,(r_D + 1)\frac{y_F - x_W}{x_D - x_W} \tag{5-4}$$

The quantity of the top product is:

$$D = \frac{V_D}{r_D + 1} \tag{5-5}$$

from which the quantity of bottom product becomes:

$$W = F - D \tag{5-6}$$

By Eqs. (3-48) and (3-49) and with $x_F = y_F = y_i$

$$V_e = F\,(r_D + 1)\,\frac{1 - \frac{y_D}{K}}{1 - \frac{y}{K}}\,\frac{y_F - x_W}{x_D - x_W} \tag{5-7}$$

To find the largest column diameters for the enriching and the stripping sections, the results of Eqs. (5-1), (5-2), (5-4), and (5-7) for V are inserted in Eq. (3-39) if the admissible vapor velocities u_V are known.

It is recommended that, first, the available quantity of mixed vapors V (kmol/h) be established according to the conditions of the reaction involved. This is identical with the column feed F(kmol/h) and has a concentration of the low boiler of y (mol-%). If the mixed vapors are formed at a virtually constant thermal load Q_R (kcal/h) of the reactor, $F = f(Z)$ can be found from

$$F = \frac{Q_R}{\lambda_F} \tag{5-8}$$

where $\lambda_F = f(y_F)$ (kcal/kmol) = molar heat of evaporation for the known function $y_F = f(Z)$. V_W as a function of Z is known from Eq. (5-4), V_D from Eq. (5-5), D and V_W from Eq. (5-5). Furthermore, the reflux quantity at the column top is

$$L_D = r_D D \tag{5-9}$$

and that at the column base is

$$L_W = V_W - W \tag{5-10}$$

as a function of the time.

Analogously with the derivation of Eq. (3-62), and with $x_F = y_F = y_i$, the heat requirement Q of the column becomes

$$Q = F\left\{\frac{y_F - x_W}{x_D - x_W}\left[(r_D + 1)\lambda_D + h'_D - h'_W\right] + h'_W - h''_F\right\} \tag{5-11}$$

where h''_F is the enthalpy of the vapor fed into the column. The other variables have been discussed in Chapter 3.

Example 1 An esterification reactor delivers, under virtually constant heat load of $Q_R = 8.52 \cdot 104$ kcal/h, a methanol-monoethylene glycol vapor mixture whose composition is given in Fig. 5.3 as a function of operating

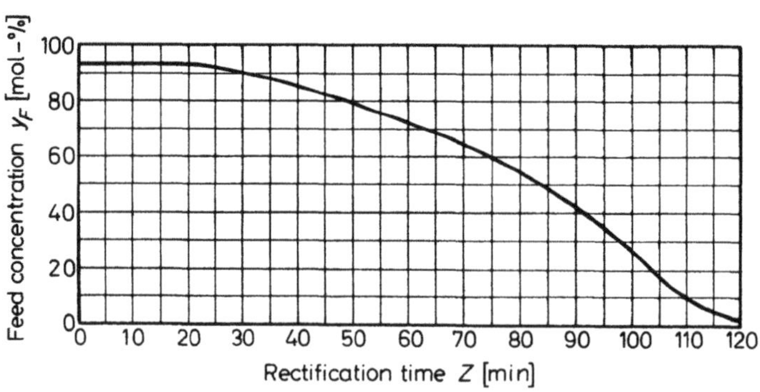

Fig. 5.3 Feed concentration as a function of time. Conditions of Example 1.

time. This mixture is to be decomposed in a setup shown in Fig. 5.1 in such a way that the top product contains no less than x_D = 99.5 mol-% of methanol and the bottom product no more than x_W = 0.4 mol-% of methanol. How must the reflux ratio at the column top be regulated during operation, how much will vapor and liquid quantities vary at the top and at the bottom of the column, and what are the production quantities?

Answer Following the procedure outlined, from Eq. (5-8) and the molar heat of evaporation λ_F, there results

$$\lambda_F = \lambda_l \frac{y_F}{100} + \lambda_h \frac{100 - y_F}{100} \tag{5-12}$$

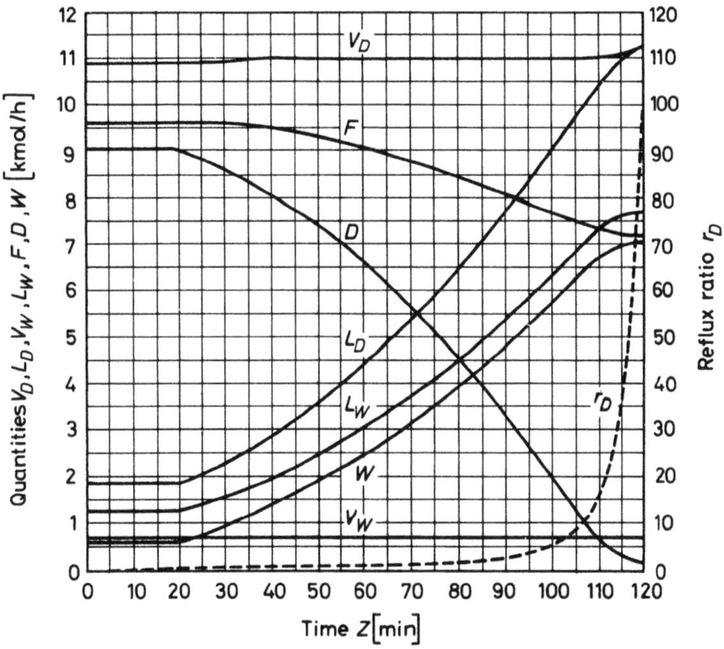

Fig. 5.4 Semicontinuous enrichment. Reflux ratio and characteristic molar quantities as functions of the data in both Example 1 and Fig. 5.3.

from the vaporization heats, λ_h and λ_l, of the higher and lower-boiling components; from Fig. 5.3 is found the relation $y_F = f(Z)$ for the quantity F entering the column during the time Z.
From Eq. (5-4), and with $y_i = y_F$, Fig. 5.2 can be used to show, from the equilibrium curve for the methanol–monoethylene glycol system at normal pressure, that the reflux ratios r_D are slightly larger than the ones dependent

on y_F, with values V_D and V_W from Eq. (5-1) as a function of Z. $y_i = y_z = f(Z)$ are known from Eq. (5.3).

The calculation yields a maximum V_W = 0.65 kmol/h. If this is held constant in order to achieve a more uniform column load, the corresponding reflux ratios will follow from Eq. (5-3) – r_D. As seen from Fig. 5.3, V_D follows from Eq. (5-4), D from Eq. (5-5), W from Eq. (5-6), L_D from Eq. (5-9), and L_W from Eq. (5-10) as functions of time. All these have been shown as molar values in Fig. 5.4; also, reflux ratio r_D has been included.

Fig. 5.5 is the same presentation, converted to a kg basis.

It should be noted that toward the end of the run the intended separation requires very large reflux ratios.

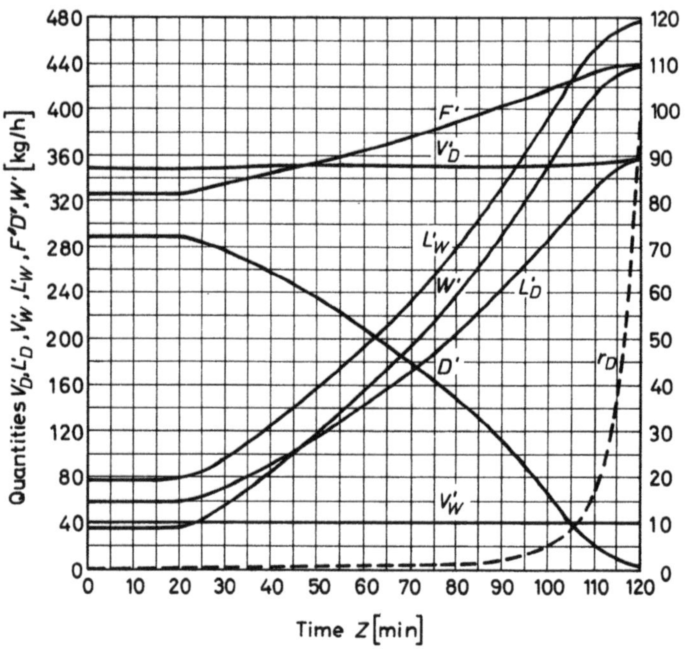

Fig. 5.5 Curves of Fig. 5.4 converted to kilograms.

VI Engineering Data, Optimization of Costs, and Selection of Column Internals

1. GENERAL

The height and diameter of a column for a specified duty depend on the type of internals under consideration. The purpose of such internals is to promote an intimate contact of vapor and liquid to ensure separation by rectification. There are various types of packing, and mass transfer trays or plates. While packing usually takes all of the available space inside a column, plates or trays are spaced evenly apart. Since efficiencies and capacities vary for different designs, differences result in the construction heights and in the diameters of columns.

The separation effect of internals may be judged from the characteristic curves which are usually found by experimentation. These show the interrelation of the separation effect and the capacity of the internals. Together with the influence of pressure loss upon capacity when the vapor flows across the internals, such curves serve as guides in design planning.

The variety of the demands made in the chemical and allied industries upon separating equipment stimulates the development of novel designs and the perfection of conventional details. Besides, equipment builders often possess exclusive rights covering proprietary arrangements, under the pressure of competition, in view of licensing agreements, or to maintain some monopoly in the field. As a result, the engineer planning a separation system will have an ever widening choice of devices that may be of use. Which one will finally be adopted will depend on a critical comparison of costs to be expected in view of the probable performance of the equipment.

Engineering criteria that must be met in order to minimize the investment in a given project include a high specific separation effect over a wide band of loads, and high specific throughput. When sensitive mixtures that tend to deteriorate must be split up they must be treated in a vacuum. Therefore, the pressure loss at maximum load, and the holdover time in the plant should be relatively low.

Thus in the ideal separation unit all of these features would be present to their best advantages, and furthermore the cost per unit separation and unit throughput, referring to the column volume, should be minimal. Whether, and to what extent such properties can be realized is investigated in this chapter.

A. Packing types

All the various packings are based on the realization that an effective mass transfer between vapor and liquid requires an ample surface. To achieve good separation, the vapor and liquid should be spread uniformly over the entire column cross section along the flow path, and redistribution and intermixing should be frequent, with the film thickness remaining as uniform as possible. The pressure loss of the vapor flowing through the packing should be small, especially when high-boiling and temperature-sensitive substances are rectified under vacuum. The packing elements should be mechanically and chemically resistant and low-weight. No known packing offers a maximum of all these characteristics. Therefore, it must be decided which type of packing will be most suitable in this instance.

No matter which kind of packing is chosen, the separation effect will depend largely on a uniform distribution of the reflux liquid across the column cross section. Available liquid distributors will be discussed here and in Chapter 13.

Packing sizes of existing installations vary generally between 8 and 80 mm, and the preferred materials are metal and ceramics. Plastics have been used in special cases only.

When a column is charged with bulk packing, the elements may arrange themselves in a more or less orderly pattern next to the inside of the column shell, with a larger, unobstructed cavity volume than is formed within the bulk of the packing. In operation, this will cause a wall effect and consequently inadequate distribution at the core of the packing that damages the separating power. To counteract that effect, the shell may be fitted with internal baffles, properly spaced, which return the liquid to the packing. An undulated shell has also been suggested.[1] Conical redistributors having passages for vapor and liquid located at different levels of the column are useful for preventing the formation of rivulets, which takes place because of surface tensions wherever packing elements stay unwetted.

The term "packing" shall apply to all suitable kinds of elements that fill the internal column space and are characterized essentially by the relative size of the cavities formed and in the extension and wettability of their surfaces. Mostly, they are regularly shaped bodies formed from metallic webbing of different sizes, or from metal strips of many shapes. Nonmetallic elements also exist.

The action and capacity of a body of packing is generally expressed by a load curve as shown in Figure 6.1. Its saddle shape is typical of many packed columns. It is found by experiment. The vapor load factor

$$F_V = u_V \rho_V^{1/2} \tag{6-1}$$

is derived from the velocity, u_V (m/s), of the vapor, related to the free column section, and the square root of its density [ρ_V(kg/m^3)]. The corresponding number of theoretical steps n_t is related to a unit depth of H(m), for a constant liquid/vapor ratio L/V. The function

$$\frac{n_t}{H} = f(u_V \rho_V^{1/2})_{L/V=\text{constant.}} \tag{6-2}$$

is then plotted. Its outline depends on the physical data for the mixture and on the packing design.

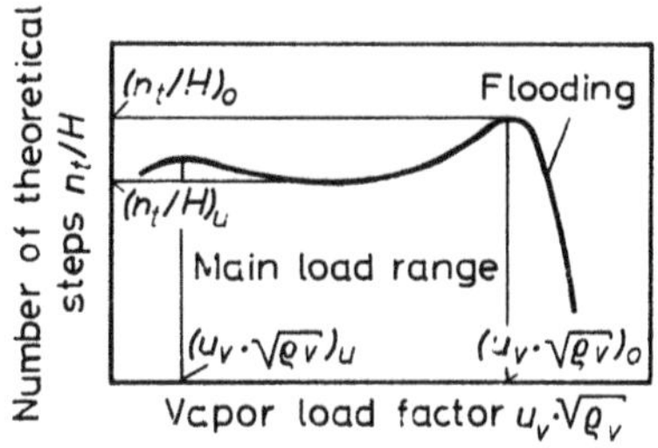

Fig. 6.1 General configuration of the separation versus the load curve of a packed column.

Thus n_t/H is a yardstick for the separation effect, being the number of theoretical steps equivalent in its action to a packing of 1-m depth. Of practical significance is that branch of the load curve that, in Fig. 6.1, is located between the value $(n_t/H)_u$ and the maximum $(n_t/H)_o$ and is bounded by the load factors $(u_V \rho_V^{1/2})_u$ and $(u_V \rho_V^{1/2})_o$. This last term is particularly important in the operation of a packed column since it fixes its maximum load, and its value may not be exceeded, as otherwise the liquid would be blocked by the vapor flow and move upward, meaning that the column would be flooded and the separation impaired.

Configurations of load curves vary for different types. Frequently, the number of transfer units is employed to state the effectiveness of a packing. Assuming the concentration gradient to exist on the vapor side, then[3]

$$NTU = \int_{y_W}^{y_D} \frac{d_y}{y_E - y} \tag{6-3}$$

If all of the mixed vapor is condensed $y_D = x_D$. y_W is the phase equilibrium concentration corresponding to x_W. In Fig. 6.2 the difference $y - y_E$ is the vertical

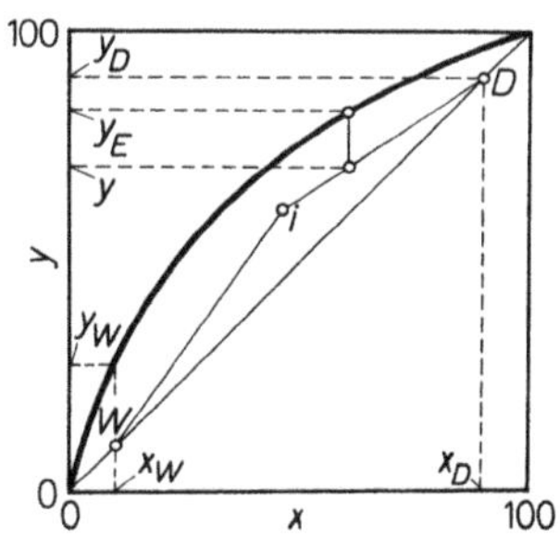

Fig. 6.2 Finding the number of transfer units of a packed column.

distance between the equilibrium curve and the operating line. For a given equilibrium curve, the integral term may be determined graphically by a y/x plot.

With the height h_V of a transfer unit known, the packing depth will follow from

$$H = h_V NTU \tag{6-4}$$

For a determination of h_V see the bibliography.[4]

In systems with small relative volatilities, the equilibrium curve and the operating lines are virtually parallel in the y/x plot,

$$NTU(\text{for small } \alpha \text{ value}) \sim n_t \tag{6-5}$$

and h_V will approximate the reciprocal of the number of theoretical steps per unit of packing depth

$$(h_V)\ (\text{for small } \alpha \text{ value}) \sim \frac{H}{n_t} = HETP, \tag{6-6}$$

the height equivalent per theoretical plate.

Among available packing elements are the Raschig rings, Berl saddles, Pall rings, Intalox saddles, and Interpack elements. Also, the miscellaneous grid and web types should be noted and are discussed in Sections 2 and 3 of both chapters 6 and 8.

B. Plates and trays

The contact surface necessary in rectification may also be obtained by means of interchange plates mounted inside columns at suitable intervals. On these elements, the reflux liquid circulates and encounters the vapor which must pass through it.

The usual rectifying plate is a circular sheet, perforated to offer vapor passages. The simplest embodiment is the sieve plate, with holes of 2 to 12 mm in size. Conventional bubble-cap plates, however, have much larger openings, equipped with uptakes or chimneys covered by bell caps that deflect the vapor and cause it to enter the mass of liquid laterally. On both types of plates, the liquid coming from the influx end moves across the plate to the downpipe and arrives on the influx side of the plate below. The interchange area proper is situated between the influx end and the exit end delimitated by a weir that maintains a certain depth of liquid.

Other tray types that have become popular are: the cascade tray,[5] the turbogrid,[6] the tunnel plate, the expanded metal tray, the dual flow tray, the valve tray, the slotted sieve tray, and the multidowncomer tray (note Chapters 6 and 8, Sections 2 and 3, and 4, respectively.)

By comparing the enrichment Δy_p of the low boiler in the vapor phase with the corresponding enrichment Δy_t by a theoretical step one obtains the relation

$$E = \frac{\Delta y_p}{\Delta y_t} \tag{6-7}$$

which is a yardstick for the separating ability of a plate. It was said in Chapter 3 that a theoretical plate works properly if the vapor leaving is at phase equilibrium with the liquid leaving. Thus, the mean efficiency of a tray in a column performing with n_p actual plates with a rectification of n_t theoretical stages is

$$E = \frac{n_t}{n_p} \tag{6-8}$$

By analogy with column packing, Eq. (6-2), one has, for the load curves of separation plates, the relation

$$E = f(u_V \rho_V^{1/2})_{L/V=\text{constant}} \tag{6-9}$$

It is usually controlled by the physical properties of the mixture on the one hand, and by the tray design on the other.

Figure 6.3 presents qualitatively the shape of a load curve, and Fig. 6.4 gives the generalized graph defining the separation by a plate for a binary mixture.

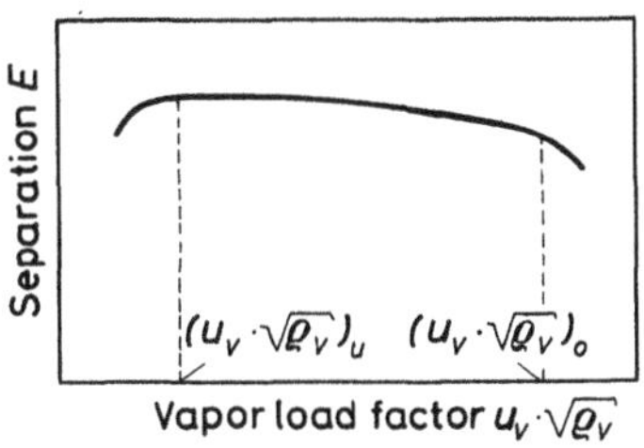

Fig. 6.3 General outline of the separation versus the load curve for a plate column

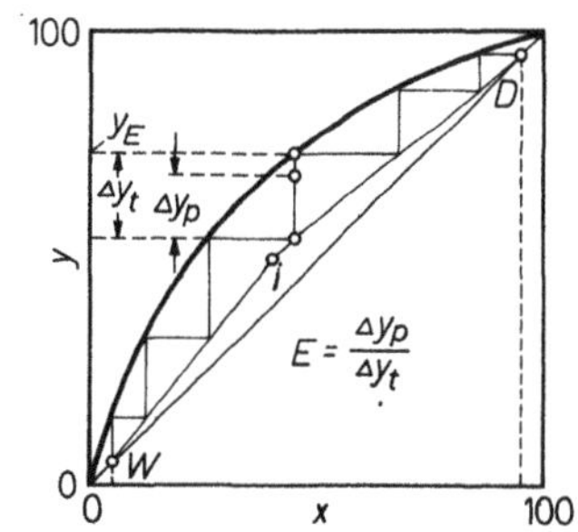

Fig. 6.4 Separation effect of a plate column

Fig. 6.3 shows how the working range of a tray is delimitated by a lower load $(u_V \rho_V^{1/2})_u$, below which the separating effect falls off because the liquid rains through, and by an upper load $(u_V \rho_V^{1/2})_o$, above which the separating effect also diminishes because of flooding or entrainment of the liquid by the vapor. In general, commercial columns are designed for and operated at the vicinity of the upper load range. The lower range becomes significant only where low pressure loss is mandatory, or where capacities required are variable, as in discontinuous operation. For many tray types the load ranges between the limits stated are fairly inflexible.

Along with the load factor of Eq. (6-1), which is basic in showing the dependence of the separating effect on the load, the term

$$F_V^* = u_V \left(\frac{\rho_V}{\rho_L - \rho_V} \right)^{1/2} \tag{6-10}$$

which has been extended by the density, ρ_L (kg/m^3), is shown.[7] It is particularly useful for high-pressure rectification where the specific gravity of the vapor is relatively large, and the difference $(\rho_L - \rho_V)$ is no longer of the order of ρ_L.

2. DESIGNS AND FUNCTIONS

In recent years, various column internals have been developed whose design features and performances differ from conventional ones of those encountered in the chemical and allied industries, the ones investigated by researchers are discussed at greater length. They are presented, as far as packings go, in Figs. 6.5. A–E and, where

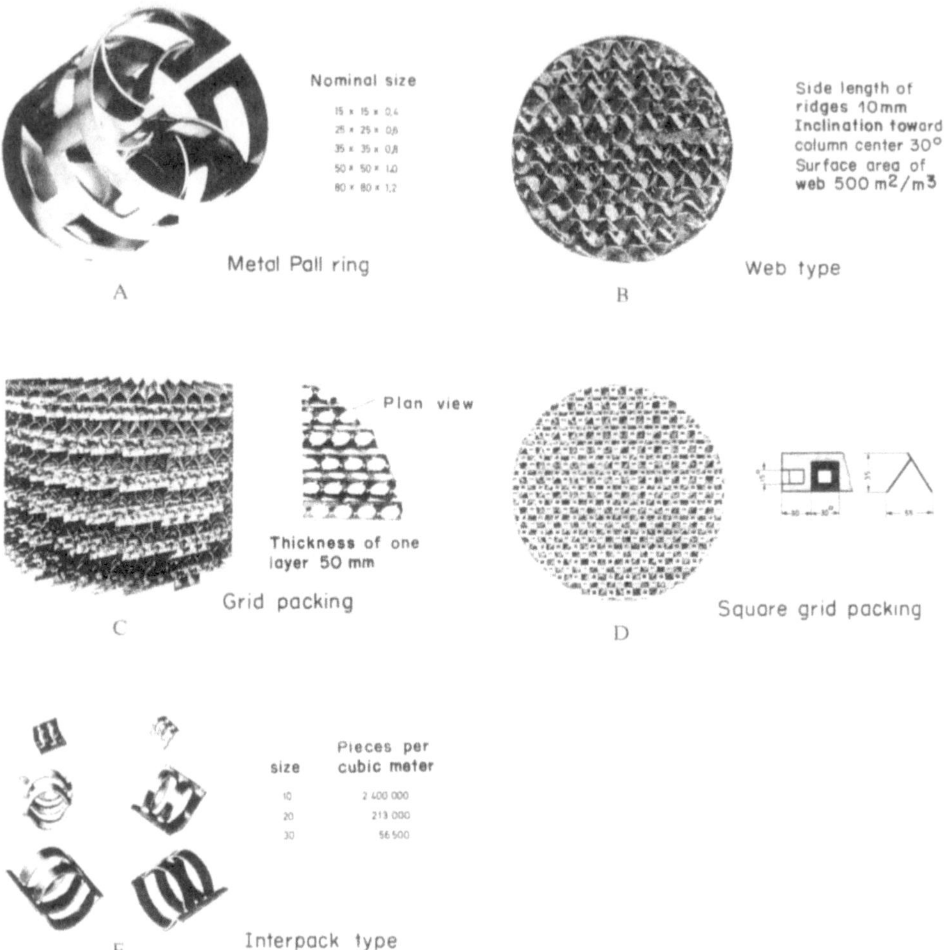

Fig. 6.5 Packing elements of columns

plates, are concerned in Figs. 6.6 A–1, along with standard makes and principal dimensio

All of these devices endeavor to create conditions favorable to materials separation. To what extent that has been achieved will be seen for various load applications.

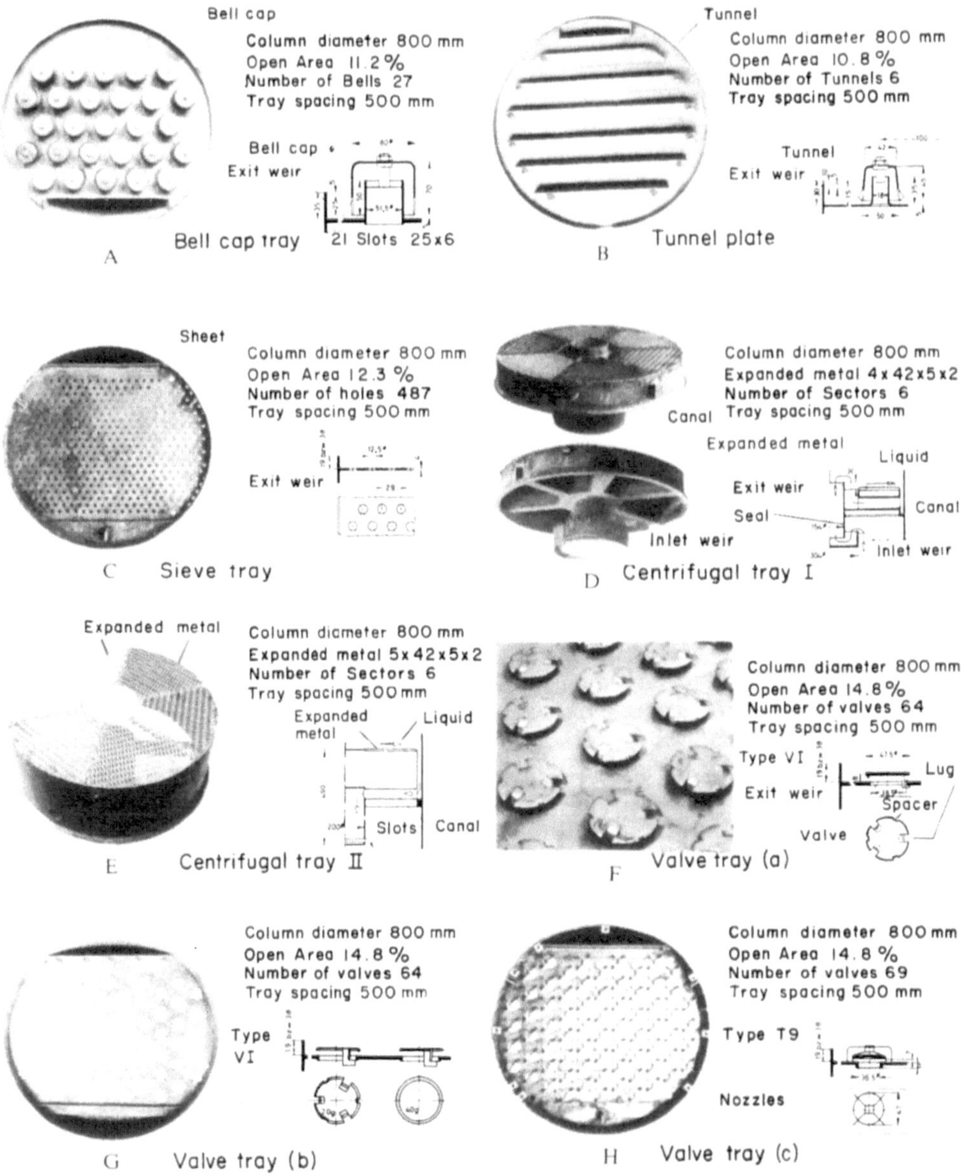

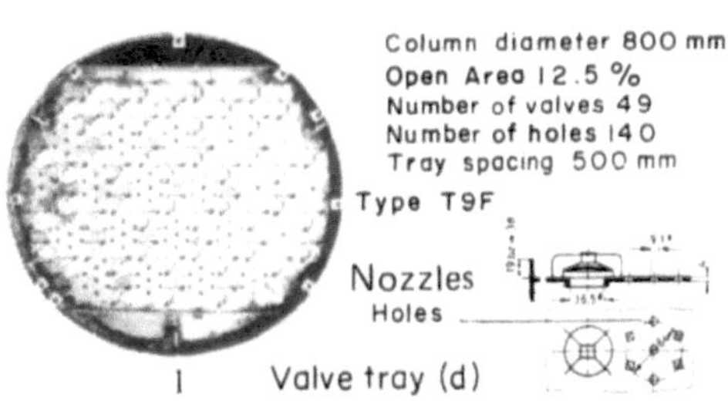

Fig. 6.6 Types of trays.

A. Packed towers

In splitting up thermally unstable systems, columns should have low base temperatures and low pressure loss per theoretical step. That requirement is best met by equipment in which the reflux is broken up into liquid films and passed through the column counter to the vapor. Refer to the elements shown in Figs. 6.5. A–E.

Pall Rings Ceramic Pall rings have been successful over the last 20 years for various separations, and recently metallic Pall rings have been offered in 15, 25, 35, 50, and 80 mm diameters equalling height. Careful investigations[8-14] revealed their suitability for vacuum operation. They are made from tubular stock cut to lengths equalling the diameter, with rectangular cutouts that are bent inward (Fig. 6.5.A)

Web Packing. This type is fairly novel, and its understanding is based on the facts of formation and the stability of liquid films. Its elements[15-17] are parallel and vertical serrated interwoven ribbons of stainless steel, combined into a cylindrical packing 100 to 200 mm high. The inclinations of the serrations of adjoining ribbons are opposite, which allows for mutually crossing, parallel flow paths of a triangular section of about 10 mm side length.

Vapor and liquid flow countercurrently and zigzag through the column. Because of the strong capillary action of the surfaces, even relatively small reflux quantities will be uniformly dispersed into a liquid film. The inclination of the channels formed provides intense crosswise mixing in parallel paths, the 90° stagger of the elements provides good intermixing of the phases across the horizontal column section. As a result, good separation may be expected.

Two different variations on this concept have gained preference, different in the size of the channels formed and their inclination. Fig. 6.5.B is a top view of type BX. Its geometry yields an extremely low-pressure loss per theoretical step and is therefore suited to vacuum separations of thermally unstable mixtures. The other type, CY, offers a very large specific surface and is preferred for rectifications of many steps, such as the separation of H_2O and D_2O. What follows refers to the BX.

In process equipment, normally a loss of 0.1 mm per year is customary, but that figure may not be applied since the wire of the web is only 0.16 mm thick. Hence this packing is made of stainless steel and phosphor bronze only. Corrosive mixtures that will attack these metals cannot be processed with this kind of packing. Mechanical cleaning of web packing is not feasible. Neither can this packing be considered when backwashing by a solvent cannot be applied.

Most organic liquids will readily wet clean wire web. Therefore, good separation may be expected if the packing is properly designed and installed. A simple test will disclose the wetting potentiality.

Grid Packing. The shape of its elements[18] results from the metal remaining from the production of valve trays which is discussed later[19]. Members formed in the stamping process, 50-mm high, are placed horizontally and crosswise into a column section. Their free volume is relatively large. High throughput with moderate pressure loss results, which makes the packing suitable for vacuum rectification. There is a coarse and a fine type of grid. The members of a grid may be up to 650 mm wide comprising a total width of several meters, depending on column size (Fig. 6.5 C).

Square Grid Packing. This type, developed from the one above,[20] consists of flat bars coiled in such a way that the end faces have, alternatingly, square openings of 15-mm side length, and square cutouts of 30 mm side length—30 mm apart. The perforations provided on angular stock yield a staggered arrangement. The spacing of the element edges is approximately 55 mm. The members are so installed in the column section that 45° angles are formed between adjacent layers; this arrangement is supposed to equalize concentrations in the column, and achieve better routing and rerouting of the vapor, and proper contact with the liquid surface (Fig. 6.5.D).

Interpack Type. This is a novel design of a packing made of steel at moderate production expense. The ample passages warrant adequate throughput at low pressure loss. Rectangular pieces of sheet metal are given parallel slots, and the remaining interconnected strips are bent outward and inward, alternatingly (Fig. 6.5. E).

B. Plate columns

To optimize mass transfer between vapor and liquid in thermal separations, various types of trays have been redesigned and improved. Compared to conventional types as described in Section 2A of this chapter, the bubble cap plate and sieve tray, functions have been improved and the cost of the equipment has been reduced, both as compared to throughput and to effectiveness, which are the two essential viewpoints. Figs. 6.6. A to 6.6.I give the principal data and design characteristics.

Bell-Cap Tray. As stated before, the vapor rising through the nozzles is deflected by the bell caps and by their slots distributed in the liquid outside the cap, and therefore some turbulence occurs. In the design shown, the liquid will flow from the entering end crosswise toward the downcomer, and hence to the next tray below. The submergence, which is the vertical distance between the upper edge of the weir and the top of the slot edge of the caps is small, in this case about 5 mm, to reduce the pressure loss. The clearance, defined as the ratio of the column cross section and the total of all nozzle sections, is 11.2%, which is typical of conventional plate trays (Fig. 6.6A).

Tunnel Tray. This is somewhat special (Fig. 6.6. B). The plate proper is composed of U-shaped elements[22,23] in such a way that vapor chimneys are formed that are covered by elongated covers to return the vapor flow. In between space remains for liquid flow. Lateral slots in the covers let the vapor escape tangentially so that the liquid runs along without being choked off. Since the slot directions alternate on both side of the covers, the directions of flow will alternate likewise. Flow paths and velocities of vapor and liquid can be adjusted within limits by varying numbers and dimensions of the constructional elements. The design characteristics, such as submergence, cap area, coverage, and clearances are about the same as in the standard bell cap described above.

Fig. 6.7 presents part of a tunnel plate and its attachment to the shell. For large column diameters, the tray components may be subdivided as shown in Fig. 6.8.

Sieve Plate. Conventional sieve trays have perforations of 2 to 4 mm. However, the plate in Fig. 6.6. C has holes of 12.3 mm size and a clearance of 12.5%, a more recent configuration. Again, its operation is that of simple crosswise flow of the liquid on the plate. The vapor ascending to the sieve plate penetrates the liquid, causing intense mixing of the phases and good conditions for separation.

Efficiency of operation, pressure loss and capacity can be adjusted by proper selection of correlated hole sizes and clearances.

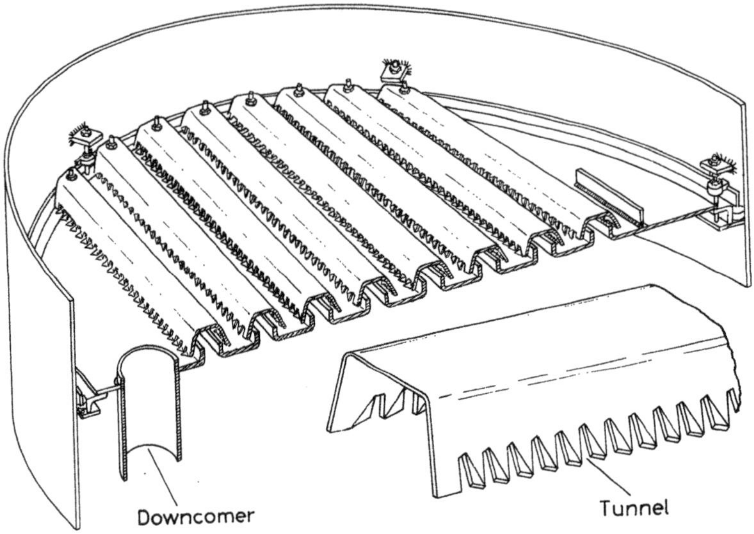

Fig. 6.7 Details of the Thormann System tunnel tray, made by Montz.

Centrifugal Trays. Figs. 6.6. D and 6.6. E show the components of this system[24] which was developed for large throughputs. The liquid is fed in at the center of the tray, and passes via horizontal segments of expanded metal fitted with slots. These are so arranged that the vapor leaves the slots tangentially and blows upward, thus rotating the liquid and forming a spiral path. The liquid eventually drops into an annular channel and is returned to the center by radial passages and finally flows down to the center of the next tray below. By proper dimensioning of the central downcomer repeated recirculation of the liquid can be achieved, which may be desirable for exceptionally slight liquid loadings (Fig. 6.6. D).

The two centrifugal trays, I and II in Figs. 6.6. D and E, differ in the slot sizes of the tray segments made of expanded metal, and in the design of the downcomers. In type I they cause the reflux liquid to recirculate several times because of the tall duct, while in type II the liquid passes over the plate only once.

Obviously, the design of large-size centrifugal trays becomes difficult in view of the liquid gradients that vary along the flow path.

The Valve Tray.[25,26] This has found wide acceptance in the chemical and allied industries. It is appreciably less expensive than other trays and offers operational advantages.

Figs. 6.6. F and 6.9 explain the design and the action of valve plate. Valves *b* over tray *a* will be lifted as the vapor pressure rises. Three guides *c* hold the valves in place. The stops at the lower guide ends limit the lift. For satisfactory flow, the lift should roughly equal the hole diameter. Because of the presence of rest *d*, a lateral

Fig. 6.8 Tray with subdivided tunnels, Thormann system.

exit equalling about $\frac{1}{4}$ of the full opening is always available. The sloping rim *e* of the valves enforces a flow pattern of the exiting vapor. The arrow marks the formation of a turbulence zone for phase mixing.

Investigations[26,27] have shown that the valve plate *a*, manufacturers' type V1, will outperform conventional bell cap plates in efficiency, low-pressure loss, and load capacity. However, actual figures are affected by the shape and weight of the valve plates and on the design of the vapor passages. This makes a careful selection of the design variables mandatory.

Fig. 6.6.G, the *b* type of valves V1, weighting 20 g and 40 g apiece, features simple circular plates with a bent rim and welded guides but without a rest, so that the disks will rest on the plate at low loads. Differently from type *a* (Fig. 6.6.F), the holes have inward rims, to reduce pressure losses at high loadings.

Two different kinds of valves are available. The former includes type A[28] and the aforementioned type V1, which are essentially alike. The other kind is manufacturers' type T.[29] Fig. 6.10 presents the basic designs. Type A has guiding legs, with the plate opening functioning as a component, while in type T the valve disks are guided in a sort of yoke, which makes the valve mechanisms independent of the plate proper. A

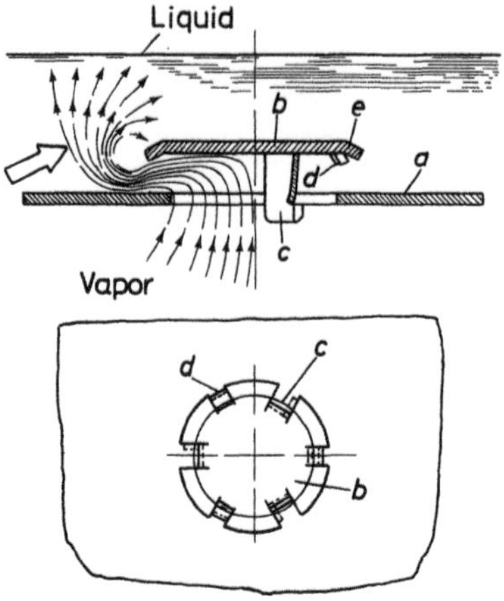

Fig. 6.9 Details of valve Type V1: a Plate, b Valve, c Guides, d Stops, e Rim. (Glitsch System).

decade's experience has been gained with valve trays in all branches of separation plants.

The pressure loss is reduced sensibly by modifying the shape, the weight and the guiding of the valve disks, and the height of the weir. The vapor passages in the tray may be made into Venturi-shaped channels[30-33] as seen in Fig. 6.10. It has been proven that[26] valve trays with down collars have less pressure loss than those with

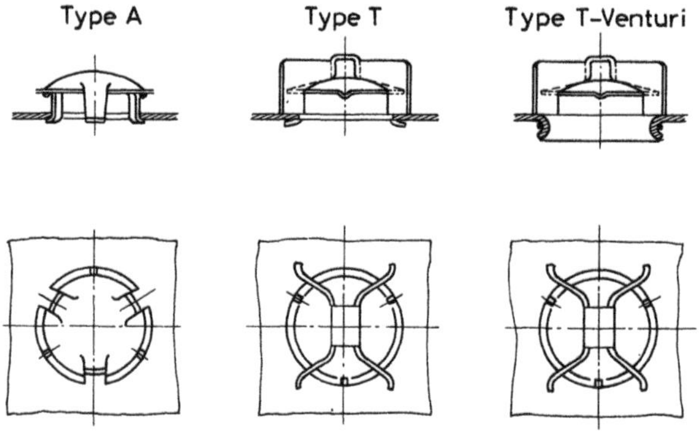

Fig. 6.10 Details of various valve types. (Koch System).

smooth holes. That design will be suitable mainly with the T type because of the favorable guiding of the valve disks, while the A and the V1 types require fairly long guide legs that are more expensive to produce and may jam in operation. Also it has been found that the reliability of these types varies; blocking by incrustations is less likely to take place for the T type than for the A; and there can be no inside wear with the former.

Type *c* (Fig. 6.6.H,) shows a valve plate with T disks and Venturi type holes, manufacturers' Type T 9.

One characteristic of types *a*, *b*, and *c*, in Fig. 6.6 F, G, and H, is the relatively large clearance ratio of 14.8 %, which is desirable from the angles of pressure loss and capacity.

Another, and relatively simple, approach to a reduction of the pressure loss is in combining type T 9 disks with perforations, as seen in type *d* of Fig. 6.6. I, manufacturers' type T 9 F.

3. EVALUATION OF RECTIFYING COLUMNS AND BEST MODE OF OPERATION

When working on separation problems, the planning engineer must often select, from among a great number of column devices, the one most likely to give best results. A rational comparison must include cost optimization based on a knowledge of processing factors, which in turn includes data on investment called for on one hand and limitations on effectiveness, loads, flexibility, pressure loss, and sensitivity to fouling during operation. However, to optimize a process involving column internals, constant service conditions have to be provided.

In the evaluation procedure outlined here, a knowledge of the separation principle of a given device and its pressure loss under given loads for a given expense will yield some minimum for the unit column cost per unit separation per unit mass of product. That minimum cost is a function of the load characteristic and of the construction material for a given separation device and is not necessarily associated with any maximum load, however.

A. Evaluation and calculations, separating effect, pressure loss, load, specific column volume, and specific column cost.

Fig. 6.11 presents the scheme for the engineering and commercial evaluation of a separating device, for two different separation versus load characteristics. Numbers 1 to 8 indicate the sequences of evaluating stages by individual plots as follows:[34,35]

1. Determination of separating efficiency E of mass transfer plate, or of the equivalent theoretical stages equal to a packing height H (m), n_t/H (m^{-1}) as a function of the vapor load factor $u_V \rho_V^{1/2}$ ($m^{-1/2}$ s^{-1} $kg^{1/2}$) with the vapor velocity based on the free column area, the specific weight of the vapor being ρ_V (kg/m^3)

2. Determination of the pressure loss p/n_p (mm water) of an actual plate, or, for packed columns, of the pressure loss per unit height p/H (mm water) as a function of the vapor-load factor.

3. Calculation of the pressure loss $\Delta p/n_p$ (mm water) per theoretical stage, a critical variable in the rectifiers for heat-sensitive mixtures, as a function of the vapor-load factor.

4. Calculation of the specific tower volume as a function of the vapor-load factor, or of the specific pressure loss as a basis for the cost optimization. The specific tower volume[10,11,36] is that specific volume v_V ($m^3/m^3 s$) required per unit

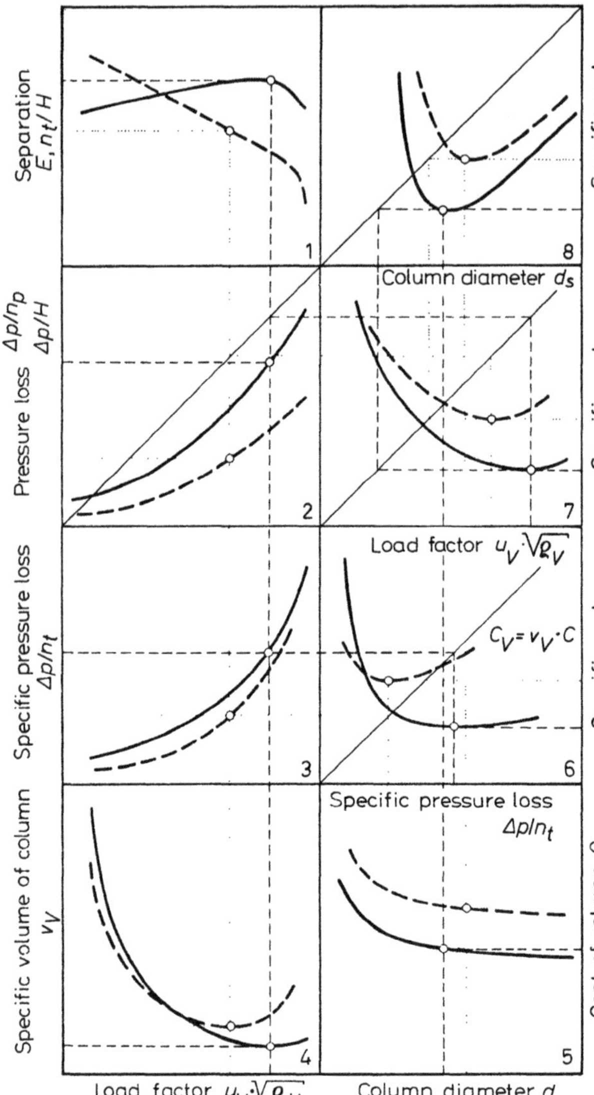

Fig. 6.11 Succession of steps in the evaluation and optimization of columns.

of vapor throughput and of separation. With tower diameter d_s (m) and

$$V^+ = u_V d_S^2 \frac{\pi}{4} \tag{6-11}$$

the vapor flow through the tower for the specific tower volume

$$v_V = \frac{d_S^2 \frac{\pi}{4} H}{V^+ n_t} = \frac{H}{u_V n_t} \tag{6-12}$$

For column packing the ratio H/n_t may be replaced with the reciprocal of the theoretical stages referred to the packing height. For plate towers, with T_s (m) plate spacing, and E = separation factor,

$$\frac{H}{n_t} = \frac{T_S}{E} \tag{6-13}$$

and with Eq. (6-12)

$$v_V = \frac{T_S}{u_V E} \tag{6-14}$$

In comparing separation devices, sometimes the reference load becomes the corresponding air velocity[31] u_A in m/s

$$u_A = u_V \left(\frac{\rho_V}{\rho_A} \right)^{1/2} \tag{6-15}$$

which is relative to atmospheric air of density $\rho_A \approx 1.2$ kg/m^3. The specific tower volume v_A (m^3/m^{3}s) is then

$$v_A = \frac{d_S^2 \frac{\pi}{4} H}{u_V \left(\frac{\rho_V}{\rho_A} \right)^{1/2} d_S^2 \frac{\pi}{4} n_t} = \frac{\rho_A^{1/2} H}{u_V \rho_V^{1/2} n_t} \tag{6-16}$$

The relative air velocity is a convenient measure of the load on a rectifier when test stand results are available for air-and-water experiments on various loadings.

5. Determination of the tower cost per m^3 of construction volume, C (monetary units/m^3), considering shell cost related to column diameter.
6. Calculation of the specific tower cost

$$C_V = v_V C \tag{6-17}$$

$$C_A = v_A C \tag{6-18}$$

as functions of the specific pressure loss $\Delta p/n_t$ if the latter is relevant.

7. Determination of C_V or C_A as functions of the vapor-load factor $u_V \rho_V^{1/2}$ when the pressure loss is irrelevant.
8. Calculation of the specific tower cost as a function of column diameter. For a diameter of d_S (m)

$$d_S = \left(\frac{4}{\pi} V^+ \frac{1}{u_V}\right)^{1/2} \tag{6-19}$$

and with Eq. (6-12)

$$d_S = \left(\frac{v_V n_t 4 V^+}{H\pi}\right)^{1/2} \tag{6-20}$$

Replacing vapor quantity V^+ (m^3/s) by a term connecting the molar vapor quantity V (kmol/h) for the pressure p (mm water) and the absolute temperature T (K)

$$V^+ = V \frac{22.4}{3600} \frac{T}{273} \frac{760}{p} \tag{6-21}$$

will give:

$$d_S = 0.149\left(\frac{T}{p} V \frac{n_t}{H} v_V\right)^{1/2} \tag{6-22}$$

Substituting the specific gravity ρ_V (kg/m^3) of the vapor having a molecular weight μ_V (kg/kmol) finally will yield[38,39]

$$d_S = 0.053\left[\frac{V}{u_V \rho_V^{1/2}}\left(\mu \frac{T}{p}\right)^{1/2}\right]^{1/2} \tag{6-23}$$

Thus the optimum tower diameter can be found which is related to the minimum specific tower cost for a given vapor throughput, using that vapor load factor $u_V \rho_V^{1/2}$ for which, in Fig. 6.11, Part 7, the specific cost C_V is at a minimum. That load factor need not be identical with the top load; it depends on the load characteristic of the separation on the one hand, and on the material of construction on the other.

B. Numerical evaluation for packed towers.

A prerequisite for an objective comparison of columns is a constant basis of reference. The throughput, separating power, and pressure loss will be affected by the physical properties of the mixture to be processed, and hence by the operating pressure, the liquid-vapor relationship inside the column, and upon a uniform distribution of these two phases.

a. Separating Effect, Pressure Loss, and Loading of Packed Towers

In vacuum rectifications, especially in the case of thermally unstable materials, the pressure loss in the bulk of the packing must be low, in order to maintain a moderate column-base temperature. Also, compared with atmospheric processes, the throughput will be less for a given load factor, and the packing that will be favored is the one which allows maximum throughput at minimum pressure loss per theoretical step[40] – the thin-walled metallic element.

At this point, the Interpack design will be compared with the Pall ring, which has given exceedingly good results in many applications.

The mixture was ethyl benzene and styrol, the column top pressure was 100 mm Hg, and the liquid/vapor ratio was unity (for the sake of simplicity). In practice, in continuous rectifiers, the liquid/vapor ratio will exceed unity in the enriching section and will be below unity in the stripping section.

The ethyl benzene–styrol mixture was selected, first because its boiling pattern has been studied widely and is essentially ideal (note Chapter 2) it could thus be accepted as standard. Second, the separation of this mixture is commercially important, and large quantities of styrol are being produced, mostly in plate type columns over 10 m in diameter. The reasons why packing is unsuited for that size column are discussed later. However, styrol tends to polymerize at higher temperatures, and therefore the pressure loss per theoretical step must be held down when distilling the ethyl benzene–styrol mixture. A decisive point here is whether the separating column can be designed for straightforward operation, or whether a repetitious procedure involving intermediate condensation and reevaporation would be indicated, which would be more costly in investment and service (note Chapter 7).

Fig. 6.12 presents a dimensioned view of the packing in question. They were tested

Pall ring elements

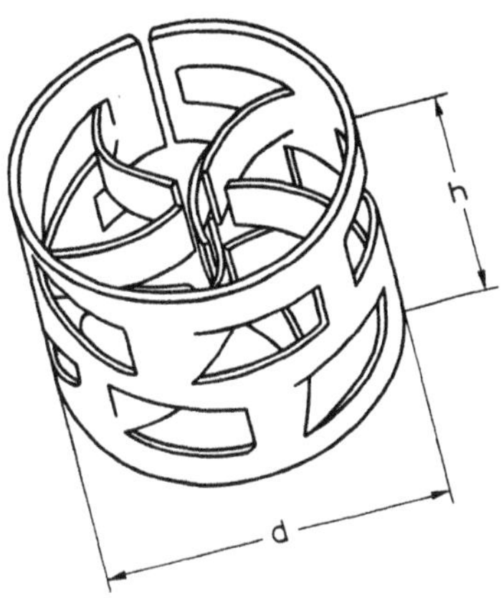

d = h

Interpack elements

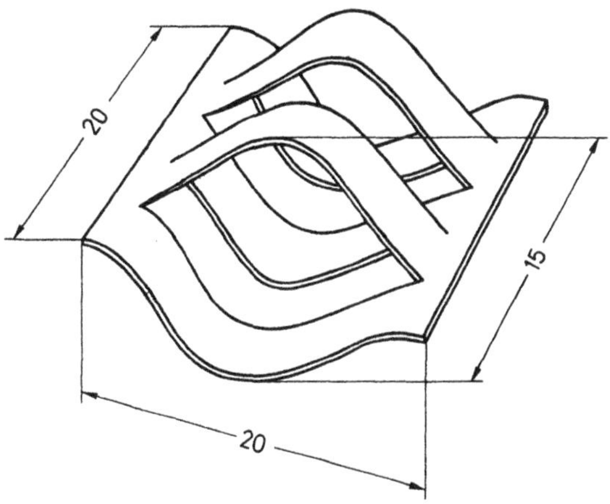

Basic dimension mm	Number per m³	Clearance volume %
50 x 50 x 1	6100	95.2
25 x 25 x 0.6	47500	94.2
15 x 15 x 0.4	224000	93.5

Number per m³	Clearance volume %
213000	95

Fig. 6.12 Principal dimensions of packing elements.

in the semiindustrial installation of Fig. 6.13 under the conditions described. Part of the equipment is the liquid redistributor shown in Fig. 6.14. Since the distribution of the reflux liquid has a marked effect upon the separating effect, its nature must be stated in order to have a complete description of the test.

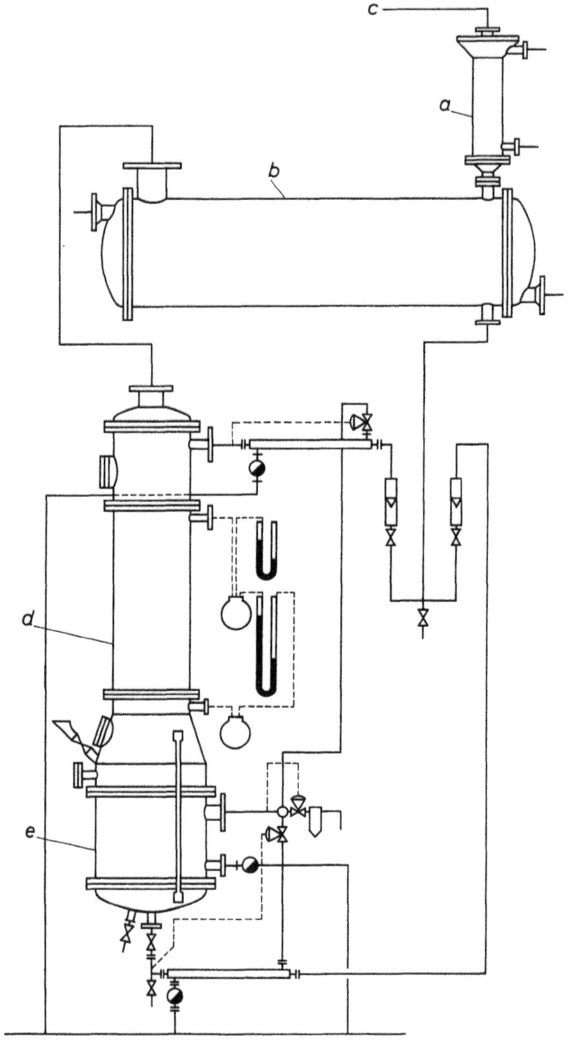

Fig. 6.13 Setup for testing packing elements, diameter of column 500 mm: a gas cooler, b reflux condenser, c vacuum connection, d rectifier, e reboiler.

The distributor in Fig. 6.14 of two tube plates connected by vapor tubes forms a weldment, together with a short shell and an annular feeder arrangement. The lower tube plate is also fitted with 2 mm nozzles, yielding six supply points per dm^2 of column cross-sectional area.

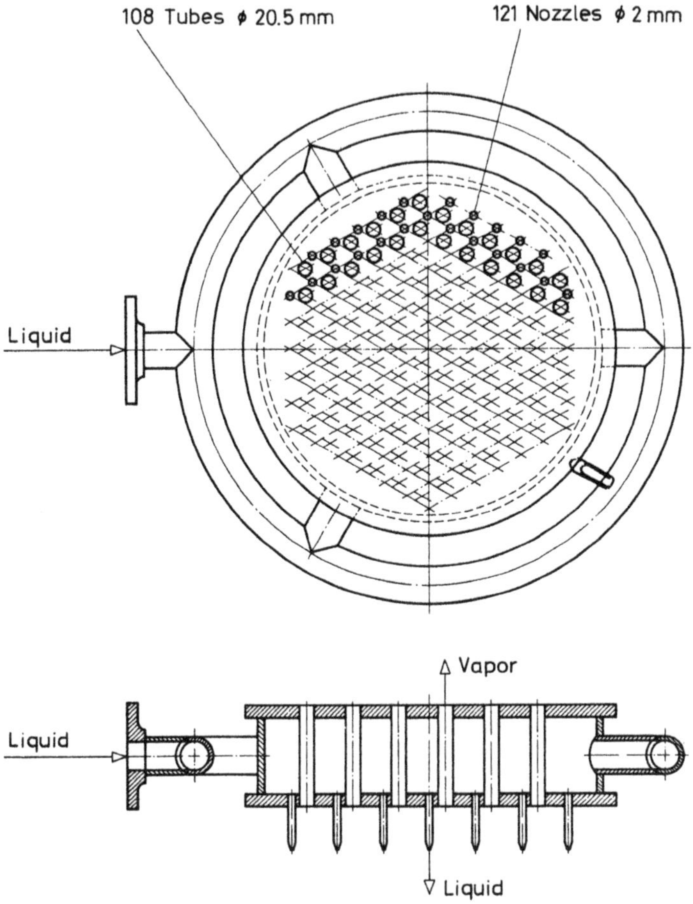

Fig. 6.14 Liquid distributor.

Fig. 6.15 presents the test results, as described in Section A. of this Chapter and outlined in plots 1, 2, and 3. The separation curves for the Pall rings are as expected inasmuch as for small-size rings the separation is more thorough, while the load limit is lower than for large-size rings. For the Interpack, at low loads the effectiveness is the equivalent of that of the 15 mm Pall rings, at average loads equivalent to 25 mm Pall rings, while at high loads it is less than that of 50 mm Pall rings. The highest loads relate to that flow condition at which the column starts flooding. It is seen that the maximum admissible loading for a 20 mm Interpack type is even below that of a 25 mm Pall-ring layer, but above the loadability of a 15 mm Pall-ring layer. In the main load range, the pressure loss per unit layer depth is much less than for the Interpack layer. The same is true for the pressure loss per unit of separation, disregarding the load range of the 15 mm Pall rings. This evaluation is definitely in favor of the Pall ring. However, a test made with a rectification of some other mixture may possibly come up with slightly different results.

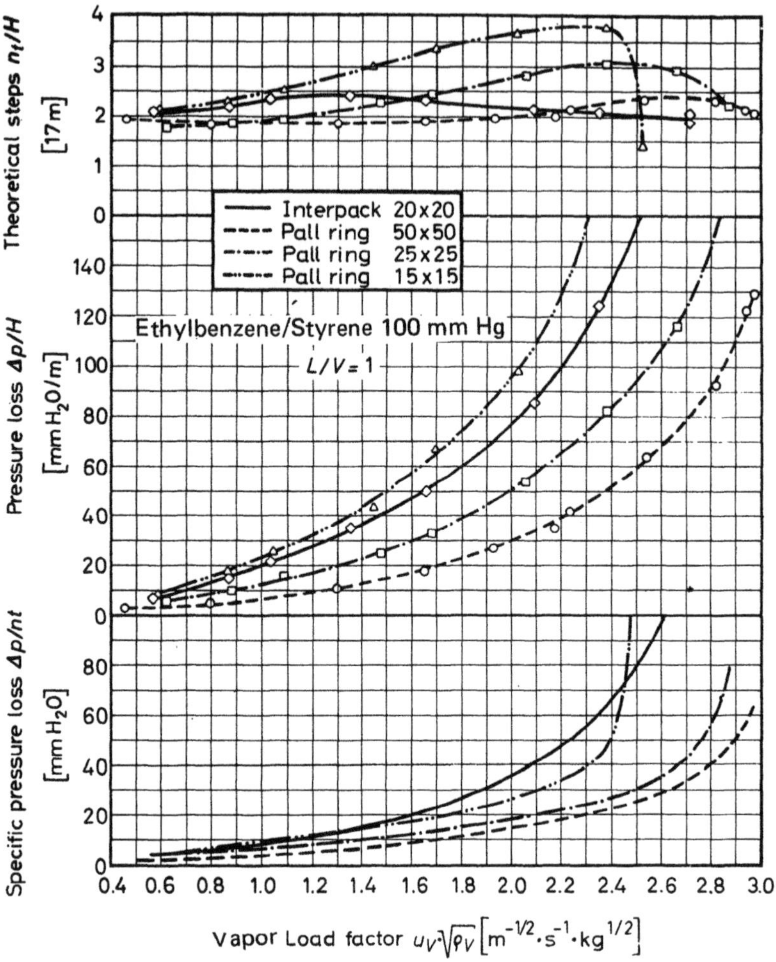

Fig. 6.15 Separation and pressure loss of different packings and loads.

b. Specific Column Volume and Specific Column Cost of Packing Tower

In the total evaluation of a packed tower, the processing indices reviewed above are not entirely comprehensive; from the view of the responsible planner the end cost of a tower suitable for a given separation will decide. To that end, the evaluation stages 4 and 5 of Fig. 6.11 must furnish the specific tower volume and the relative tower cost C (cost units per m^3). Since, however, the resulting tower diameters will differ for different packings, and the price of the shell per m^3 of tower volume will change with the tower diameter, the determination of the relative column cost must consider the share of the shell.

Fig. 6.16 is the result of a determination of the tower cost per m^3 as a function of the diameter for carbon steel construction. The lower cost curve applies to the shell only and represents averages. For any given tower diameters, the relative tower costs are slightly higher with 20 mm Interpack than with a 50 mm Pall ring layer, but much less than with 25 mm or 15 mm Pall rings.

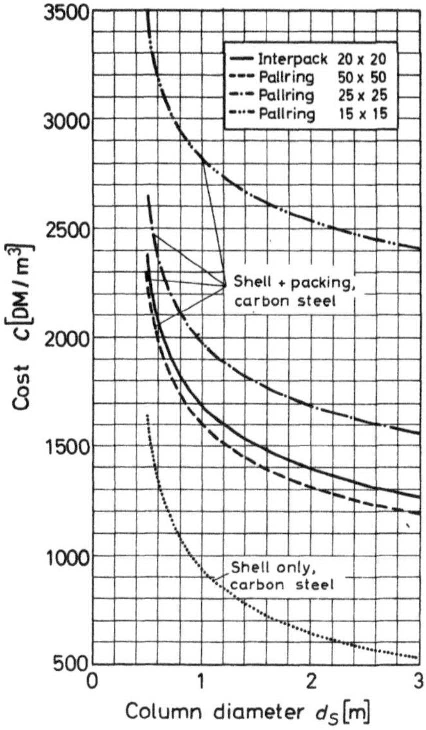

Fig. 6.16 Packed columns. Cost per cubic meter for various types and diameters. Prices of 1970.

The specific column volume found by Eqs. (6-12) and (6-16) is presented as a function of the specific pressure loss in the upper part of Fig. 6.17. By multiplying the values in these curves by the corresponding data from Fig. 6.16, referring to Eqs. (6-17) and (6-18), stage 6 in Section A of this chapter, yields the specific tower costs, which are to be met per unit separation and throughput for a given application. These are plotted in the lower part of Fig. 6.17 against the specific pressure loss, based on the cost of an Interpack tower of $\tilde{d}_S = 0.5$ m size.

In this connection, the diameters of the Pall-ring columns operated at different load factors $u_V\rho_V^{1/2}$ follow from

$$d_S(\text{Pall}) = d_S(\text{Interpack})\left[\frac{[u_V\rho_V^{1/2}(\text{Interpack})]_{\Delta p/n}}{[u_V\rho_V^{1/2}(\text{Pall})]_{\Delta p/n_t}}\right]^{1/2} \qquad (6\text{-}24)$$

the cost per m^3 of construction, required for the calculation of the specific cost by Eqs. (6-17) and (6-18) taken from Fig. 6.16.

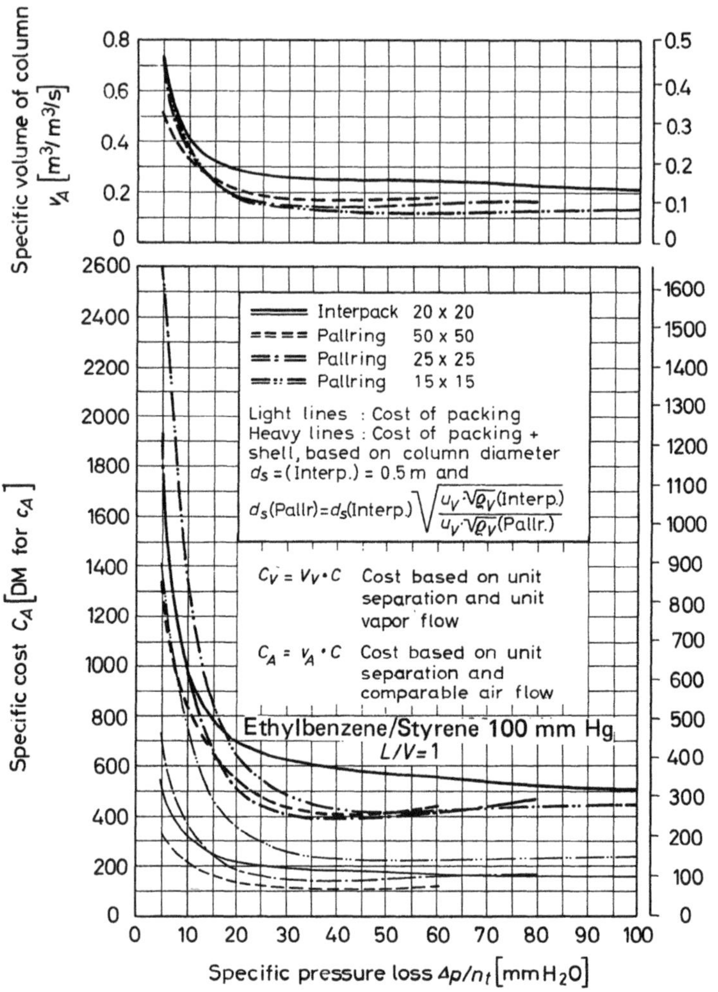

Fig. 6.17 Specific column volumes and specific column costs for various packings and specific pressure losses.

In the lower part of Fig. 6.17 the light lines give the packing cost, while the heavy lines apply to the packing plus the shell. Obviously, neglecting the shell cost when figuring the tower cost will cause errors.

From the heavy lines in the lower part of Fig. 6.17, which make allowance for the shell cost, the Pall ring appears superior for the main load range of a given separation, as far as specific column cost goes. One characteristic of the cost curves of Pall rings is the existence of minimums that reflect specific pressure losses coordinated to maximum separation.

The upper part of Fig. 6.17 also indicates that specific tower volumes for Pall rings are usually less than for Interpack.

Determining the percent cost of Pall-ring towers

$$C_p = \frac{C_V \text{ (Pall rings + Shell)}}{C_V \text{ (Interpack + Shell)}} (100) \tag{6-25}$$

as a percentage of the cost, based on separation and throughput, of comparable Interpack columns of 0.5 m diameter, one obtains the dependence on the specific pressure loss reflected in Fig. 6.18. The corresponding diameters of the Pall towers again follow

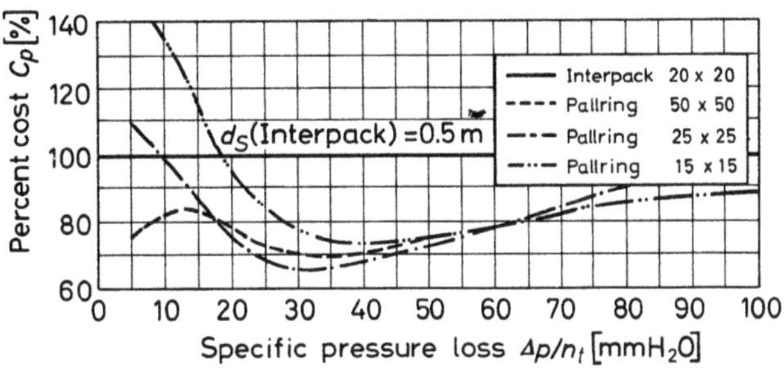

Fig. 6.18 Cost of packed columns for different specific pressure losses.

from Eq. (6-24), where $u_V \rho_V^{1/2}$ are the load factors for equal specific pressure losses plotted in the lower part of Fig. 6.15.

Transferring results found on a pilot tower of 0.5 m diameter to towers of larger diameters and heights in order to gain a precise cost optimization is difficult. Unknowns are the effects of irregular vapor and liquid flows, blowby and rivulet formation, and the resulting reduction in separating power. However, by assuming that the percent decrease in separation of diverse packings in larger size columns are about equal, one can find the percent cost of Pall towers by Eq. (6-25) and Figs. 6.15 and 6.16, using Eqs. (6-12), (6-17), and (6-24), and plotting this as shown in Fig. 6.19 over the tower diameter, with the specific pressure loss as the parameter.

For specific pressure losses of ordinary magnitude, the Pall ring will be more profitable even with larger columns. As for the data in Fig. 6.19, a use of small-size packing in a large column would be exceptional.

c. Evaluation of Highly Loaded Packed Columns

Table 6.1 lists the engineering and commercial indices for top loadings that still offer adequate safety against flooding. Again, Pall rings show advantages. Using a 0.5 m reference diameter, the specific tower volume of the Interpack column is appreciably larger than the column volumes of the Pall design, and despite virtually equal clearances the Interpack layer will have a much greater pressure loss and much more liquid retention. However, in the working of temperature-sensitive substances, low liquid holdover and small liquid content are desirable.

Table 6.1

Packed columns operating at maximum loads. Reference design Interpack $d_S = 0.5$ m, carbon steel construction, ethylbenzene/styrene system, at 100 mm Hg, $L/V = 1$.

Packing		Interpack 20 X 20	Pallring 50 X 50	Pallring 25 X 25	Pallring 15 X 15
$u_V \rho_V^{1/2}$ ($m^{-1/2} s^{-1} kg^{1/2}$)		2.4	2.65	2.48	2.27
n_t/H (m^{-1})		2.02	2.4	3.1	3.9
$\Delta p/n_t$ (mmH_2O)		68	31.5	30.5	38
v_V ($m^3/m^3/s$)		0.1438	0.1095	0.0907	0.0788
C_V(DM)	Packing	107	73	94	153
	Shell + Packing	346	273	258	277

d. Comparison of Packings and Grids in Working a Given System

To juxtapose results obtainable with Pall-ring layers, wire-cloth packings, and grid packings, tests were run in 500 mm columns for wire-cloth packing, and 800 mm columns for grids and Pall rings. The equipment is shown diagrammatically in Figs. 6.13 and 6.20. In the experiments on Pall rings and grids, the reflux was recirculated by means of a redistributor (Fig. 6.21). As in the redistributor in Fig. 6.14, there are about six feeder points per dm^2 of tower section. A special device was used for wire-cloth packing, shown in Fig. 6.22, with about 10 feeder points per dm^2 of tower section.

As outlined in Section A of this chapter, an investigation of the effectiveness of mass transfer will come first. The essential indices are presented in Fig. 6.23, being the separation efficiency n_t/H and pressure losses $\Delta p/H$ and $\Delta p/n_t$ as functions of load $u_V \rho_V^{1/2}$. The plot reveals how wire-cloth packing improved separation and specific pressure loss per theoretical step over the principal load range. No other separation device is competitive in this respect. Differently from other columns of the dynamic type, separation improves as the liquid load diminishes because of capillary action. However, there is the drawback that no mechanical cleaning of the packing is feasible. In terms of separation the, next best are the 50 mm Pall ring, the 80 mm Pall ring, and the square grid, the fine grid, and the coarse grid, if the zone next to the upper load limit is disregarded.

As for the performance characterized by falling pressure loss per theoretical step, the important range is between $2.0 < u_V \rho_V^{1/2} < 2.7$ $m^{-1/2}$ s^{-1} $kg^{1/2}$. The succession is: wire mesh first, then coarse grid, square grid, 80 and 50 mm Pall, and fine grid.

Because of its relatively large clearances, the coarse grid has the best throughput per unit column section. Note that, because of the limited cooling capacity of the reflux condenser (Fig. 6.20), the maximum loads for the fine and the coarse grid could not be stated accurately. However, the maximum loads for these two will be close to the limits expectable from Fig. 6.24. Since in that plot the reflux quantity

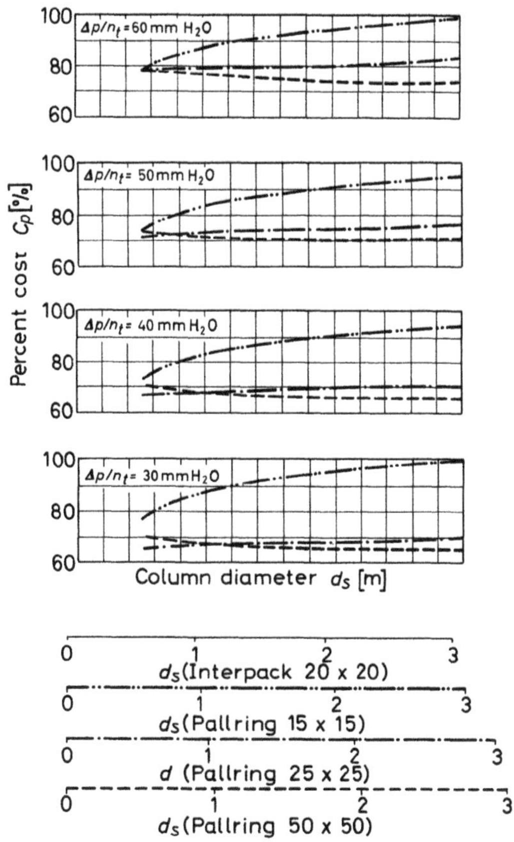

Fig. 6.19 Cost of packed columns at varying column diameters.

measured at the column top (figuring in Fig. 6.23) has been plotted as a function of the corresponding vapor load; the difference between the curve values and the diagonal values represents the liquid entrained by the vapor at the top of the column.

The tower cost referred to a unit volume of the equipment (Fig. 6.25) is expressed as a function of the column diameter. Relatively heavy, or relatively light shells would affect the correctness of that relationship. In general, the relative cost for all columns becomes less as the diameter increases, because the share of the relative cost of the shell is greater for small column sizes. Hence, the thinner the column, the greater the influence of the column size upon cost optimization. The left-hand part of Fig. 6.25 has no curve for wire-net packing, because this is only available in stainless steel.

When the specific pressure loss is decisive in cost optimization, again the specific column volumes μ_V and μ_A are plotted over $\Delta p/n_t$; note the upper part of Fig. 6.26. These volumes are found from the curves in Fig. 6.23 by Eqs. (6-12) and (6-16). All volume curves thus obtained have minima that must be identical with the minimum cost, because of Eqs. (6-17) and (6-18). This is seen from the curves in the lower part

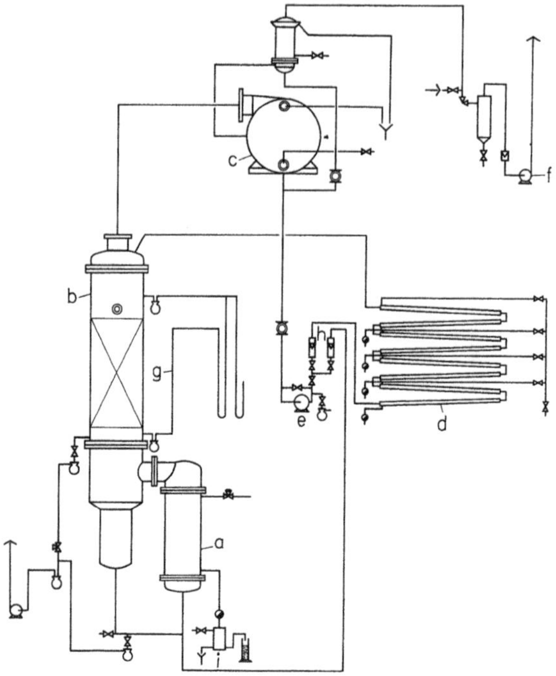

Fig. 6.20 Packing test setup, diameter of column 800 mm: *a* reboiler, *b* rectifier, *c* reflux condenser, *d* reflux heater, *e* reflux pump, *f* vacuum pump.

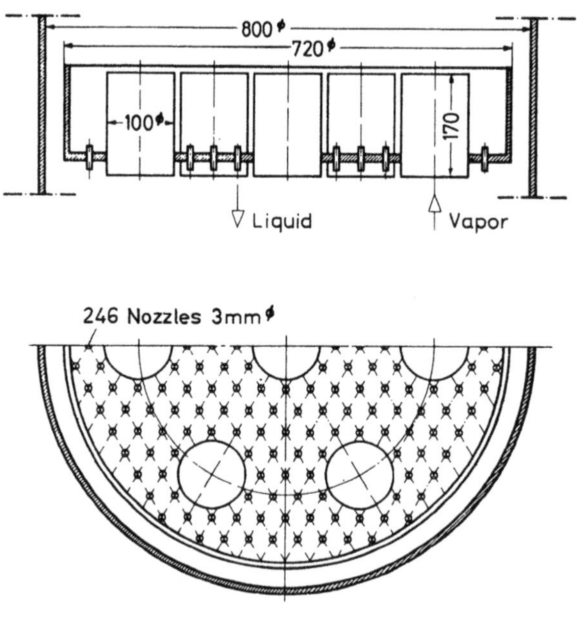

Fig. 6.21 Distributor plate.

a)

b)

Fig. 6.22 **Distributor** made by Sulzer Co. of Winterthur, Switzerland: *a* parts before assembly, *b* assembly of device.

of Fig. 6.26. These were obtained by the procedure of Fig. 6.11, by using the relative cost from Fig. 6.25 and multiplying by the volume curves in the upper part of Fig. 6.26. The reference diameter was 0.8 m = d_S for the coarse-grid type. For the diameters of other columns of equal $\Delta p/n_t$

$$d_S = d_S\ (\text{coarse-grid}) \left[\frac{[u_V \rho_V^{1/2}(\text{coarse-grid})]_{\Delta p/n_t}}{(u_V \rho_V^{1/2})_{\Delta p/n_t}} \right]^{1/2} \qquad (6\text{-}26)$$

where

$[u_V \rho_V^{1/2}(\text{coarse-grid})]_{\Delta p/n_t}$ = vapor load factor of the reference column, and

$(u_V \rho_V^{1/2})_{\Delta p/n_t}$ = that of the column compared having the same specific pressure loss $\Delta p/n_t$

Obviously, the wire-net packing is the most compact. Price advantages, however, will appear only for that load range with specific pressure losses below 7-mm water; therefore it is recommended for rectifications of thermally unstable mixtures under high vacuum where low temperatures are desirable to prevent spoilage – meaning low specific pressure losses.

In the load ranges with specific pressure losses above 10 mm water, however, the tower with 50-mm Pall rings will be attractive in price. Its minimum cost is located at a specific column pressure of 35 mm water. There, follow the column with 80-mm Pall rings, the coarse-grid column, and farther off the wire-net column, the square grid, and the fine grid types.

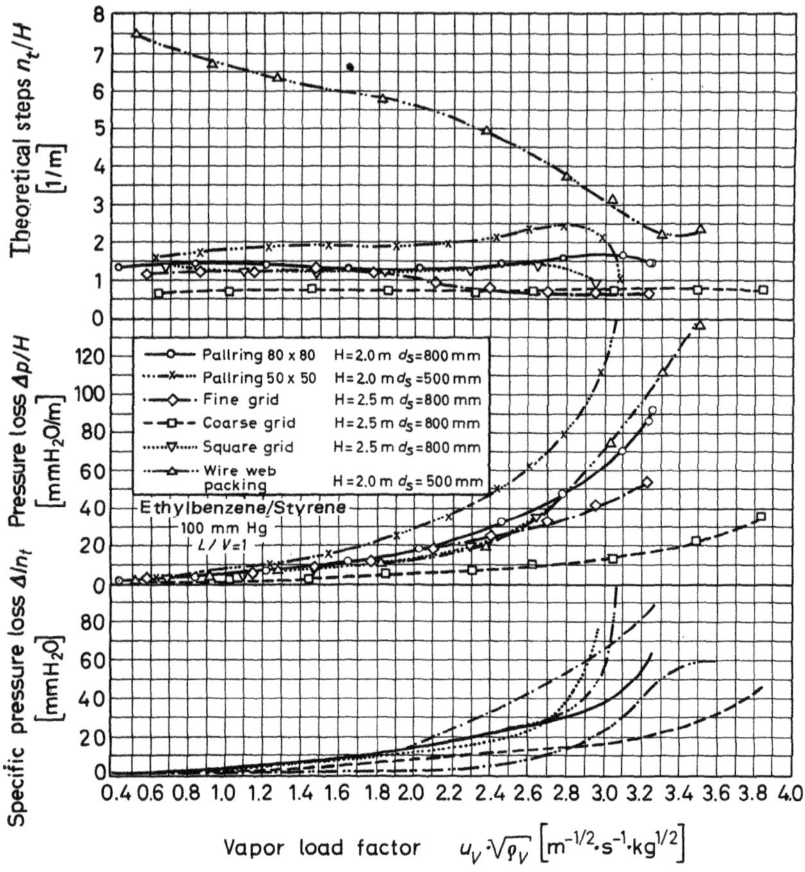

Fig. 6.23 Separation and pressure drop of packing versus vapor load factor.

As the relative column cost depends on the diameter, its cost differential may be expected to vary as the size increases. For a coarse-grid column the percent costs

$$C_p = \frac{C_V\ (\text{packing} + \text{shell})}{C_V\ (\text{coarse-grid} + \text{shell})}\ (100) \qquad (6\text{-}27)$$

were determined as a function of the reference d_S of the coarse-grid column. They are the ratio of the cost of the column investigated and the coarse-grid type serving as a reference, expressed in percent. Since, in Fig. 6.23 the results of the test comprised only columns of 0.5 and 0.8 m size, it was assumed that operational variations due to uneven vapor and liquid distribution have comparable effects for all packings considered.

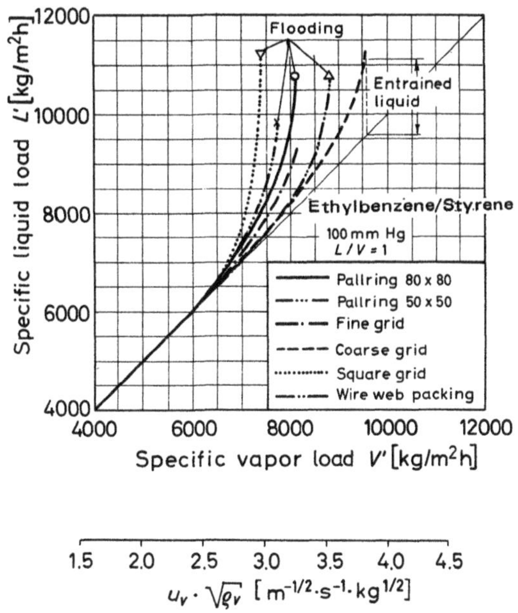

Fig. 6.24 Packed columns. Specific reflux load versus vapor load at column top.

Fig. 6.27 thus reveals the percentage costs according to Eq. (6-27) over the reference diameter of the coarse-grid column plotted for the range between 0.5 and 3 m. Although the diameter somewhat affects the cost, the sequence of Fig. 6.26 is retained, on the whole.

When the nature of the intended separation does not impose any particular restrictions of the specific pressure loss, the packing may be selected on the basis of Fig. 6.28, where the specific tower volume is shown as a function of the vapor load factor. This figure reveals that the load that would cause a minimum construction volume and cost will not necessarily be the same as that for a maximum loading capacity. This especially applies to packing types for which increasing load causes loss of separation.

The middle part of Fig. 6.28 presents the proper cost curves according to phase 7 of Fig. 6.11 for stainless steel construction, and the lower part of Fig. 6.28 applies to carbon steel. The column diameter is 0.8 m. For the minimum cost, the conclusion is again the same as for Figs. 6.26 and 6.27: 50-mm size Pall rings give the best economy.

One characteristic of the Pall rings is that the cost minimum is located at a vapor load $u_V\rho_V^{1/2}$ slightly below the maximum. Here, the fine-grid and wire-net packings have their optimum loads much below the maximum, because their separation drops off with increasing load (Fig. 6.23, top part).

The absolute specific cost minimum for a given vapor throughput of a column is at the same level as the minimum specific construction volume and that of the specific cost for a given column diameter. This is because the change in the cost dC/dd_S per m^3 of the column volume over the range considered is negligible as compared to the

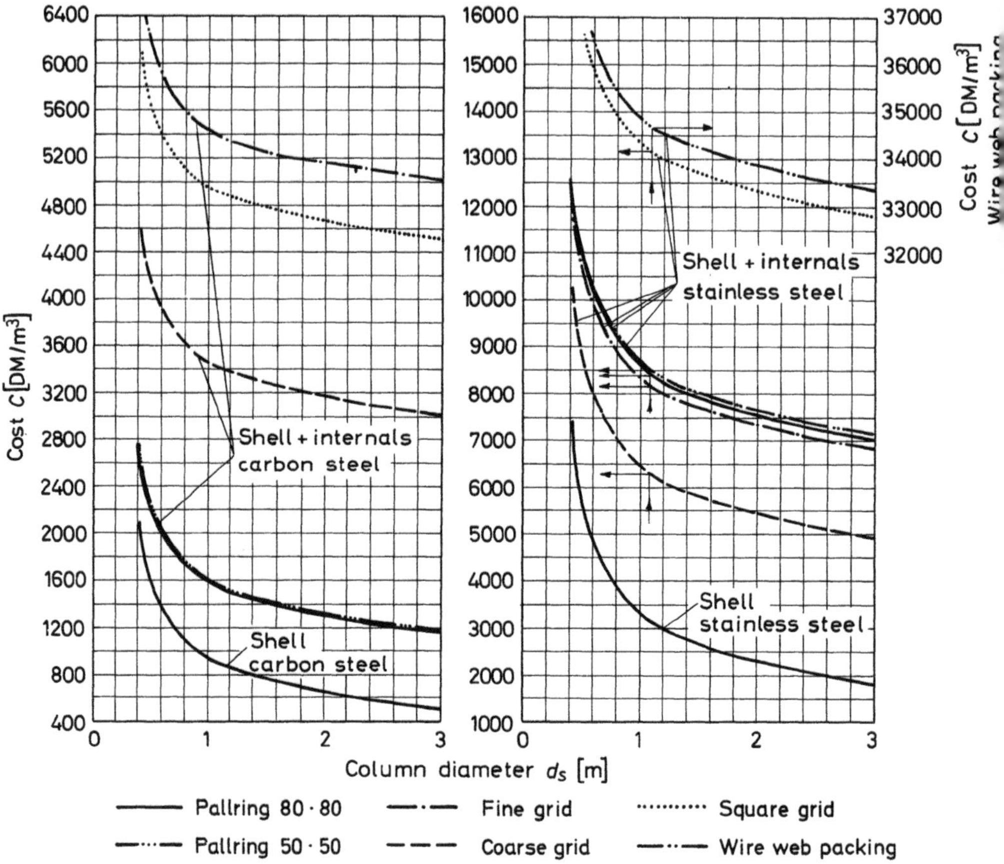

Fig. 6.25 Packed columns. Cost per unit volume versus column diameter. Prices of 1970

corresponding change in specific column volume $dv/d(u_V\rho_V^{1/2})$. This is demonstrated in Fig. 6.29 for the rectification of ethyl benzene and styrol at 100 mm Hg in a column with 50-mm Pall rings of stainless steel for the vapor rates $V^* = 0.82$ and $2.09\ m^3/s$.

By evaluation of phase 8 of Fig. 6.11, the minimum tower volume v_{Vm} corresponds to diameters d_S of 0.5 and 0.8m.

For loads $u_V\rho_V^{1/2}$, below or above the ones correlated to minimum volume v_{Vm}, Fig. 6.29 yields, in its lower part, the heavily drawn cost curves for constant diameter d_S at variable V^* rates, and the lightly drawn curves for constant throughput V^* at variable diameter d_S. Thus, cost differentials become less as diameter d_S increases and the dC/dd_S values decrease.

A column layout for $u_V\rho_V^{1/2}$ ratings deviating from the optimum for the minimum tower volume ρ_{Vm} may cause a much higher tower cost, from phase 8 in Fig. 6.11. This is seen from Fig. 6.30.

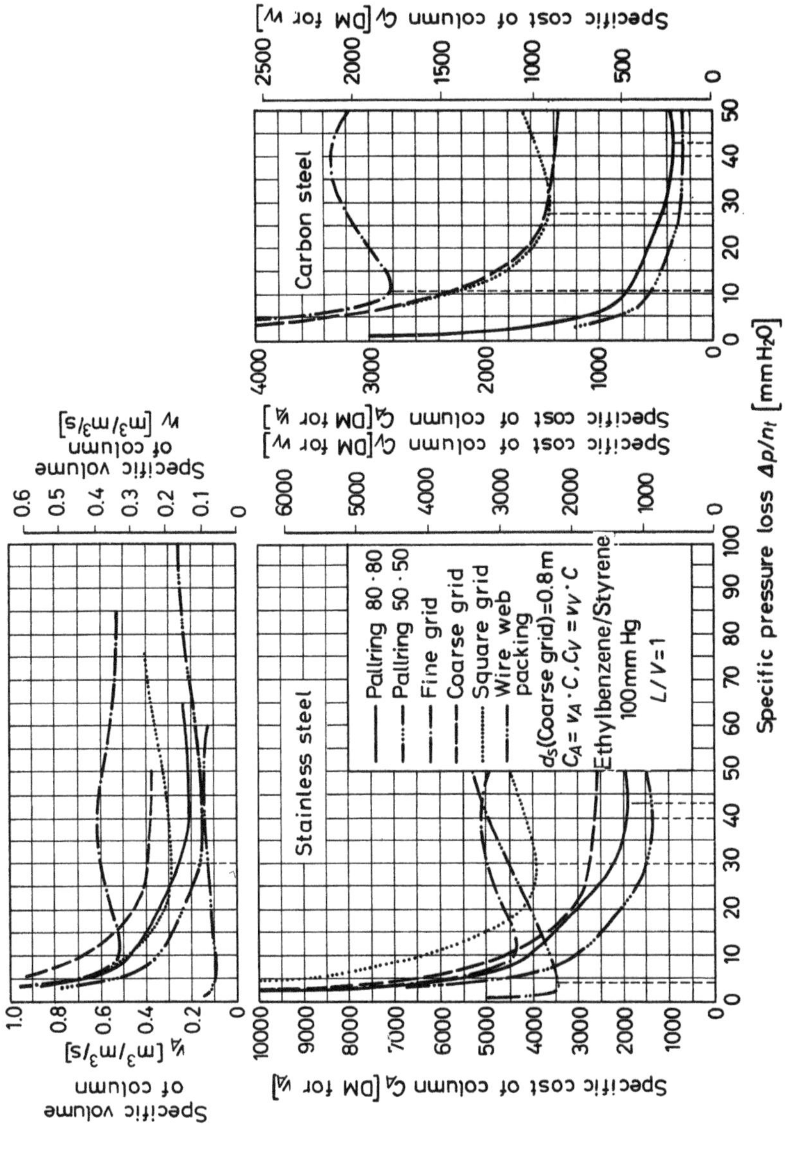

Fig. 6.26 Packed columns. Specific column volume and specific cost versus specific pressure loss.

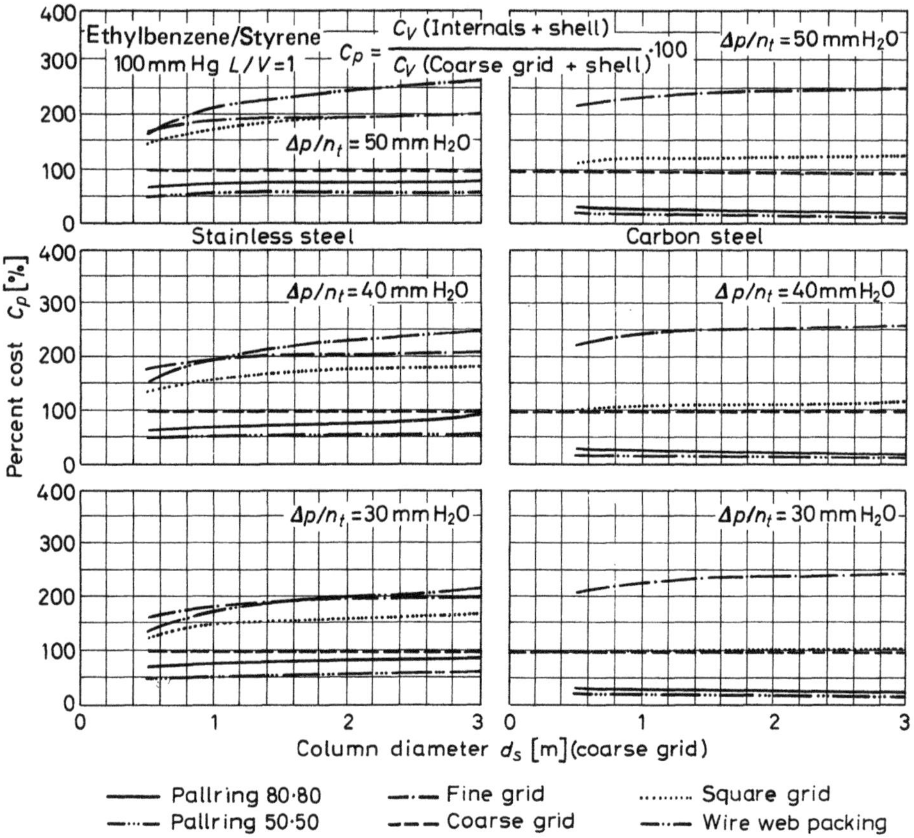

Fig. 6.27 Packed columns. Percentual cost versus column diameter, for various pressure losses per theoretical step.

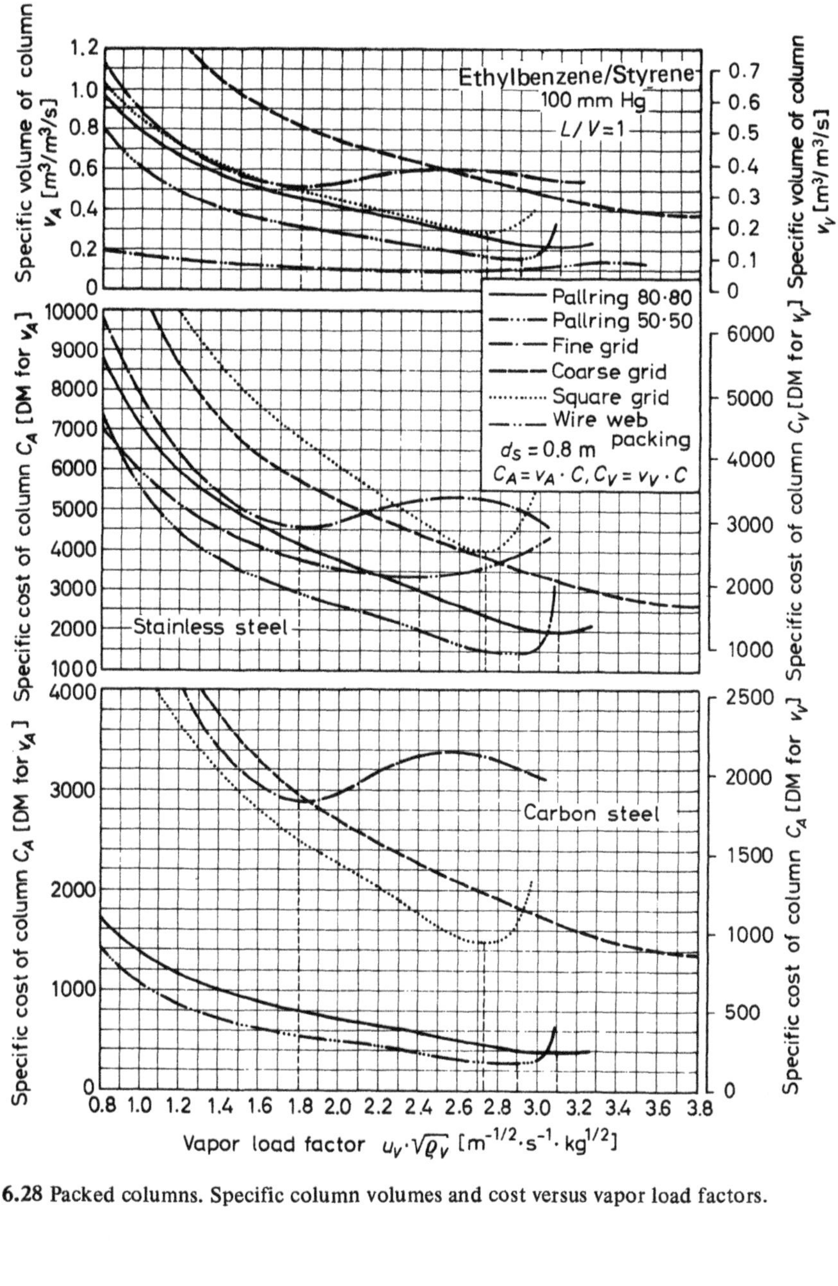

Fig. 6.28 Packed columns. Specific column volumes and cost versus vapor load factors.

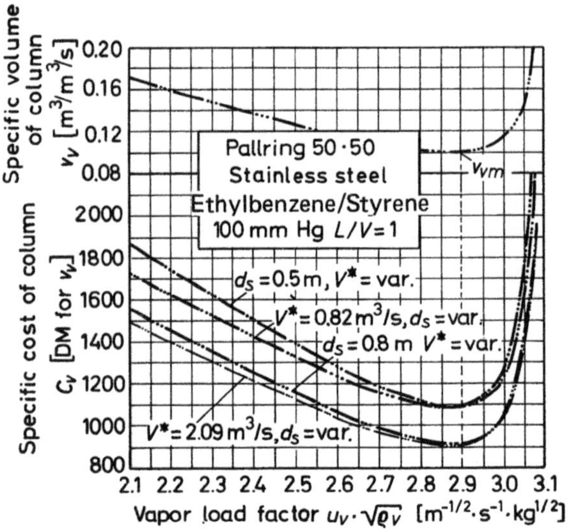

Fig. 6.29 Specific column volumes and cost for Pall ring packing 50 • 50 mm versus vapor load factors. Constant vapor flow at V* = 0.082 and 2.09 m³/sec. Constant diameters d_s = 0.5 and 0.8 m. Mixture of ethyl benzene and styrene at 100 mm Hg.

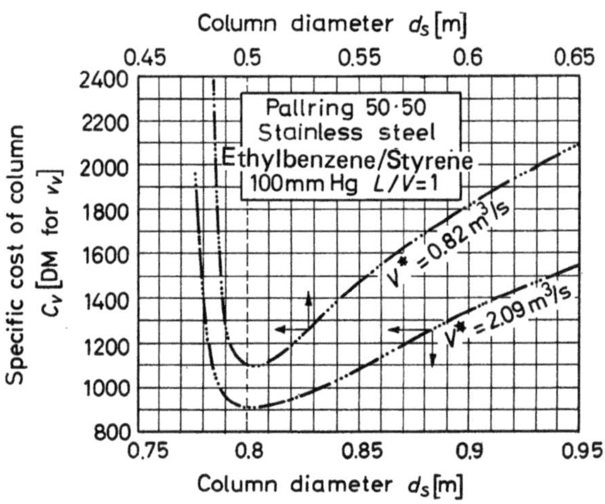

Fig. 6.30 Optimum column diameter for the conditions of Fig. 6.29.

e. Performance of Packed Columns Handling Various Systems

It has been seen that the performances of 50-mm Pall rings and wire cloth stand out in some ways, and these two will be compared to find how the nature of a mixture will affect operation and expenses. Along with the ethyl benzene/styrene mixture at 100 mg Hg, the same mixture will be investigated at 50 mm Hg, as will a 1, 2-propylene glycol– ethylene glycol mixture at 10 mm Hg, and methanol–ethanol at atmospheric pressure.

Fig. 6.31 shows the process properties of the Pall ring. Generally, separating effect and pressure loss increase with greater pressure and liquid loads. However, the specific pressure losses do not differ much. The separation curves pass through a maximum at a certain load. This maximum characterizes rings of all sizes, and is more marked the smaller the ring size. Note Chapter 8, Section 2.

At vapor rates giving the best separation, considerable liquid entrainment by the vapor stream may occur, especially in vacuum operation, but there will be no flooding of the packing proper. Fig. 6.32 shows this for some typical tests, with the reflux at the column top plotted over the vapor load. The open dots are for loads at optimum separation, while the black dots indicate the beginning of flooding and the lessening of separation. That flooding condition corresponds to the points of the maximum load in Fig. 6.31, disregarding the curve for the system 1, 2-propylene glycol–ethylene glycol.

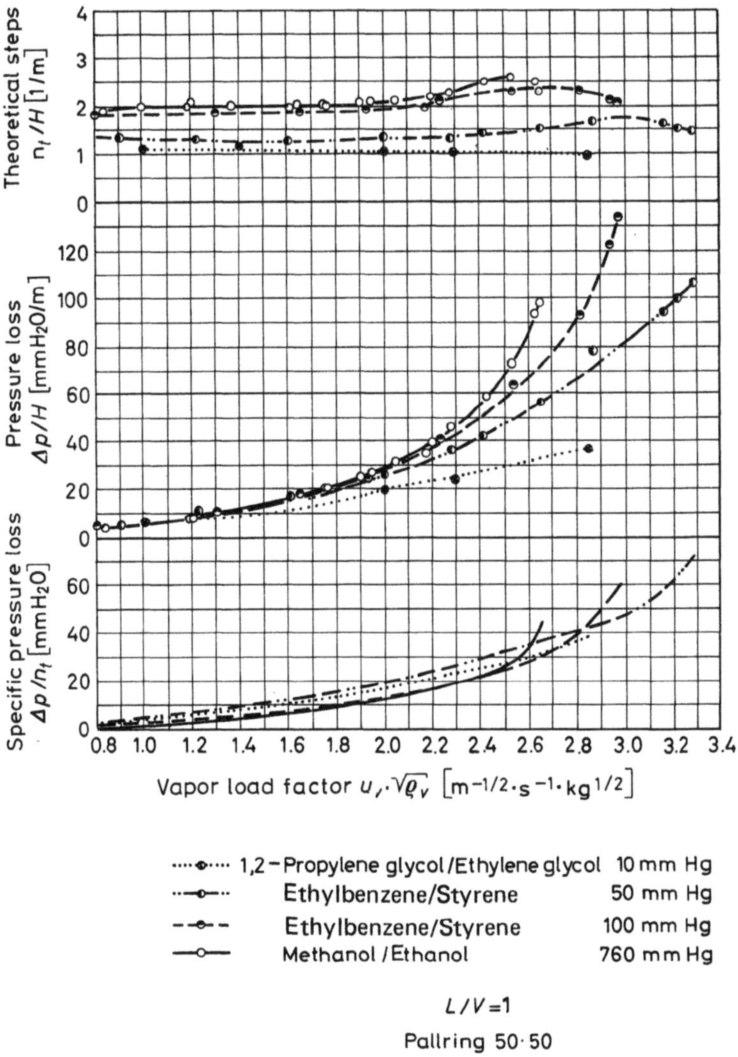

Fig. 6.31 50-mm Pall rings. Separation and pressure loss for different systems at various vapor loads.

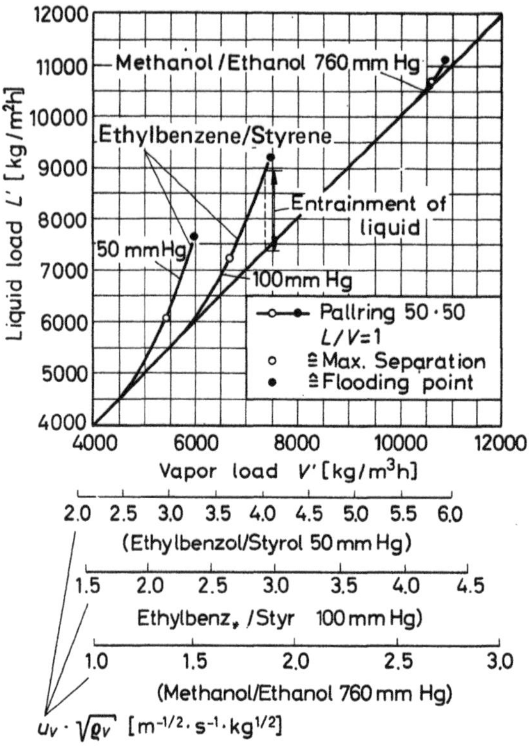

Fig. 6.32 50-mm Pall ring columns, different systems. Specific reflux loads versus vapor loads at column top.

Fig. 6.33 presents analogous findings for wire-cloth packing. A favorable situation seems to prevail for the 1, 2-propylene glycol–ethylene glycol mixture at the lowest pressure; at lower vapor-load factors it yields the higher specific number of steps if the zone of lowest loads is disregarded, while the Pall-ring column gives the lowest number of steps at low loads. This brings out the separation characteristics of these two packings; as opposed to the action in the Pall-ring tower, the separation in the wire-net packing improves as the liquid load diminishes (note Fig. 6.34) and it fades only when the liquid film ruptures when the liquid is insufficient in quantity to form a continuous film.

A special advantage of the wire-net packing is that even relatively viscous mixtures with relatively high surface tensions, such as 1, 2-propylene glycol and ethylene glycol, are well distributed across the packing prior to separation, while in the usual dumped packing they would become stringy.

Figs. 6.31, 6.33, and 6,34 present the separation curves for the system 1, 2-propylene glycol and ethylene glycol. Note that because of reboiler limitations the vaporization did not reach the flood point. For most systems, the maximum vapor-load factor at the end of the curve locates the flood point, and this is higher up the less the system pressure.

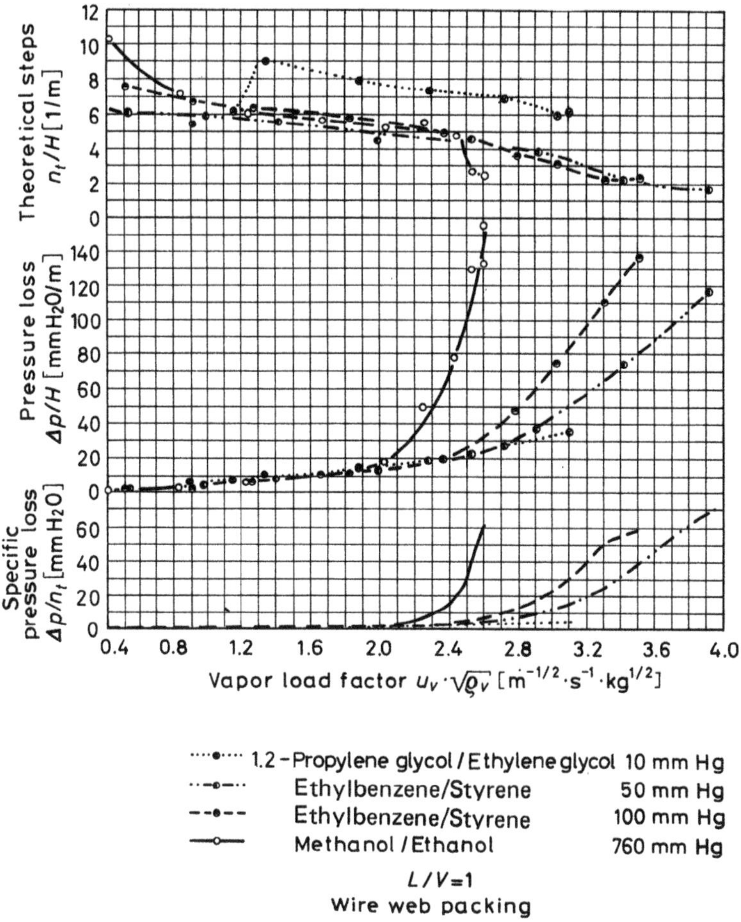

Fig. 6.33 Wire web packing. Separation, pressure loss and loads for various systems.

In Fig. 6.35 the reflux measured at the top of the wire-net packing is plotted over the correlated vapor load; the difference of ordinate and diagonal values determines the amount of entrainment at the column top. Especially in vacuum systems, much liquid is carried over below flood loading occurs. The specific tower volume is given by Eq. (6-12), from Eq. (6-17), and the cost curves from Fig. 6.25. The specific cost is determined as a function of the specific pressure loss, for a reference diameter of the Pall-ring column of $d_S = 0.8$ m. Then, the corresponding diameter for the wire-web packing used in the cost calculation becomes

$$d_S(\text{cloth}) = d_S\,(\text{Pall}) \left[\frac{u_V \rho_V^{1/2}\,(\text{50-mm Pall})_{\Delta p/n_t}}{u_V \rho_V^{1/2}\,(\text{web})_{\Delta p/n_t}} \right]^{1/2} \tag{6-28}$$

The $u_V \rho_V^{1/2}$ values in the numerator and denominator of the root in Eq. (6-28) both refer to the same pressure loss $\Delta p/n_t$.

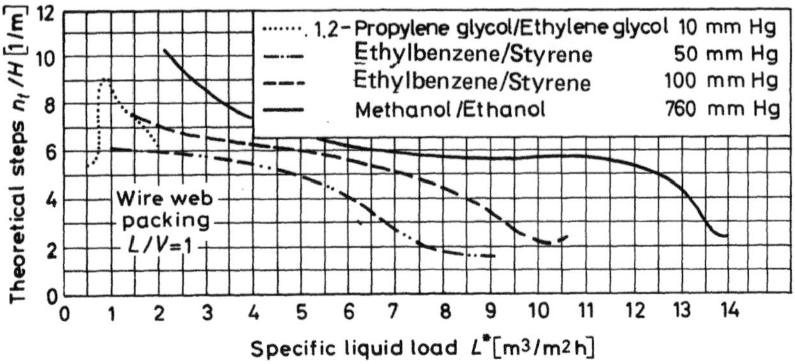

Fig. 6.34 Wire web packing, theoretical steps versus specific liquid load, for various systems.

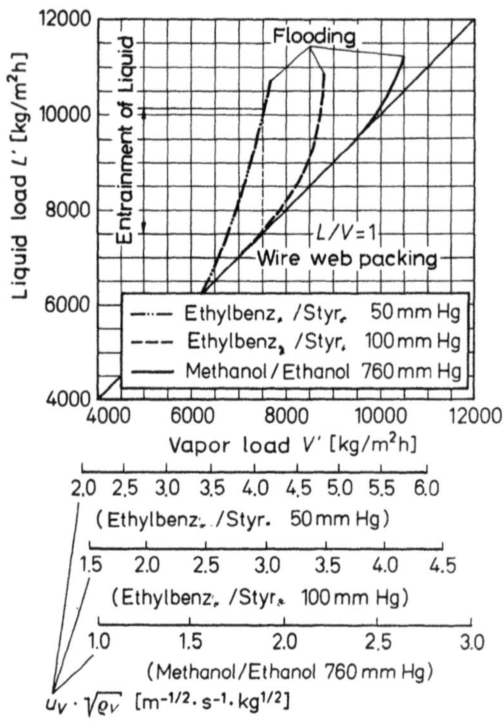

Fig. 6.35 Wire web packing. Specific reflux load versus vapor load at column top, for various systems.

Fig. 6.36 presents the results of these calculations. The upper part shows the specific tower volumes v_V, and the lower part the specific cost C_V as a function of the specific pressure loss $\Delta p/n_t$. For the wire-web packing, v_V and C_V will increase with the system pressure; for the Pall column this does not necessarily apply. To determine when the wire-web packing will be more economical than the Pallring, percent cost

$$C_p = \frac{C_V\,(\text{web + shell})}{C_V\,(\text{50-mm Pall rings + shell})}\,(100) \tag{6-29}$$

is plotted as a function of the specific pressure loss $\Delta p/n_t$ (Fig. 6.37)

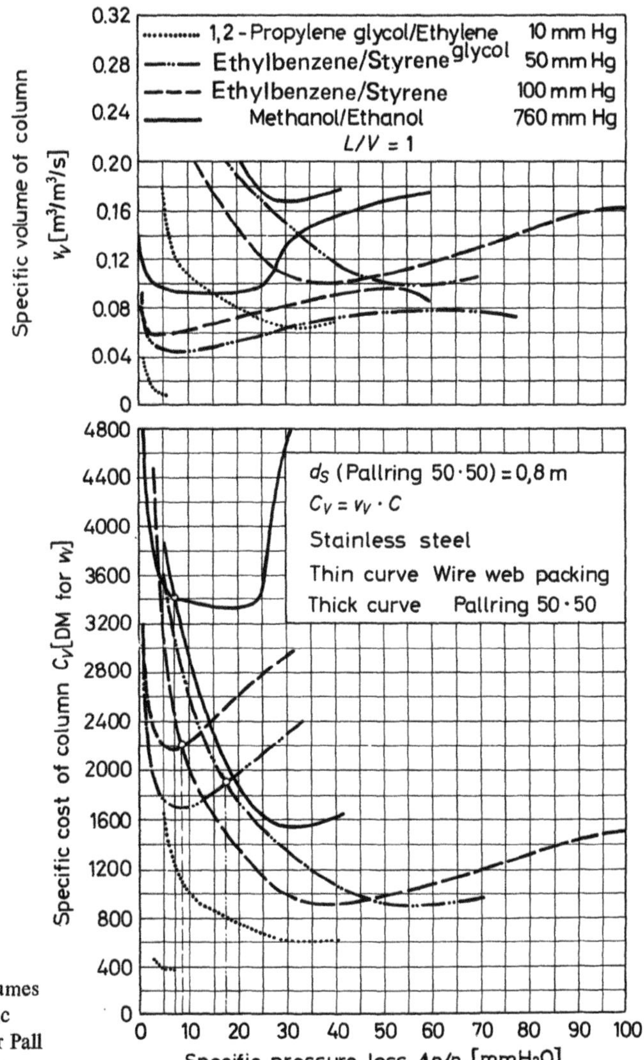

Fig. 6.36 Specific column volumes and column cost versus specific pressure loss, for wire web, for Pall rings, and for different systems.

From this, separations having specific pressure losses $\Delta p/n_t < 7$–18mm water can be run more economically on wire-web packing than in Pall-ring columns, especially in vacuum operation. The lower the pressure, the higher is that specific pressure loss for which the column costs become equal.

The comparison made in Figs. 6.36 and 6.37 is based on a reference column with only one theoretical step. For a real column having several theoretical steps, the

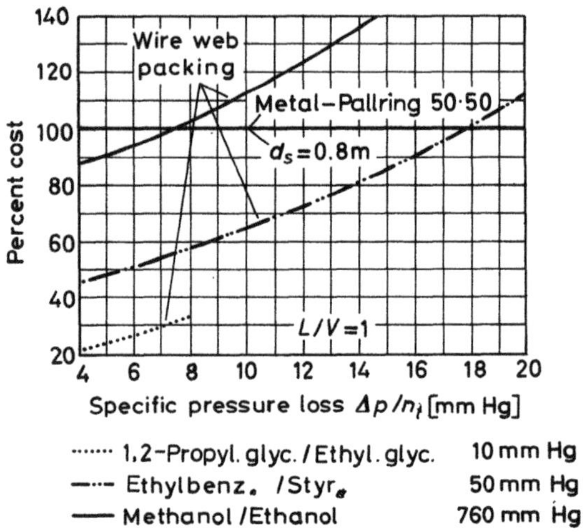

Fig. 6.37 Web packed column versus Pall ring column. Percent cost versus specific pressure for various systems.

calculation must be made by stages. Then, the p value will vary along the column, and from Eq. (6-9) derived from Eq. (6-23), the vapor load factor $u_V \rho_V^{1/2}$ will be:

$$u_V \rho_V^{1/2} = 2.8 \cdot 10^{-3} \frac{V}{d_S^2} \left(\mu \frac{T}{p} \right)^{1/2} \tag{6-30}$$

This allows for the changes of $n_t/H = f(u_V \rho_V^{1/2})$ and $\Delta p/n_t = f(u_V \rho_V^{1/2})$.

In vacuum rectifications of thermally unstable substances, the highest admissible base temperature, and hence the pressure loss $\Delta p/n_t$ as well as the pressure p_W at the column are meaningful for the cost calculation (Fig. 6.38).

The plot is based on a vapor throughput of 4000 kg/h; it gives the total cost C_T as a function of the column-base pressures for wire web and for Pall rings and for a different number of theoretical steps, n_t. For guidance, there are also lines connecting the cost points for constant top pressures p_T of the Pall-ring columns. The intersections of correlated cost curves C_T for equal n_t values indicate at what base pressure p_W both column types will have the same cost. Along the column height, n_t/H was assumed constant. Also, for the wire-mesh packing $\Delta p/n_t$ was considered constant. For the Pall-ring column, however, $\Delta p/n_t$ along the column was divided into five theoretical steps and adjusted according to the changes in $u_V \rho_V^{1/2}$.

Fig. 6.38 permits a comparison of the two columns operating at optimum conditions, which in this case are: for wire-web packing $u_V\rho_V^{1/2} = 2.1\ \mathrm{m}^{-1/2}\ \mathrm{s}^{-1}\ \mathrm{kg}^{-1/2}$, for Pall rings $u_V\rho_V^{1/2} = 2.7\ \mathrm{m}^{-1/2}\ \mathrm{s}^{-1}\ \mathrm{kg}^{-1/2}$, which are fairly close.

If v_V and C_V are determined as functions of the load factor $u_V\rho_V^{1/2}$, a column $d_S = 0.8$ m will give the results of Fig. 6.39. The dependence on pressure is given by Fig. 6.39. Again, for a web packing the minimum cost does not coincide with the maximum load.

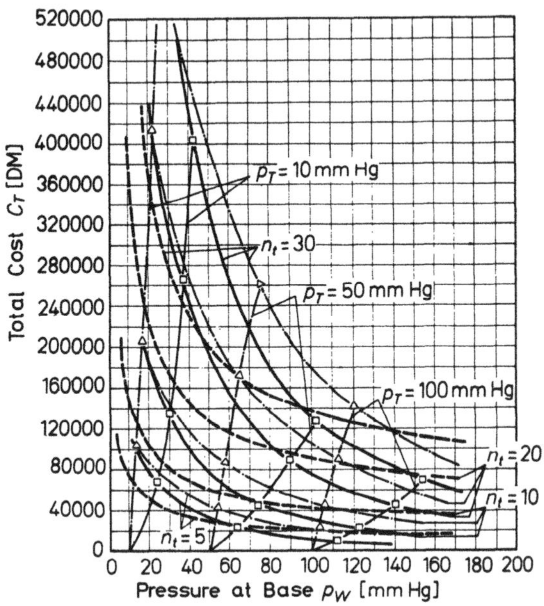

Type of column	$u_V\cdot\sqrt{\rho_V}$ (Top) $[\mathrm{m}^{-1/2}\cdot\mathrm{s}^{-1}\cdot\mathrm{kg}^{1/2}]$	$\Delta p/n_t$ (Top) [$\mathrm{mmH_2O}$]	n_t/H [1/m]
Wire web packing	2.3	4 — —	5
Pallring 50·50	2.74 1.6	35 —— 10 —·—	1.5
Vapor load 4000 kg/h			

Fig. 6.38 Web packed column versus Pall ring column. Total cost versus column base pressure for separations by different numbers of theoretical steps n_t, and for different column top pressures p_T. Vapor load 4000 kg/h.

For comparison, the upper part of Fig. 6.40 is a plot of the percent cost, as defined by Eq. (6-29), over the load factor. From this, Pall-ring columns are attractive pricewise for higher system pressures, and web columns are favorable for very low system pressures. Form Fig. 6.40 (lower part), it follows that the web-type column is superior to the Pall-ring column for all practical applications, as the specific pressure loss of the web packing is only 10 to 25% of that of Pall rings.

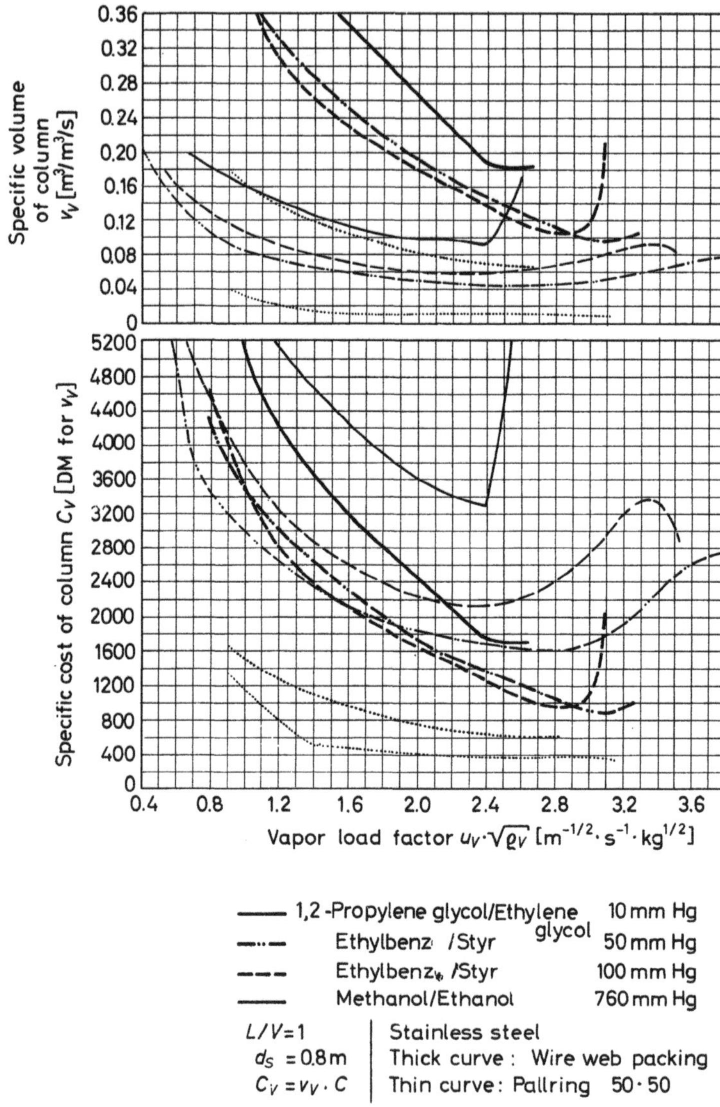

Fig. 6.39 Web packed column versus Pall ring column. Specific volumes and cost versus vapor load factors, for different systems.

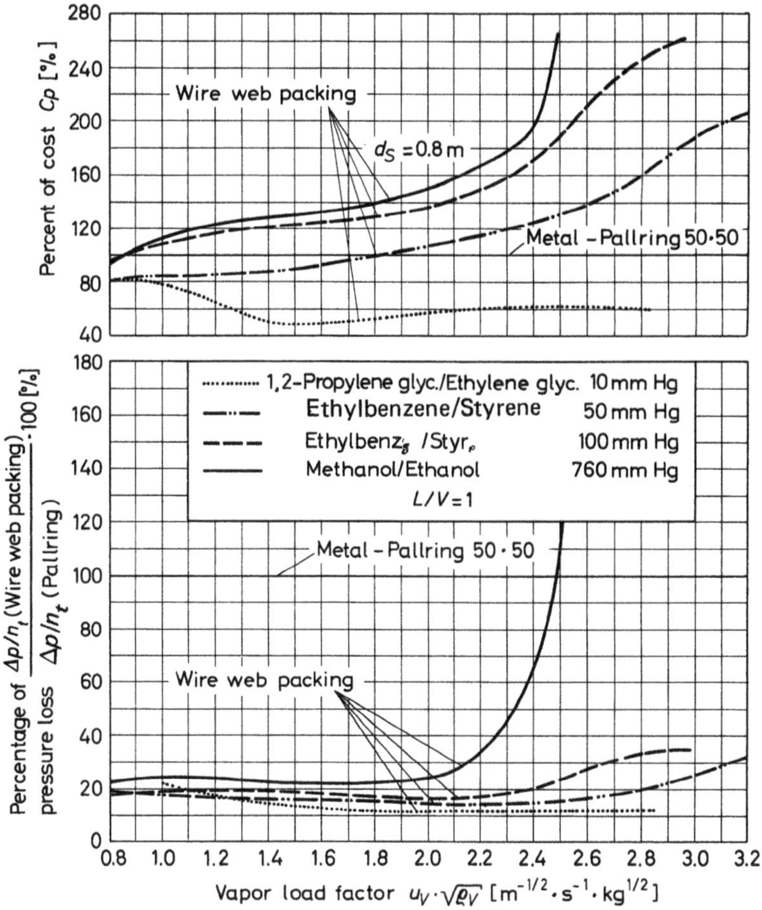

Fig. 6.40 Web packed column versus Pall ring column. Specific pressure losses and percent cost, and specific cost versus vapor load factor, for different systems.

C. Quantitative evaluation for plate-type columns

As discussed in Section B of this Chapter, for packed columns, the evaluation will be treated for plate-type columns, based on pilot plant tests, for the same system, and under the same conditions as before.

a. Separation, Pressure Loss, and Load Capacity for Bell-Cap Trays, Tunnel Trays, Valve Trays, and Centrifugal Trays

The mass transfer trays shown in Fig. 6.41 have different processing characteristics. Shown here are a conventional bell-cap tray, a special tunnel tray, a standard valve tray *a* (as in Fig. 6.6. F), and a centrifugal tray, type I (as in Fig. 6.6. D). The principles of all these have been explained in Section 2 of this chapter.

This evaluation again follows Section 3.A of this chapter and Fig. 6.11, phases 1,

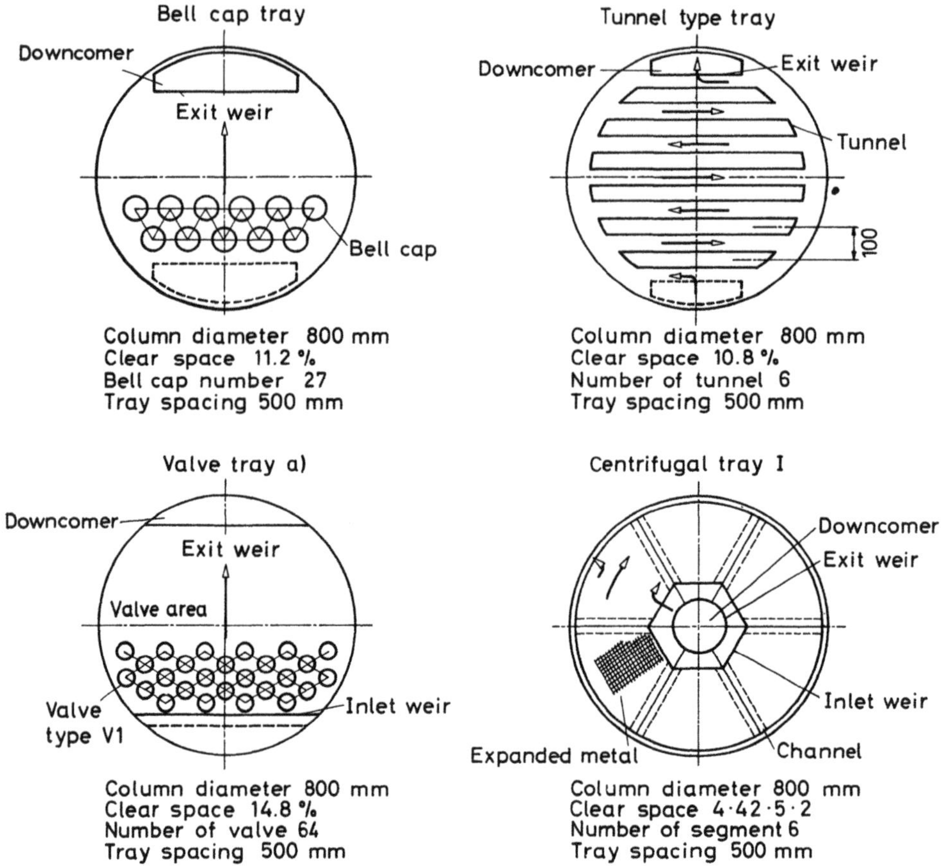

Fig. 6.41 Principal dimensions of trays investigated.

2, and 3; results are presented in Fig. 6.42. Most of the curves are for total reflux, $r = \infty$. Only for the bell-cap and for the tunnel tray are some results based on a reflux ratio $r = 4$, which indicates that the separation is generally unaffected by reflux ratios between 4 and ∞; only at high loads is the tray with the smaller reflux ratio inferior.

Separation, upper load range, and specific pressure loss are all more favorable for tunnel trays and for valve trays than for bell-cap trays. The performance of centrifugal trays was much improved; with good separation and low pressure loss it supports a remarkably high load beyond the capacities of tunnel trays and valve trays.

While this performance is excellent, one should bear in mind that it cannot be taken for granted with larger sizes, as long as the hydraulics of the radial-flow component of the liquid have not been investigated and a proper design for optimum liquid and vapor distribution has not been established.

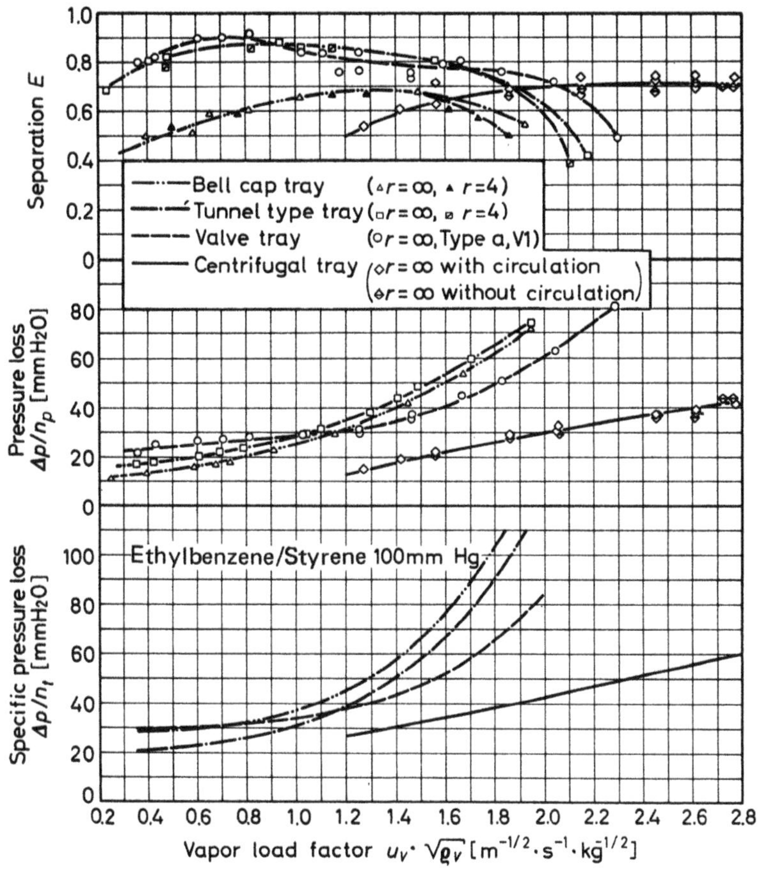

Fig. 6.42 Separation and pressure loss of different trays versus vapor load factor for a tray spacing of 0.5m.

Means that are open to designers in order to reduce the pressure loss include a reduction of submergence, more clearance, and a lowered liquid gradient. However, along with plate hydraulics, mass transfer is always a factor. Investigations[42] reveal that clearances of bell caps have certain limitations to maintain separation at higher loads. When centrifugal trays are laid out for larger sizes, clearances and slot orientation must be chosen to prevent a laying bare of part of the plate, which would impair separation.

b. Specific Tower Volume and Specific Column Cost

The scheme applied in reviewing packed columns and their modifications will be retained hereafter. Again the first step will be to find the tower cost per m^3 of column volume C (cost units per volume unit) as presented in Fig. 6.43 as a function of the

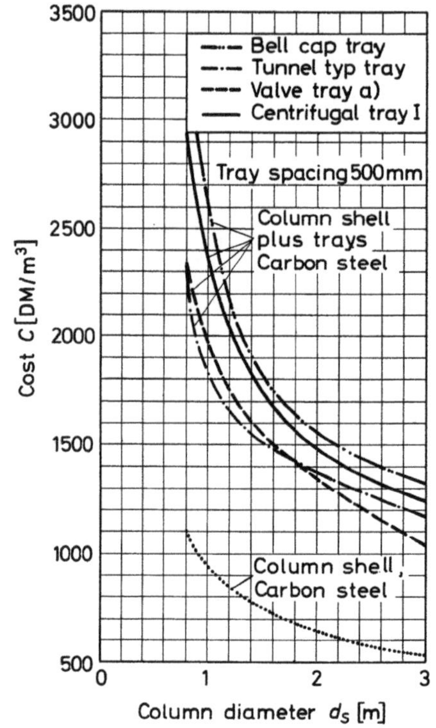

Fig. 6.43 Cost per unit volume of different tray type columns versus column diameter. Prices of 1970.

column diameter. The data apply to carbon steel and to prices as of 1970. Note the relatively low price of tunnel plates, and valve plates as compared to bell caps, especially for small diameters

Referring to the method outlined in Section 3.A of this chapter, the specific column cost, per unit vapor throughput and per unit separation, can be found by multiplying the specific column cost of Fig. 6.43 by the specific tower volume. The latter is computed by Eqs. (6-14) and (6-16) and the upper part of Fig. 6.44 as a function of the specific pressure loss.

The result is presented in the lower part of Fig. 6.44. It is based on a reference diameter of the bell-cap plate of d_S = 1.5 m. For other plates, the diameters are:

$$d_S \text{ (tray)} = d_S\text{(for bell-cap plate)} \left[\frac{u_V \rho_V^{1/2} \text{(bell-cap plate)}_{\Delta p/n_t}}{u_V \rho_V^{1/2} \text{(tray)}_{\Delta p/n_t}} \right]^{1/2} \qquad (6\text{-}31)$$

with Fig. 6.42 supplying the proper load factors for the different $\Delta p/n_t$ data.

By Fig. 6.44 the bell-cap plate investigated is definitely the least economic; it yields the largest and hence the most expensive column. On the whole, the tunnel plate and the valve plate are equivalent, and so are their cost. The least tower volume results for the centrifugal tray, and the cost difference is slight.

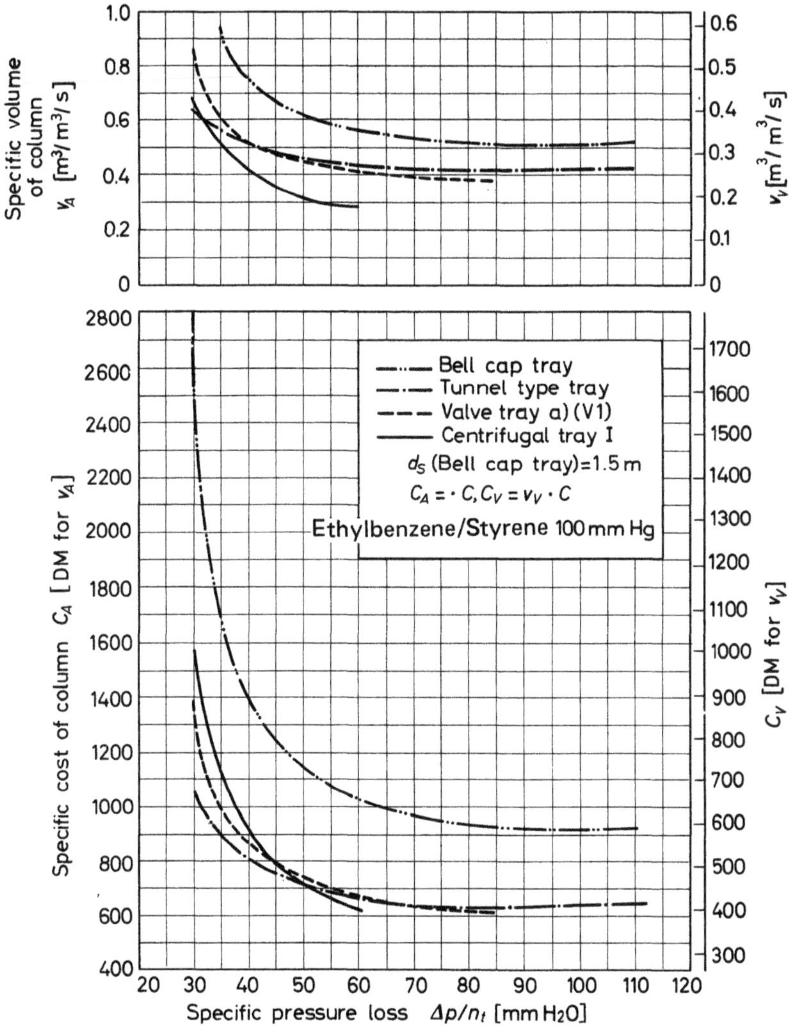

Fig. 6.44 Plate type columns. Specific column volume and cost versus specific pressure loss. Plate spacing 0.5 m.

Assuming that the same relative differences exist in separation, in load and in pressure loss for plates within the 0.8 to 3m range, as found for the 0.8 m size, one can compute the dependence of the percentage cost from the column diameter; from Fig. 6.45:

$$C_p = \frac{C_L \text{ (tray + shell)}}{C_L \text{ (bell-cap tray + shell)}} (100) \tag{6-32}$$

Even allowing for minor errors, this cost curve does indicate that differences in cost are also caused by the specific pressure loss, meaning the load, and the tendency of the cost curves is not uniform inasmuch as the differences, if any, may favor one or the other type of tray.

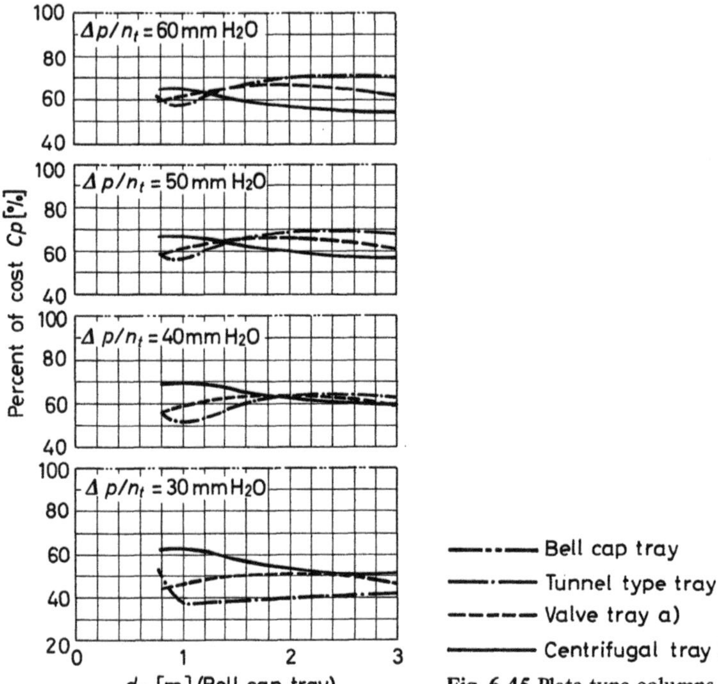

Fig. 6.45 Plate type columns. Cost versus diameter.

That leads to the conclusion that, among the types investigated, there is no ideal plate offering only advantages at all diameters and all load conditions. For large size centrifugal designs, refer to Section 3.C.I of this chapter, which mentions the hazards of transferring test results.

c. Evaluations at Top Column Loads

Table 6.2 sums up process and cost data for a bell-cap tower of 1.5 m operating a maximum load. While for tunnel trays, valve trays, and centrifugal trays the process data differ, the costs are roughly the same. The bell-cap tower, however, is the most expensive by far.

d. Evaluation of Valve Trays

The use of valve trays in processes has increased greatly in recent years. This is due to the relatively low cost and the operating advantages of this construction.

Table 6.2

Plate columns operating at maximum loads. Related to bell cap plate $d_S = 1.5$ m, carbon steel construction, ethyl benzene/styrene system, at 100 mm Hg, $L/V = 1$.

Tray	Bell cap	Tunnel	Valve	Centrifugal
$u_V\rho_V^{1/2}$ ($m^{-1/2} s^{-1} kg^{1/2}$)	1.82	1.89	2.08	2.80
E	0.6	0.7	0.7	0.7
$\Delta p/n_p$(mm H_2O)	63	71	66	43.5
v_V[$m^3/m^3/s$]	0.318	0.2638	0.239	0.1776
C_V[DM] Shell + Tray	582	402	390	396

Returning to Section 2.B of this chapter, the variants shown in Figs. 6.6F–6.6.I. are treated at greater length. The separation data were obtained on a pilot installation with a 0.8-m column, working the system ethyl benzene-styrene at 100mm Hg. The tests results were again evaluated by the method described in Section 3.A of this chapter. The separation E for a tray, found from Eq. (6-7), the pressure loss $\Delta p/n_p$ of the vapor passing a plate, and the specific pressure loss $\Delta p/n_t$ for one theoretical step, are plotted over the load factor $u_V\rho_V^{1/2}$ as a yardstick for the vapor load of the column.

Figs. 6.46 and 6.47 present these evaluation data. Across the main load range, and with the exception of the extreme high, the separation is about constant, however there are marked differences in the specific pressure losses. Near the maximum load, they amount to 40–50% of that characteristic for the standard design *a*, as far as the *c* design with venturi openings, and the combined sieve-valve design *d* are concerned. Hence the two designs labeled *c* and *d* are advantageous in vacuum rectifications of thermally unstable systems.

Now with designs *a, b,* and *c* the clearances amount to 14.8%, while design *d* has only 12.5%. Possibly the specific pressure could be reduced even farther by increasing the clearance to 14.8% as well.

Flood loadings by which the column becomes choked with liquid, either through entrainment or by accumulation in downcomers, are higher than the maximum loads figuring in Figs. 6.46 and 6.47. This, however, is not significant since separations below 0.5 are impractical; the corresponding load would be around $u_V\rho_V^{1/2} = 2.7\ m^{-1/2}\ s^{-1}\ kg^{1/2}$.

The measurement points are plotted in Figs. 6.48 and 6.49 for designs *c* and *d*, and the effect of the weir elevation upon the separation and the pressure loss in the main load range are clearly seen: the higher the weir, the better the separation, but the greater also the pressure loss per plate and the specific pressure loss.

A more pronounced drop in separation, below 0.7, sets in at load factor $u_V\rho_V^{1/2} = 2.3\ m^{-1/2} s^{-1}\ kg^{1/2}$. This drop is probably due to increased entrainment as may be noted from Fig. 6.50, also it is still remote from the flooding limit.

For designs *c* and *d*, Fig. 6.50 shows the amounts of liquid ΔL entrained by the vapor from the top plate into the reflux condenser as a percentage of the reflux, L, against the vapor load factor. These quantities are not, however, identical with those that effectively move from one plate to the one above because the latter causes a partial separation, meaning that only a portion controlled by the plate design, the physical properties of the system, and by load situation passes the next plate, reducing the separation. The measured and clearly recognizable amounts of entrainment begin

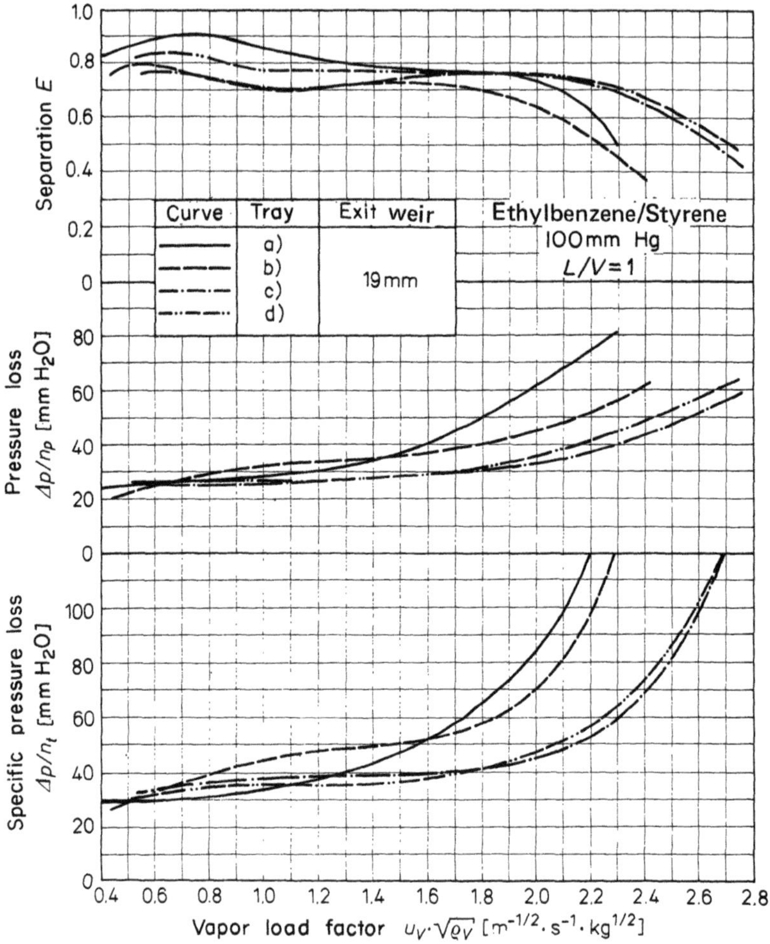

Fig. 6.46 Valve trays. Separation and pressure loss versus vapor load factor. Height of exit weir 19 mm. Tray spacing 0.5 m.

to show for load factors above $u_V \rho_V^{1/2} = 1.2\ \text{m}^{-1/2}\ \text{s}^{-1}\ \text{kg}^{1/2}$. The higher the weir, the less the entrainment.

The question as to which of the valve plates will give the optimum column cost requires a cost optimization. According to the method presented in Section 3.A of this chapter, two cases may be considered. First, the columns will serve in vacuum rectifications of heat-sensitive mixtures; the important criterion is the pressure loss per theoretical step; cost becomes a function $\Delta p/n_t$. Second, the columns will be evaluated regardless of pressure loss.

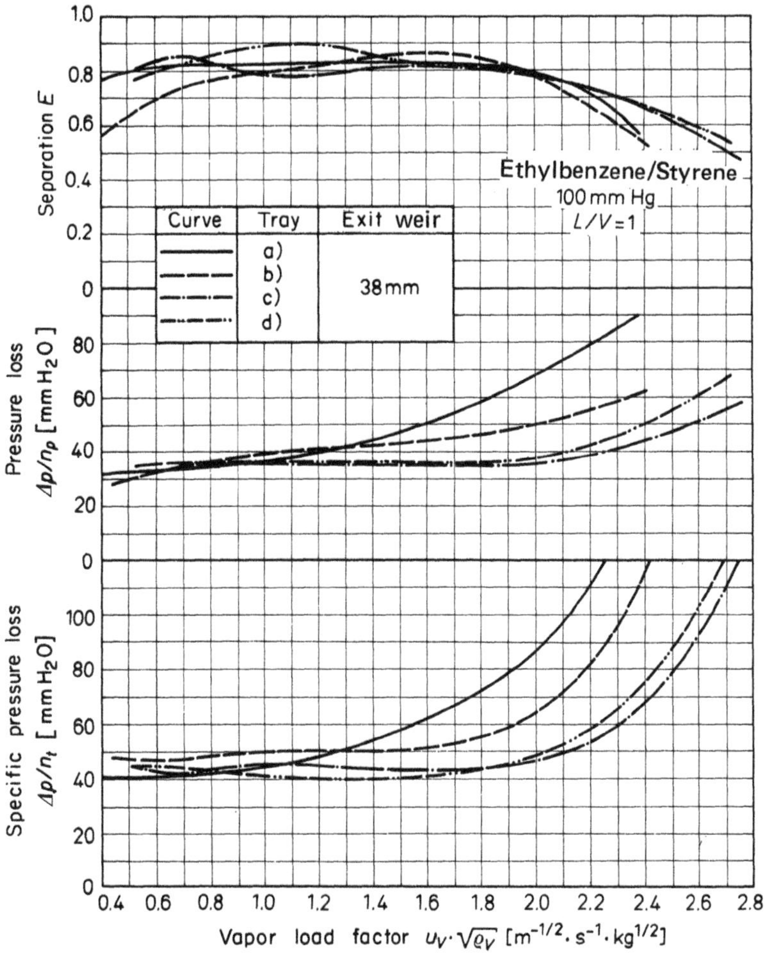

Fig. 6.47 Valve trays. Separation and pressure loss versus vapor load factor. Height of exit weir 38 mm.

Thus, ν_V is found from Eq. (6-12), using Figs. 6.46 and 6.47 for the $\Delta p/n_t$ functions. Thus one obtains the curves $\nu_V = f(\Delta p/n_t)$ of Fig. 6.51, upper part, for 19-mm weir height. The cost $C_V = \nu_V\, C$ follow from the $C = f(d_S)$ curves (not shown).

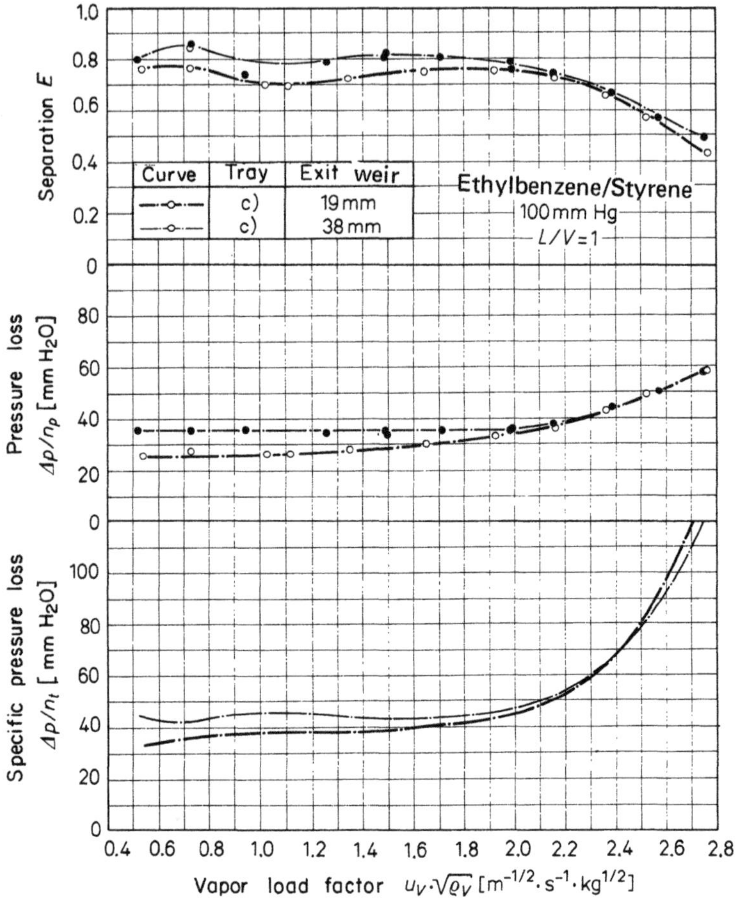

Fig. 6.48 Valve trays, Venturi type. Separation and pressure loss versus vapor load factor. Tray spacing 0.5 m.

In Fig. 6.51, the two lower plots give the results of cost optimization for carbon-steel construction and for stainless steel, with the percent cost

$$C_p = \frac{C_V\ (\text{ tray } b,\ c,\ d + \text{shell})}{C_V\ (\text{ tray } a + \text{shell})}(100) \tag{6-33}$$

and design a as the reference at 100% of the cost.

This evaluation is based upon a column diameter of $d_S(a) = 1.5$ m and design a. Then, for the comparable designs b, c, and d the diameters become

$$d_S(b,\ c,\ d) = d_S(a)\left[\frac{[u_V\rho_V^{1/2}(a)]_{\Delta p/n_t}}{[u_V\rho_V^{1/2}(b,\ c,\ d)]_{\Delta p/n_t}}\right]^{1/2} \tag{6-34}$$

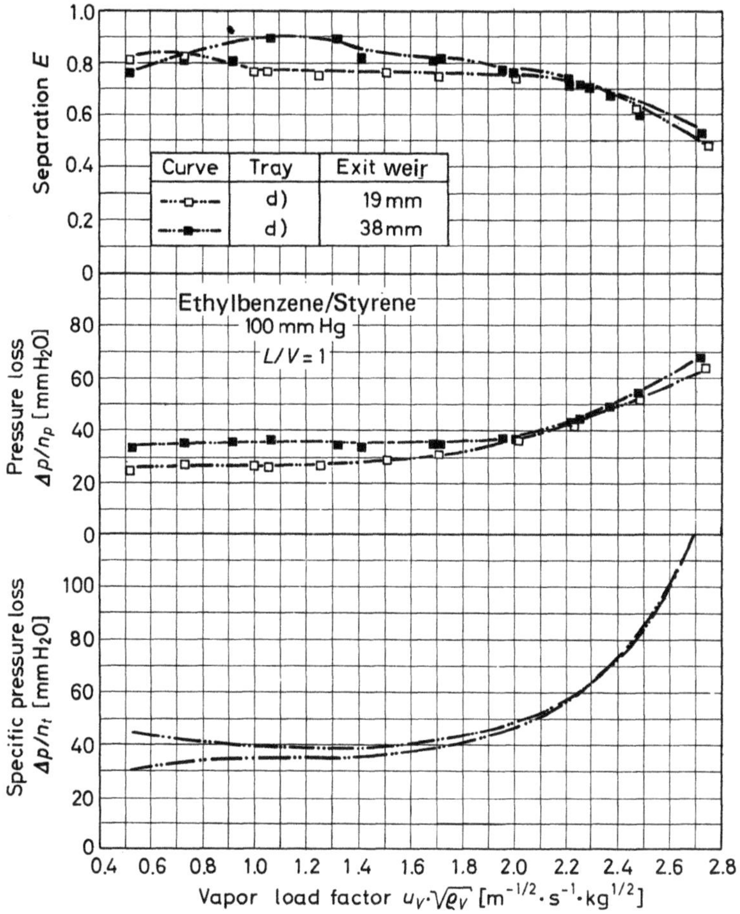

Fig. 6.49 Sieve-and-valve trays, 0.5 m spacing. Separation and pressure loss versus vapor load factor.

In Eq. (6-34) the load factors $u_V\rho_V^{1/2}(a)$ and $u_V\ \rho_V^{1/2}(b, c, d)$ refer to the same specific pressure losses $\Delta p/n_t$, and are taken from Figs. 6.46 and 6.47.

Designs *c* and *d* are also economical. The cost savings over design *a* are, for designs *c* and *d*, up to 16 and 17%, with a specific pressure loss around $\Delta p/n_t$ = 45 mm water, at a load factor around 1.9 $m^{1/2}\ s^{-1}\ kg^{-1/2}$ There is not much difference between the curves for carbon steel and for V2A.

Fig. 6.52 shows the effect of the column diameter upon the cost. The percent cost found from Eq. (6-33) has been plotted as a function of the column diameter of design *a* for diameters of 1 - 3m, a weir height of 19 mm, and carbon steel. The specific pressure losses $\Delta p/n_t$ = 40, 50, and 60 mm water, occurring in the main load ranges, were the parameters.

Results show that, with increasing column diameter, designs *c* and *d* become more economical for constant process conditions. For the entire range design *b* is the most expensive.

Fig. 6.53 is a comparison of costs based on 38-mm-weir height and carbon steel

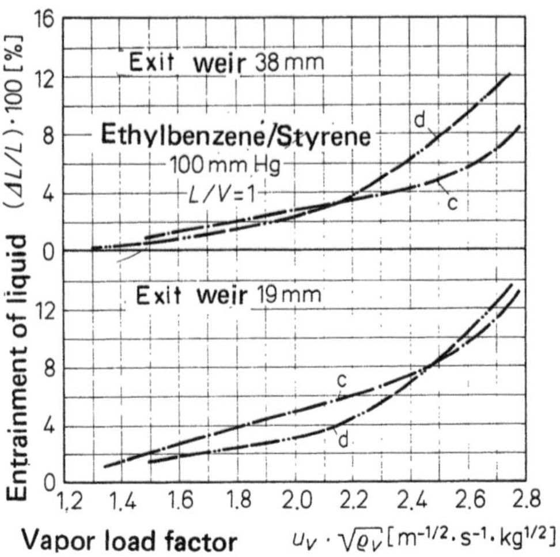

Fig. 6.50 Valve trays, and sieve-and-valve trays, Venturi type, 0.5 m spacing. Liquid entrainment at column head versus vapor load factor.

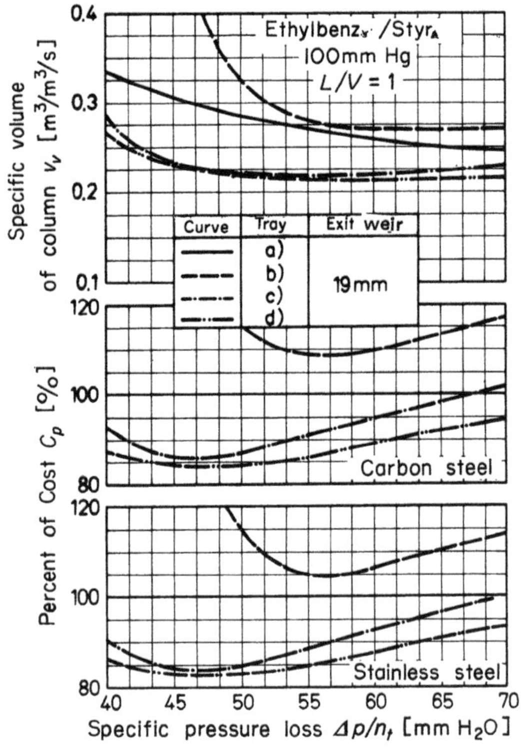

$H = 0.5\,\text{m},\ d_S(a) = 1.5\,\text{m}$

$$d_S(b,c,d) = d_S(a) \cdot \sqrt{\frac{u_V \cdot \sqrt{\varrho_V}\,(a)}{u_V \cdot \sqrt{\varrho_V}\,(b,c,d)}}$$

Fig. 6.51 Valve trays spaced 0.5 m. Specific column volume and cost versus specific pressure loss.

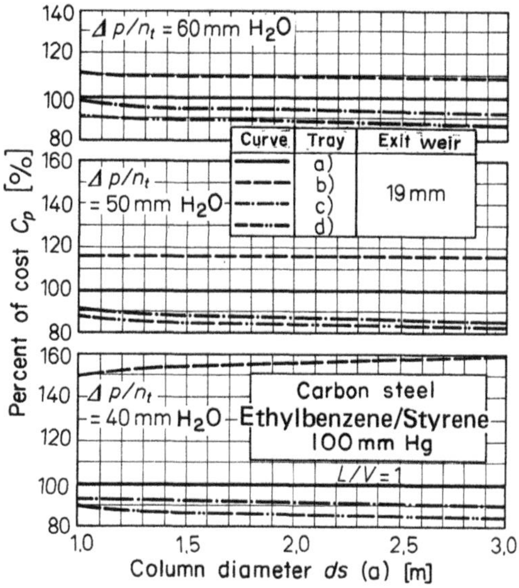

Fig. 6.52 Valve trays of various diameters, made of carbon steel, exit weirs 19 mm high, tray spacing 0.5 m. Percent cost.

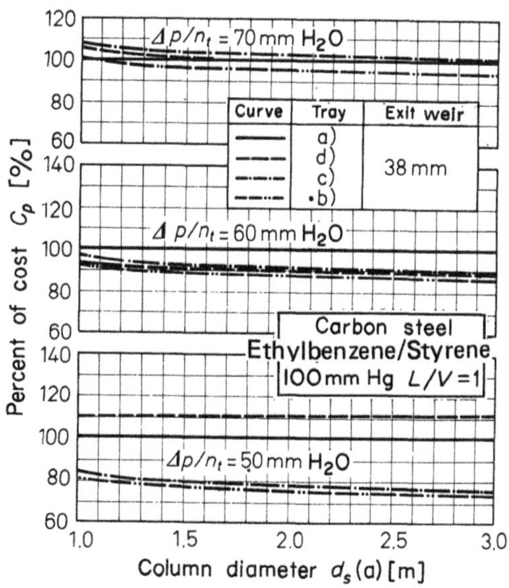

Fig. 6.53 Valve trays of various diameters. made of carbon steel, exit weirs 38 mm high, tray spacing 0.5 m. Percent cost.

construction, with three different specific pressure losses, $\Delta p/n_t$ = 50, 60, and 70 mm water as the parameters. Again design *d* is the most economical; the saving over design *a* is, for *d* and *c*, 21 and 23% for a reference size of $d_S(a)$ = 1.5 m, and with $\Delta p/n_t$ = 50 mm water.

No plots are given for results with stainless-steel construction as the plots would be practically the same as Figs. 6.52 and 6.53.

Table 6.3

Valve plates at maximum loads, E = 0.6 and h_W = 19 mm weir. Ethyl benzene/styrene at 100 mm Hg, L/V = 1.

Plate-type				*a*	*b*	*c*	*d*
$u_V\rho_V^{1/2}(m^{-1/2}s^{-1}kg^{1/2})$				2.22	2.08	2.49	2.53
$\Delta p/n_p(mm\ H_2O)$				76	47.3	47.3	54
$\Delta p/n_t(mm\ H_2O)$				126.8	78.9	78.9	90
$v_V(m^3/m^3/s)$				0.2613	0.279	0.233	0.2295
C_P (%)	d_S(m) (a)	0.8	St	100	110.9	91.7	87.2
			V2A	100	107.3	90.5	87.1
		1.5	St	100	110	92	87.5
			V2A	100	107.3	90.3	86.9
		2.0	St	100	108.8	91	86.8
			V2A	100	107.9	90.9	87.5
		3.0	St	100	109.7	91.8	87.2
			V2A	100	107.2	90.6	87.0

Table 6.4

Valve plates at maximum loads, E = 0.6 and h_W = 38 mm weir. Ethyl benzene/styrene 100 mm Hg, L/V = 1

Plate-type				*a*	*b*	*c*	*d*
$u_V\rho_V^{1/2}(m^{-1/2}s^{-1}kg^{1/2})$				2.35	2.31	2.53	2.58
$\Delta p/n_p(mm\ H_2O)$				88.3	58.8	48.7	59.2
$\Delta p/n_t(mm\ H_2O)$				147.2	98	81.2	98.7
$v_V(m^3/m^3/s)$				0.2468	0.251	0.2293	0.2248
C_P (%)	d_S (m) (*a*)	0.8	St	100	105.3	95.5	90.4
			V2A	100	101.9	94	90
		1.5	St	100	104.9	96	90.9
			V2A	100	102.2	94.1	90.1
		2.0	St	100	103.6	94.8	90.4
			V2A	100	102.7	94.9	90.8
		3.0	St	100	104.2	95.6	90.3
			V2A	100	100.6	94.1	90.3

When pressure loss is not important, the column may be simply designed for that load that yields, from Eq. (6-12), the minimum construction volume. This gives a separation that according to Figs. 6.46 and 6.47 is at least $E = 0.6$, which corresponds to a load well below the admissible. Results for $E = 0.6$ are presented in Tables 6.3 and 6.4. Again, designs *c* and *d* are superior, with *d* appearing preferable to *c* and preferable on all counts.

e. Columns Fitted with Sieve-Valve Plates, Sieve Plates, and Centrifugals

Once the combined sieve and valve tray, design *d* in Fig. 6.6.I, had proven its superiority[43] an evaluation of additional designs became interesting.

Thus, Figs. 6.54 and 6.55 compare a sieve tray of recent design[44], illustrated in Fig. 6.6.C. Note the excellent capacity and separation of design *d*. Also the specific pressure losses over the main load range, between 1.3 and 2.4 $m^{-1/2}\ s^{-1}\ kg^{1/2}$, are

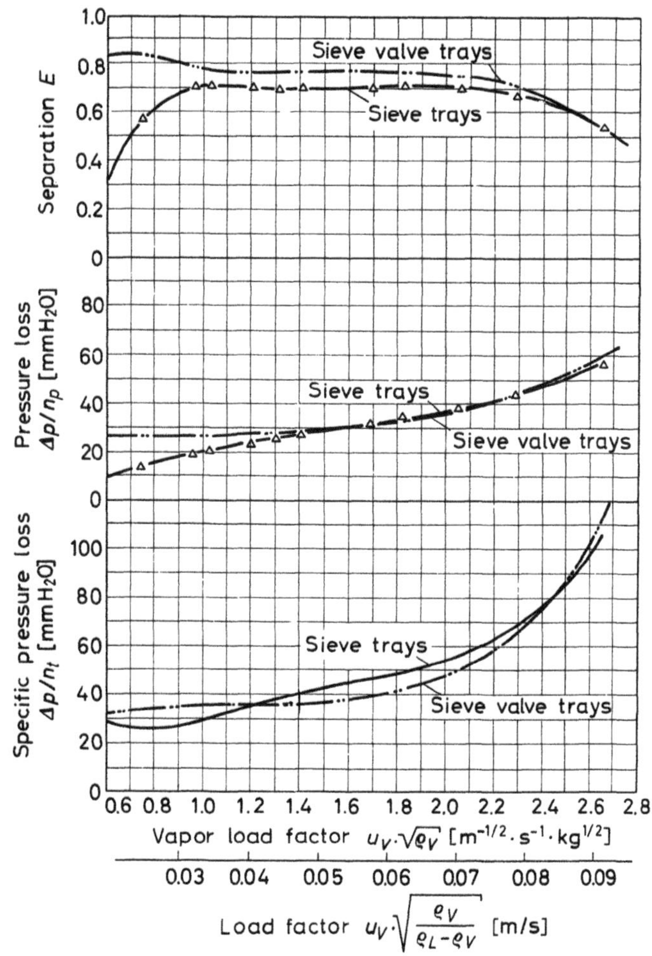

Fig. 6.54 Valve trays and combined sieve-valve trays. Separation and pressure loss versus vapor load factor. 19-mm weirs, 0.5 m tray spacing. Ethylbenzene/Styrene 100 mm Hg.

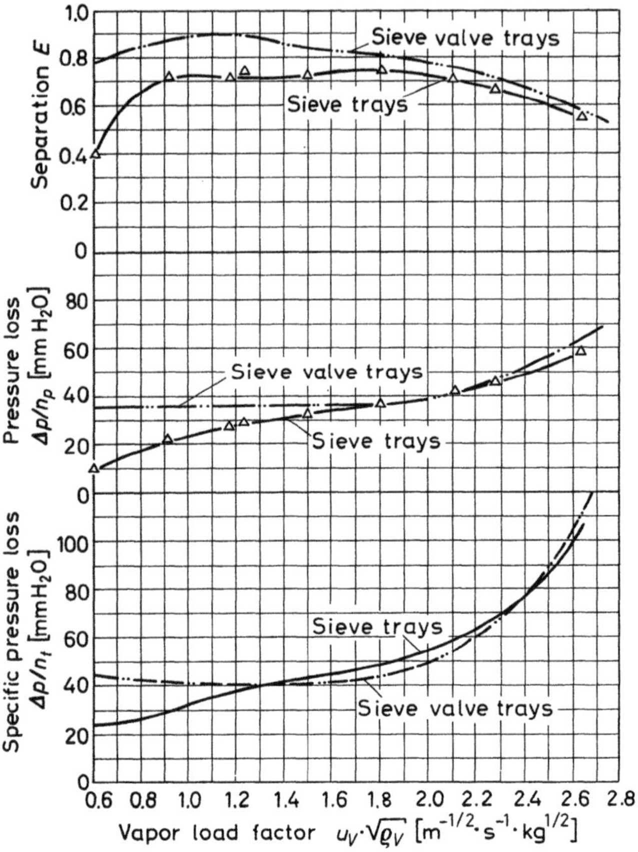

Fig. 6.55 Sieve plates and combined sieve-and-valve trays, exit weirs 38 mm high, 0.5 m spacing. Separation and pressure loss versus vapor load factor. Ethylbenzene/Styrene 100 mm Hg.

slightly less for the combined sieve and valve tray than for an ordinary sieve tray. The marked drop in separation by the sieve plate for load factors below 1 $m^{-1/2}$ s^{-1} $kg^{1/2}$ is due to raining through, typical of sieve trays.

From Fig. 6.56, the amount of liquid entrained by vapor does not vary much in the upper load range.

Again taking the specific pressure loss as a basis, the upper part of Fig. 6.57 shows how the specific column size depends on the specific plate design. The lower part of Fig. 6.57 gives the cost relation, specific cost (in percent) for the sieve-valve combination versus sieve plate, as a function of specific pressure loss. The sieve-valve combination is slightly more economical.

Finally, Fig. 6.58 compares performances of the two centrifugal trays shown in Fig. 6.6., I and II, which differ in height of exit weir and in sizes of slots in the expanded metal.

As compared to the valve and sieve designs, centrifugal trays work at higher vapor-load factors, but their separation effect is much less, especially for design II. The

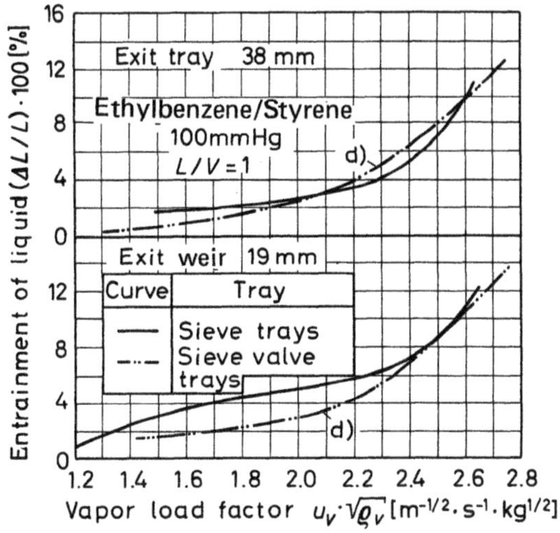

Fig. 6.56 Sieve plates and combined sieve-and valve trays, 0.5 m spacing. Percentage entrainment versus load factor.

maximum admissible load of centrifugals exceeds the highest measured loads in Fig. 6.58.

A cost optimization conceived in analogy with Fig. 6.51, based on the specific pressure loss, yields the curves in Fig. 6.59, in the lower part of which the percent costs are plotted for carbon steel and stainless steel

$$C_p = \frac{C_V \text{ (tray + shell)}}{C_V \text{ (sieve tray + shell)}} (100) \tag{6-35}$$

and the upper part giving the specific tower volumes.

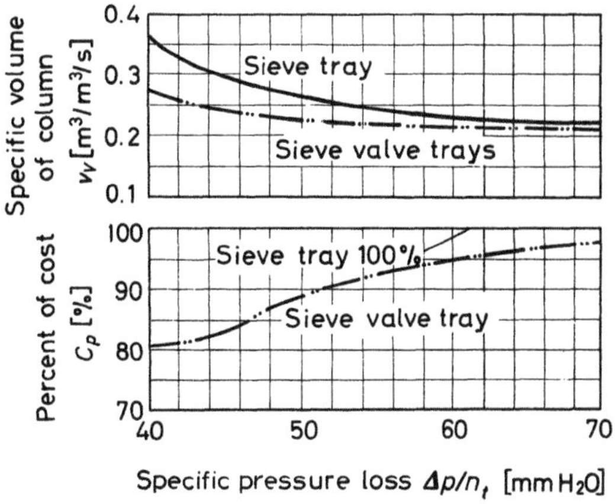

Fig. 6.57 Sieve plates and combined sieve-and-valve trays; 0.5 m spacing. Specific volume and cost versus specific pressure loss.

In Eq. (6-35) the numerator contains the term C_V for the cost of that column under consideration with which the sieve-type column is compared. Fig. 6.59 is based on a reference column with sieve plates having 1.5-m diameter. The diameter, at the same pressure loss, $\Delta p/n_t$ of the plate compared becomes:

$$d_S\,(\text{tray}) = d_S\,(\text{sieve}) \left[\frac{u_V \rho_V^{1/2}(\text{sieve tray})_{\Delta p/n_t}}{u_V \rho_V^{1/2}(\text{tray})_{\Delta p/n_t}} \right]^{1/2} \tag{6-36}$$

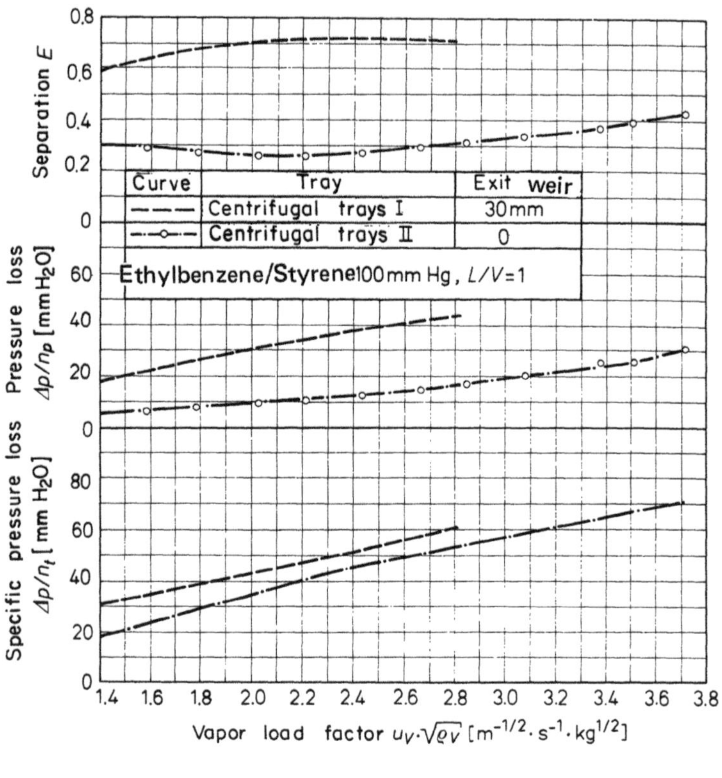

Fig. 6.58 Centrifugal trays, 0.5 m spacing. Separation and pressure loss versus vapor load factor.

The price advantages of the sieve plate and of the combined sieve-valve plate under the given conditions are obvious. The shape of the curves for the centrifugal trays reveals that these should be considered only where maximum throughput at maximum pressure loss is admissible, in view of the high cost involved.

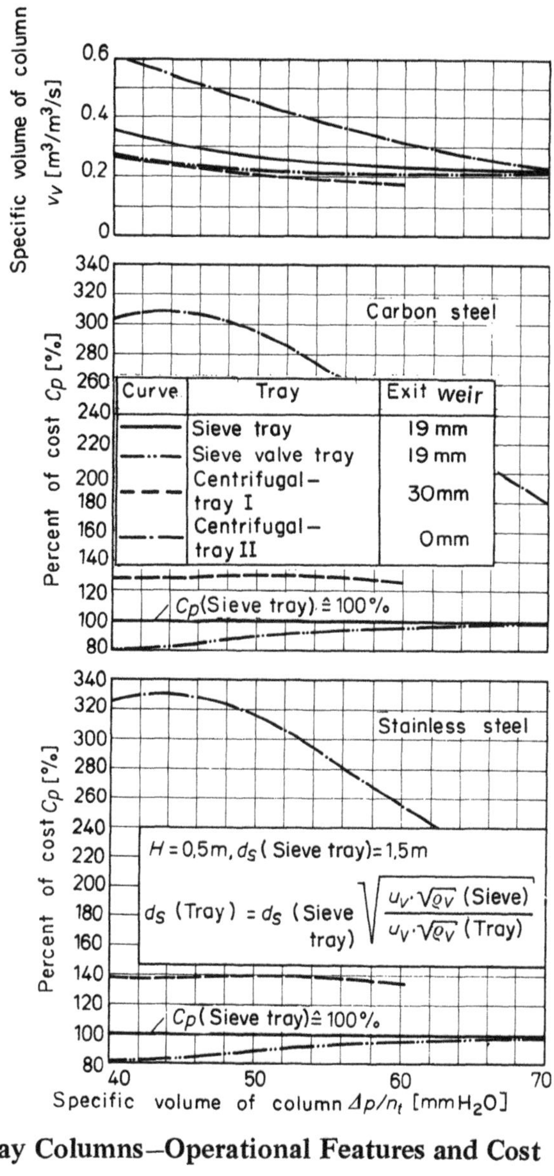

Fig. 6.59 Sieve trays combined sieve and valve trays, and centrifugal trays, spaced 0.5 m. Specific column volumes and cost, referred to specific pressure loss.

D. Packed Columns versus Tray Columns—Operational Features and Cost

The findings submitted in the foregoing open up fresh viewpoints and approaches in column layout. The earlier view, still widely held, is that the highest capacity column that affords good separation is the best, or that the internals offering least cost per unit volume of construction would be the optimum; these views must be adjusted through a combination of functional evaluation and a corresponding analysis of cost.

For the precise determination of the specific cost of a separating column, based on unit separation and unit throughput, one prerequisite is the adequate insight into the interdependence of separating power and the load, allowing for the pressure loss. This interaction must, at this time, be found by tests for most column internals, and to that end different columns must be subjected to uniform conditions and work the same systems.

Tables 6.5 to 6.11[34] are intended to facilitate column selection for a given separation. They list the principal operating data and relative cost factors of the columns discussed in this work – all of which were investigated by the same ethylene benzene–styrene system, at 100 mm Hg, under identical test conditions.

Table 6.5 presents the values n_t/H, $\Delta p/n_t$, and v_V as functions of that load $u_V \rho_V^{1/2}$ that equals about 80% of the maximum. The sequence follows the descending orders of the load.

Table 6.6. shows the dependence $n_t/H = f(u_V \rho_V^{1/2})$, the sequence being that of descending n_t/H values at load factors of 80% of the maximum.

Table 6.7 gives the $p/n_t = f(u_V \rho_V^{1/2})$ values. Sequence is by increasing $\Delta p/n_t$ at 80% of the maximum load.

Tables 6.8 and 6.9 review the columns investigated in the order of increasing

Table 6.5

Typical performances of various column internals at 80% of the flooding load. Ethyl benzene/styrene, 100 mm Hg, $L/V = 1$

Column internal	$u_V \rho_V^{1/2}$ ($m^{-1/2} s^{-1} kg^{1/2}$)	n_t/H (1/m)	$\Delta p/n_t$ (mm H_2O)	v_V ($m^3/m^3/s$)	T_s (m)
Coarse grid	3.04	0.75	17.1	0.30	
Centrifugal tray II, 5 × 42 × 5 × 2; $h_W = 0$	2.98	0.65	55.8	0.36	0.5
Wire web packing	2.82	3.69	14.0	0.068	
Pallring 80 × 80 × 1.5	2.61	1.47	26.1	0.18	
Fine grid	2.59	0.71	41.4	0.38	
Pallring 50 × 50 × 1.0	2.47	2.18	23.8	0.13	
Pallring 35 × 35 × 0.8	2.40	2.71	25.9	0.11	
Square grid	2.38	1.29	17	0.226	
Pallring 25 × 25 × 0.6	2.30	3.02	24.2	0.10	
Centrifugal tray I, 4 × 42 × 5 × 2; $h_W = 30$ mm	2.24	1.42	47.6	0.219	0.5
Valve tray c), $\varphi = 14.8\%$	2.22	1.43	53.5	0.219	0.5
Valve tray d), $\varphi = 12.5\%$	2.20	1.46	56.9	0.217	0.5
Interpack 20 × 20 × 0.4	2.18	2.08	46.1	0.15	
Sieve plate $\varphi = 12.3\%$	12	1.40	57.6	0.234	0.5
Valve tray b), $\varphi = 14.8\%$	2.12	1.16	83.5	0.283	0.5
Pallring 15 × 15 × 0.4	2.02	3.66	26.9	0.094	
Valve tray a), $\varphi = 14.8\%$	1.84	1.53	69.0	0.25	0.5
Tunnel tray, $\varphi = 10.8\%$	1.74	1.52	93.6	0.26	0.5
Bell cap trays, $\varphi = 11.2\%$	1.54	1.34	69.4	0.34	0.5

Table 6.6

Separation in terms of equivalent theoretical steps versus vapor load. Ethyl benzene/styrene at 100 mm Hg, $L/V = 1$.

	80% of flooding load		$n_t/H(1/m)$						
			$u_V\rho_V^{1/2}(m^{-1/2}s^{-1}kg^{1/2})$						
Column internal	n_t/H (1/m)	$u_V\rho_V^{1/2}$ ($m^{-1/2}s^{-1}kg^{1/2}$)	1.0	1.5	2.0	· 2.5	3.0	3.5	T_s (m)
Wire web packing	3.69	2.82	6.62	6.10	5.59	4.60	3.13	2.32	
Pallring 15 × 15 × 0.4	3.66	2.02	2.42	3.05	3.65	2.84	–	–	
Pallring 25 × 25 × 0.6	3.02	2.30	1.89	2.22	2.74	3.06	–	–	
Pallring 35 × 35 × 0.8	2.71	2.40	1.97	2.13	2.37	2.86	1.67	–	
Pallring 50 × 50 × 1.0	2.18	2.47	1.80	1.90	1.91	2.22	2.09	–	
Interpack 20 × 20 × 0.4	2.08	2.18	2.30	2.35	2.14	1.99	–	–	
Valve tray. a), $\varphi = 14.8\%$	1.53	1.84	1.70	1.56	1.46	–	–	–	0.5
Tunnel tray, $\varphi = 10.8\%$	1.52	1.74	1.75	1.64	1.27	–	–	–	0.5
Pallring 80 × 80 × 1.5	1.47	2.61	1.40	1.35	1.30	1.42	1.65	–	
Valve tray d), $\varphi = 12.5\%$	1.46	2.20	1.54	1.53	1.50	1.23	–	–	0.5
Valve tray c), $\varphi = 14.8\%$	1.43	2.22	1.41	1.48	1.50	1.18	–	–	0.5
Centrifugal tray I, 4 × 42 × 5 × 2; $h_W = 30$ mm	1.42	2.24	–	1.24	1.39	1.44	–	–	0.5
Sieve plate, $\varphi = 12.3\%$	1.40	2.12	1.42	1.39	1.41	1.21	–	–	0.5
Bell cap tray, $\varphi = 11.2\%$	1.34	1.54	1.31	1.35	–	–	–	–	0.5
Square grid	1.29	2.38	1.30	1.27	1.24	1.37	–	–	
Valve tray b), $\varphi = 14.8\%$	1.16	2.12	1.43	1.45	1.28	–	–	–	0.5
Coarse grid	0.75	3.04	0.73	0.75	0.73	0.71	0.75	0.77	
Fine grid	0.71	2.59	1.24	1.27	1.03	0.72	0.65	–	
Centrifugal tray II, 5 × 42 × 5 × 2; $h_W = 0$	0.65	2.98	–	0.60	0.53	0.56	0.65	0.77	0.5

Table 6.7

Specific pressure loss versus vapor load. Ethyl benzene/styrene at 100 mm Hg, $L/V = 1$.

Column internal	80% of flooding load		$\Delta p/n_t$ (mm H_2O)						
	$\Delta p/n_t$ (mm H_2O)	$u_V\rho_V^{1/2}$ ($m^{-1/2}s^{-1}kg^{1/2}$)	$u_V\rho_V^{1/2}$ ($m^{-1/2}s^{-1}kg^{1/2}$) 1.0	1.5	2.0	2.5	3.0	3.5	T_s (m)
Wire web packing	14.0	2.82	0.8	1.3	2.1	5.9	22.3	59.7	
Square grid	17.0	2.38	3.1	6.3	11.3	19.4	–	–	
Coarse grid	17.1	3.04	1.8	4.3	7.8	11.3	16.8	27.9	
Pallring 50 × 50 × 1.0	23.8	2.47	4.2	8.1	14.8	24.2	55.0	–	
Pallring 25 × 25 × 0.6	24.2	2.30	6.3	11.7	18.3	30.2	–	–	
Pallring 35 × 35 × 0.8	25.9	2.40	4.9	9.9	18.0	28.0	10.2	–	
Pallring 80 × 80 × 1.5	26.1	2.61	3.7	7.0	13.7	23.8	37.0	–	
Pallring 15 × 15 × 0.4	26.9	2.02	9.3	16.0	26.2	–	–	–	
Fine grid	41.4	2.59	3.2	6.6	15.9	36.9	65.8	–	
Interpack 20 × 20 × 0.4	46.1	2.18	8.1	17.8	35.6	79.3	–	–	
Centrifugal tray I, 4 × 42 × 5 × 2; h_W = 30 mm	47.6	2.24	–	32.3	42.5	53.3	–	–	0.5
Valve tray c), φ = 14.8%	53.5	2.22	37	39	45	81	–	–	0.5
Centrifugal tray II, 5 × 42 × 5 × 2; h_W = 0	55.8	2.98	–	20.0	33.6	47.1	56.0	65.8	0.5
Valve tray d), φ = 12.5%	56.9	2.20	35.2	37.4	47.7	85.5	–	–	0.5
Sieve plate, φ = 12.3%	57.6	2.12	28.3	41.8	53.3	83.5	–	–	0.5
Valve tray a), φ = 14.8%	69.0	1.84	33.8	47.0	83.8	–	–	–	0.5
Bell cap tray, φ = 11.2%	69.4	1.54	37.3	65.9	–	–	–	–	0.5
Valve tray b), φ = 14.8%	83.5	2.12	44.3	50.2	70	–	–	–	0.5
Tunnel tray, φ = 10.8%	93.6	1.74	30.9	57.2	–	–	–	–	0.5

Table 6.8

Specific cost factors of various columns made of carbon steel. Ethyl benzene/styrene at 100 mm Hg, $L/V = 1$

Column	C_S	$u_V \rho_V^{1/2}$ ($m^{-1/2} s^{-1} kg^{1/2}$)	v_V ($m^3/m^3/s$)	
Interpack 20 × 20 × 0.4	0.934	1.09	0.271	80% of flooding load
Pallring 50 × 50 × 1	1.00	1.24	0.305	
Pallring 35 × 35 × 0.8	1.035	1.20	0.287	
Valve tray d)	1.12	1.10	0.411	
Valve tray a)	1.185	0.92	0.432	
Sieve plate	1.238	1.06	0.462	
Pallring 80 × 80 × 1.5	1.263	1.31	0.387	
Valve tray c)	1.27	1.11	0.448	
Valve tray b)	1.31	1.06	0.463	
Pallring 25 × 25 × 0.6	1.32	1.15	0.308	
Tunnel tray	1.658	0.87	0.455	
Pallring 15 × 15 × 0.4	1.853	1.01	0.282	
Centrifugal tray I	2.88	1.12	0.703	
Bell cap tray	3.135	0.77	0.743	
Centrifugal tray II	3.238	1.49	0.79	
Coarse grid	4.58	1.52	0.61	
Square grid	5.00	1.19	0.46	
Fine grid	5.23	1.29	0.431	
Pallring 35 × 35 × 0.8	0.93	2.40	0.11	80% of flooding load
Pallring 50 × 50 × 1	1.00	2.47	0.13	
Pallring 25 × 25 × 0.6	1.04	2.30	0.10	
Interpack 20 × 20 × 0.4	1.21	2.18	0.15	
Pallring 80 × 80 × 1.5	1.37	2.61	0.18	
Valve tray d)	1.39	2.20	0.217	
Pallring 15 × 15 × 0.4	1.449	2.02	0.094	
Valve tray c)	1.452	2.22	0.219	
Sieve plate	1.467	2.12	0.234	
Valve tray a)	1.618	1.84	0.25	
Valve tray b)	1.88	2.12	0.283	
Centrifugal tray I	2.10	2.24	0.219	
Tunnel tray	2.24	1.74	0.26	
Bell cap tray	3.363	1.54	0.34	
Centrifugal tray II	3.46	2.98	0.36	
Coarse grid	5.28	3.04	0.30	
Square grid	5.81	2.38	0.226	
Fine grid	10.81	2.59	0.38	

Table 6.9

Specific cost factors of various columns made of stainless steel. Ethyl benzene/styrene 100 Hg, $L/V = 1$, $d_S = 1.5$ m.

Column	C_S	$u_V \rho_V^{1/2}$ $(m^{-1/2} s^{-1} kg^{1/2})$	v_V $(m^3/m^3/s)$
40% of flooding load			
Valve tray d)	0.563	1.10	0.411
Valve tray a)	0.598	0.92	0.432
Sieve plate	0.624	1.06	0.462
Valve tray c)	0.629	1.11	0.448
Valve tray b)	0.644	1.06	0.463
Tunnel tray	0.799	0.87	0.455
Interpack 20 × 20 × 0.4	0.853	1.09	0.271
Pallring 50 × 50 × 1	1.00	1.24	0.305
Pallring 35 × 35 × 0.8	1.088	1.20	0.287
Wire web packing	1.138	1.41	0.0798
Pallring 80 × 80 × 1.5	1.25	1.31	0.387
Fine grid	1.357	1.29	0.431
Pallring 25 × 25 × 0.6	1.40	1.15	0.308
Bell cap tray	1.438	0.77	0.743
Coarse grid	1.444	1.52	0.61
Pallring 15 × 15 × 0.4	1.50	1.01	0.282
Centrifugal tray I	1.568	1.12	0.703
Centrifugal tray II	1.76	1.49	0.79
Square grid	2.39	1.19	0.46
80% of flooding load			
Valve tray d)	0.698	2.20	0.217
Valve tray c)	0.721	2.22	0.219
Sieve plate	0.749	2.12	0.234
Valve tray a)	0.812	1.84	0.25
Valve tray b)	0.924	2.12	0.283
Pallring 35 × 35 × 0.8	0.977	2.40	0.11
Pallring 50 × 50 × 1.0	1.00	2.47	0.13
Pallring 25 × 25 × 0.6	1.068	2.30	0.10
Tunnel tray	1.07	1.74	0.26
Interpack 20 × 20 × 0.4	1.11	2.18	0.15
Centrifugal tray I	1.143	2.24	0.219
Pallring 15 × 15 × 0.4	1.172	2.02	0.094
Pallring 80 × 80 × 1.5	1.36	2.61	0.18
Bell cap tray	1.542	1.54	0.34
Coarse grid	1.668	3.04	0.30
Centrifugal tray II	1.882	2.98	0.36
Wire web packing	2.275	2.82	0.068
Square grid	2.758	2.38	0.226
Fine grid	2.805	2.59	0.38

Table 6.10

Specific cost factors of various columns made of carbon steel. Ethyl benzene/styrene 100 mm Hg, $L/V = 1$, d_S (Pallring 50 × 50) = 1.5 m.

Column	C_S	$u_V \rho_V^{1/2}$ $(m^{-1/2} s^{-1} kg^{1/2})$	v_V $(m^3/m^3/s)$	$\Delta p/n_t$ (mm H_2O)
Pallring 50 × 50 × 1	1.00	1.69	0.22	
Pallring 35 × 35 × 0.8	1.072	1.506	0.216	
Interpack 20 × 20 × 0.4	1.22	1.11	0.266	
Pallring 80 × 80 × 1.5	1.36	1.77	0.298	
Pallring 25 × 25 × 0.6	1.43	1.355	0.244	10
Pallring 15 × 15 × 0.4	2.42	1.036	0.273	
Coarse grid	4.46	2.31	0.418	
Square grid	4.50	1.90	0.293	
Fine grid	5.53	1.77	0.327	
Pallring 35 × 35 × 0.8	0.899	2.586	0.0901	
Pallring 50 × 50 × 1.0	1.00	2.646	0.1107	
Pallring 25 × 25 × 0.6	1.072	2.485	0.0911	
Pallring 80 × 80 × 1.5	1.448	2.77	0.16	
Interpack 20 × 20 × 0.4	1.55	1.88	0.169	
Pallring 15 × 15 × 0.4	1.58	2.107	0.0882	
Sieve plate	2.66	1.06	0.462	
Valve tray a)	2.98	0.70	0.55	30
Tunnel tray	3.425	0.98	0.405	
Centrifugal tray I	4.13	1.37	0.437	
Valve tray d)	4.39	0.49	0.876	
Valve tray b)	4.40	0.53	0.826	
Coarse grid	5.36	3.56	0.254	
Square grid	5.58	2.74	0.184	
Centrifugal tray II	7.06	1.86	0.691	
Bell cap tray	8.10	0.63	0.986	
Fine grid	12.60	2.355	0.379	
Pallring 35 × 35 × 0.8	0.961	2.833	0.0941	
Pallring 50 × 50 × 1.0	1.00	2.894	0.1078	
Pallring 80 × 80 × 1.5	1.11	3.174	0.1194	
Pallring 25 × 25 × 0.6	1.128	2.762	0.0931	
Pallring 15 × 15 × 0.4	1.42	2.401	0.0771	
Interpack 20 × 20 × 0.4	1.432	2.235	0.1504	
Valve tray d)	1.581	2.064	0.224	
Valve tray c)	1.615	2.15	0.2215	
Sieve plate	1.755	1.89	0.2605	
Valve tray a)	1.87	1.565	0.285	50
Valve tray b)	2.17	1.48	0.326	
Centrifugal tray I	2.26	2.35	0.2066	

Table 6.10 cont.

Column	C_S	$u_V \rho_V^{1/2}$ ($m^{-1/2} s^{-1} kg^{1/2}$)	v_V ($m^3/m^3/s$)	$\Delta p/n_t$ (mm H_2O)
Tunnel tray	2.715	1.40	0.2975	
Bell cap tray	3.81	1.28	0.396	
Coarse grid	5.16	3.885	0.2385	
Centrifugal tray II	5.25	2.65	0.452	
Square grid	6.60	2.89	0.211	
Fine grid	12.75	2.75	0.372	

Table 6.11

Specific cost factors various columns made of stainless steel. Ethyl benzene/styrene 100 mm Hg, $L/V = 1$, d_S (Pallring 50 × 50) = 1.5 m.

Column	C_S	$u_V \rho_V^{1/2}$ ($m^{-1/2} s^{-1} kg^{1/2}$)	v_V ($m^3/m^3/s$)	$\Delta p/n_t$ (mm H_2O)
Pallring 50 × 50 × 1	1.00	1.69	0.22	
Interpack 20 × 20 × 0.4	1.112	1.11	0.266	
Pallring 35 × 35 × 0.8	1.122	1.506	0.216	
Wire web packing	1.261	2.695	0.0634	
Pallring 80 × 80 × 1.5	1.337	1.77	0.298	10
Coarse grid	1.42	2.31	0.418	
Fine grid	1.428	1.77	0.327	
Pallring 25 × 25 × 0.6	1.516	1.355	0.244	
Pallring 15 × 15 × 0.4	1.96	1.036	0.273	
Square grid	2.12	1.90	0.293	
Pallring 35 × 35 × 0.8	0.934	2.586	0.0901	
Pallring 50 × 50 × 1	1.00	2.646	0.1107	
Pallring 25 × 25 × 0.6	1.251	2.485	0.0911	
Pallring 15 × 15 × 0.4	1.27	2.107	0.0882	
Sieve plate	1.37	1.06	0.462	
Interpack 20 × 20 × 0.4	1.418	1.88	0.169	
Pallring 80 × 80 × 1.5	1.421	2.77	0.16	
Valve tray a)	1.497	0.70	0.55	
Tunnel tray	1.568	0.98	0.405	
Coarse grid	1.70	3.56	0.254	30
Valve tray d)	2.125	0.49	0.876	
Valve tray b)	2.138	0.53	0.826	
Centrifugal tray I	2.235	1.37	0.437	
Bell cap tray	2.62	0.63	0.986	
Square grid	2.63	2.74	0.184	
Wire web packing	3.16	3.11	0.0807	
Fine grid	3.24	2.355	0.379	
Centrifugal tray	3.83	1.86	0.691	

Table 6.11 cont.

Column	C_S	$u_V \rho_V^{1/2}$ $(m^{-1/2} s^{-1} kg^{1/2})$	v_V $(m^3/m^3/s)$	$\Delta p/n_t$ (mm H_2O)
Valve tray d)	0.797	2.064	0.244	
Valve tray c)	0.816	2.15	0.2215	
Sieve plate	0.896	1.89	0.2605	
Valve tray a)	0.955	1.565	0.285	
Pallring 50 × 50 × 1	1.00	2.894	0.1078	
Pallring 35 × 35 × 0.8	1.002	2.833	0.0941	
Valve tray b)	1.088	1.48	0.326	
Pallring 80 × 80 × 1.5	1.099	3.174	0.1194	
Pallring 15 × 15 × 0.4	1.143	2.401	0.0771	
Pallring 25 × 25 × 0.6	1.192	2.762	0.0931	50
Centrifugal tray I	1.223	2.35	0.2066	
Tunnel tray	1.248	1.40	0.2975	
Interpack 20 × 20 × 0.4	1.31	2.235	0.1504	
Coarse grid	1.649	3.885	0.2385	
Bell cap tray	1.845	1.28	0.396	
Centrifugal tray II	2.775	2.65	0.452	
Square grid	3.10	2.89	0.211	
Fine grid	3.285	2.75	0.372	
Wire web packing	3.815	3.305	0.0946	

specific cost factors for loads of 40 and 80% of the maximum. By the term "specific cost factor" C_S means the ratios of the specific costs C_V of the column in question and the corresponding cost of a column packed with 50-mm Pall rings adequate for the separation intended.

$$C_S = \frac{C_V \text{ (internals + shell)}}{C_V \text{(50 mm pall rings + shell)}} \tag{6-37}$$

The reference column with 50-mm Pall rings always has a specific costs factor of $C_S = 1$.

Tables 6.10 and 6.11 show, in sequence, the increasing specific cost factors for specific pressure losses of 10 mm, 30 mm, and 50 mm of water.

Although for some of the towers listed in Tables 6.5–11 certain structural details may be revised, because of operational advantages, it is felt that there would be only minor changes in the results of the evaluation. However, differences in the figures of merit might be caused by experimenting with systems behaving unlike the ethyl benzene–styrene mixture employed before; note the curves in Fig. 6.36 and 6.39.

Tables 6.8–11 were compiled from manufacturers' estimates and are subject to changes. The market may become unstable, and manufacturing methods are subject to rationalization.

The tables reflect only the cost of the internals and the shell and disregard the accessories. Whether the latter would cause the cost factors to vary depends much on the application. However, the tables should be useful for project work under average conditions. Pronounced fouling tendency, or specific material problems, might bar the use of some data. Rectification at elevated pressures would demand heavy shells, and, for example, the cost pattern would be affected.

E. Special designs for vacuum operation

As stated before, a rectification of temperature-sensitive components in vacuum equipment demands rectifiers of the lowest loss of vapor pressure, with a maximum of separation effect and load capacity. One instance is the various distillation stages in the production of plastics, and special equipment has been developed recently, such as the rotary spray column,[45] the nonadiabatic thin-layer rectifier[46] and the trickling-film column.[47] Test results are presented, and they are compared to those of a Pall-ring column as the conventional vacuum rectifying device.

a. The Rotary Spray Column

Fig. 6.60 is a diagrammatic representation of the construction elements of a rotary column. A shaft revolving inside the column is fitted with superimposed separating

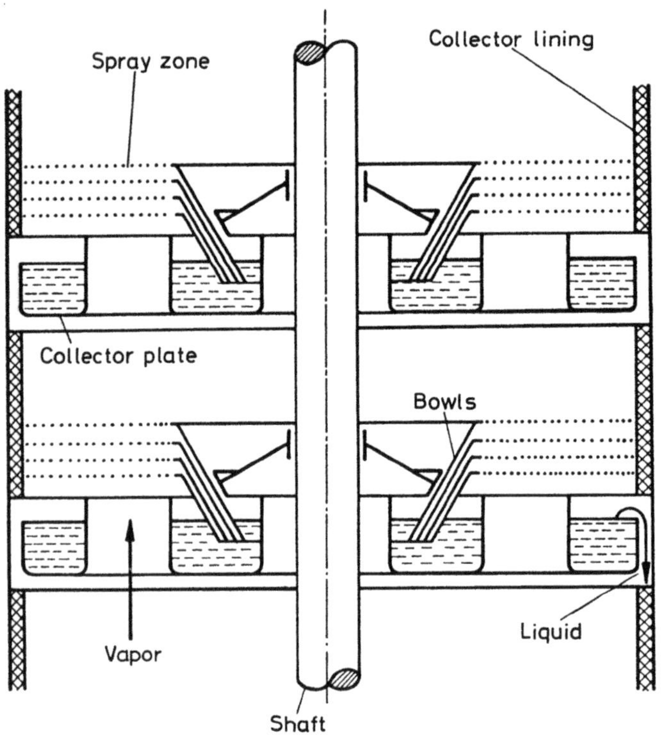

Fig. 6.60 Components of rotary spray column.

bowls that lift the reflux liquid from stationary collectors and spray it against the shell by centrifugal action. The liquid is gathered by a web and returned to the tray by flowing downward together with reflux dripping from collector to collector tray. Each tray has passages for the ascending vapor phase in exchange with the flowing liquid.

The horizontal layers of spray form parallel zones that may be relatively close to each other. In the pilot column of Fig. 6.61 the spacing is only 150 mm for a tower

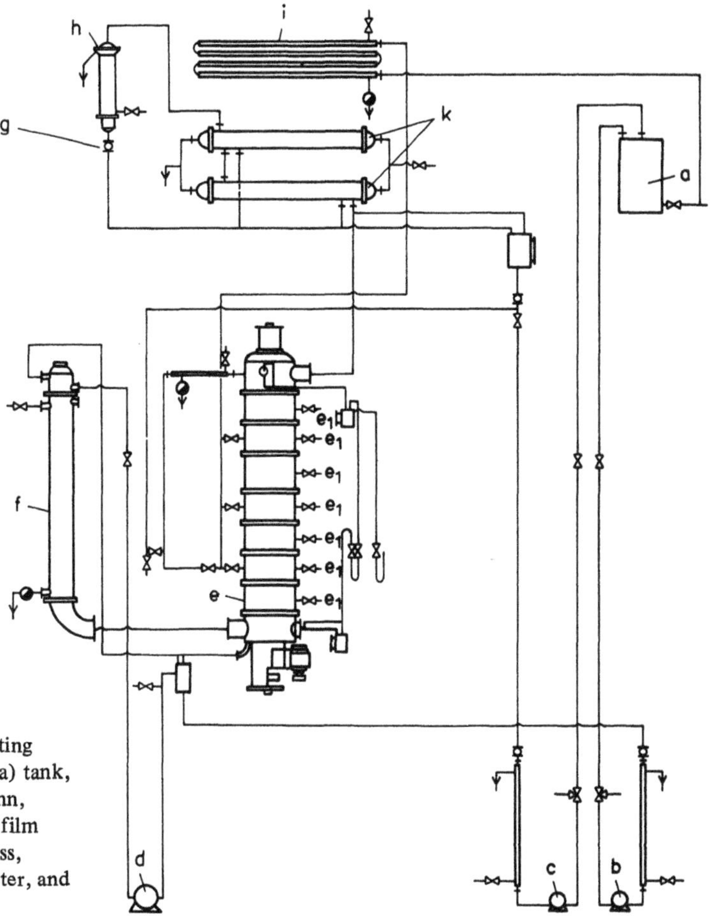

Fig. 6.61 Setup for testing rotary spray columns: a) tank, b)c)d) pumps, e) column, e_1) samplers, f) falling film evaporator, g) sight glass, h) gas cooler, i) preheater, and k) reflux condenser.

diameter of 500 mm. Thus for 15 separation stages, the column height is only 2.25 m. In other designs[51] the bowls spray the liquid upward at an angle. This gives a lower pressure loss per theoretical stage but a greater spacing than for horizontal sprays.

For simple spray columns with no liquid recycling[52] and having the film column, often the separation effect slackens off as the vacuum rises because the mass of reflux dwindles and becomes insufficient for maintaining a dense spray layer or liquid film. The column in Fig. 6.60 avoids that drawback by systematic recirculation of the reflux, creating a spray layer virtually independent of the vacuum with a fluid surface that is renewed permanently.

Fig. 6.61 presents a pilot installation where the rotor is driven from below. It is mounted at both ends on antifriction bearings and is properly sealed by the process liquids. At column top and bottom, and at every other of the 15 collecting trays, liquid samples may be drawn. To avoid the static liquid pressure typical of thermosyphon-type reboilers a falling-film evaporator is provided which is described in Chapter 9. The necessary vacuum is maintained by a water ring pump and ejector combined with a Roots blower connected to the back of the reflux condensers. The vacuum is controlled by a mercury seal and a magnetic valve.

The test results that follow were obtained with total reflux; all of the condensate formed was returned to the top of the column. The boiling temperature of the reflux could be maintained by a heat exchanger located between the reflux condenser and the column top.

The test mixture was *trans*-decaline and 1, 2-propylene glycol/ethylene glycol. These were rectified at a column-top pressure of 10 mm Hg and a rotor speed of 900 rpm. Furthermore, comparative tests were run at pressures of 3, 5, 10, and 15 mm Hg. To determine the pressure loss, every other stage was dismantled, which gave a stage spacing of 300 mm.

Fig. 6.62 gives mean values of separation and pressure loss for one separation stage as a function of column top condition and comparative air velocity as shown in Eq. (6-15).

The separation stages were evaluated analogously to those of the rectification plates; Section 1.B of this chapter. The separations thus determined were compared with those obtained on other rotary columns using the same principle[51,52] (Fig. 6.62). By this, the separation of one stage is identical with that of a conventional vacuum tray only at small loads.

As opposed to the conventional rectification plates, the available liquid surface is almost constant for a given system in a rotary spray column of constant speed. With that condition, the liquid quantity L circulated per unit time remains the same even though the column load varies, while on trays the liquid surface provided grows with the vapor load. Therefore, the separation effect of a rotary spray column must decline faster than that of a plate column with increasing vapor load and diminishing contact time between vapor and liquid.

The liquid surface in a rotary spray column is

$$A = \frac{1000}{3600} \frac{L\mu_L}{\rho_L} \frac{d^2 \pi t_R}{\frac{d^3 \pi}{6}} = \frac{5}{3} \frac{L\mu_L}{\rho_L} \frac{t_R}{d} \qquad (6\text{-}38)$$

where:

t_R = residence time (s)
d = mean diameter of liquid drops in spray layer of recycled liquid L, (mol/h)

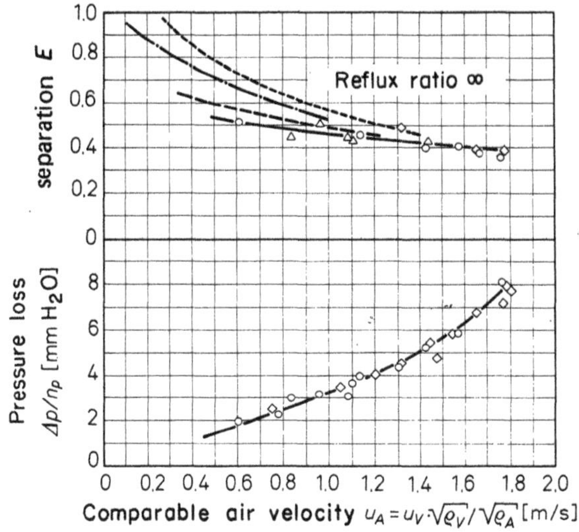

Mixture	p [mm Hg]	Column diameter [mm]	Spacing of steps [mm]	Speed [1/m]
○ △ Trans-/cis-Dekalin	10	500	150	900
◇ 1,2-Propyl.glyc./Ethyl.glyc.	10	500	150	900
--- Chlorbenz. / Ethylbenz.	50	500	150	765
—— 1-/2-Methylnaphtalin	5	500	150	765
—·— Ethanol/Water	100	900	650	300

Fig. 6.62 Separation and pressure loss versus comparable air velocity, for rotary spray columns.

u_L = molecular weight (kg/kmol)
ρ_L = density (kg/m^3)

The pressure loss per separation stage of a rotary column is, according to Fig. 6.62, less than one-tenth of that of a comparable bell-cap plate, if the reflux is sprayed horizontally.

Even less of a pressure loss would result if the spray from the bowls were directed at an upward angle instead of horizontally. The spacing would become larger as a result. The pressure loss measured with the shaft moving would be less than with the shaft at rest, which indicates a suction effect of the liquid spray.[51]

At higher loads the measured separation effects per set of internals is almost the same regardless of the system investigated (Fig. 6.62). Recirculation of the sprayed liquid will yield a relatively large surface; therefore, the separations achieved per transfer unit are superior to those of simple rotary devices without the liquid recycle.[53] Few data are known regarding the effect of the rotor speed and of the spray action upon the separation and loadability (6-54).

Further evaluation demonstrates that an increase of the spacing from 150 mm to

300-mm will hardly affect the pressure loss. Fig. 6.63 shows the results for 150 mm spacing, using the pressure-loss findings in Fig. 6.62. The differences are slight. It was noted that the entrainment was less than in tests for 150-mm of spacing.

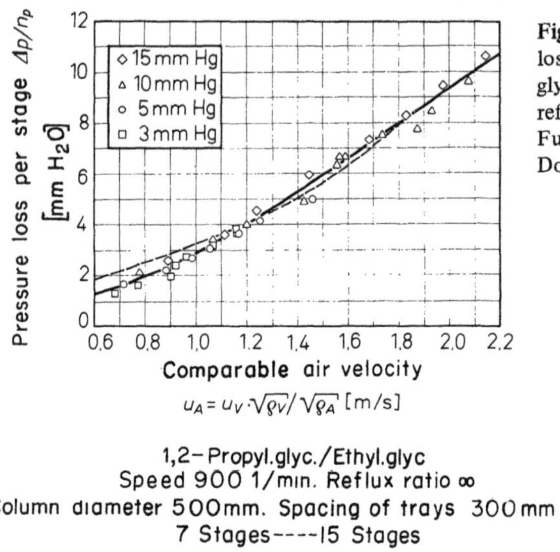

Fig. 6.63 Rotary spray columns. Pressure loss versus load. Mixture 1.2 propylene glycol/ethylene glycol, speed 900 rpm, reflux ratio ∞, column diameter 0.5 m. Full line 300 mm spacing, 7 stages. Dotted line 150 mm spacing, 15 stages.

The power requirement N of the rotor was measured at a constant rate of 900 rpm. Above a reflux quantity of 500 kg/m^2 h it varies little with increasing liquid loading. Measurements revealed a power requirement of $N = 1.6$ kW per m^2 of column cross section per stage. The liquid content per stage was 0.012 m^3 per m^2 of plate area,[54] that value can be reduced by employing tapering collectors.

The largest industrial rotary column has a 1400-mm diameter and a maximum of 15 stages. If more stages are needed, several columns must be used in series because of shaft-length limitations, which means that other methods must be adopted.

b. Thin-Layer Rectifier

A characteristic of the thin-layer rectifier is its nonadiabatic operation. The reflux liquid forms a thin film on the inside of a vertical cylinder heated from the outside, counter to the ascending product vapor. It is held in motion by a cooled coaxial rotor equipped with wipers (Fig. 6.64). The components of this construction differ from those of a simple thin-layer evaporator, described in Chapter 9 for continuous partial separation, only by the presence of the cooled rotor. The shape of the cooled rotor is such that any liquid formed on its cooling surface is immediately returned to the heated wall. Thus, in the case of the thin-layer rectifier, the basic actions in rectification are physically separate, being distributed over an evaporation area and a condensing area. The vapor ascending in the annular space between evaporating area and cooling area undergoes only a slight pressure loss, which permits working at very low pressures.

According to the theory[55] of the thin-layer rectifier the optimum performance may be expected if the ratio of the material evaporated and the material condensed becomes unity. The mass of product vapor may be condensed and reevaporated several times. The number of reevaporations will affect the separation.

Fig. 6.64 Section through a film type rectifier (Luwa of Zürich, Switzerland)

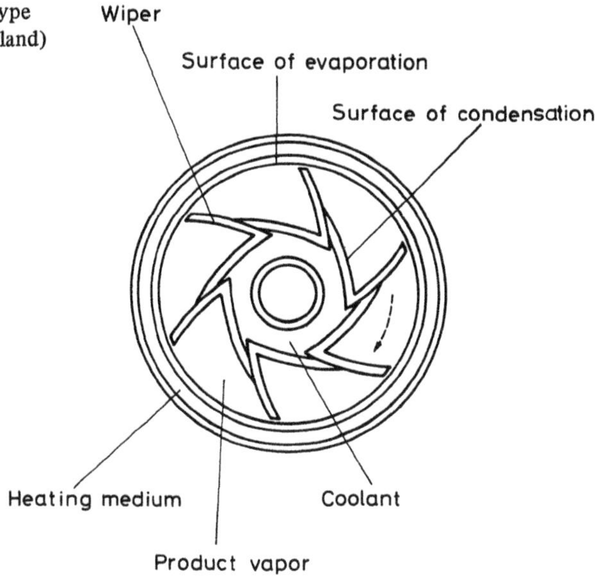

Fig. 6.65 is the diagram of a pilot plant as used in this investigation. The heating surface was 0.96 m^2, the heated length 1.7 m, and the ID 180 mm. The cooling surface of the rotor is 0.35 m^2, its speed is 750 rpm, based on a circumferential wiper velocity of around 7 m/s (a supposed optimum).

To achieve a low bottom boiling point, the evaporator is of the thin-layer type with a revolving wiper. The reflux liquid is formed in a condenser connected to a vacuum setup. The reflux, the bottom product, the evaporate, and the feed are handled by separate proportioners.

A mixture of 1, 2-propylene glycol and ethylene glycol was rectified at total reflux with column top pressures of 100, 50, and 10 mm Hg. The ratio of evaporate and condensate was always unity. For best separation, the rectifier rotor was always supplied with a maximum of coolant.

Fig. 6.66 gives the test results scaled as follows: for the load – the comparable air velocity, equivalent to the vapor flow, related to atmospheric conditions; for the separation – the theoretical number of stages equivalent in effect to the unit height of the rectifier, n_t/H ; for the pressure loss–the corresponding $\Delta p/H$ value. The separation curves for 50 and 100 mm Hg resemble each other, but should differ from the curve applicable for 10 mm Hg, on principle.

The better separations found for higher pressures in the zones of less vapor load are presumably due to better wetting of the rectifier at higher pressures, and the cooling of the rotor improved at higher pressures, which in turn improved re-

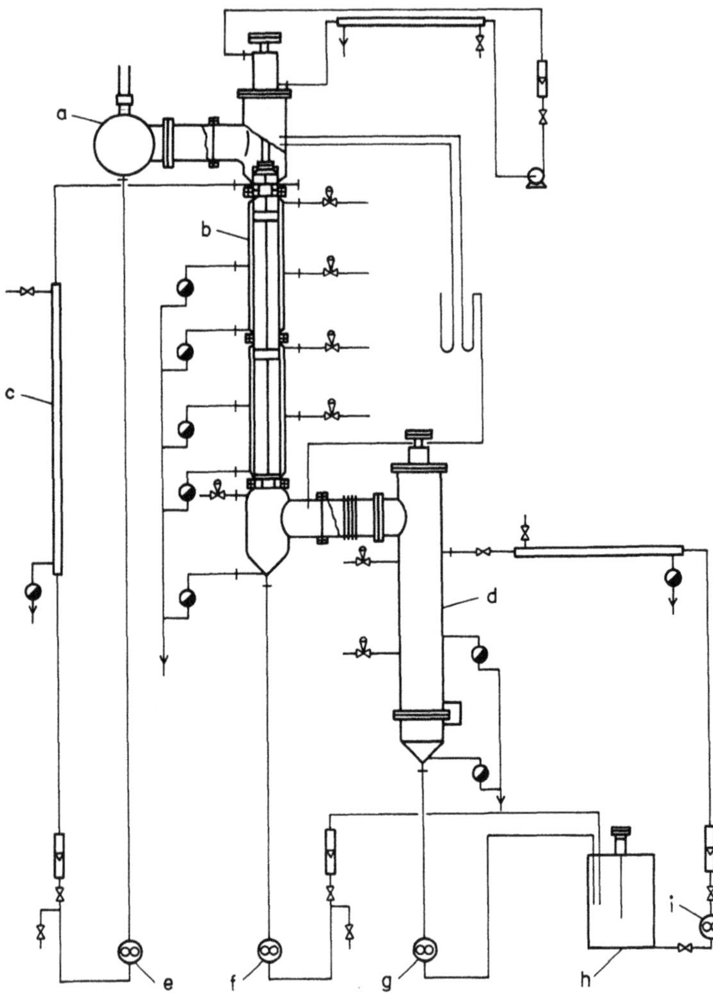

Fig. 6.65 Test stand for a film-type rectifier equipped with film-type vaporizer: *a* reflux condenser, *b* rectifier, *c* preheater, *d* vaporizer, *e*, *f*, and *g* pumps, *h* tank, *i* pump.

evaporation. These data are compiled in Table 6.12 as a function of the specific heat flow q. The product of the vapor flowing through the rectifier is based on a free passage area and on unit time multiplied by the heat of evaporation.

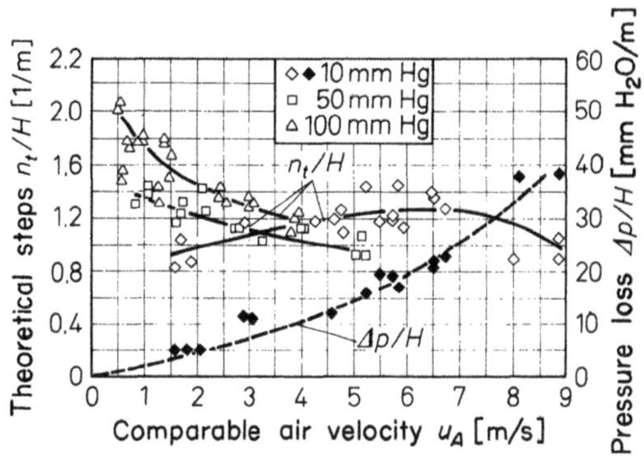

Fig. 6.66 Film rectifier. Separation and pressure loss versus load. Mixture 1.2 propylene glycol/ethylene glycol, reflux ratio–∞, speed–750 rpm, column height–1.7 m, column diameter–180 mm, re-evaporation data in Table 6.12.

The effectiveness of a thin-layer rectifier at total reflux may be expressed by the term[55]:

$$n_{RK} = \frac{\alpha - 1}{\alpha + 1} \frac{y_D - y_W}{100} + \ln \frac{y_D\,(100 - y_W)}{y_W\,(100 - y_D)} \frac{\alpha}{\alpha^2 - 1} \tag{6-39}$$

where:

y_D = vapor concentration at the rectifier top (mol-%)
y_W = vapor concentration at the rectifier bottom (mol-%)
α = relative volatility of the binary mixture for the concentration range in question

based on tests, rectifier characteristic n_{RK} is related to the number of steps by[55]

$$n_t \cong 1.8 \cdot n_{RK} \tag{6-40}$$

The separation in a thin-layer rectifier can be set within wide limits by increasing the reevaporation number.

Table 6.12

Reevaporation coefficients of an experimental film rectifier versus pressure at various specific heat flows. Coolant water.

q(kcal/m²h)	Reevaporation coefficients		
	10 mm Hg	50 mm Hg	100 mm Hg
$4 \cdot 10^5$	3.4	4.1	5.1
$8 \cdot 10^5$	1.7	2.2	2.7
$12 \cdot 10^5$	1.2	1.6	1.9

For the system hexanedecane/heptyl benzoate, and for more than 10 reevaporations at 3 mm Hg, characteristics over 2 were found per meter of rectifier height[56]

The heat transfer coefficients of the cooled rotor, at low vapor loads, were around 1400 kcal/m^2 h °C, and at high vapor loads, around 1700 kcal/m^2 h °C.

The largest existing thin-layer rectifiers are 9-m long and have 850-mm diameters. Operational advantages of the thin-layer rectifier are the low liquid content, brief treatment time of the product, and the insensitivity to fouling media. It would primarily be suitable for stripping, and relatively viscous liquids could be handled without incurring a risk of plugging. The small content means short startup periods. As thin-layer rectifiers have a limited height, they cannot be used for separations with many stages. A separation of encrusting materials in a conventional device would have to be an enriching operation to obviate plugging, and a high reflux rate would be mandatory for continuous work. Also, a dirt-resisting heater would be required to supply the vapor, such as a thin-layer evaporator with revolving internals. In that event, the steam requirement for the special concept might well exceed that of a conventional rectifier, with a thin-layer stripper and a thin-layer evaporator.

The thin-layer rectifier is characterized by a very low pressure loss per theoretical stage, high specific vapor rates, and brief process times, all of which makes it suitable for handling heat sensitive materials in a vacuum. Examples include the making of pure caprolactam, the rectification of vitamins, and of methacrylic acid.

Fig. 6.67 shows the field of application of thin-layer rectifiers in the recovery of thermally sensitive substances by giving the vapor-pressure curves for some specific organic materials.

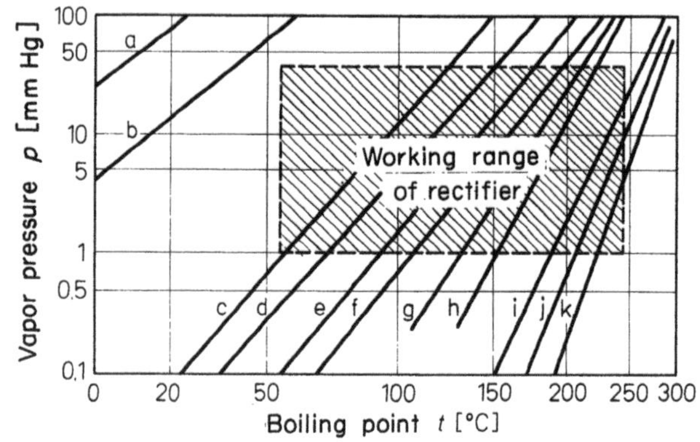

Fig. 6.67 Area of application of film rectifiers made by Luwa of Zurich, Switzerland. Vapor pressure curves of a) vinyl acetate, b) water, c) ethylene glycol, d) butyl phenol, e) caprolactam, f) C–10 fatty acid, g) C–12 fatty acid, h) C–14 fatty acid, i) C–18 fatty acid, j) C–20 fatty acid, k) vitamin E.

c. *Film-Type Columns*

The demand for equipment in vacuum rectification with low pressure loss, good separation, and ample capacity is frequently met by the various types of trickling film columns. In these, the liquids descend on walls counter to the ascending vapor. Although the principle has long been known, industrial applications of film rectification have been few. Experiences with film-type columns cover designs employing internals of flat and parallel sheets, sheet packages and wire web,[57–60] vertical wire helices,[61] and research work[62–64] permits the following conclusions. The efficacy of film type columns depends largely on even liquid distribution. The supply of liquid is a decisive design factor. Distribution must be uniform across the entire column length, and no blowby is permissible. Furthermore, to ensure a constant vapor-liquid ratio, the vapor distribution must likewise be uniform, and the vapor passages must be interconnected to avoid concentration and pressure gradients.

The trickling film column in Fig. 6.68[47] largely meets these requirements. Its internals are made of expanded metal forming flat and parallel layers mounted

a)

Fig. 6.68 Elements of a trickling film column, plane and side view. Jantzen design. Made by Schmidding of Cologne, Germany. Expanded metal.

vertically and 20 mm apart crosswise. This yields vertical passages of square section in which vapor and liquid flow counter to each other. The slots in the expanded metal allow an equalization of the pressures and concentrations of the phases; also the slot openings promote intermixing which assists in improving the specific separation.

Since the pressure loss per theoretical stage in a film-type column is very slight, its use in rectifications of limited pressure range may be expected to be of considerable economic advantage. Within the boundaries of a specified overall pressure drop, trickling film columns can provide more theoretical steps than plates, which means a lowered reflux ratio. Furthermore, the throughput of a trickling film column is larger in view of the high specific rate, and that reduces the effective diameter as compared to a plate tower for the same capacity.

The packing elements of the pilot installation in Fig. 6.69 have a diameter of 200 mm.

Four aggregates, each 1-m long, are superimposed, which yields an effective separation height of 4 m. The evaporator is a falling film type with liquid recirculation as described in Chapter 13. Other components include the reflux condenser with gas cooler and vacuum equipment, preheaters, pumps for the feed and the reflux, and instrumentation and controls.

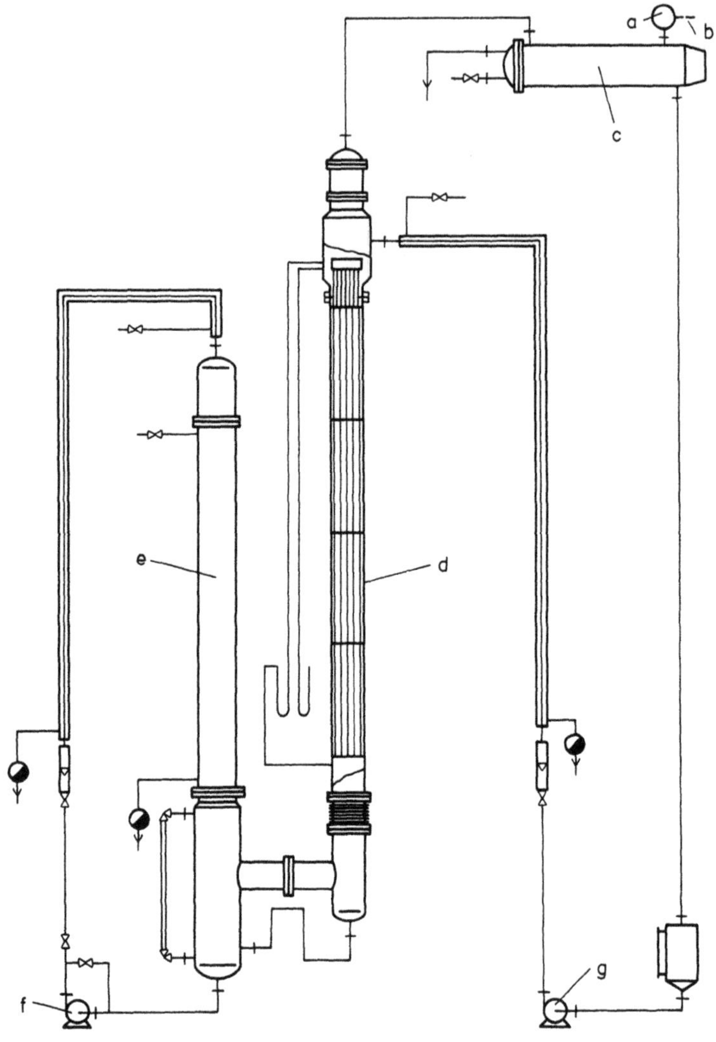

Fig. 6.69 Test stand for trickling film columns, with falling film evaporator serving as reboiler: *a*) gas cooler, *b*) vacuum connection, *c*) reflux condenser, *d*) trickling column, *e*) evaporator, *f*) and *g*) pumps.

In the tests on the mixture 1, 2-propylene glycol–ethylene glycol at 15 mm Hg, the liquid redistribution[65] of Fig. 6.70 was employed. It consists of a plate fitted with tube stubs extended at both ends, and expanded in place. Syphoning wires are pinned into the upper rims and can be adjusted to move 80 to 85% of the liquid overboard. The quantity thus transferred depends on the physical properties of the system, the number of the clips, and the levels set on the plates.

The transfer of liquid controls the separating action of the column packing. To find a quantitative relationship, the distributor in Fig. 6.70 was compared to a

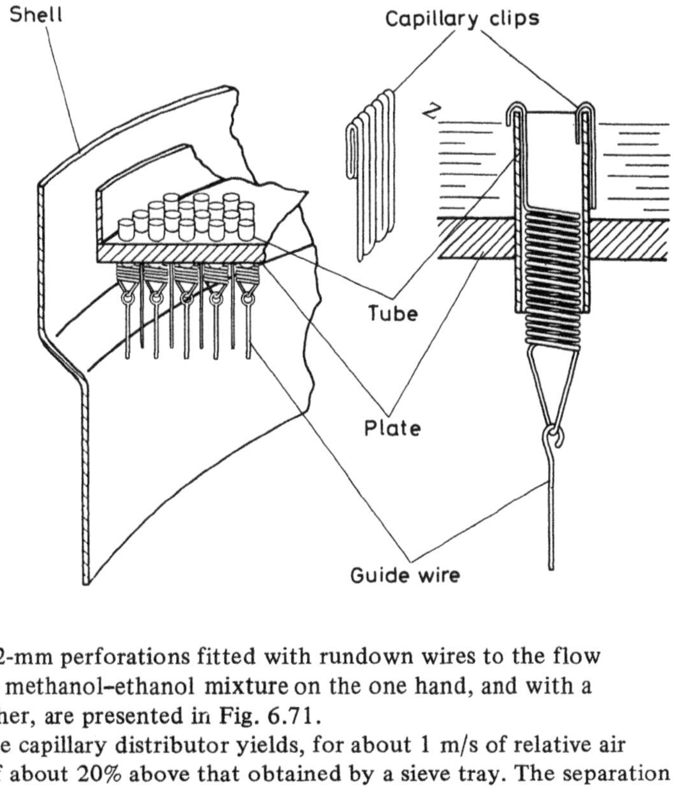

Fig. 6.70 Liquid distributor for trickling film column. Kloss design, made by Montz of Hilden, Germany.

simple sieve plate with 2-mm perforations fitted with rundown wires to the flow passages. Results with a methanol–ethanol mixture on the one hand, and with a glycol system on the other, are presented in Fig. 6.71.

According to this, the capillary distributor yields, for about 1 m/s of relative air velocity, a separation of about 20% above that obtained by a sieve tray. The separation is much better at ordinary pressure; in a vacuum the separation per meter of packing height is only 0.8 to 1 theoretical steps. Therefore, wall spacings exceeding 20 mm are impractical; according to the theory the separation effect decreases by the reciprocal of the square of the wall spacing.[58,59]

Other tests in which the reflux liquid was distributed across the packing by a nozzle gave a much inferior separation.

In Fig. 6.71 the separation decreases with the mounting load. However, at no time were the loads high enough to cause flooding.

When the column space is fitted with flat strips, parallel to each other, the simplicity of the construction and straightforward performance will make the theoretical treatment relatively uncomplicated.

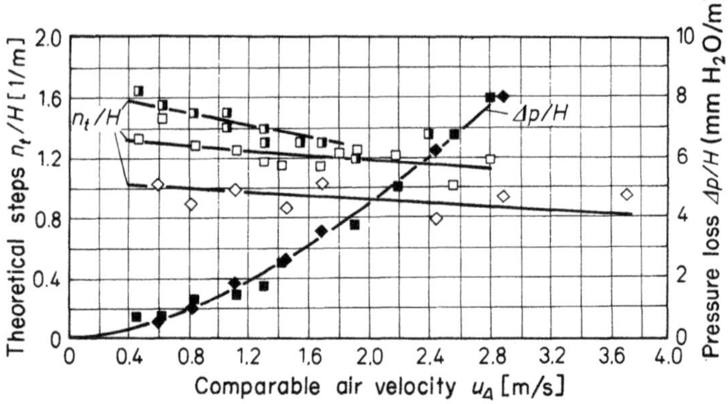

Fig. 6.71 Trickling film column 200-mm diameter, made by Schmidding of Cologne, Germany, with 20·20 mm expanded metal packing 4000 mm deep. Separation and pressure loss versus vapor load. ◇◆ 1, 2 propylene glycol–ethylene glycol, 15 mm Hg, Montz capillary feeder plate, □■ methanol–ethanol, 760 mm Hg, feeder sieve plate and guide wires. □■ methanol–ethanol, 760 mm Hg, Montz capillary feeder plate, ∞ reflux ratio.

With a laminar flow of vapor the velocity profile is parabolic, and the materials exchange is through a purely molecular diffusion. The plate height H for a trickling column separating a binary mixture becomes[58]

$$H = \frac{17}{35} \frac{s_a^2}{D} u_{V_m} \int \frac{dy}{(y - x) - \frac{1}{r}(x_D - y)} \tag{6-41}$$

where:

s_a = distance of the liquid film from the symmetry plane between two walls, or half the distance between two neighboring liquid films, (m)
D = the diffusion coefficient (m^2/s)
u_{V_m} = the average vapor velocity (m/s)
y = the equilibrium concentration of the low boiler in the vapor at phase equilibrium with a liquid of concentration x
r = reflux ratio
x_D = concentration of low boiler in the top product

The integral in Eq. (6-41) can be evaluated from the equilibrium curve. For turbulent vapor flow the equation for the plate height becomes more involved. Tests conducted with a 350-mm column packed with flat, parallel, smooth metal strips 6 m high, using the binary methanol–water mixture, gave for 24-mm plate spacing a

60% separation, and for 12-mm spacing, 40% of the perfect separation.[58] A reduction of plate spacing improved separation, but the discrepancy between theoretical values and test results increased. The equivalent number of steps obtained was between 0.5 and 1 theoretical step per meter of plate height. Other tests on film type columns with bundled flats of 18- and 20-mm spacing gave comparable results.[57]

For film type columns as described, fitted with flat strips, the arrangement of Fig. 6.72 for collecting the vapor and supply of the liquid may be adapted.[66]

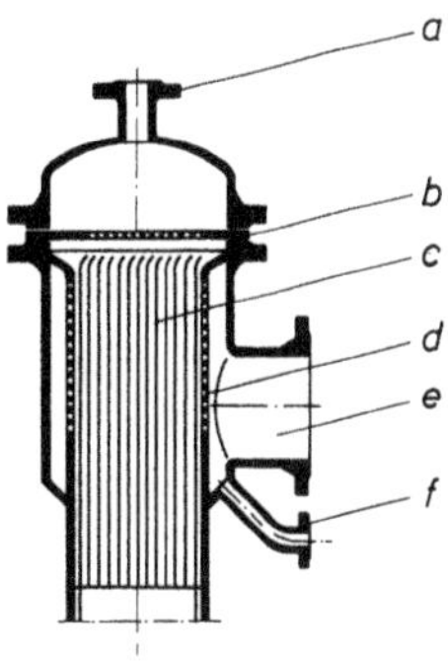

Fig. 6.72 Top of a trickling film column with parallel and flat baffling: *a* Liquid feed, *b* Perforated plate, *c* Baffles, *d* Vapor passages, *e* Vapor exhaust, *f* Drain for entrained liquid.

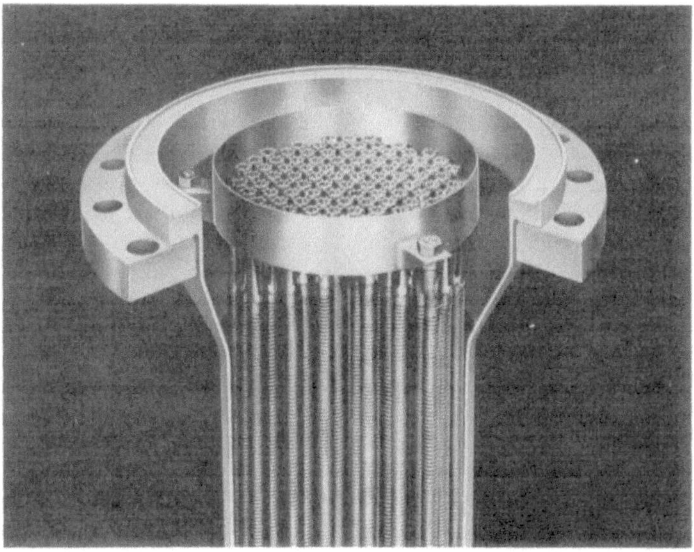

Fig. 6.73 Film column with helical packing. Kloss design, made by Montz of Hilden, Germany.

Another means for conservative rectification of a heat-sensitive material is offered by a column fitted with a packing of helical wires[67] as in Fig. 6.73. The liquid would be supplied according to Fig. 6.70.

An even spreading of the liquid as a film on the surfaces of springs along a properly designed path is conducive to good separation, and the relatively large clearances allow large vapor volumes at low pressure loss.

d. Special Designs versus Pall-Ring Columns

In the past, packed columns having low pressure loss were frequently preferred to plate columns when low process temperatures were indicated. A comparison may not be out of place. Separation, loadability, and pressure loss of a packed column depend largely on the shape, dimensions, and material of the internals. It has been seen in Section 3.B of this chapter that the Pall ring is highly loadable and gives a low pressure loss. Two makes were selected: 50-mm metallic thin-wall Pall rings, and 25-mm ceramic Pall rings. For a 1, 2-propylene glycol–ethylene glycol mixture, separation and pressure loss were determined as a function of load at total reflux for 10 mm Hg column-top pressure.

In the rectification of unstable, heat-sensitive materials, a low pressure loss of the vapor and the best separation are desirable, and the pressure loss per theoretical step is especially significant. Fig. 6.74 presents that specific pressure drop versus the

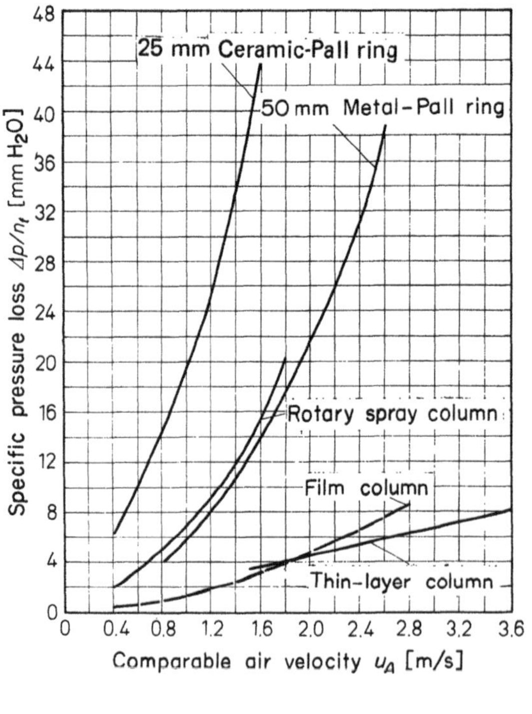

Fig. 6.74 Pressure loss per theoretical step versus comparable air velocity measuring load condition, for various vacuum rectifiers and Pall ring dump packing, 1,2 propylene glycol–ethylene glycol at 10 and 15 mm Hg.

corresponding air velocity. Because of their construction and design characteristics, thin-layer rectifiers and film-type columns will stand higher loads per unit column area than rotary spray columns and Pall-ring columns, for a given pressure drop per theoretical step and a given throughput; hence the required column area will be less in the former case. Comparing the height per theoretical step at low specific pressure losses from Fig. 6.75 demonstrates the advantage to the rotary spray if the power requirement is disregarded.

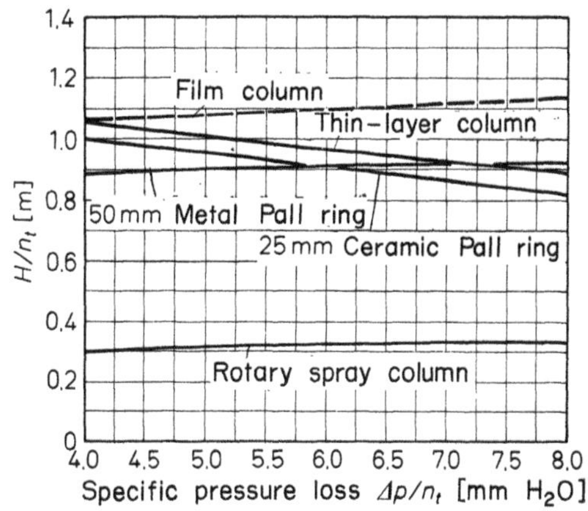

Fig. 6.75 Height per theoretical step versus pressure drop per theoretical step. 1,2-propylene glycol methylene glycol at 10 and 15 mm Hg.

As far as the specific tower volume goes, Fig. 6.76, the special designs are preferable to the Pall columns. For the thin-layer rectifier and for the rotary spray, that advantage is lost if the towers must be built in two stages to acquire the indicated number of theoretical steps.

The dwell type of the process material is less for the thin-layer rectifier because of the low liquid content. Hence, that device recommends itself for working temperature-sensitive materials. On the other hand, there is the drawback of heat losses due to lateral partial reevaporation.

One advantage of the rotary spray is that fluctuations in the feed rate will not impair the separation mechanism. Conceivably, the separation by rotary spray will not be affected at a low pressure, such as 3-mm Hg, given the relatively large contact area between the vapor and the liquid.

Both the rotary-spray and the thin-layer rectifier have limitations as to construction height and diameter because of their rotating parts, and they are more likely to malfunction. Towers with rotary internals are especially costlier than packed towers, and in many instances of vacuum distillation, large thin-walled Pall rings or some equivalent packings will hold their own.

No attempt is made to present column cost data for special designs. It may be noted that a wire-web column suitable for vacuum work, as discussed in Section 3.B of this Chapter, is much less expensive than a thin-layer rectifier for comparable duty.

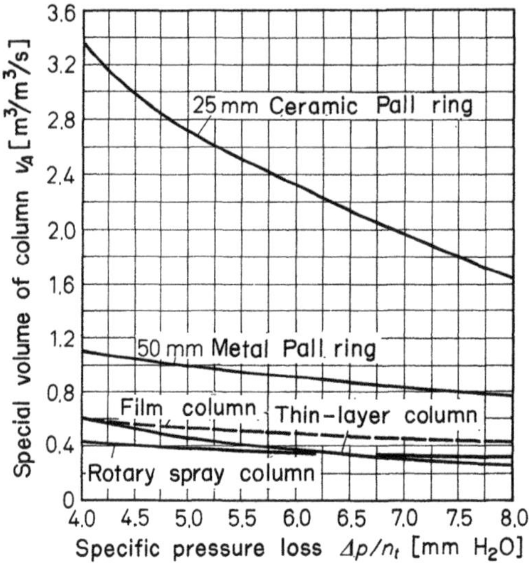

Fig. 6.76 Specific column volume versus pressure loss per theoretical step. 1,2 propylene glycol–ethylene glycol at 10 and 15 mm Hg.

4. TESTS OF FULL-SIZE TOWER INTERNALS

To arrive at a proper choice and eventually determine the cost optimum of the internals for a rectifier being planned for a given job, knowledge of process data such as separation effect and pressure loss as a function of load is a prerequisite. As long as no dependable calculation data were available, one had to use excessively large safety factors; also, extensive research could be had at prohibitive cost only. In the past, it was often necessary to use excessively large safety factors due to the lack of dependable calculation data, and extensive research also was only available at prohibitive cost. Today, however, Fractionation Research Inc. in South Pasadena, California, provides test facilities to which its members have access (Fig. 6.77). The tower in the foreground handles high-pressure rectifications, the one in the middle is for normal to low pressures, and the rear column is for overall and blank tests.

The central tower in Fig. 6.77 consists of an upper part that is 2.44 m in diameter, and a lower part with a diameter of 1.22 m. The upper part can accommodate up to 6 plates at 600-mm spacing and serves primarily for recirculation tests. The lower part can take up to 10 trays at 600-mm spacing in a single-pass arrangement. The shells have 50 or more taps for pressure, temperature, and for sampling. The vaporizer part can transmit $3.78 \cdot 10^6$ kcal/h, referred to butane at 11 kg/cm^2 gage. The reflux condenser can handle the same throughput. The feed can be supplied from a storage tank either to the column base or to the column top, depending on the magnitude of the liquid-vapor ratio. The condensate of the top vapor is collected in a special tank. With total reflux, all of the condensate is recycled to the column top. However, for top product recovery, part is drawn off with the balance going to a storage tank that also receives the bottom product. The various streams inside the plant can be controlled

Fig. 6.77 Overall view of test stand of Fractionation Research Inc., South Pasadena, California, is an organization for the experimentation and promotion of fractionation techniques, and has about 70 member- companies, chiefly chemical, oil, and engineering firms.

and measured. Among these there is the heating steam, its condensate (from which the heat flow is computed) the coolant, the feed, the top product, the bottom product, and the reflux. All quantities are recorded; the corresponding temperatures, along with those on the trays, are measured by copper-constantan thermoelements and are recorded by an electronic, multiple precision potentiometer. The temperatures of selected streams at chosen points on the columns are continuously recorded by multiple potentiometers.

Pressure losses and hydrostatic variables are measured by gages with nitrogen-charged connecting lines. The valving is such that the pressure loss of individual plates, or of groups of trays can be read off. Also the total loss for the entire column can be noted. With overpressure distillation, the excess pressure is controlled by a layer of inert gas in the condenser, or by coolant control of quantity. With vacuum distillations the pressure is controlled by a regulating valve in the ejector steam.

To determine the influence of the load on the separation, the load is varied, at total reflux, between 10 and 100% of the flooding load, in addition to measurements of the flooding load itself. The flood tests proper are made at several constant liquid loads so that liquid-vapor ratios are established as in actual enrichment and stripping processes. Alongside with pressure tests at total reflux, special tests can demonstrate pressure losses at various constant liquid loads as functions of vapor loads. Furthermore, in tray tests the pressure loss of the topmost dry plate can be found after sealing the downcomer by passing all of the condensate formed by steaming to the reflux tank and the column bottom.

This testing equipment is operated on systems such as cyclohexane–*n*-heptane, *n*-butane–isobutane, isopropanol–water, *o*-xylene–*p*-xylene, isooctane-toluene, *n*-heptane–absorption oil, Freon 11 and the water-water vapor phases. In that way, performances of column internals can be investigated, and calculations for new units can be refined.[68] Research work mostly covers recent designs that appear promising. If necessary, pretesting is undertaken.

The equipment provided to that end (Fig. 6.77, rear) comprises a column 5.5 m tall, in which up to 3 trays of 1.67-m diameter can be installed. Up to 27,000 m^3/h of a gas and up to 125 m^3/h of liquid can pass through. The gases used are air, carbon dioxide, and natural gas; the liquids used are water, kerosene, oils, and other noncorrosives. As in the rectifier installation, the flow in the trays can be observed through looking glasses. The measurements reveal the pressure losses, liquid gradients, vapor and liquid distribution, intensity of entrainment, and flooding conditions. Finally, the properties of the gas and liquid phases are investigated for the conditions stated. The plate activity is judged by means of optical evaluations.

VII Optimum Separation

Once a decision has been made in the choice of a rectifying column, from the criteria outlined in Chapter 6, the overall equipment must be dimensioned for optimum operation. A different situation exists if an existing setup must be reequipped for a new and unforeseen separation. Again, the question of the optimum operation arises. In any event, the interlinking of the various cost factors, and their dependence upon the basic separation parameters must be investigated. This is especially true for multicolumn systems where the fixing of desiderata may lead to complexities.

1. OPTIMIZATION OF SIMPLE COLUMNS

In plant planning the choice of the column internals and proper loading must be correlated with best overall design. Thus, a low pressure in vacuum operation would avail little if unfavorable reflux conditions and staging causes an unnecessary pressure rise. Rather, for a minimum column-bottom temperature those dimensions and operational conditions must be found for which a minimum pressure loss can be maintained. An example will explain this for a case of continuous operation.

A. The Theory and Its Application

Note the flowsheet, in Fig. 7.1, the scheme in Fig. 7.2, and the sequence in Fig. 7.3. From the project data, find the concentration of the low boiler x_F (mol-%) in the feed, x_D (mol-%) in the top product, and (x_W) mol-% in the bottom product.

Graphical or analytical methods will yield, for the rectifier and the stripper, the number of theoretical stages n_{tT} versus reflux ratio r (phase 1 of Fig. 7.3).

Then, assume various specific feed loads f (kmol/m^2 h) for the column of diameter d_S (m) and feed F (kmol/h)

$$f = \frac{F}{d_S^2 \frac{\pi}{4}} \tag{7-1}$$

which yields the vaporload factor $u_V \rho_V^{1/2}$ versus the reflux ratio r by means of the relations:

$$u_v = K_1 f \frac{VT}{Fp} \tag{7-2}$$

$$K_1 = 1.73 \cdot 10^{-2} \text{h m}^3 \text{ mm Hg/s kmol K} \tag{7-3}$$

$$\frac{V}{F} = (r+1)\frac{x_F - x_W}{x_D - x_W} \tag{7-4}$$

Strictly speaking, in Eq. (7-4), V (kmol /h) is the molar vapor quantity at the column head. This, however, is for almost equal molar heats of evaporation roughly the same as the molar heat quantity at any given rectifier section. With the feed at boiling temperature, V is also the molar quantity at any given stripper section.

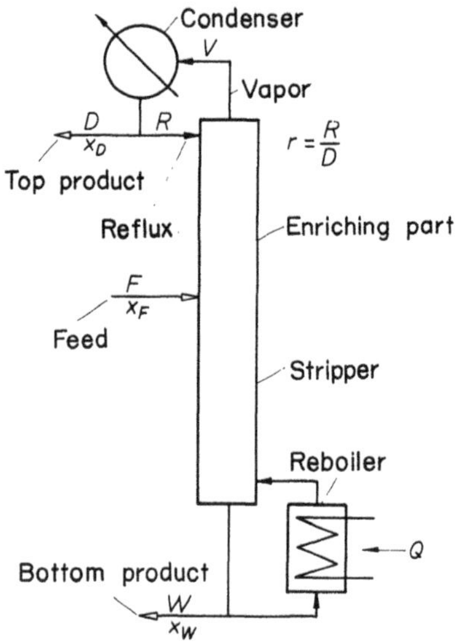

Fig. 7.1 Diagram of a continuous rectifier, and typical quantities and concentrations.

The calculation of V for different molar heats of evaporation and for different feed conditions is described in Chapter 3, and an example is given. Accordingly, for a given point in the rectifier, with y (mol-%) of the low-boiling component and the reflux ratio $r = r_D$ referred to the top of the column,

$$\frac{V}{F} = \frac{1 - \frac{x_D}{K}}{1 - \frac{y}{K}} (r + 1) \frac{x_F - x_W}{x_D - x_W} \tag{7-5}$$

Furthermore, for a given point of the stripper where: y = low boiler in vapor, (mol-%) y_i = coordinate of intersection (mol-%) (in y/x plot Fig. 7.2.) of the stripping line and the enriching line

$$\frac{V}{F} = \frac{1 - \frac{y_i}{K}}{1 - \frac{y}{K}} \left[(r + 1) \frac{x_F - x_W}{x_D - x_W} \frac{1 - \frac{x_D}{K}}{1 - \frac{y_i}{K}} + f_i - 1 \right] \tag{7-6}$$

In eqs. (7-5) and (7-6) K is the coefficient computed from the molar heats of evaporation by relations (3–11) while the factor f_i that characterizes the thermal condition of the feed is found from relations (7-3)–(7-3?

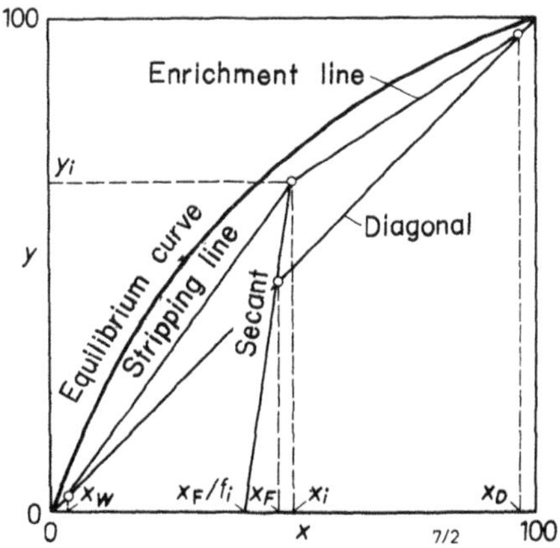

Fig. 7.2 Separation of a binary. Typical concentrations and operating lines. y/x plot; diagrammatic.

Starting from the equation of state for ideal gases, the specific weight ρ_V(kg/m^3) can be estimated from

$$\rho_v = \frac{\mu}{K_2} \frac{p}{T} \tag{7-7}$$

where:

μ = the molecular weight (kg/kmol)
K_2 = 62.3 m^3 mm Hg/kmolK and a constant (7-8)
p = pressure, mm Hg
T = absolute temperature, K.

Especially for a low vacuum at the column top, the vapor-load factor at constant column section may vary greatly along the column. Therefore, the enriching and stripping parts of the columns should be subdivided; the loads can be checked by means of Eq. (6-30). Also, a first approach may be obtained by using an average load factor, or its maximum to be on the safe side.

In Fig. 7.3., second phase, points on curve $u_V \rho_V^{1/2} = f(r)$ for f = constant represent

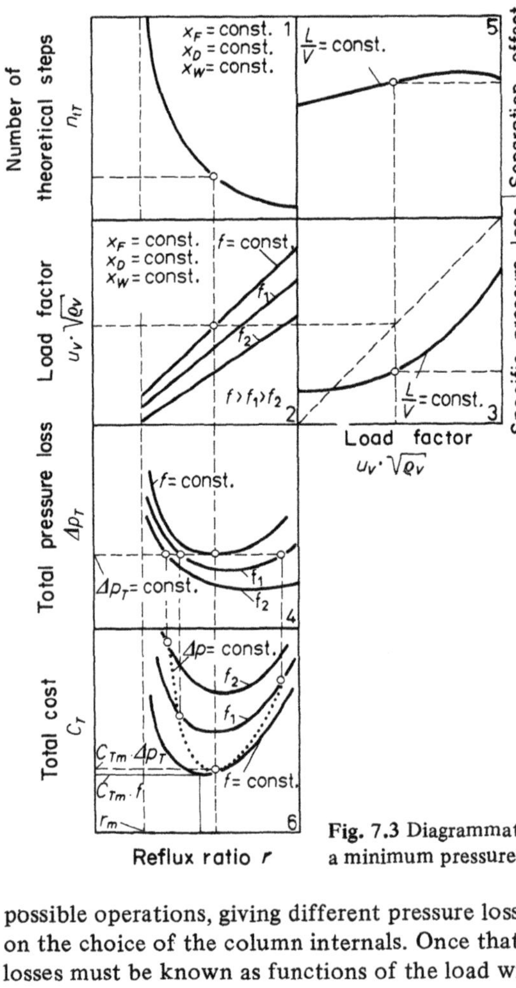

Fig. 7.3 Diagrammatic sequence of phases in the optimization for a minimum pressure loss rectification.

possible operations, giving different pressure losses for given conditions, depending on the choice of the column internals. Once that choice has been made, the pressure losses must be known as functions of the load within the pressure range in question, and for the liquid-vapor ratios incurred (third phase in Fig. 7.3). Typical curves are presented in Chapter 6, for L/V values of unity, in some cases for $r = 4$ (Fig. 6.42) and for $L/V = 0.8$, in accordance with Eq. (3-7) for enriching columns

$$L/V = \frac{r}{r+1} \tag{7-9}$$

In many rectifications, the molar L/V-values in the enriching and stripping parts of the column will differ from unity to the extent of −30 to +30 %. Moreover the specific pressure losses $\Delta p/n_t$ are widely independent of L/V for $u_V \rho_V^{1/2}$ (for many packings), the p/n_t values, that are valid for $L/V = 1$ will often be satisfactory; note the figures of Chapter 6.

Referring to Fig. 7.3 it can be seen that for different reflux ratios r the correlated values of the theoretical steps n_{tT} and of the load factors $u_V \rho_V^{1/2}$ for a specific feed load f can be found; also for the latter the corresponding specific pressure losses $\Delta p/n_t$ can be found. Then, by forming the products of the correlated n_{tT} and p/n_t the total pressure losses of the column, correlated to various reflux ratios r becomes

$$\Delta p_T = (\Delta p/n_t)\, n_{tT} \tag{7-10}$$

The fourth phase of Fig. 7.3. is a general view of that curve versus the reflux ratio for constant f-values. Since the total pressure loss increases for increasing theoretical steps and for an increasing reflux ratio, there is for any f-value a certain reflux ratio r and a number of stages n_{tT} that offer a minimum total pressure loss in rectification. The curves $\Delta p_T = f(r)$ will pass through a minimum Δp_{Tm}.

Referring to phase 5 in Fig. 7.3, the curves for the separation $n_t/H = f(u_V \rho_V^{1/2})$ determined by measurement, $E = f(u_V \rho_V^{1/2})$, and those of the theoretical number of steps n_{tT} corresponding to the minimum pressure loss Δp_{Tm} determined by optimization will supply the depth of packing

$$H = \frac{n_{tT}}{n_t/H} \tag{7-11}$$

or the actual plate number

$$n_p = \frac{n_{tT}}{E} \tag{7-12}$$

One advantage of using the specific feed load f is that the plates or packings of existing columns can be optimized through exchange, by which means the pressure drop is minimized. Ultimately, a correction must be provided in the number of plates or the packing height, and feedpoint location.

For a new plant, the admissible pressure at the column base is a deciding factor. If the use of a certain type of insert is mandatory, there exists, for a given f, a pair of

n_{tT} and r that yield a minimum total pressure loss Δp_{Tm}. The reflux ratio correlated to the minimum pressure loss of a column having a given specific feed rate will not necessarily agree with that giving the lowest overall cost at the same feed rate; note phase 6 in Fig. 7.3. Instead, it is a question of the selection of the construction material and the length of the amortization period allowed in the cost optimization.

To achieve the cost optimization for a new column, one determines, for various f-values, a family of curves $\Delta p_T = f(r)$ and the corresponding total cost lines $C_T = f(r)$. By drawing a section through the family of $p_T = f(r)$ curves from the point of highest admissible pressure loss Δp_T = constant, the intersects (for reflux r) with the curves $C_T = f(r)$ will supply the proper cost. The connecting line passing through these specific cost points yields the cost $C_T = f(r)$, which will have a minimum for a group of n_{tT}, r, and f. The reflux ratio that goes with this minimum cost is identical with that of the pressure loss minimum of the $\Delta p_T = f(r)$ curve valid for the same specific feed load f, but it will not in general coincide with the reflux ratio for the minimum cost of curve $C_T = f(r)$ for the same f value.

B. Quantitative Evaluation

The numerical evaluation by this method will be demonstrated by an example of a continuous rectification of a binary mixture.

> **Example 1** A continuous rectifier is to separate F = 10 kmol/h of a binary mixture having a mean relative volatility of α = 1.54 and a low boiler concentration x_F = 58.6 mol-% in such a way that the top product contains x_D = 98.9 mol-% and the bottom product only a residual x_W = 0.19 mol-% of the low boiler. The specific weight of the vapor at the column head is ρ_V = 0.0935 kg/m^3, while the other physical properties of the system coincide with those of ethyl benzene-styrene at approximately 17 mm Hg.
>
> **Answer** For clarity of exposition, the method of Fig. 7.3 will be applied, with some simplifications. The vapor load factor is assumed constant across the entire column length. The relative volatility is considered constant. It is assumed that the stripper and rectifier will have the same diameter.

Fig. 7.4 reveals how the number of theoretical steps and the vapor load factor vary for different specific rates and reflux ratios. Also, the dependence of the heat consumption on the feed quantity Q/F and the reflux ratio are shown. However, the latter can be found from Eq. (3–62) only after the pressure losses have been determined, since the pressures called for by the enthalpies must be known.

The total pressure loss is found readily if the $\Delta p/n_t$ curves for L/V = 1 may be considered as the median (lower part of Fig. 7.5). For the extremes of this example, r = 3 and r = 9, the L/V values in the enriching part are 0.75 and 0.9, and for the stripper they are 1.26 and 1.1.

Fig. 7.6. shows the total pressure losses as functions of the reflux ratio. All curves have minima, reflux ratios being the lower the higher the f-values, the column loads. Pall-ring columns show much lower overall pressure losses than plates under equal

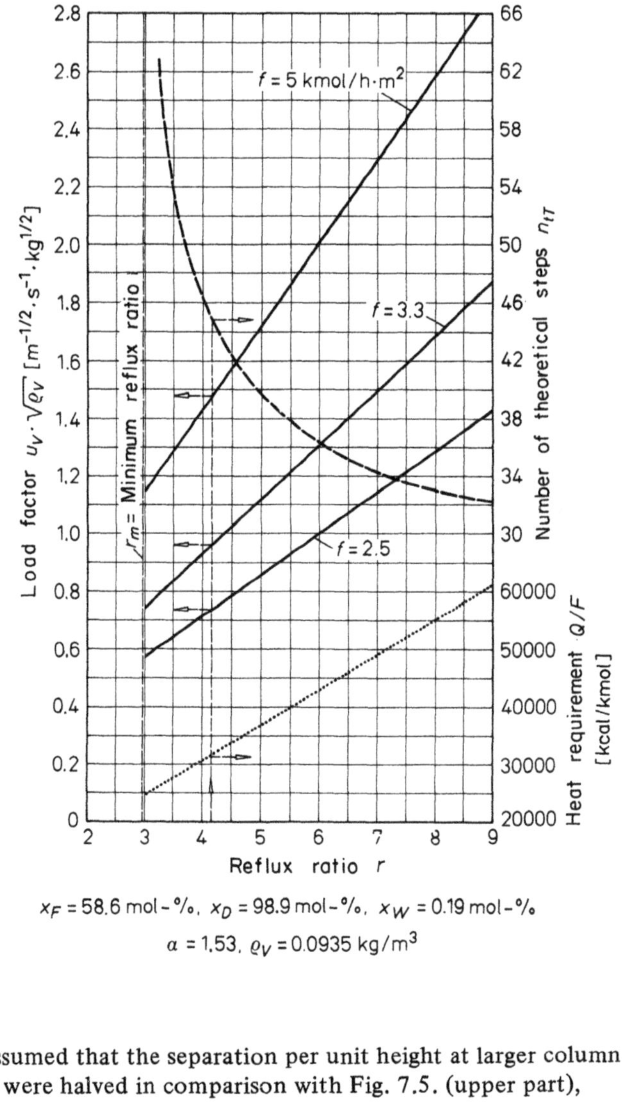

$x_F = 58.6$ mol-%, $x_D = 98.9$ mol-%, $x_W = 0.19$ mol-%

$\alpha = 1.53$, $\varrho_V = 0.0935$ kg/m^3

Fig. 7.4 Load factor, theoretical number of steps, and heat requirements versus the reflux ratio; note Example 1.

conditions. Even if it were assumed that the separation per unit height at larger column sizes (diameters and heights) were halved in comparison with Fig. 7.5. (upper part), the Pall ring would offer advantages in terms of pressure loss.

The calculation of the overall cost is based on a 5-yr amortization, a heating steam cost of 9 DM per ton of steam of 5 atm, with cooling water at 0.2 DM per ton.

The column cost was the average of recent years, as plotted in Fig. 7.7; the exchanger cost used was of the same period. Some of the calculations were also made for a tunnel-tray column in stainless construction; the cost of the columns and the heat exchanger cost used was of the same period. Some of the calculations were also made for a. ponding pressure losses are also plotted. The tiny squares in the cost curve indicate the cost minimum. For the carbon steel equipment the minimum overall cost is located at virtually the same reflux ratio as the minimum pressure loss; for stainless construction the minimum of cost and of pressure loss is correlated to almost the same reflux conditions at all loads.

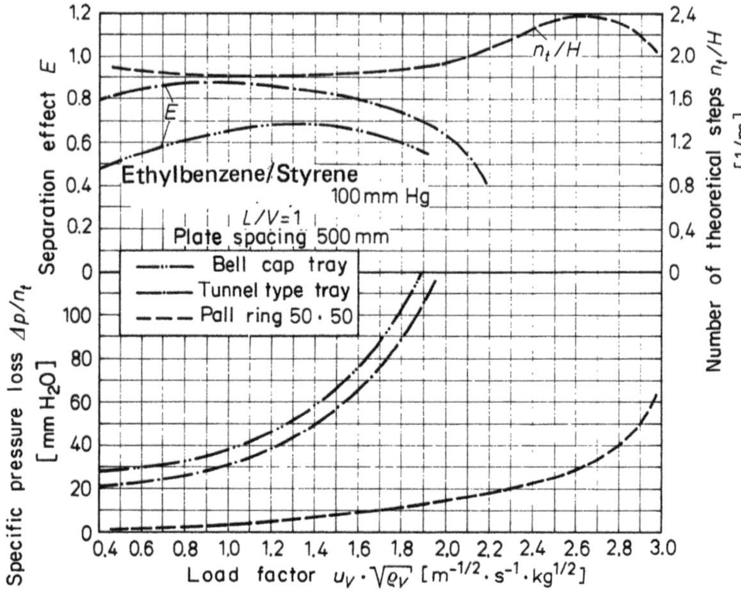

Fig. 7.5 Separation and pressure loss versus vapor load factor, for different column internals.

To determine the conditions under which minimum cost can be achieved with a given limiting pressure loss refer to Fig. 7.3. The proper cost curve is given for Δp_T = constant and Fig. 7.9 shows the example of a tunnel-type tray and two values for Δp_T. The results are almost the same for larger f-values and for Δp_T = constant.

Fig. 7.10 sums up the overall calculation results. The minimum pressure losses are plotted versus the specific feed loads, as are the correlated theoretical numbers of steps, reflux ratios, and end costs. At high specific feed loads, the minimum pressure losses are obtained at the same reflux ratio and hence at the same theoretical steps for all the equipment reviewed. At low specific feed loads, however, the minimum pressure loss of the column will yield more theoretical stages and hence a lower reflux ratio than for the column with internals causing a higher specific pressure loss (Fig. 7.5).

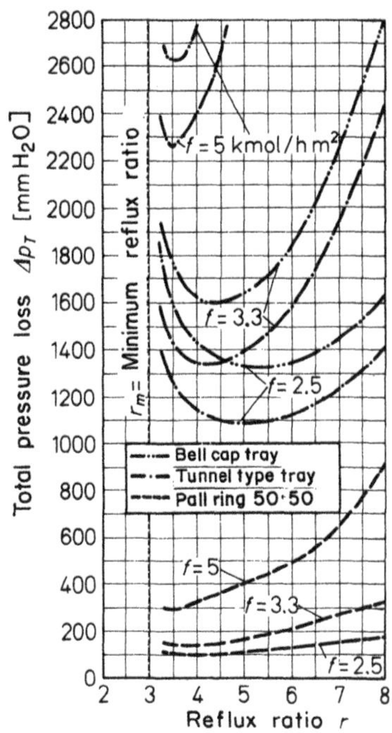

Fig 7.6 Total pressure loss versus the reflux ratio for different columns. Operating conditions of **Fig. 7.4** and Example 1.

Across the entire range of loads, the Pall-ring column offers advantages with regard to both pressure loss and investment. Note that the assumption of a constant relative volatility α simplifies calculations. However, when comparing towers of large specific pressure losses, this will falsify results if the column internals differ largely in pressure loss. In that event, many steps are required, and different internals give different averages of volatility α. Even for a single rectifier, the mean relative volatility may fluctuate through the optimization range because of the load dependence of the pressure loss. A test run with ethyl benzene–styrene at loads customary with valve-type columns will reveal that an α value of 1.57 at the column top will drop to 1.34 at the bottom. Such a variation will affect the correlation of steps and reflux and suggests design of the plates for minimum pressure loss per step. If optimization were based on a uniformly average volatility, the result would discourage the trend toward minimum pressure loss per step.

If the pressure loss in the separation is being limited, Fig. 7.10 will indicate the optima of the theoretical number of steps n_{tT} reflux ratio r, total cost C_T and specific feed load f; dotted lines and closed arrows apply to bell-cap plates, and open arrows apply to tunnel plates. By these f-values, the tower diameter d_S (m) becomes:

$$d_s = \left(\frac{4}{\pi}\frac{F}{f}\right)^{1/2} \tag{7-13}$$

Eqs. (7-2), (7-3), and (7-4) will give the corresponding vapor velocity u_V, Eq. (7-7) will supply the specific weight of the vapor, ρ_V, which in turn will yield the vapor-load factor $u_V\rho_V^{1/2}$; Fig. 7.5 will supply separation E, and from this and Eq. (7-12) there follows the number of actual plates, n_p.

Thus certain prerequisites must be met for an objective investigation of the interchange equipment in vacuum rectification.

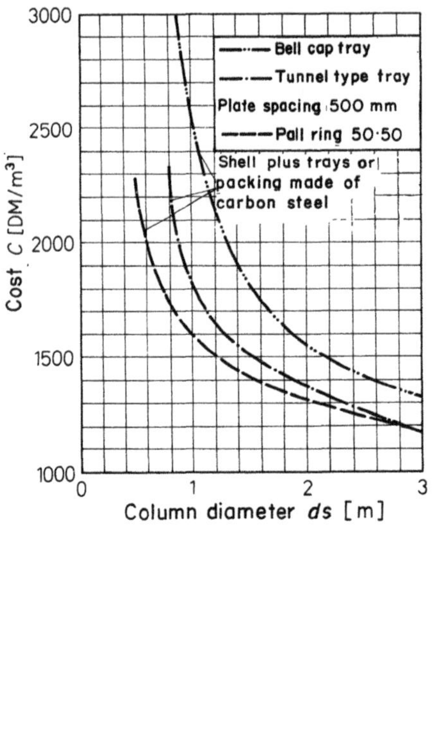

Fig 7.7 Cost per unit column volume versus column diameter, for different column types.

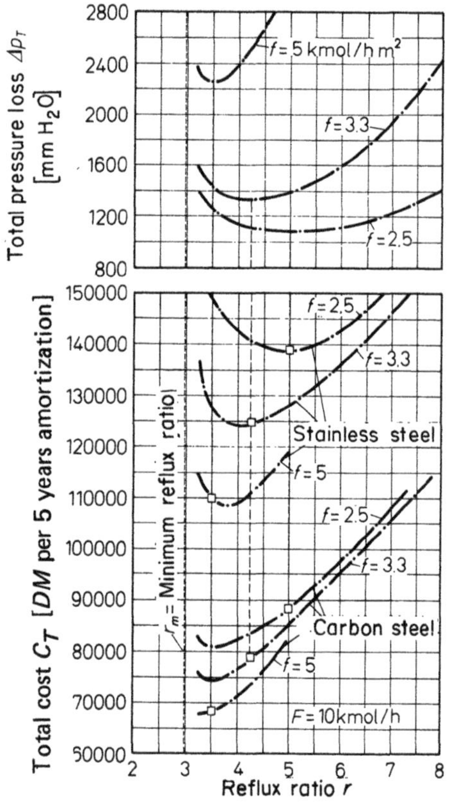

Fig. 7.8 Total pressure loss and total cost versus the reflux ratio for a tunnel type column. Operating conditions of Fig. 7.4 and Example 1.

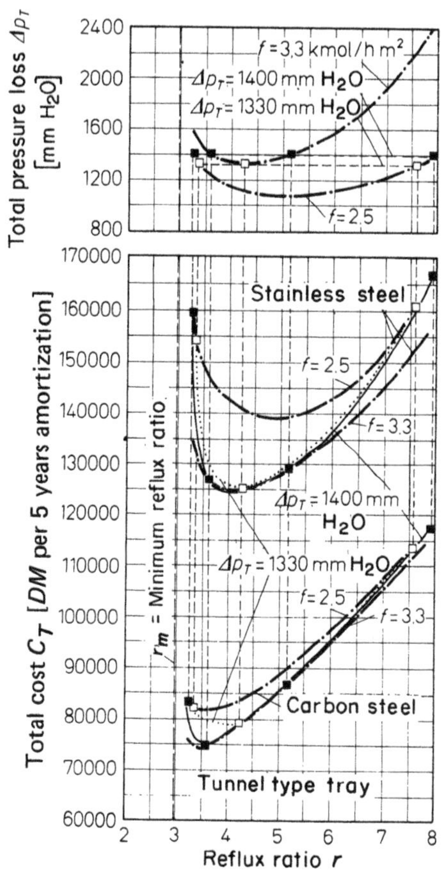

Fig. 7.9 Tunnel type tray for operating conditions of Fig. 7.4 and Example 1. Total pressure loss and total cost versus reflux ratio

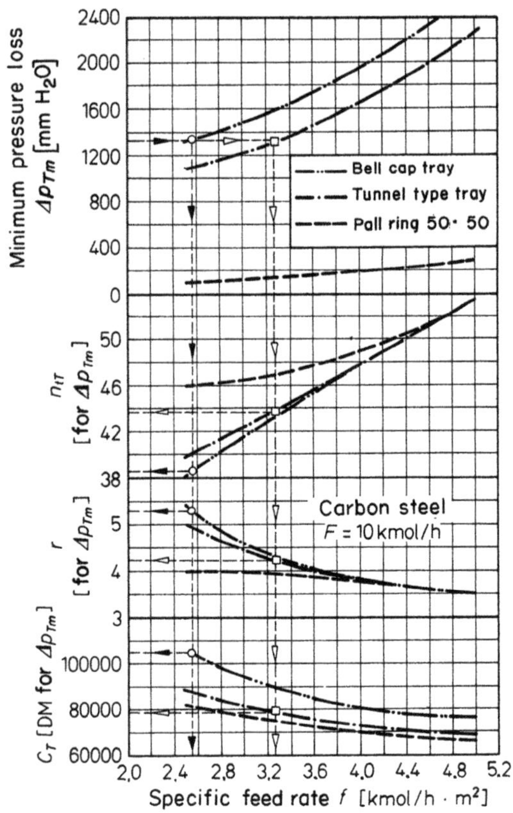

Fig. 7.10 Operating conditions of Fig. 7.4. and Example 1. Optimization of various minimum-pressure-loss columns.

2. OPTIMIZATION OF MULTIPLE COLUMNS

A rectification of liquid mixtures that become chemically unstable at elevated temperatures may become impracticable for a single column, inasmuch as the presence of many separating stages may cause considerable pressure loss and, in turn, the bottom temperature becomes excessive. In that event, multiple columns are employed. Furthermore, subdivision of a column becomes inevitable if high purity is demanded for the top and the bottom product; use of a single column would result in excessive height.

In arranging multiple columns, a variety of hookups is available, depending on design freedom and availability of heat. Some typical arrangements will be presented.

A. Duplex Columns: Number of Theoretical Steps, Reflux Ratios, and Vapor Loads

Fig. 7.11 gives two typical hookups of duplex columns a) and b). A binary mixture is to be separated into a top product of concentration x_D and a bottom product x_W, the concentration of the mixture being x_F. In hookup a), the top product of column I having a concentration x_T is the feed for column II; the bottom product of both columns is the high boiler, and the top product of column II is the low boiler. Both columns might also be regarded as forming one unit that could be so split up that equal numbers of steps would result for both columns [hookup b)].

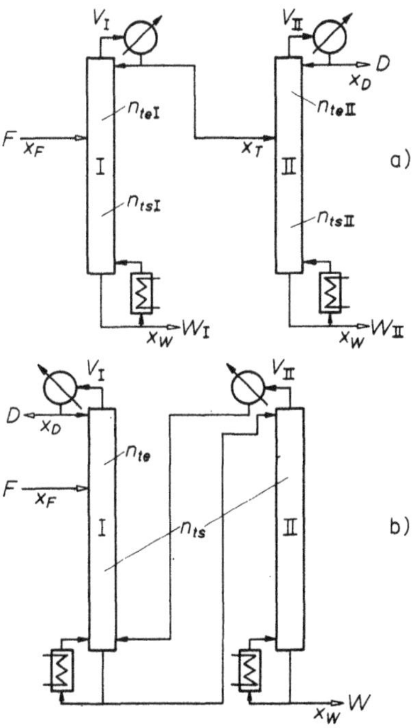

Fig. 7.11 Compound column hookup, diagrammatic.

The condenser at the top of column II and the vaporizer at the bottom of column I, hookup b), can be suppressed if the subdivision of the column is due to excessive column height rather than to pressure loss considerations. If the calculation yields a lesser number of steps for the enriching part than for the stripper, column I will assume the enriching and stripping function, and column II will be a stripper only. If the calculation yields more steps for the enriching part, column I becomes a rectifier, and column II becomes a stripper with an enriching section.

Fig. 7.12 shows the location of the operating lines for the two hookups, from which the number of theoretical steps is derived.[5] The lines will be labeled as follows:

n_{tsI} = number of theoretical steps in stripper of column I, hookup a)
n_{teI} = number of theoretical steps in rectifier of column I, hookup a)
n_{tsII} = number of theoretical steps in stripper of column II, hookup a)
n_{teII} = number of theoretical steps in rectifier of column II, hookup a)
n_{ts} = number of theoretical steps in stripper, hookup b)
n_{te} = number of theoretical steps in rectifier, hookup b)

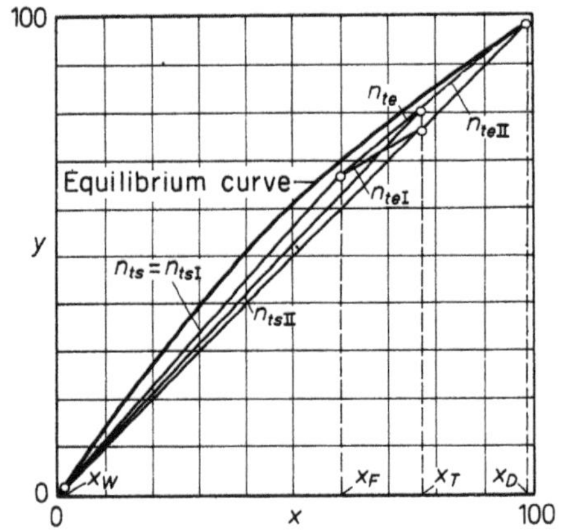

Fig. 7.12 Concentrations of Fig. 7.11 in y/x plot, and operating lines.

For convenience, identical reflux conditions will be assumed for the rectifier of column II, hookup a), and the rectifier of column I, hookup b):

$$r_{IIa} = r_{Ib} = r \tag{7-14}$$

From this follows the slopes of the rectifying lines. Furthermore, let the liquid/vapor ratio, and hence the slope of the stripping line of column I in hookup a) be identical with those for the stripper in hookup b):

$$n_{tsI} = n_{ts} \tag{7-15}$$

This fixes the location of the enriching line and the reflux ratio in column I, hookup a). The calculation of the vapor quantities follows.

Hookup a). For the vapor quantity V_I (kmol/h) in column I, referred to the unit feed quantity F (kmol/h)

$$\frac{V_I}{F} = \frac{x_F - x_W}{x_T - x_W}(r_{Ia} + 1) \tag{7-16}$$

and for vapor quantity V_{II} (kmol/h) in column II, referred to the unit feed quantity F (kmol/h)

$$\frac{V_{II}}{F} = \frac{x_F - x_W}{x_D - x_W}(r_{IIa} + 1) \tag{7-17}$$

Hence, the total vapor quantity referred to the unit feed quantity is

$$\sum \frac{V}{F} = \frac{V_I + V_{II}}{F} = \left(\frac{r_{Ia} + 1}{x_T - x_W} + \frac{r_{IIa} + 1}{x_D - x_W}\right)(x_F - x_W) \tag{7-18}$$

Hookup b): The molar vapor quantity V_I (kmol/h) in column I equals the molar vapor quantity V_{II} in column II, hence

$$\frac{V_I}{F} = \frac{V_{II}}{F} = \frac{x_F - x_W}{x_D - x_W}(r + 1) \tag{7-19}$$

In view of the validity of Eq. (7-14), Eq. (7-19) will yield the same result as Eq. (7-17); the vapor quantity V_{II} of hookup a) is the same as $V_I = V_{II}$ in hookup b). When pressure loss considerations enforce condensation and reevaporization, the total vapor per unit of feed becomes, for hookup b)

$$\sum \frac{V}{F} = \frac{V_I + V_{II}}{F} = (2)\frac{x_F - x_W}{x_D - x_W}(r + 1) \tag{7-20}$$

Example 2 A binary mixture, relative volatility $\alpha = 1.54$, is to be separated. Find the theoretical number of steps, vapor throughput, and reflux ration, for a low boiler concentration in the feed $x_F = 60$ mol-%, in the top product $x_D = 99$ mol-%, and bottoms $x_W = 0.5$ mol-%.

Since the mixture is thermally unstable, a duplex column is proposed, along with hookups a) and b) of Fig. 7.11. Calculations for hookup a) consider four cases: 70, 77, 83.5, and 90 mol-% of the low boiler in the intermediate product at the top of column I, which is the feed to column II. The molar vapor quantities in the column are virtually constant because of equal molar heats of vaporization of the individual components. The feed entering the column is at its boiling point.

Answer Equations (7-14) and (7-15) are considered applicable. They give, for the theoretical steps and vapor quantities in the column parts, the values in Fig. 7.13 as functions of the reflux ratio $r = r_{IIa} = r_{Ib}$. For hookup a), in view of the operating line in Fig. 7.12, the number of theoretical steps n_{teI}

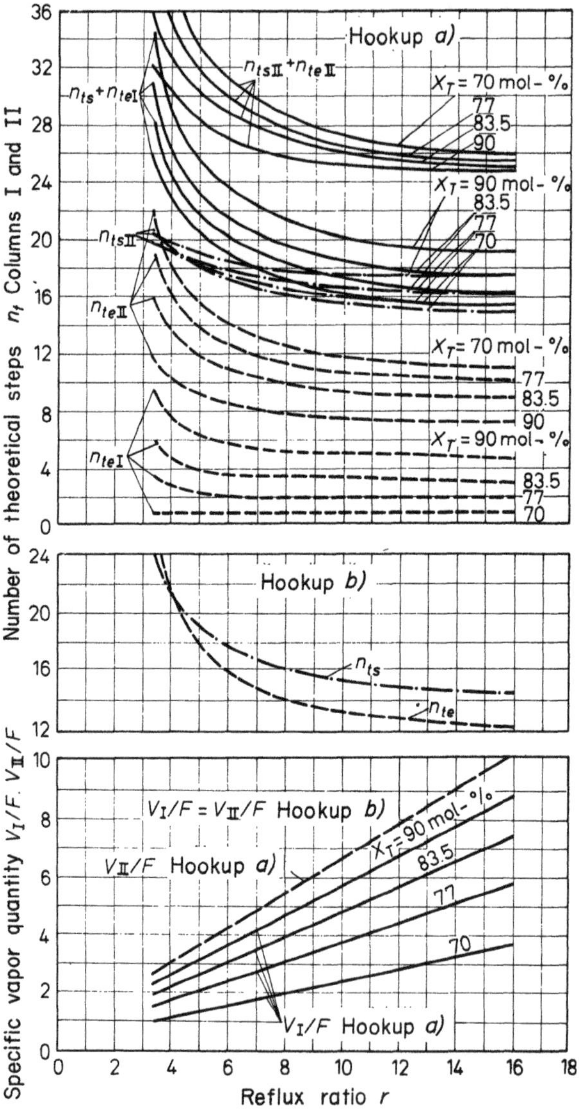

Fig. 7.13 Theoretical steps and vapor quantities vs. the reflux ratio of Fig. 7.11, hookups *a*) and *b*). Operating conditions of Example 2.

in the rectifier of column I, and in the stripper of column II n_{tsII} will be larger, and those in the rectifier of column II will be less as concentration x_T at the top of column I rises. Also, for greater x_T values the vapor load V_I/F in column I must be greater than at the lower concentrations x_T.

The upper portion of Fig. 7.14 compares the total number of steps for both hookups. The number of theoretical steps is always larger for hookup a) than for b).

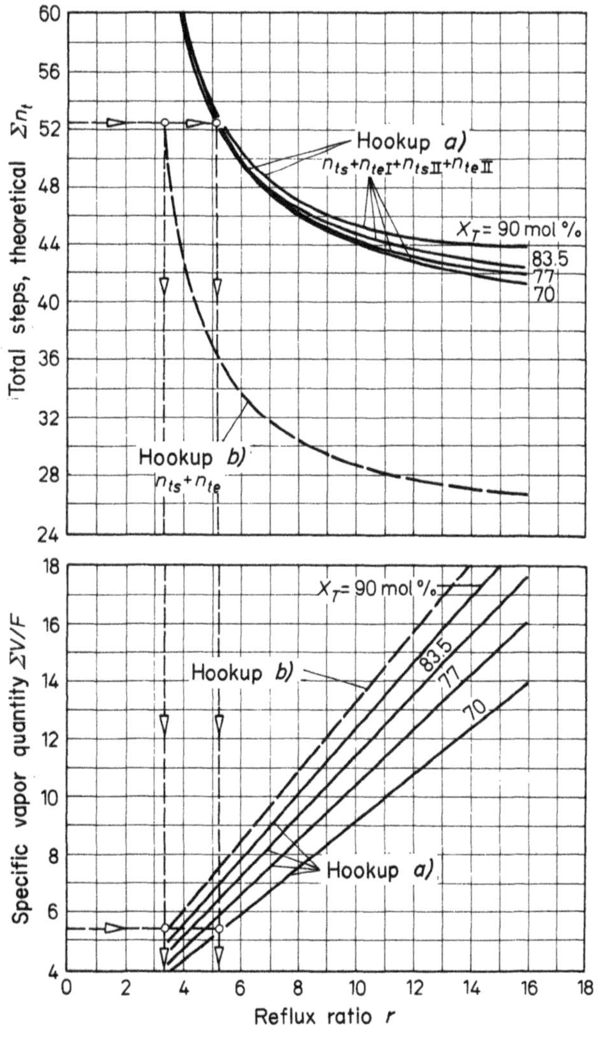

Fig. 7.14 Total theoretical steps and vapor quantities versus the reflux ratios. Hookups a) and b) of Fig. 7.11. Operation of Fig. 7.13.

The lower portion of Fig. 7.14 presents the total vapor quantities versus the reflux ratio. Hookup a) demands less vapor than b) if, for the latter, intercondensation and reevaporation become mandatory. If hookup b) requires a column subdivision only because of height limitation, so that intercondensation and reevaporation can be waived, the ordinate values in Fig. 7.14 (dotted lines) are halved. In that event, hookup b) would be more favorable than a) from the viewpoint of vapor quantity as well.

By plotting, for constant reflux, the total vapor quantities from the lower portion of Fig. 7.14 versus the corresponding values (upper portion of Fig. 7.14) of the total theoretical steps, Fig. 7.15 results. It presents the correlation of energy expense and

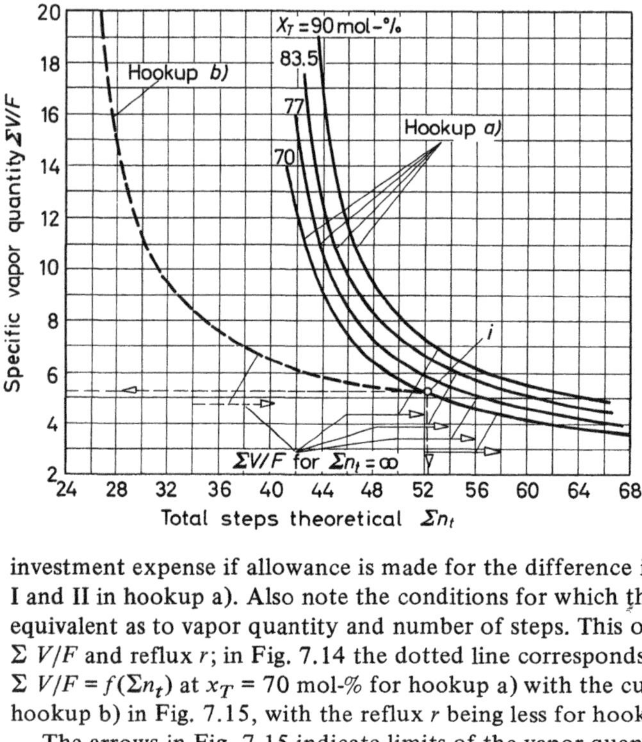

Fig. 7.15 Total vapor quantities versus theoretical steps. Hookups of Fig. 7.11., operation of Fig. 7.13.

investment expense if allowance is made for the difference in diameters of columns I and II in hookup a). Also note the conditions for which the two hookups are equivalent as to vapor quantity and number of steps. This occurs at low vapor loads $\Sigma V/F$ and reflux r; in Fig. 7.14 the dotted line corresponds to intersection i of curve $\Sigma V/F = f(\Sigma n_t)$ at $x_T = 70$ mol-% for hookup a) with the curve $\Sigma V/F = f(\Sigma n_t)$ for hookup b) in Fig. 7.15, with the reflux r being less for hookup b) than for a).

The arrows in Fig. 7.15 indicate limits of the vapor quantities that would give an infinity of steps because of the approach to the minimum reflux r_m.

It is clear that rectification in duplex columns in series will always be more costly than working in a single column, as far as energy demand and equipment are concerned. A different picture might emerge when the two columns can operate at different pressures, so that the overhead vapors from the column of higher pressure would heat the reboiler of the column of lesser pressure.

By means of a cost analysis, the optimum operating and planning conditions can be found for both hookups. If in the example above the cost were more favorable for fewer steps, hookup b) would be preferable. For the latter the cost optimization for minimum pressure loss might be handled as shown in Section 1. of this Chapter. Both columns could be regarded as one unit, and the total pressure loss for optimum conditions and the steps could be distributed over both columns. For hookup a), however, the optimizing procedure would be applied for different concentrations x_T individually, and the results would be summarized.

B. Vapor Loads of Multiple Columns Subdivided Because of Limited Height

In the planning for increased capacity sometimes the question arises how available columns can be reused for larger production. Frequently, this new demand can be met by the judicious addition, to the existing installation, of a suitable column which is no longer in operation. If the number of steps of available towers is adequate, that arrangement is optimum which corresponds to a minimum of the sum of the vapor loads of the individual columns.[5]

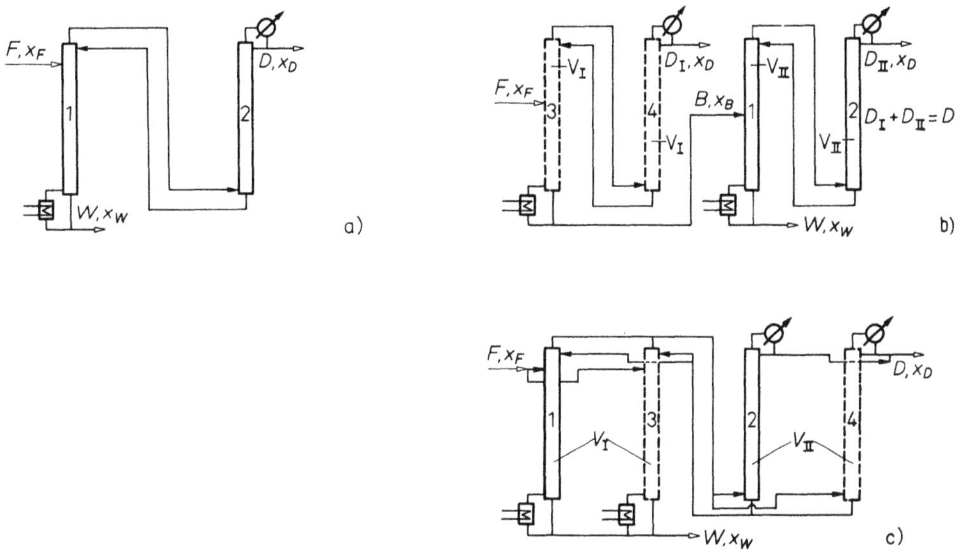

Fig. 7.16 Hookup schemes in subdividing columns to reduce height.

Hookup a). The question of best arrangement of auxiliary columns might arise in a rectification performed in columns 1 and 2, placed in series without intermediate condensation as diagrammatically shown in Fig. 7.16, hookup a).

With columns 1 and 2 no longer able to handle the throughput, two available spare columns 3 and 4 with enough stages to relieve the condition are to be added. These may be arranged in series, b), or in parallel, c). A synopsis will reveal which of the two offers advantages as to vapor load and heat consumption.[5,8] It is supposed that the column feed streams are preheated to boiling, and that the molar quantities are constant.

Hookup b). The binary mixture F (kmol/h) with x_F (mol-%) of the low boiler is first rectified in the new columns 3 and 4 arranged in series without intermediate condensation; part of the top product D_I (kmol/h) reaches the top concentration x_D (mol-%). The bottom-product of column 3, section, B, reaches the proper concentration x_B (mol-%) for columns 1 and 2 arranged in series without intercondensation. Columns 1 and 2 then achieve further separation into the partial quantity D_{II} (kmol/h) of the top concentration x_D (mol-%) and the main product W (kmol/h) of the desired end concentration x_W(mol-%). For the partial product D_I (kmol/h) per unit of feed F (kmol/h) we must have

$$\frac{D_I}{F} = \frac{x_F - x_B}{x_D - x_B} \tag{7-21}$$

The quantity of vapor V_I (kmol/h) per unit of feed F (kmol/h) in columns 3 and 4 becomes

$$\frac{V_I}{F} = \frac{D_I}{F}(r_I + 1) \tag{7-22}$$

and

$$\frac{V_I}{F} = \frac{x_F - x_B}{x_D - x_B}(r_I + 1) \tag{7-23}$$

where the reflux ratio is

$$r_I = \frac{V_I - D_I}{D_I} \tag{7-24}$$

The partial top product quantity D_{II} (kmol/h) per unit of feed F (mol/h) for column 2 becomes

$$\frac{D_{II}}{F} = \frac{B}{F}\frac{x_B - x_W}{x_D - x_W} \tag{7-25}$$

where

$$\frac{B}{F} = 1 - \frac{D_I}{F} \tag{7-26}$$

the quantity of bottom product B (kmol/h) per unit of feed F (kmol/h) for column 3, which by Eq. (7-21) then becomes

$$\frac{B}{F} = \frac{x_D - x_F}{x_D - x_B} \tag{7-27}$$

Eq. (7-25) now reads

$$\frac{D_{II}}{F} = \frac{x_D - x_F}{x_D - x_B} \frac{x_B - x_W}{x_D - x_W} \tag{7-28}$$

Vapor quantity V_{II} based on feed F (kmol/h) for columns 1 and 2 follows from

$$\frac{V_{II}}{F} \quad \frac{D_{II}}{F}(r_{II} + 1) \tag{7-29}$$

by

$$\frac{V_{II}}{F} = \frac{x_D - x_F}{x_D - x_B} \frac{x_B - x_W}{x_D - x_W}(r_{II} + 1) \tag{7-30}$$

with the reflux ratio

$$r_{II} = \frac{V_{II} - D_{II}}{D_{II}} \tag{7-31}$$

The total vapor quantity becomes

$$\sum \frac{V}{F} = \frac{V_I + V_{II}}{F} = \frac{x_F - x_B}{x_D - x_B}(r_I + 1) + \frac{x_D - x_F}{x_D - x_B} \frac{x_B - x_W}{x_D - x_W}(r_{II} + 1) \tag{7-32}$$

Hookup c). The replacement columns 3 and 4 are arranged in parallel with main columns 1 and 2. The calculations for the two column pairs without intercondensation are identical, as far as the vapor quantity V (kmol/h) is concerned, with those for the original hookup a).

The total quantity of the top product, D (kmol/h), per feed quantity F (kmol/h) comprising the partial amounts of columns 2 and 4, then becomes

$$\frac{D}{F} = \frac{x_F - x_W}{x_D - x_W} \tag{7-33}$$

and

$$\frac{V}{F} = \frac{x_F - x_W}{x_D - x_W}(r + 1) \tag{7-34}$$

Concentration x_B will be located between x_F and x_W

$$x_F > x_B > x_W \tag{7-35}$$

and a correlation of Eqs. (7-32) and (7-34) will show that under the limiting condition $x_B = x_F$ the first term in Eq. (7-32), the vapor quantity, becomes $V_I = 0$, and Eq. (7-32) merges with Eq. (7-34) ($r_{II} = r$). Under the limitation $x_B = x_W$ the second term of Eq. (7-32), vapor quantity $V_{II} = 0$, and Eq. (7-32) merges with Eq. (7-34) ($r_I = r$).

For x_B within limits x_F and x_W, Eq. (7-32) will always supply larger figures than Eq. (7-34). Viewed from the vapor load, the parallel arrangement is preferable to the series arrangement.

A numerical example will explain the difference between hookups b) and c).

Example 3 A mixture of meta-/paradiisopropyl benzene is to be rectified continuously at ordinary pressures. There is a choice between column system b) and c) of Fig. 7.16. The desiderata are: $x_F = 66$ mol-%, $x_D = 99.6$ mol-%, $x_W = 0.3$ mol-% metadiisopropyl benzene. To secure the same basis of reference for the two hookups, the minimum reflux ratios r_{Im}, r_{IIm}, and r_m shall be used, and the correlated vapor quantities V_{Im}/F, V_{IIm}/F, and V_{IIm}/F shall be computed. Find the best distribution of the vapor load.

Answer By the method outlined in Section 3 of Chapter 3, find for $r_{Im} = r_m = 8.3$, the minimum reflux with a series hookup (Fig. 7.17, upper part). From

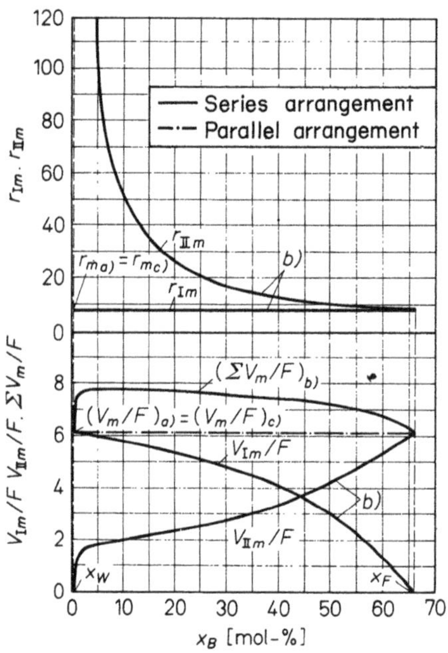

Fig. 7.17 Minimum reflux ratios and correlated vapor loads for hookups of Fig. 7.16. Note conditions of Example 3.

the lower part of Fig. 7.17, find the corresponding partial vapor loads, by Eq. (7-23) and (7-30), and their sums by Eqs. (7-32) versus the bottoms concentrations x_B. The dash-dotted line gives the value of the total vapor quantity for a parallel hookup by Eq. (7-34). It is seen that the series hookup is always inferior to the parallel hookup, with the exception of the two limit conditions for $x_B = x_F$ and $x_B = x_W$, for which one pair of columns becomes useless while the other carries an extra specific load in terms of overall vapor. With parallel hookup, however, each pair of columns can be operated with half the molar overall vapor load.

C. Optimizing Duplex Rectifiers for Minimum Pressure Loss

It has been stated that a rectification of thermally unstable mixtures requires subdivision of columns, with intercondensation and reevaporation, in the event that the total pressure loss of the vapor in one column only would cause an excessive rise in bottoms temperature. Optimum operation of subdivided columns can be calculated for minimum pressure loss conditions. In the continuous rectification of binary systems, several different column hookups may be considered.

Hookup a). In Fig. 7.18, let partial columns I and II have equal diameters and equal molar vapor loads. That situation results if the calculated optimum-pressure-loss column of constant diameter in the rectifying and stripping sections is, for example cut in half.

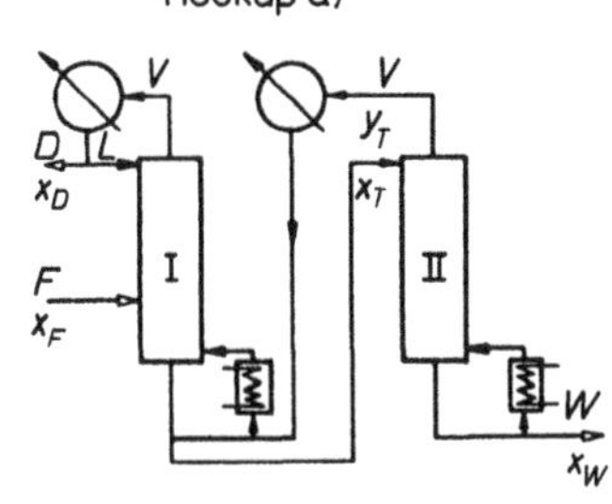

Fig. 7.18 Column hookup for divided height and intercondensation.

For the hookup in Fig. 7.18, the characteristic concentrations in the separation of a binary mixture are presented diagrammatically in the y/x plot (Fig. 7.19). The column subdivision required on the grounds of pressure loss, involving vapor intercondensation, gives at the breaking point, the pair of concentrations of y_T (mol-%) in the vapor and x_T (mol-%) in the liquid. The calculation of this hookup is the same as the one in Section 1 of this Chapter, except that presently the computed total pressure loss and the plate numbers are split up and shared by columns I and II.

Hookup b). Both columns, I and II, are fitted with enriching and stripping sections (Fig. 7.20). In that scheme, column sizes and vapor quantities of both columns may differ, depending on the operation.

In column I, feed F (kmol/h) with x_F (mol-%) is separated into the top product D (kmol/h) with x_D (mol-%) and the bottom product B (kmol/h) with x_B (mol-%). The

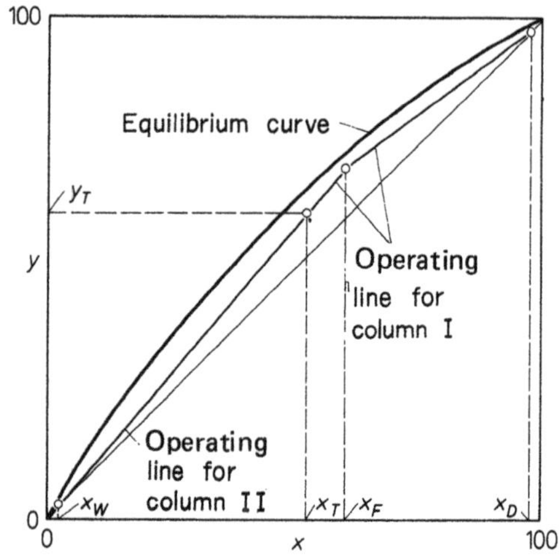

Fig. 7.19 Column hookup of Fig. 7.18. Typical concentrations and operating lines in y/x plot for separation of a binary.

latter is the feed for column II, from which the runoff W (kmol/h) with x_W (mol-%) and the top product T (kmol/h) with x_T (mol-%) is withdrawn. The latter is added to the column reflux at the point of equal concentration. V_I (kmol/h) and V_{II} (kmol/h) are the overhead vapor quantities, and L_I and L_{II} are the correlated reflux quantities.

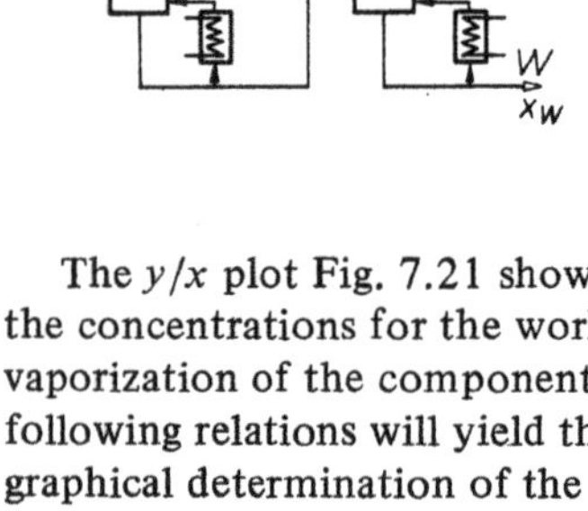

Fig. 7.20 Two-stage hookup with intercondensation.

The y/x plot Fig. 7.21 shows qualitatively the location of the operating lines and the concentrations for the working of a binary mixture. For virtually equal heats of vaporization of the components, and assuming the feed at a boiling temperature, the following relations will yield the slopes $\tan \beta$ of the interchange lines 1 to 5 for graphical determination of the number of steps:

$$\text{Column I, Enrichment line 1, } \tan \beta_1 = \frac{L_I}{V_I} = \frac{r_I}{r_I + 1} \qquad (7\text{-}36)$$

Operating line, 2 $\tan \beta_2 = \dfrac{L_I + F}{V_I}$ (7-37)

Stripping line 3, $\tan \beta_3 = \dfrac{L_I + F + T}{V_I}$ (7-38)

Column 2, Enrichment line 4, $\tan \beta_4 = \dfrac{L_{II}}{V_{II}} = \dfrac{r_{II}}{r_{II} + 1}$ (7-39)

Stripping line 5, $\tan \beta_5 = \dfrac{L_{II} + B}{V_{II}}$ (7-40)

Mass balances will demonstrate that the extension of the operating line 2 in Fig. 7.21 must pass through the intersection of the diagonal and the vertical through x_W.

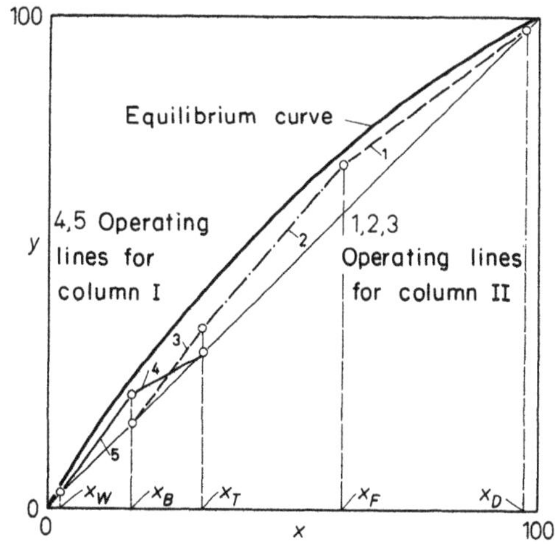

Fig. 7.21 Column hookup of Fig. 7.20. Typical concentrations and operating lines in y/x plot for separation of a binary.

The load, pressure loss, heat requirement, and number of steps in column II will be controlled by the top concentration, x_T (mol-%), and hence by the overhead quantity T (kmol/h) which is

$$T = W \frac{x_B - x_W}{x_T - x_B} \tag{7-41}$$

Furthermore, according to the hookup in Fig. 7.20 the following mass balances must be satisfied.

$$B = T + W \tag{7-42}$$

$$F = D + W \tag{7-43}$$

$$D = F \frac{x_F - x_W}{x_D - x_W} \tag{7-44}$$

Analogous to the definition of the specific feed load, Eq. (7-1), the specific runoff load

$$w^* = \frac{W}{d_{sII}^2 \frac{\pi}{4}} \tag{7-45}$$

may serve to visualize the condition of column II. For a given w^* a load diagram may be devised in which, for example, the top of column II, and for various reflux conditions, r_{II}, the vapor load factor $u_{VII}\,\rho_V^{1/2}$ is plotted versus the concentration x_T. u_{VII} may be found from the mass and concentration balances

$$u_{VII} = K_1\, w^* \frac{V_{II}}{W} \frac{T}{p} \tag{7-46}$$

and

$$\frac{V_{II}}{W} = (r_{II} + 1) \frac{x_B - x_W}{x_T - x_B} \tag{7-47}$$

Constant K_1 follows from Eq. (7-3), and the specific weight of the vapor ρ_V can be found from Eq. (7-7).

However, u_{VII} may also be determined by the method in Section 1 of this chapter, if the corresponding values for column II are inserted into Eqs. (7-2); thus $F = B$, $x_F = x_B$, and $x_D = x_T$.

For a given column section, greater column-top concentration can be achieved with a given reflux ratio and otherwise constant conditions only through an increased number of steps, with reduced heat requirement and hence less vapor load and pressure loss per theoretical step. For a given reflux ratio, r_{II}, there is a number of steps, n_{tII}, and a feasible top concentration x_T yielding a minimum overall pressure loss. In other words, for each top concentration there is a definite reflux ratio and a definite number of steps, together with a minimum total pressure loss. The location of the minimum is a function of the load dependence of the pressure loss per theoretical step, as discussed in Chapter 6 for different plates, packings, and special designs.

Hookups c1 and c2). In hookups a) and b), column II was to perform as a stripper only.

If the conditions of Example 2 are applied to hookup b) (Fig. 7.20), it will be found that column II must function as a stripper only; this follows from the use of the given α, x_F, x_D, and x_W values, and the requirement of least pressure loss in the column. That mode of operation is always given for hookup a) if the column division is located below the feedpoint. In view of the data for hookup a) the diameters of columns I and II would be the same. For hookup b) that restriction will not exist.

A summary will follow for the instances in which columns II must be run as strippers only. Fig. 7.22 gives flowsheets, and Fig. 7.23 shows concentrations and operating lines.

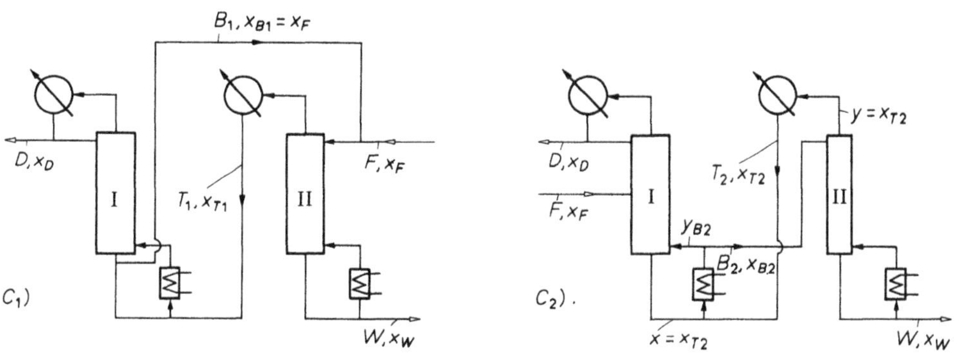

Fig. 7.22 Two-stage hookups for straight stripper function of a secondary column.—C1) Straight enriching function, the primary and secondary columns having same diameter. C2) Primary column combines enriching and stripping function, primary and secondary columns have different diameters.

In hookup c_1), diameters d_{SI} and d_{SII} of columns I and II are identical; in hookup c) the diameter of column II is less than that of column I. For comparison it will be assumed that the number I columns in both hookups, c_1) and c_2), have roughly the same diameter.

Mass balances establish the relationship of diameters d_{SI} and d_{SII} as follows: According to Eq. (7-1),

$$F = f d_{SI}^2 \frac{\pi}{4} \tag{7-48}$$

Further, from Eqs. (7-43) and (7-44)

$$W = F\frac{x_D - x_F}{x_D - x_W} \tag{7-49}$$

With this, a combination of Eqs. (7-45), (7-48), and (7-49) yields

$$d_{SI} = d_{SII}\left(\frac{w^*}{f}\frac{x_D - x_W}{x_D - x_F}\right)^{1/2} \tag{7-50}$$

For $d_{SI} = d_{SII}$ there follows

$$(f)_{d_{SI} = d_{SII}} = f' = w^*\frac{x_D - x_W}{x_D - x_F} = w^*\frac{F}{W} = \frac{F}{d_{SII}^2\frac{\pi}{4}} \tag{7-51}$$

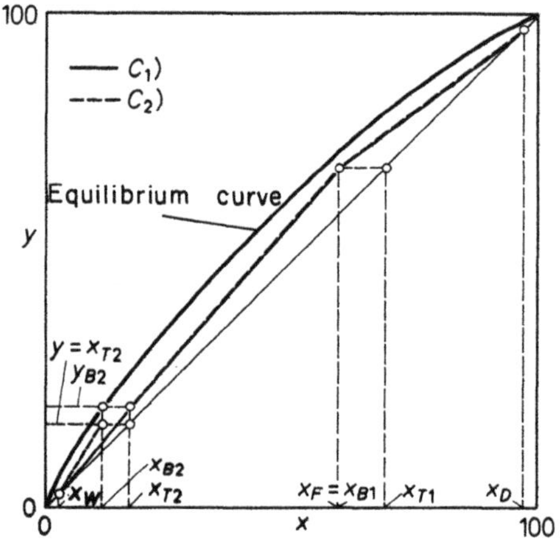

Fig. 7.23 Column hookups of Fig. 7.22. Typical concentrations and operating lines for the separation of a binary.

Example 4 Hookups a) and b) are evaluated, and hookups c_1) and c_2) are compared for the data of Example 2: x_F = 60 mol-%, x_D = 99 mol-%, x_W = 0.5 mol-%, and α = 1.54; other data is from ethyl benzene/styrene at 17mm Hg, note Example 1. The molar heats of evaporation are assumed equal, and the curves of Fig. 7.5, resulting from the rectification of the above mixture, are to apply. Calculations should be based on columns fitted with bell-cap plates, and columns packed with 50-mm Pall rings.

Answer Results for hookup a): Referring back to Sec. 1, Fig. 7.24 shows the typical curves for the total pressure loss $\Delta p_T = \Delta p_I + \Delta p_{II}$ versus different

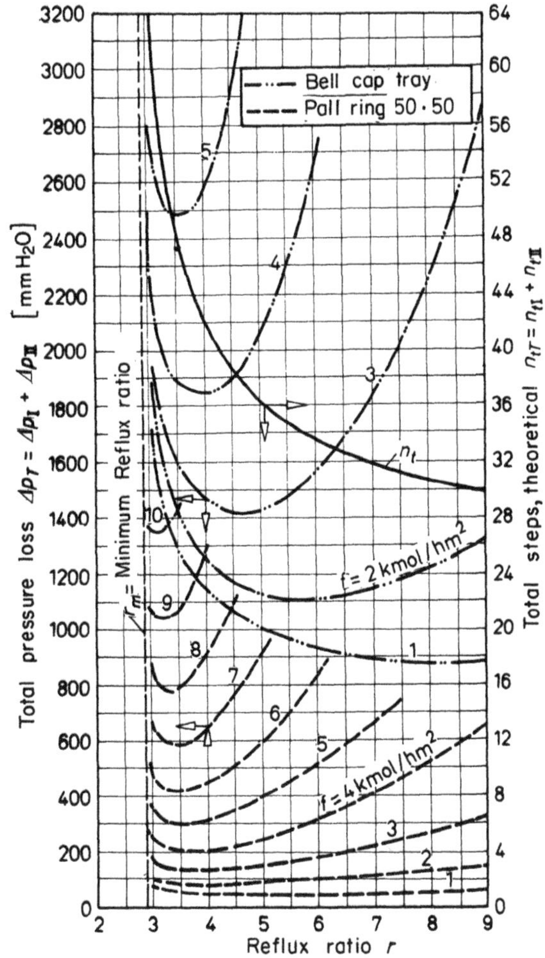

Fig. 7.24 Hookup of Fig. 7.18. Total pressure loss and theoretical steps versus the reflux ratio at various specific feed loads. Conditions of Example 4.

specific feed loads f and for the total number of theoretical steps $n_{tT} = n_{tI} + n_{tII}$ versus reflux ratio r. For a discussion and summary, note Sec. 1 of this Chapter.

Results for hookup b). Referring to the original data, let a concentration $x_B = 11.8$ mol-% with concentration x_T varying, and verify operating conditions. According to the optimization procedure discussed in Section 1 of this Chapter, first the number of theoretical steps for column I is determined as a function of the reflux ratio (Fig. 7.25). It is seen that there is only a slight dependence of the number of steps of concentration x_T within the range in question—15 mol-% $\leqslant x_T \leqslant$ 50 mol-%.

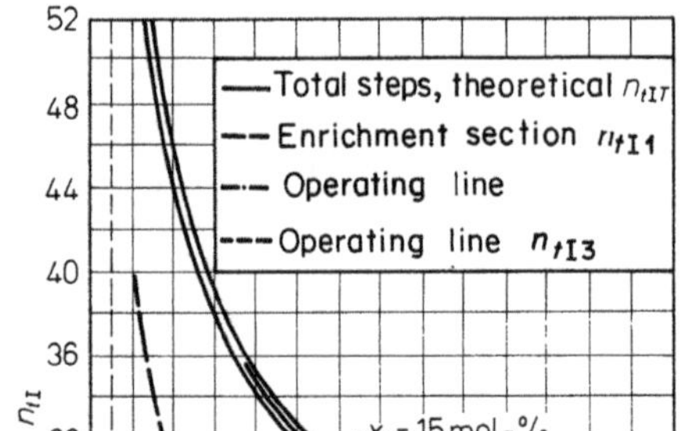

Fig. 7.25 Column I of Fig. 7.20. Number of theoretical steps versus the reflux ratio. Conditions of Example 4 with variations.

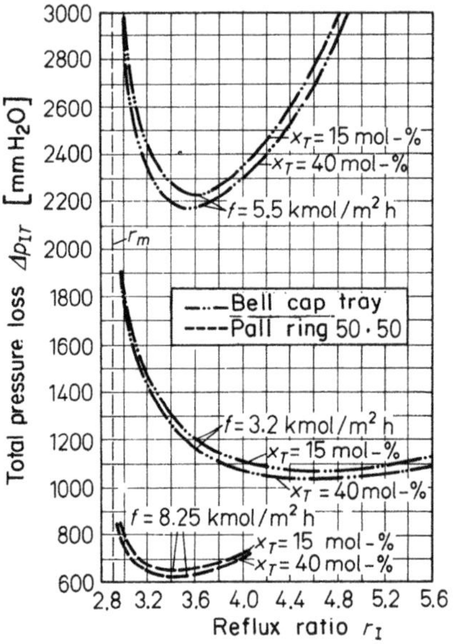

Fig. 7.26 Column I of Fig. 7.20 Total pressure loss versus reflux ratio, for various specific feed loads and top concentrations in column II. Conditions of Fig. 7.25.

Again, the following steps are as outlined in Sec. 1 of this Chapter. Thus Fig. 7.26 shows the total pressure loss p_{IT} of column I versus reflux ratio r_I for various concentrations x_T and several quantitative ratios f, for bell-cap plates and for 50-mm Pall rings. These calculations are based on the curves of Fig. 7.5. Note the small effect of concentration x_T upon the pressure loss.

A calculation of the feed quantity referred to the heat requirement Q_I/F of column I, based on heat balances, shows that here again the dependence upon x_T is slight.

Fig. 7.27 indicates how concentration x_T affects quantities B and T if at a constant feed F = 1 kmol/h concentration x_B = 11.8 mol-%.

Fig. 7.28 reveals the number of theoretical steps in the rectifier and stripper of column II for the same range of top concentrations, 12 mol-% $\leqslant x_T \leqslant$ 60 mol-%, and identical feed and drain concentrations.

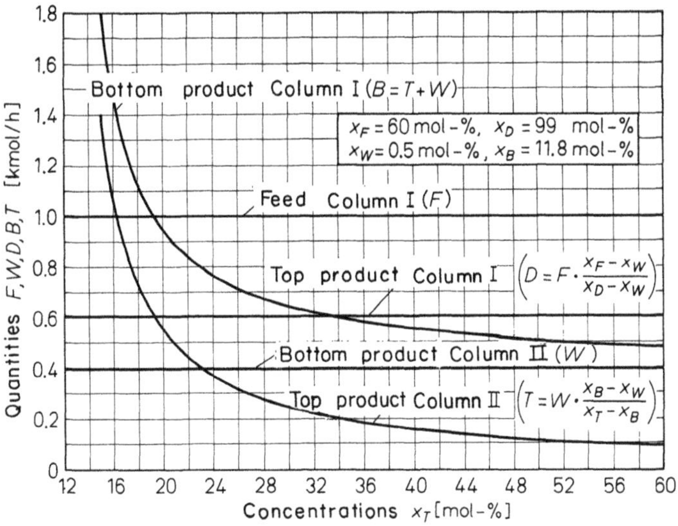

Fig. 7.27 Hookup of Fig. 7.20, conditions of Example 4. Quantities versus concentrations.

Fig. 7.29 shows the number of steps versus the reflux ratio.

Constant molar heats of evaporation were always assumed. It is seen that as the necessary steps in column II increase, the reflux ratio becomes less. Hereafter, the

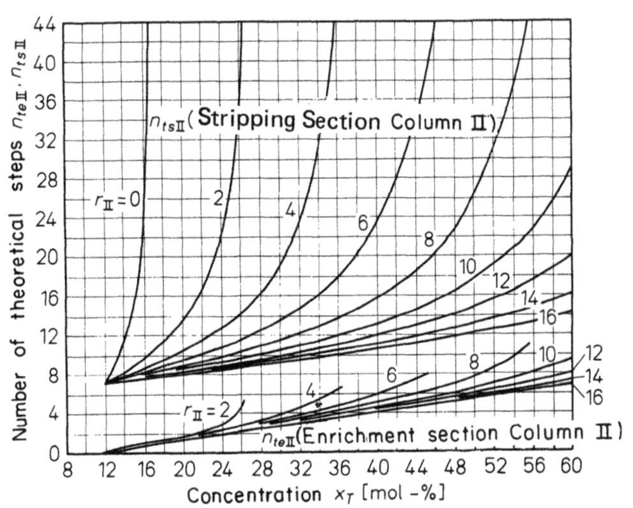

Fig. 7.28 Column II of hookup Fig. 7.20. Theoretical steps in the stripper and the rectifier versus top concentration of column II. Conditions of Example 4.

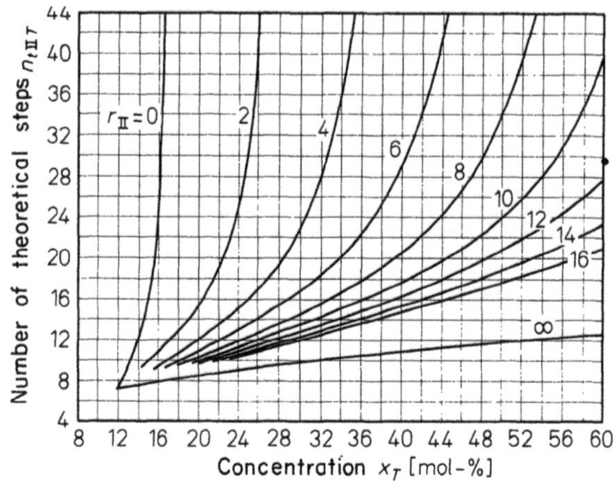

Fig. 7.29 Column II of hookup Fig. 7.20. Theoretical steps versus top concentration of column II. Conditions of Fig. 7.28.

specific runoff load is assumed, by Eq. (7-45), to equal $w^* = 3.27$ kmol/m^2 h. From this and by Eqs. (7-3), (7-7), (7-8), (7-46), and (7-47), the curves $u_{VII}\,\rho_V^{1/2} = f(x_T)$ versus different reflux ratios, r_{II}, can be determined, (Fig. 7.30).

For a given reflux ratio, r_{II}, and a given top concentration x_T, Fig. 7.29 will supply the theoretical number of steps n_{tII}. Fig. 7.30 will yield the corresponding load factors $u_{VII}\,\rho_V^{1/2}$, and Fig. 7.5 will supply the proper pressure losses per step. The product of steps times specific pressure loss then gives the total pressure loss Δp_T.

The total pressure can be plotted versus the top concentration for different reflux ratios (Fig. 7.31). It is seen that the pressure loss minima are much lower for the

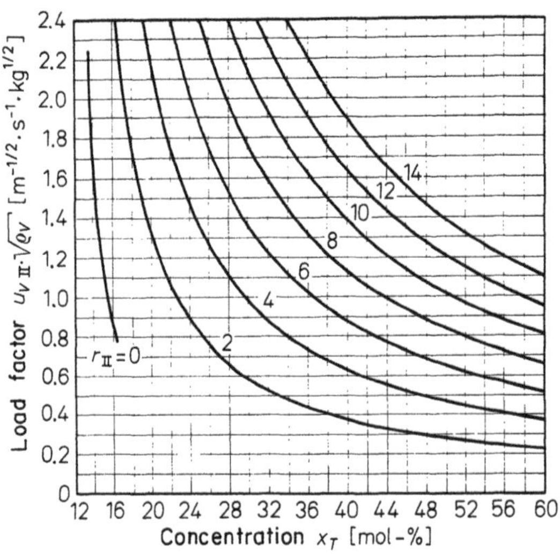

Fig. 7.30 Column II of hookup Fig. 7.20. Load factor at top of column II versus top concentration at various reflux ratios. Conditions of Example 4.

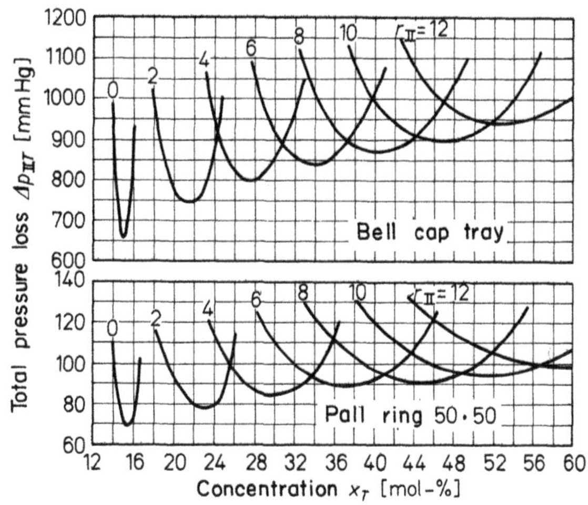

Fig. 7.31 Column II of hookup Fig. 7.20. Total pressure loss versus top concentration at various reflux ratios. Conditions of Fig. 7.30.

Pall-ring column, with higher top concentrations, than for the bell-cap column, for an identical reflux ratio r_{II}. The minimum, with the lowest total pressure loss, is obtained with $r_{II} = 0$, when column II functions as a straight stripper.

Finally, Fig. 7.32 presents the minimum pressure losses Δp_{IIm} and the correlated

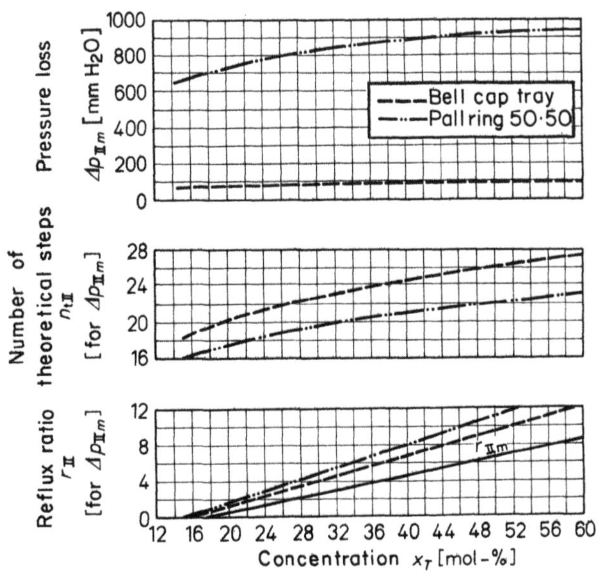

Fig. 7.32 Column II of hookup Fig. 7.20. Minimum total pressure loss, correlated theoretical numbers of steps and reflux ratios versus top concentrations. Conditions of Fig. 7.30.

numbers of steps and reflux conditions for column II. As opposed to column I, these are markedly dependent on x_T. There is also a line indicating the change in minimum reflux versus the top concentration. An adequate safety margin is thus given between the reflux ratio at minimum pressure loss and the corresponding minimum reflux ratio for the same top concentration. Note that for a different concentration x_B, different r, n_t, and Δp values may result for either column. The x_B value should be chosen according to the admissible level of the bottoms temperatures in both columns.

Results for hookups c_1) and c_2) Table 7.1 summarizes, for a bell-cap-type column, calculation results in optimization. In case c_1), corresponding values of f, Δp, n_t, and r are drawn from Fig. 7.24. In case c_2), these are taken from Figs. 7.25, 7.26, 7.28, 7.31, and 7.32. Also, the heat requirements based on feed quantities Q/F (kcal/kmol)

Table 7.1.

Example 4. Comparison of column arrangements.

<table>
<tr><td>Arrangement</td><td colspan="2">c_1)</td><td colspan="2">c_2)</td></tr>
<tr><td>Column</td><td>I</td><td>II</td><td>I</td><td>II</td></tr>
<tr><td>d_S</td><td>d_{S_I}</td><td>$d_{S_{II}} = d_{S_I}$</td><td>d_{S_I}</td><td>$0.62 \cdot d_{S_I}$</td></tr>
<tr><td>f(kmol/m²h)</td><td>3.2</td><td>–</td><td>3.2</td><td>–</td></tr>
<tr><td>f'(kmol/m²h)</td><td>–</td><td>3.2</td><td>–</td><td>8.25</td></tr>
<tr><td>w^*(kmol/m²h)</td><td>–</td><td>1.27</td><td>–</td><td>3.27</td></tr>
<tr><td>Δp_m(mm H_2O)</td><td>740</td><td>740</td><td>1070</td><td>660</td></tr>
<tr><td>n_t</td><td>19.3</td><td>19.3</td><td>27.5</td><td>16.1</td></tr>
<tr><td>r</td><td>4.53</td><td>–</td><td>4.60</td><td>–</td></tr>
<tr><td>Q/F(kcal/kmol)</td><td>32,400</td><td>36,000</td><td>32,900</td><td>16,350</td></tr>
<tr><td></td><td colspan="2">68,400</td><td colspan="2">49,250</td></tr>
</table>

are included. It is seen that case c_2) is more favorable than case c_1) as far as energy consumption is concerned. Minimum pressure loss is offered for column II by hookup c_2); if column size d_{S_I} is enlarged, other conditions being unchanged, the pressure loss of column I might be reduced. A total cost summary would decide which of schemes c_1) and c_2) would ultimately prevail.

3. OPTIMUM OPERATION OF COMBINED COLUMNS OF DIFFERENT TYPES UNDER SPECIAL CONDITIONS

Columns designed for a given set of variables may under changed conditions become surplus, and their reuse for new applications may be contemplated. While reuse of a single column may be a relatively simple matter, reuse of a set of columns entails difficulties in establishing the optimum, as has previously been demonstrated. Having a plurality of different units on hand will introduce additional viewpoints. Best utilization of a pair of columns, one trickling film column and one plate-type unit, in a special application, is treated next.

A. Parallel Arrangement

In the flowsheet of Fig. 7.33, the top product T of the plate-type column is recycled to feed F entering the trickling film column. Note that the feedpoint of the trickling film column may not be modified, which is different from that of the plate column.

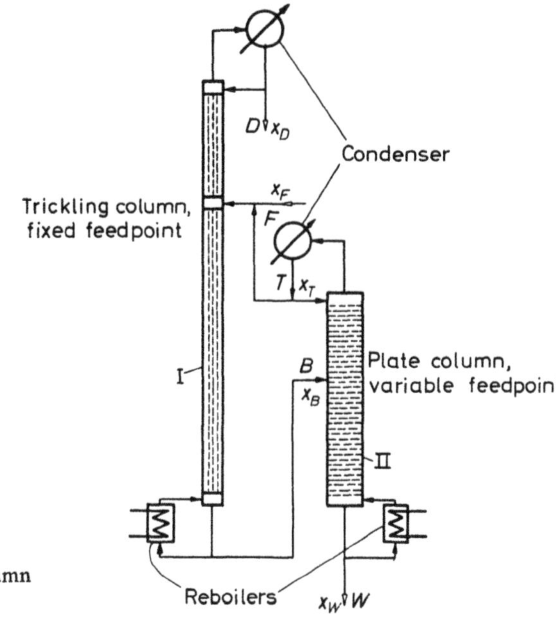

Fig. 7.33 Flowsheet of a trickling film column combined with a plate column.

Now find the optimum reflux ratios r_{I} and r_{II}, and vapor loads V_{I} and V_{II} (kmol/h) for a given quantity of feed F (kmol/h), and for given low boiler concentrations in the feed x_F (mol-%), in the top product, x_D (mol-%), and in the runoff x_W (mol-%). Which plate in the plate-type column will be the feedplate in order to obtain a minimum $V_{\mathrm{I}} + V_{\mathrm{II}}$? Because of the variables x_B and x_T there will be multiple answers.[8]

To find the optimum load conditions, first the interrelation of quantities and concentrations for operating columns must jointly be expressed mathematically. We have:

F = Feed to trickling column (kmol/h)
x_F = Concentration of feed to trickling column (mol-%)
D = Top product of trickling column (kmol/h)
x_D = Concentration of top product of trickling column (mol-%)
B = Bottom product of trickling film column (kmol/h)

x_B = Concentration of bottom product of trickling column (mol-%)
r_{I} = Reflux ratio of trickling column
V_{I} = Vapor in trickling column (kmol/h)
T = Top product in plate column (kmol/h)
x_T = Concentration of top product of plate column (mol-%)
W = Bottom product of plate column (kmol/h)
x_W = Concentration of bottom product of plate column (mol-%)
r_{II} = Reflux ratio of plate column
V_{II} = Vapor in plate column (mol/h)

Vapor quantities V_{I} and V_{II} are assumed constant inside their columns, which is true if in operation the heat exchanged per mole is virtually constant across the range of concentrations in question, and amounts M and B enter at boiling temperature into their columns.

Because of balanced amounts and concentrations, for the top product of the trickling column

$$D = F\frac{x_F - x_W}{x_D - x_W} \tag{7-52}$$

for the quantity of vapor in the trickling column

$$V_{\text{I}} = F\frac{x_F - x_W}{x_D - x_W}(r_{\text{I}} + 1) \tag{7-53}$$

for the reflux ratio in the trickling column

$$r_{\text{I}} = \frac{V_{\text{I}} - D}{D} \tag{7-54}$$

for the runoff of the trickling column

$$B = F\left(\frac{x_D - x_F}{x_D - x_W}\frac{x_T - x_W}{x_T - x_B}\right) \tag{7-55}$$

for the bottom product of the plate column

$$W = F\frac{x_D - x_F}{x_D - x_W} \tag{7-56}$$

for the overhead quantity of the plate column

$$T = F \frac{x_D - x_F}{x_D - x_W} \frac{x_B - x_W}{x_T - x_B} \tag{7-57}$$

and for the vapor quantity of the plate column

$$V_{II} = F \frac{x_D - x_F}{x_D - x_W} \frac{x_B - x_W}{x_T - x_B} (r_{II} + 1) \tag{7-58}$$

Eqs. (7-53) and (7-58) will yield vapor quantity V_{II} as a function of vapor quantity V_I:

$$V_{II} = V_I \frac{r_{II} + 1}{r_I + 1} \frac{x_D - x_F}{x_F - x_W} \frac{x_B - x_W}{x_T - x_B} \tag{7-59}$$

B. Series Arrangement

For comparison, consider a flowsheet in which the two columns operate as in Fig. 7.34 and may be computed as a single unit. The entire plate column functions as

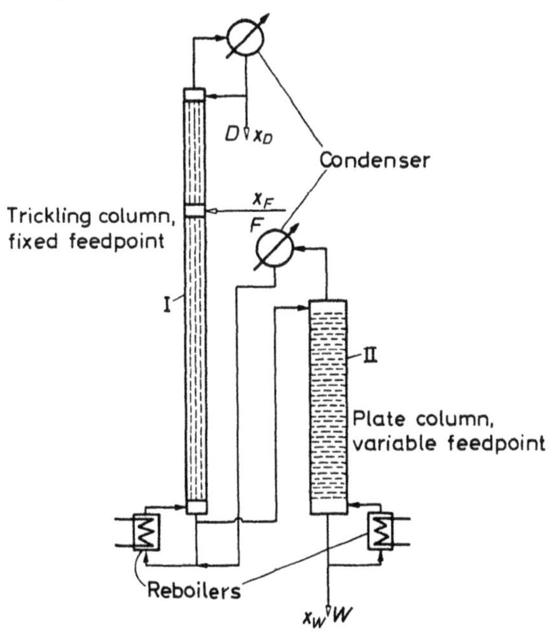

Fig. 7.34 Same as Fig. 7.33.

a stripper; the condensate formed goes to the bottom of the trickling column for reevaporation. The reflux from the stripper of the trickling column is the feed to the top of the plate column. There is no provision for utilizing the reflux heat. The vapor quantities generated in the vaporizer of the trickling column and in that of the plate column are to be equal

$$V = V_{\mathrm{I}} = V_{\mathrm{II}} = F \frac{x_F - x_W}{x_D - x_W} (r + 1) \tag{7-60}$$

Example 5 Review the flowsheets of Figs. 7.33 and 7.34. A raw material to be rectified is to yield a certain amount of first fraction. The mean relative volatility of the system First Fraction/Main Product is $\alpha = 1.5$. The first fraction is to have the following concentration

$x_F = 0.5$ mol-%
$x_D = 5$ mol-%
$x_A = 0.01$ mol-%
x_B = variable: 0.05, 0.075, 0.1, 0.15, 0.2, 0.3 mol-%
x_T = variable: 0.1, 0.2, 0.3, 0.4, 0.5 mol-%

The average separation of the available trickling column is about equivalent to that of 12 theoretical steps, and that of the available plate column is equivalent to 11 theoretical plates:

Trickling column $n_{t\mathrm{I}} = 12$ (constant)
Plate column $n_{t\mathrm{II}} = 11$ (constant)

For the trickling column, the feedpoint location is that of theoretical step 6, while for the plate column it is variable:

Trickling column $n_{tF} = 6$
Plate column n_{tF} = variable

What are the prerequisites for optimum operation of either column?

Answer Fig. 7.35 gives, for hookup a) and by calculation via Eqs. (7-53) and (7-58), the vapor quantities referred to the feed quantities F. For the trickling column, vapor quantities V_{I} (kmol/h) are given by dash-dotted curves, and for the plate column V_{II} (kmol/h) by dotted curves. The sums $(V_{\mathrm{I}} + V_{\mathrm{II}})/F$ are plotted over the bottom concentration x_B (mol-%) of the trickling column for various top concentrations $x_T = 0.1$ to 0.5 mol-% of the plate column. Similarly, the reflux ratios r_{I} and r_{II} correlated to vapor quantities V_{I} and V_{II} are plotted versus x_B for $x_T = 0.1$ to 0.5 mol-% (Fig. 7.36).

Optimization gives a minimum total vapor for a vapor load at the top of the trickling column = 3.86 kmol per kmol of feed, with a reflux ratio of 38.2. For a reflux ratio of 4.21, the vapor load at the top of the plate column is 4.39 mol per mol of feed. Theoretically, the trickling column gives a bottom concentration of 0.15 mol-%, and the plate column gives a top concentration of 0.3 mol-%.

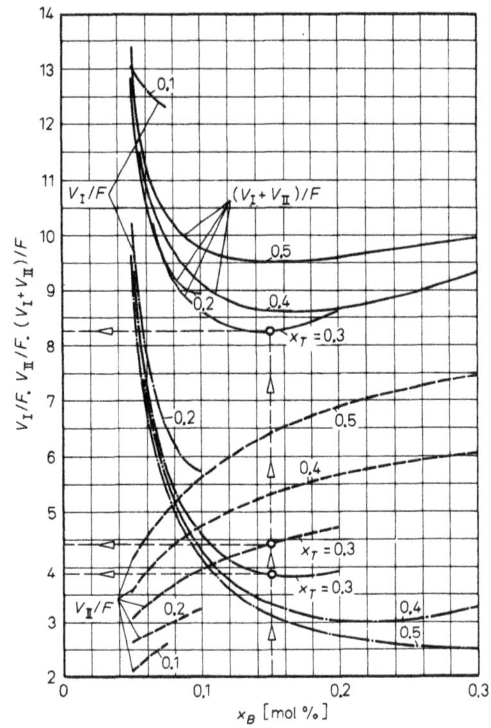

Fig. 7.35 Columns of Fig. 7.33 Conditions of Example 5, Vapor loads.

Fig. 7.37 reveals, for hookup a), the theoretical location of the feedplate for various pairs of x_B and x_T values. To meet optimum operation requirements, the runoff from the trickling column must be fed to the plate column at the level corresponding to that of theoretical step number 10, counting up from the bottom of the plate column.

Fig. 7.38 summarizes, for hookup b), results applying for a variable top concentration x_D a variable number of total steps n_t (as opposed to the optima of Fig. 7.37). The plot includes lines that show the top concentration x_D versus the reflux ratio for constant V/F values. Also there are n_t lines connecting points of constant theoretical steps.

For a comparison with the optima above, only those values that belong to $x_D = 5$ mol-% and to $n_t = 12 + 11 = 23$ are of interest. They are $V/F = 3.54$ and $r = 35$. For completeness, there is the V/F line $= f(r)$ for $x_D = 5$ mol-%, that will show $V/F = 3.54$ at $r = 35$.

The original column flowsheet in Fig. 7.33 is inferior to the one in Fig. 7.34 as far as the load goes, since, for the former, the sum of the vapor quantities $(V_I + V_{II})/F = 3.86 + 4.39 = 8.25$, and for the latter, only $(V_I + V_{II})/F = 3.54 + 3.54 = 7.08$.

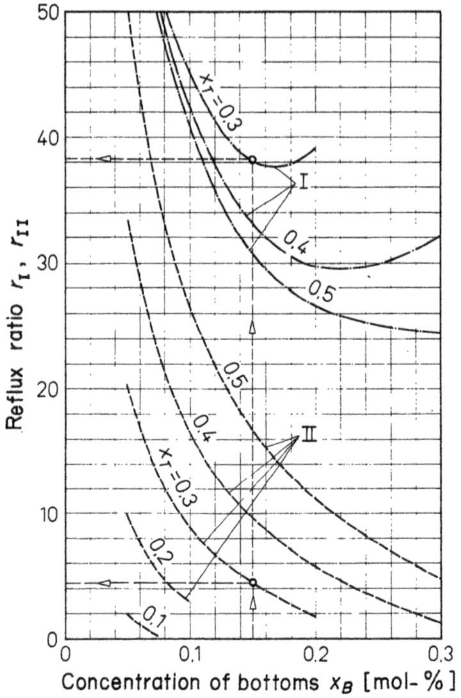

Fig. 7.36 Columns of Fig. 7.33. Reflux ratios conditions of example 5.

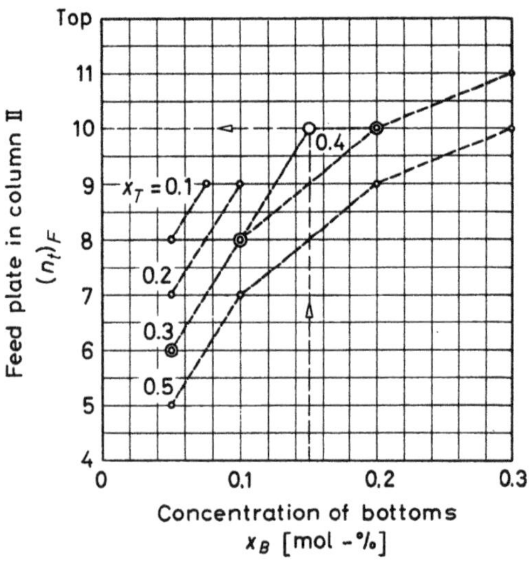

Fig. 7.37 Feed plate in column II of Fig. 7.33 Conditions of example 5

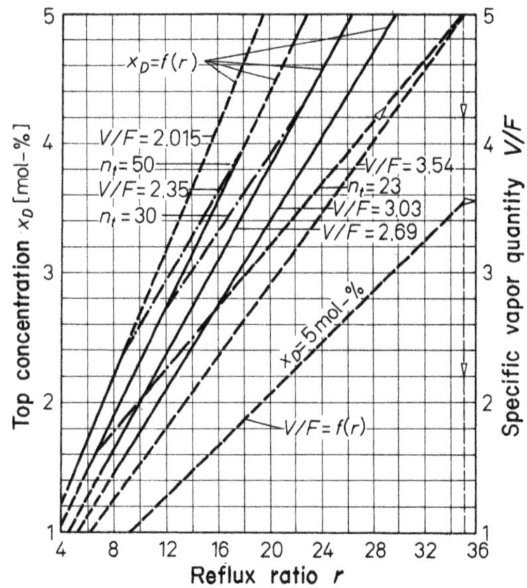

Fig. 7.38 Hookup of Fig. 7.34. Top concentration of column I, and specific vapor quantities of columns I and II. Conditions of Example 5.

4. SPECIALIZED OPERATIONS

Along with the schemes discussed, there exist others, applied according to the separation problem, operational performance, and physical properties of the mixture. The use of straight strippers or straight rectifiers is not uncommon, and the exigencies of the task in hand will ultimately decide.

A. Specialized Hookups and Their Calculation

As pointed out in Chapter 2, azeotropic mixtures cannot readily be separated beyond their azeotropic point. Once the azeotropic composition of the liquid has been reached in distillation or rectification, the vapor formed will have that composition, and no further enrichment is feasible. Nevertheless, a soluble azeotropic binary mixture can be split up simply by continuous rectification in a two-column setup, to yield the essentially pure components. To that end, advantage is taken of the fact that increasing the pressure will shift the azeotropic point toward the component of the larger heat of evaporation. The two columns must, then, work at different pressures. The original mixture is fed to the low-pressure column whose top product feeds the high-pressure unit. Under limiting conditions, the bottom product of the low-pressure column is the high-boiling component, while the top product may reach the azeotropic

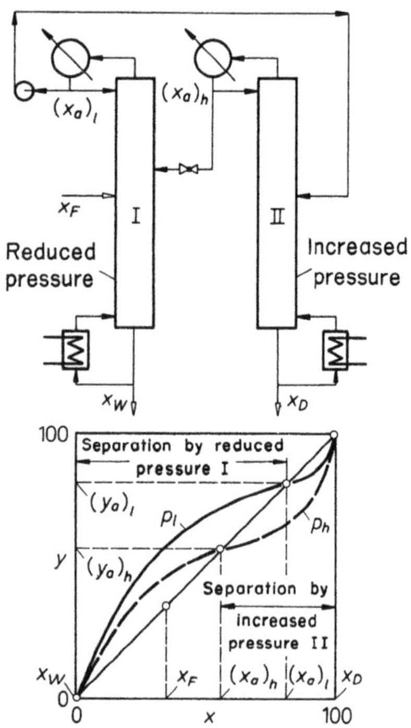

Fig. 7.39 Flowsheet for the separation by two pressure process of a binary with an azeotropic point, and y/x plot with typical concentrations (diagrammatic).

composition correlated to the pressure. The top product of the high-pressure column is recycled to the low-pressure column at the point of proper composition. Both columns may be computed from the instructions in Chapter 3.

In the range of concentrations where the equilibrium curve of the y/x plot is located below the diagonal, the operating lines must also be located below the diagonal.

Fig. 7.39 illustrates a two-pressure installation for decomposing an azeotropic binary mixture, together with a diagrammatic plot of the typical concentrations in the y/x presentation. Under certain conditions the evaporator of the low-pressure column may serve as a condenser to the high-pressure column; this may save heat. Also, the continuous separation of a partially insoluble binary mixture whose azeotropic composition is located within the range of the miscibility gap can be achieved in a two-column set-up[10]. In the limiting case, the bottom product again comprises the pure components. The azeotropic point is overcome by passing the cooled top product to a decanter where two layers are formed whose compositions in the equilibrium are on opposing sides of the azeotropic point. Each of these two layers is returned to its own column as reflux. In this way, both columns may operate at the same pressure.

Fig. 7.40 also shows the typical operating lines 1, 2, and 3, which indicate the the y/x plot. The dotted solubility curve in the lower right-hand corner indicates compositions x_{SI} and x_{SII} of the two phases in the decanter at different temperatures t_S; it must be found by experiment. If the lower boiling component is water, column

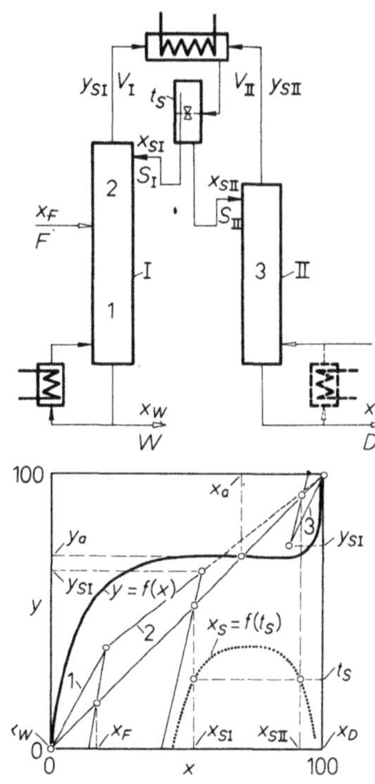

Fig. 7.40 Flowsheet for separation of a binary with an azeotropic point in the miscibility gap, and y/x plot with typical concentrations and interchange lines (diagrammatic).

II may be heated by direct steam. In practice, this is being done in the separation of water and *n*-butanol.

Fig. 7.40 also shows the typical exchange lines 1, 2, and 3, which indicate the relation of vapor and liquid concentrations in any section of the three column parts. Regarding the upper limits and the graphic orientation, all three operating lines are used like those of straight strippers. Quantities S_I and S_{II} in the decanter are governed by the vapor loads of columns I and II only, and thus do not interfere with each other in steady operation.

Example 6 From a vapor mixture of water–*n*-butanol totaling 52.4 kmol/h, with a water content of 12 mol-%, continuous rectification at normal pressure is to recover the butanol with a purity of 99.5 mol-%. The concentration of butanol in the water recovered must not exceed 0.5 mol-%. Find the principal data for the plant.

Answer The separation will take place as described above. The original mixture is first fed to a rectifier with enriching and stripping functions (Fig. 7.41). From the bottom, virtually pure butanol is drawn off, the overhead vapors are condensed totally.

The condensate formed passes to a decanter cooled to 40° C to assist separation; there will be one layer rich in *n*-butanol and one layer rich in water. The former is the reflux to the aforementioned column, while the latter is the feed to the top of a stripper from the bottom of which practically pure water is withdrawn, and whose overhead vapors are condensed in the condenser common to both columns.

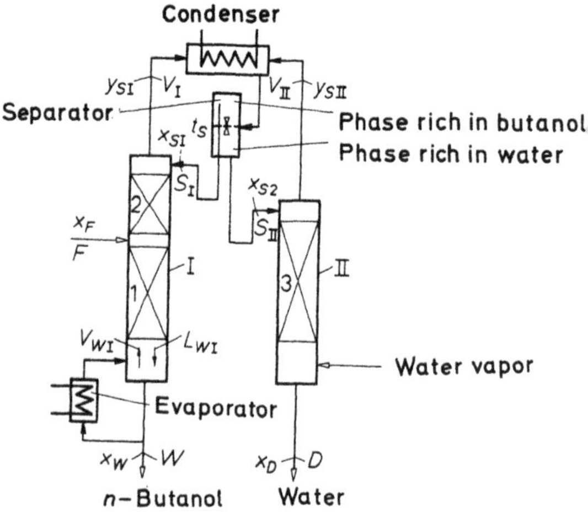

Fig. 7.41 Flowsheet for the separation of the partly insoluble azeotropic mixture of water and n-butanol, with typical concentrations and quantities.

Fig. 7.42 presents the characteristic data and curves in a y/x plot correlated to the flowsheet in Fig. 7.41. The specified values were:

$x_F = 12.0$ mol-%
$x_D = 99.5$ mol-% $\qquad F = 52.4$ kmol/h
$x_W = 0.5$ mol-%

From this, the runoff of the water column becomes

$$D = F\frac{x_F - x_W}{x_D - x_W}$$

$$= 52.4\ \frac{12 - 0.5}{99.5 - 0.5} = 6.09 \text{ kmol/h} \qquad (7\text{-}61)$$

The runoff from the butanol column is

$$W = F - D$$

$$= 52.40 - 6.09 = 46.31 \text{ kmol/h} \qquad (7\text{-}62)$$

From the equilibrium curve for the mixture water-*n*-butanol at normal pressure and in view of the vapor feed of the butanol column, the proper slope for its exchange line is "1" (in view of the number of steps and vapor loads); assuming constant molar heats of vaporization (Fig. 7.42)

$$\tan \beta_{\text{I}} = 4.9$$

Based on mass balances for the butanol column, $\tan \beta_{\text{I}}$ may also be expressed by the ratio of the molar quantities of liquid and vapor in the stripper:

$$\tan \beta_{\text{I}} = \frac{L_{W\text{I}}}{V_{W\text{I}}} = \frac{V_{W\text{I}} + W}{V_{W\text{I}}} = 1 + \frac{W}{V_{W\text{I}}} \qquad (7\text{-}63)$$

and there follows the vapor quantity to be generated by the reboiler

$$V_{W\text{I}} = \frac{W}{\tan \beta_{\text{I}} - 1} \qquad (7\text{-}64)$$

$$= \frac{46.31}{4.9 - 1} = 11.9 \text{ kmol/h}$$

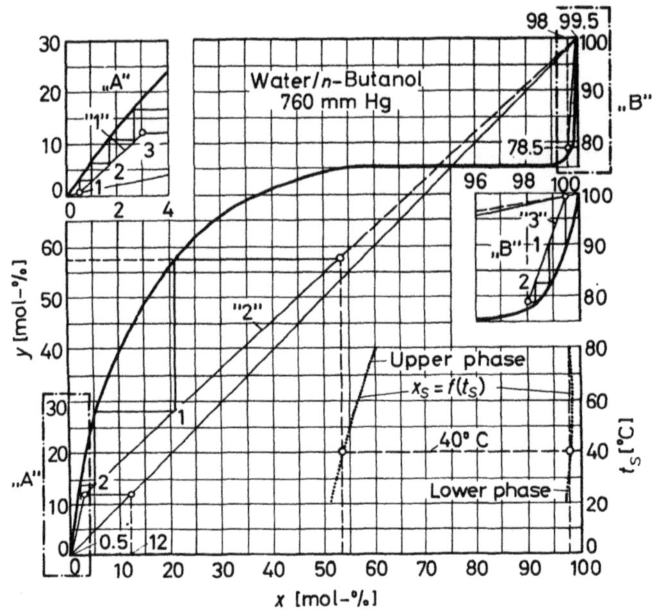

Fig. 7.42 Number of theoretical steps in an installation of Fig. 7.41, for separating the mixture of water and n-butanol. Conditions of Example 6.

The overhead vapor from the top of the butanol column, assuming equal molar heats of evaporation of the components, and for vapor feed

$$V_{\mathrm{I}} = V_{W\mathrm{I}} + F$$

$$= 11.9 + 52.4 = 64.3 \text{ kmol/h} \tag{7-65}$$

The corresponding quantity of reflux from the decanter is

$$S_{\mathrm{I}} = W + V_{\mathrm{I}} - F$$

$$= 46.31 + 64.3 - 52.4 = 58.21 \text{ kmol/h} \tag{7-66}$$

Based on the solubility curve $x = f(t_S)$ at $t_S = 40$ C, its concentration is

$$x_{S\mathrm{I}} = 53 \text{ mol-\%}$$

For this, operating line "2" in Fig. 7.42 gives the vapor concentration

$$y_{SI} = 57.5 \text{ mol-\%}$$

Hence, for the butanol column

$(n_{te})_I$ = theoretical enrichment steps, and
$(n_{ts})_{II}$ = 3 theoretical stripping steps

If the slope of the stripping line (note line "3" in Fig. 7.42) is obtained from

$$\tan \beta_{II} = \frac{S_{II}}{V_{II}} \tag{7-67}$$

$$= 14$$

the solubility curve will supply the concentration

$$x_{SII} = 78.5 \text{ mol-\%}$$

and there will be

$$(n_{ts})_{II} = 2 \text{ theoretical stripping steps}$$

Hence, the water column will supply the vapor quantity

$$V_{II} = D \frac{x_D - x_{SII}}{x_{SII} - y_{SII}} \tag{7-68}$$

$$= 6.09 \frac{99.5 - 98.0}{98.0 - 78.5} = 0.469 \text{ kmol/h}$$

The reflux quantity from the decanter is

$$S_{II} = V_{II} + D$$

$$= 0.469 + 6.09 = 6.559 \text{ kmol/h} \tag{7-69}$$

With these data the balance of the decanter is

$$V_{I} + V_{II} = S_{I} + S_{II}$$

$$64.3 + 0.469 = 58.21 + 6.559$$

$$64.769 = 64.769 \tag{7-70}$$

The azeotropic point of a partly immiscible mixture can be circumvented through separation of the liquid phase into two layers. Hence, perfectly miscible azeotropic binary mixtures can be split up into pure components by conversion into partly immiscible ternary mixtures by adding a suitable new component. For instance, pure industrial alcohol can be produced by adding benzene[11,12] or stabilized trichloroethylene to the raw spirits in a two-column setup. The bottom product of the main column is the absolute alcohol, and the top product in the limit event is an azeotropic ternary mixture that, upon cooling and possibly when water is added, will give two layers in a decanter. The layers containing ethanol-benzene or trichloroethylene are recycled to the main column, while the layer containing ethanol-water can be decomposed into water and azeotropic alcohol, which is returned to the main column. The upper part of Fig. 7.43 presents the flowsheet. The lower part is the distillation triangle with typical concentrations for the use of benzene as an additive. To the original mixture of concentration x_F enough benzene must be added

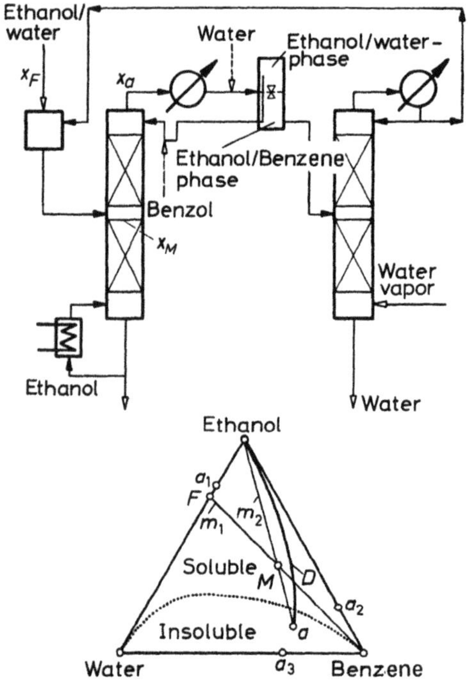

Fig. 7.43 Separation of the azeotropic mixture of ethanol and water, using benzene as an additive insoluble in water, and a distillation triangle (diagrammatic).

so that a mixture of concentration x_M is obtained at intersection M of the mixing lines m_1 and m_2, which can be split up into ethanol and an insoluble mixture of composition x_a. In the plot, a_1, a_2, and a_3 are the azeotropic points of the pairs of substances, and a is the azeotropic point of the ternary mixture.

Operating line D indicates the direction in which the vapor concentration changes relatively to that of the liquid. A precise calculation requires accurate information on the distillation plot, including an adequate number of operating lines. This is known as "azeotropic distillation." For some mixtures, however, the azeotropic point can be shifted by adding a soluble salt or other substance,[14,15] or it may be removed altogether. Thus the azeotropic behavior of the mixture isopropyl alcohol-water can be deleted, and its relative volatility increased across the entire concentration range.

If the azeotropic mixture of minimum boiling point benzene–cyclohexane is given an addition of aniline, the azeotropic point of the y/x plot is shifted toward the corner of the high-boiling component cyclohexane, and the equilibrium curve of the original components falls below the diagonal. Continuous separation of that mixture into its pure components, benzene and cyclohexane, is feasible by having two columns in series. The first column receives the original mixture and the additive and separates them into cyclohexane as the top product and benzene–aniline as the bottom product. The second column separates the latter into benzene as the top product and aniline as the bottom product, which is recycled to the first column. This procedure is known as "extractive distillation."

In discussing the heat requirements of continuous rectifiers (Chapter 3) it was stated that in the decomposition of multiple mixtures such as crude oils, the possibility of recovering byproducts as oil fractions was often exploited. In such cases, a combination of main columns and side columns is often preferred over that of main columns only (Fig. 3.13). Having merely to provide the reflux for the main column simplifies the plant. The side column then does the essential work. Its vapors are recycled directly via the tray of the main column that delivers the liquid to the side column. Finding the number of steps for the side column is the same as for a stripper. An efficient arrangement of main and auxiliary columns yields effective separations.[16] Fig. 7.44 explains the splitting up of a multicomponent mixture by one main column and several side columns.

Combining a main column and an auxiliary column is also advantageous when a pure component is to be recovered from a fraction concentrated in the main column (Fig. 7.45). The vapor fraction drawn from the main column is fed to the bottom of the side columns functioning as a rectifier without additional reboiler. The runoff from the side column is recycled to the main column.

The hookup of Fig. 7.45 is also suitable for many azeotropic and extractive distillations, with the additional component being supplied at the top of the main or auxiliary column. In extractive distillation, reflux condensation may be minor if the additive is cold enough.

Depending on the system worked and circumstances there are many choices in combinations of several columns. Other flowsheets[17-23] that should be mentioned are those which decompose binaries containing lesser amounts of high or low boilers.[24] Fig. 7.46 presents such systems, along with the basic operating lines in a y/x plot.

If the top product of a column contains traces of a low boiler, they can be removed

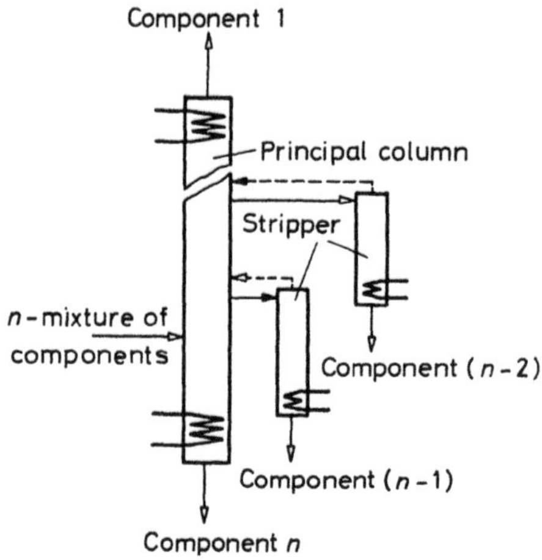

Fig. 7.44 Combination of a principal column with lateral columns in the separation of multicomponent mixtures. Lateral columns function as strippers supplied with intermediate products.

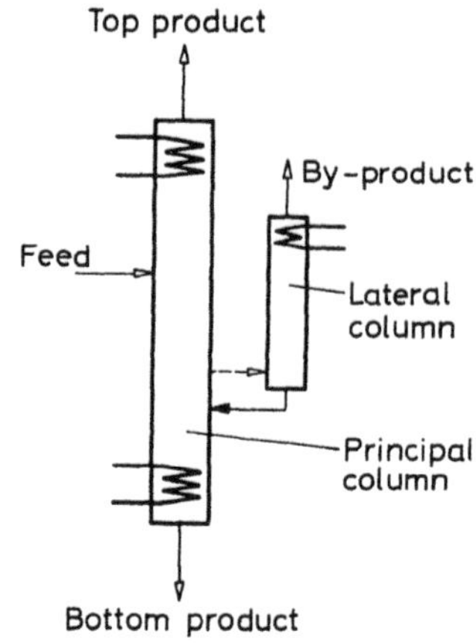

Fig. 7.45 Combination of a principal column with one lateral column functioning as a straight rectifier supplied with overhead product.

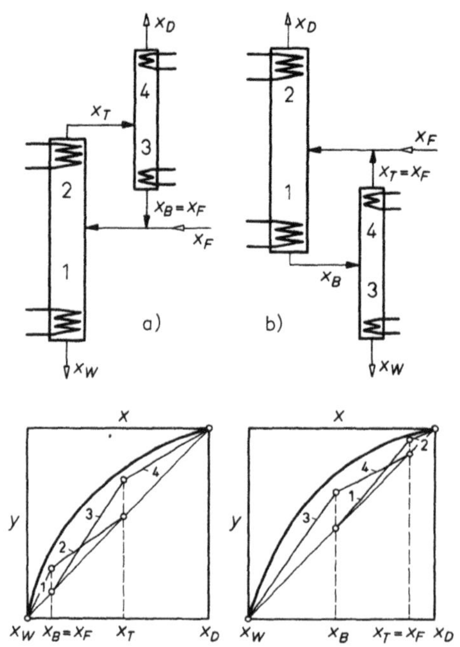

Fig. 7.46 Main column and side columns functioning as combined rectifiers and strippers; *a*) if the amount of low boiler is minor, and *b*) if the amount of high boiler is minor. Also corresponding y/x plots, diagrammatic.

by the scheme of Fig. 7.47. The low boiler is enriched in a column top extension above the runoff of main product D. If T is the amount of the low boiler drawn off from the column extension, the reflux quantity below D is, with r_D as the local reflux ratio,

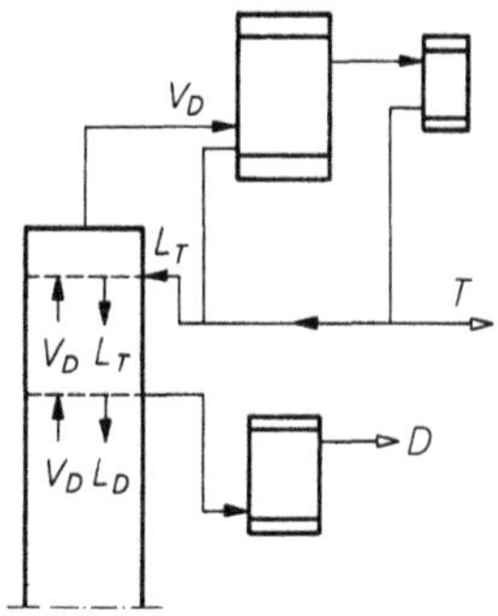

Fig. 7.47 Flowsheet for separation of traces of a component contained in an overhead product.

$$L_D = r_D (D + T) \tag{7-71}$$

T is a fraction of D:

$$T = mD \tag{7-72}$$

For the reflux quantity at the column extension, with reflux ratio r_T

$$L_T = r_T T = m r_T D \tag{7-73}$$

Furthermore,

$$L_T = D + L_D \tag{7-74}$$

From this, and Eqs (7-71) and (7-72)

$$L_T = D [1 + r_D (1 + m)] \tag{7-75}$$

By equating eqs. (7-73) and (7-75), the relation between the reflux ratios above and below the main product runoff follows

$$r_T = \frac{1 + r_D (1 + m)}{m} \tag{7-76}$$

Because of its simplicity, this flowsheet is often preferred to the more costly auxiliary column arrangement of Fig. 7.48, as in the separation of aldehydes in the production of prime spirits. An auxiliary column, while more expensive, provides a cleaner aldehyde cutoff.

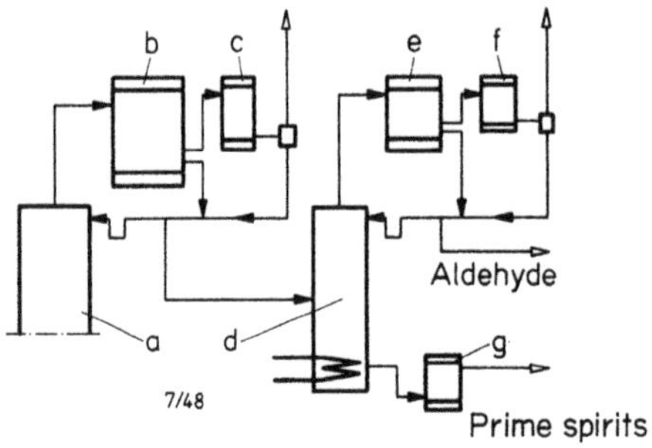

Fig. 7.48 Separation of aldehydes in making prime spirits: *a* main column, *b* reflux condenser, *c* gas cooler, *d* extra column, *e* reflux condenser, *f* gas cooler, *g* product cooler.

Example 7 In a continuous rectifier installation, traces of low boiler must be removed by enrichment in a column extension. Their quantity is 2% of the main product. At what reflux ratio should the column top operate if the column part below the main product runoff is to function at reflux ratio 5?

Answer From Eq. (7-76) the reflux ratio above the main product runoff becomes

$$r_D = \frac{1 + 5\,(1 + 0.02)}{0.02} = 30.5$$

B. Rectification in Straight Stripping Columns

In discussing combined column schemes, uses of continuous strippers have been mentioned (note Figs. 7.22, 7.41, and 7.44). They are investigated further, noting that the enthalpy of feed F as sent to the column top h_F (kcal/kmol) differs from enthalpy h'_F (kcal/kmol) at boiling.

Fig. 7.49 presents the quantities, concentrations, and operating line for this separation in a flowsheet, while Fig. 7.50 is the corresponding y/x plot of a binary mixture. Layout and optimization will be explained.

From a heat balance around the equipment, the heat requirement is

$$Q = Dh''_D + Wh'_W - Fh_F \tag{7-77}$$

That heat produces the top product quantity D (kmol/h) that leaves overhead as a vapor of enthalpy h''_D (kcal/kmol):

$$D = F - W \tag{7-78}$$

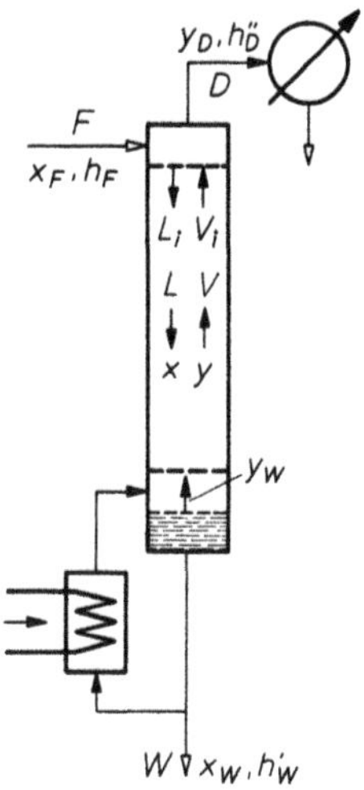

Fig. 7.49 Hookup of a straight stripper, with typical quantities, concentrations, and enthalpies.

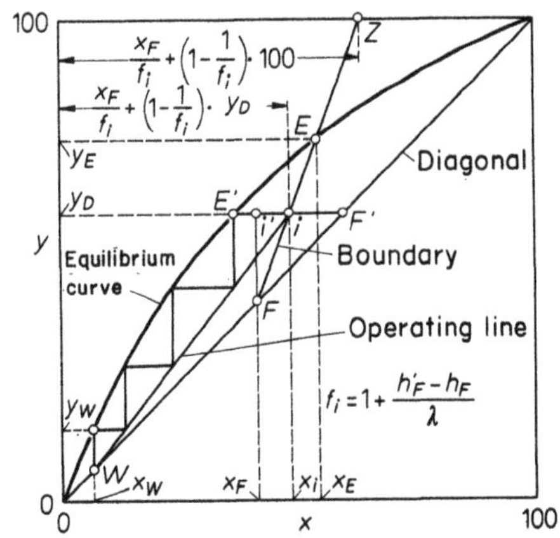

Fig. 7.50 Rectification of binaries in straight strippers according to Fig. 7.49, and y/x plot (diagrammatic).

or

$$D = \frac{Fx_F - Wx_W}{y_D} \tag{7-79}$$

Runoff W (kmol/h) with enthalpy h'_W (kcal/kmol) gives

$$W = F\frac{y_D - x_F}{y_D - x_W} \tag{7-80}$$

With heat loss Q_l Eqs. (7-77), (7-78), and (7-80) the heat requirement becomes:

$$Q = F\left[(h''_D - h_F) - \frac{y_D - x_F}{y_D - x_W}(h''_D - h'_W)\right] + Q_l \tag{7-81}$$

or, referred to the runoff quantity

$$\frac{Q}{W} = \frac{y_D - x_W}{y_D - x_F}(h''_D - h_F) - (h''_D - h'_W) + \frac{Q_l}{W} \tag{7-82}$$

The heat balance for the feedpoint is:

$$(V_i - D)\lambda = F(h'_F - h_F) \tag{7-83}$$

where V_i = the vapor quantity entering at the feedpoint (kmol/h) and λ = molar heat of condensation of vapor condensed at the feedpoint (kcal/kmol). Also, there is the mass balance

$$F + V_i = L_i + D \tag{7-84}$$

where L_i = the liquid quantity leaving at feedpoint level (kmol/h). This differs from feed quantity F(kmol/h) by a factor f_i

$$L_i = f_i F \tag{7-85}$$

Relations (7-83), (7-84), and (7-85) give

$$f_i = 1 + \frac{h'_F - h_F}{\lambda} \tag{7-86}$$

which is analogous to Eq. (3-33)

The stripping line correlates the vapor and liquid concentrations at any point of the stripper. By a balance around any column section

$$Fx_F + Vy = Lx + Dy_D \tag{7-87}$$

The number of theoretical steps is found from that part of the stripping line limited by concentrations x_W and y_D.

From $x = x_W$, and because of

$$Fx_F - D\,y_D = Wx_W \tag{7-88}$$

and

$$L - W = V \tag{7-89}$$

Eq. (7-87) will yield $y = x_W$.

Vapor concentration x_i, while fictitious, limits the stripping line at y_D, and will yield, together with Eq. (7-84) to (7-87) and $V = V_i$, $y = y_D$, $L = L_i$, and $x = x_i$ the relation

$$x_i = \frac{x_F}{f_i} + \left(1 - \frac{1}{f_i}\right) y_D \tag{7-90}$$

This is the boundary, the locus of all intersections i of the stripping line with the horizontal $y = y_D$. The boundary line intersects the horizontal y = 100 mol-% at Z, the abscissa

$$(x_i)_Z = \frac{x_F}{f_i} + \left(1 - \frac{1}{f_i}\right)100 \tag{7-91}$$

Intersection i, located between F and E in Fig. 7.50, is important in the actual operation of a stripper. At $i = E$ the theoretical number of steps becomes infinite and Eqs. (7-79) and (7-80) give the ratio

$$\frac{D}{F} = \frac{x_F - x_W}{y_D - y_W} \tag{7-92}$$

where now $y_D = y_E$.

With $i = F$, $y_D = x_F$, and from Eq. (7-92) $D/F = 1$, or $W = 0$. When i comes closer to the limit E, the heat requirement will become less, as seen from Eqs (7-81) and (7-82), but the number of steps n_t increases. The optimum location of i between limits F and G, for a given enthalpy of the feed, i_F, can only be determined from an economic analysis.

Another approach to an optimization is to first assume a certain top concentration, y_D, then vary entropy i_F of the feed by partial vaporization of the feed prior to admission to the column in a preheater by preheating only, or by cooling. If the preheat reaches the boiling point, $h_F = h'_F$ the secant becomes identical with the vertical througl F that intersects the horizontal at ordinate $y = y_D$, point i'. The connecting line $\overline{i'F'}$ is the locus for all intersections i that satisfy the conditions of feed heating, that is, $h_F < h'_F$. On the other hand, connecting line $\overline{i'E'}$ is the locus for all intersections i involving partial vaporization of feed, that is, $h_F > h'_F$. For optimization, it is also necessary to consider the cost for the purchase and the operation of the said heat exchanger in the overall picture.

If the molar heats exchanged between the low boiler and the high boiler during the rectification of the mixture are approximately equal, the stripping line will merge with connecting line $\overline{Wi}$. If, however, the molar heats of vaporization of the components are markedly different, the stripping line will follow from the equation below [note comments on Eq. (3-27)]

$$y = \frac{\left(m^* - \frac{y_D}{K}\right)x + y_D - m^* x_i}{1 + (m^* - 1)\frac{x}{K} - m^* \frac{x_i}{K}} \tag{7-93}$$

Factor m* is identical with the liquid/vapor ratio prevailing immediately below the feedpoint:

$$L_i = m^* V_i \tag{7-94}$$

which is combined with Eq. (7-85) to give

$$m^* = f_i \frac{F}{V_i} \tag{7-95}$$

From Eqs. (7-84) and (7-85):

$$V_i = D + F(f_i - 1) \tag{7-96}$$

Thus by substitution of Eqs. (7-78) and (7-80) one can find the vapor quantity V_i for fixing the column size at the top

$$V_i = F \left(\frac{x_F - x_W}{y_D - x_W} + f_i - 1 \right) \tag{7-97}$$

Inserting this into Eq. (7-95) will give:

$$m^* = \frac{f_i}{\dfrac{x_F - x_W}{y_D - y_W} + f_i - 1} \tag{7-98}$$

Factor K in Eq. (7-93) is identical with relation (3-11).

Should the molar heats of evaporation for the components differ appreciably, the corresponding quantities of vapor and liquid at a given level of the stripper may be derived from

$$V = F \frac{1 - \dfrac{y_D}{K}}{1 - \dfrac{y}{K}} \left(\frac{x_F - x_W}{y_D - y_W} + f_i - 1 \right) \tag{7-99}$$

$$L = V + F \frac{y_D - x_F}{y_D - x_W} \tag{7-100}$$

According to Chapter 8, these amounts must be accurately known in order to find the column dimensions.

Continuous rectification in straight strippers can, for nonazeotropic systems, supply a bottom product free from low-boiling components.

C. Rectification in Straight Enriching Columns

Continuous rectification in straight enriching columns, on the other hand, will for nonazeotropic systems supply a top product free of high-boiling components. This procedure sometimes offers advantages.

Fig. 7.45 shows how the pure low-boiler can be obtained, from a cut enriched in a main column, by rectification through a side column functioning as a straight enriching device.

For any mixtures, the use of a straight rectifier is the only way to obtain the low boiler on a continuous basis in regular operation. For example, the raw material may contain substances that will in time form deposits inside the columns, or even plug them. In that event, columns with stripping functions are useless, and continuous working is feasible in straight enriching columns only. Reboiler design is an important consideration. If the contaminants are crystals unlikely to adhere to metallic surfaces, a falling-film evaporator is often serviceable (Fig. 7.51). If, however, incrustation may be expected, a rotary-film evaporator is indicated. Frequently, a jacketed still kettle with an agitator will be the only choice (Fig. 7.52).

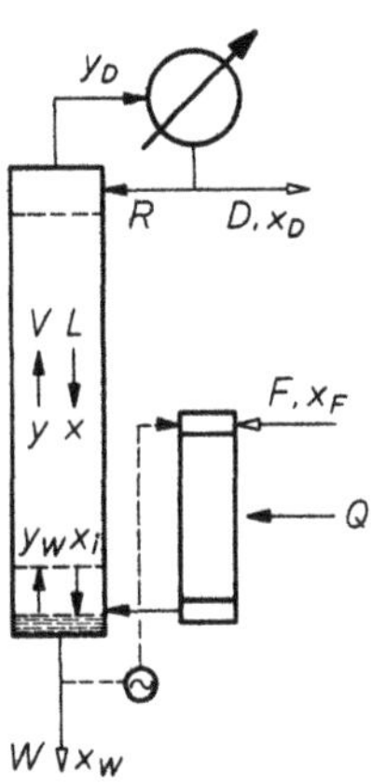

Fig. 7.51 Continuous straight enrichment column with falling film type reboiler and typical motions of phases and concentrations.

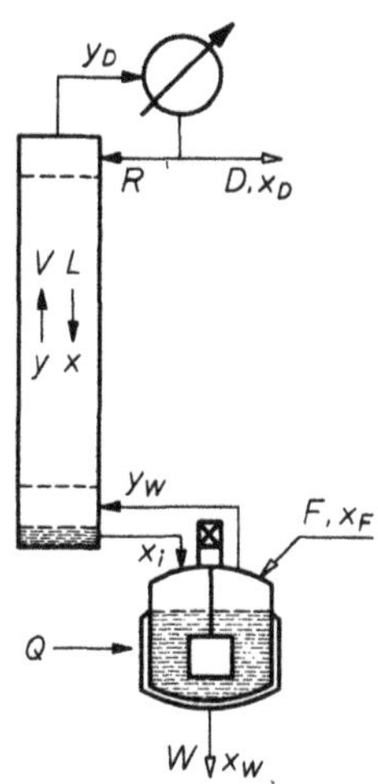

Fig. 7.52 Continuous straight enrichment column with jacketed still kettle, and typical motions of phases and concentrations. .

For continuous operation of a straight enriching column, the feed is charged into the reboiler (Figs. 7.51 and 7.52). To achieve a low concentration of the low boiler in the bottoms, the reflux ratio must be high; note the y/x plot for the continuous separation of a binary in Fig. 7.53. Since the heat requirement will be high, a thin-film rectifier may be mounted in between the enriching column and the thin-film type evaporator (Fig. 7.54). That scheme is discussed in Chapter 6. It improves thermal economy, the reflux ratio for the enriching column is less, and the reevaporation rate is also less.

The following remarks will apply to straight enriching columns. The characteristics and compositions are as presented in Figs. 7.51, 7.52, and 7.53. For the continuously operated, straight enriching column, the y/x plot is in Fig. 7.53, and Eqs. (3.10) to (3.16) apply. Again assuming that the vapor from the reboiler enters the enriching column with the phase equilibrium concentration y_W corresponding to the bottom runoff concentration x_W, the concentration of reflux x_i at the bottom of the enriching column is found in the y/x plot by making the parallel to the abscissa y_W intersect the enrichment line (point i in Fig. 7.53).

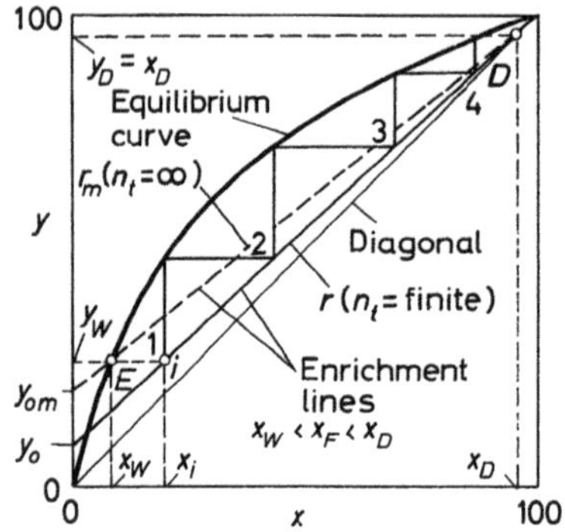

Fig. 7.53 Continuous straight rectifier in y/x plot, and determination of theoretical number of steps (diagrammatic).

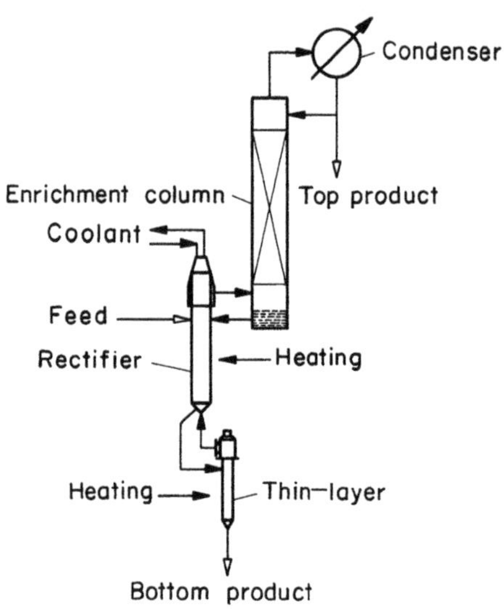

Fig. 7.54 Straight enrichment column fitted with thin-film rectifier for stripper function and thin-film evaporator as reboiler.

The analytical procedure employs Eq. (3-10)

$$x_i = \frac{\dfrac{r_D}{r_D+1}\,\dfrac{K-y_W}{x_D-y_W}\,x_D - K}{\dfrac{r_D}{r_D+1}\,\dfrac{K-y_W}{x_D-y_W} - 1} \qquad (7\text{-}101)$$

For equal molar heats of evaporation K according to Eq. (3-11), becomes infinite and Eq. (7-101) is simplified

$$x_i = \frac{r_D+1}{r_D}\,y_W - \frac{x_D}{r_D} \qquad (7\text{-}102)$$

Again, reflux ratio r_D must exceed minimum r_m. For equilibrium curves without an inflection point, Eq. (3-35) gives

$$r_m = \frac{K - y_W}{K - x_D} \frac{x_D - x_W}{y_W - x_W} - 1 \tag{7-103}$$

For equal molar heats of evaporation, $K = \infty$ and

$$r_m = \frac{x_D - y_W}{y_W - x_W} \tag{7-104}$$

The operating line correlated to the minimum reflux ratio intersects the equilibrium curve at intersection E, which is the parallel to the ordinate through x_W, $(x_i = x_W)$.

The stepping off takes place between the operating line and the equilibrium curve, between points i and D, the intersection of the operating line and the diagonal.

Quantities D, V, and L again result from Eqs. (3-23), (3-41), and (3-50). The reflux ratio at any given column level follows from Eq. (3-49) combined with Eq. (3-16).

By writing $x = x_i$ and $y = y_W$ the balance at the column base is obtained from

$$Fx_F + Lx = Wx_W + Vy \tag{7-105}$$

The liquid/vapor ratio is given by Eq. (7-9).

The heat requirement Q (kcal/h) becomes, according to Eq. (3-62)

$$Q = F\left[\frac{x_F - x_W}{x_D - x_W}\left\{(r_D + 1)\lambda_D + h'_D - h'_W\right\} + h'_W - h_F\right] + Q_l \tag{7-106}$$

where

λ_D = molar heat of condensation of the vapor condensed as reflux (kcal/kmol)
h'_D = molar enthalpy of top product (kcal/kmol)
h'_W = molar enthalpy of bottom product at boiling (kcal/kmol)
h_F = molar enthalpy of feed at entering temperature kcal/kmol
Q_l = heat losses (kcal/h)

Example 8 A straight enrichment column operating continuously is to separate a binary mixture having a relative volatility of $\alpha = 1.54$ virtually independent of its composition in such a way that the bottom product contains no more than x_W = 5 mol-%, and the top product no less than x_D = 99 mol-% of the low boiler. The molar heats of vaporization of the components are equal. Find the number of theoretical steps, and state the various quantities for feed concentrations x_F = 10, 30, 50, 70, and 90 mol-%, and a reflux ratio r_D = 1.4 times the minimum.

Answer The equilibrium curve for α = 1.54 can be calculated from Eq. (2-4). Results are: x_W = 5 mol-% yields a y_W = 7.5 mol-%. Eq. (7-104) gives a r_m = 35, or $r = r_D$ = 49. From this Eqs. (3-23), (3-41), and (3-50) yield the top product D(kmol/h), vapor V (kmol/h), and liquid L (kmol/h) as well as the runoff at the column base.

$$W = F - D \tag{7-107}$$

referred to the unit of feed F (kmol/h). The results are plotted in Fig. 7.55.

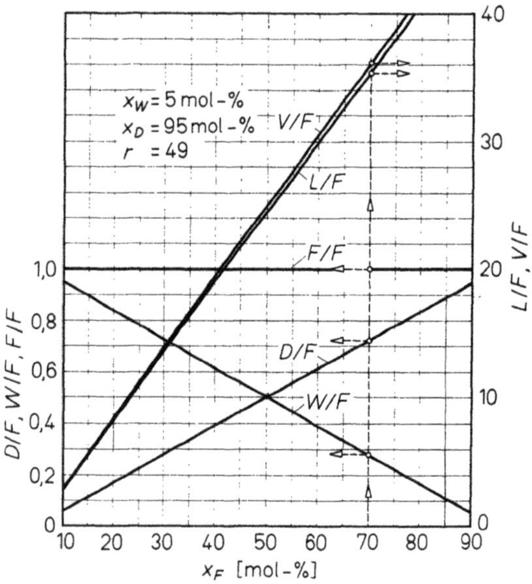

Fig. 7.55 Example 8. Amounts of vapor and liquid feed concentration, for continuous rectification in a straight enrichment column.

Fig. 7.56 shows the location of the equilibrium curve, enrichment line, and characteristic concentrations of the vapor and liquid for the range of the column base. The number of theoretical steps is around n_t = 17.5.

Eq. (7-105) together with Figs. 7.55 and 7.56, gives for a feed concentration x_F = 50 mol-%. For example,

$$(1)(50) + (24.5)(5.72) = (0.5)(5) + (25)(7.5)$$

$$190 = 190$$

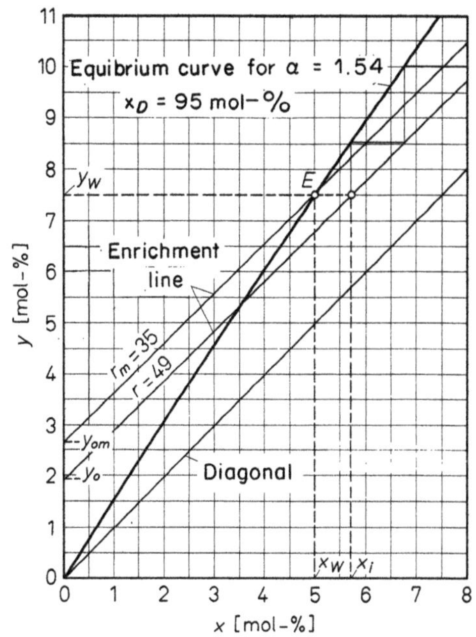

Fig. 7.56 Example 8. Concentrations in a straight enrichment column, bottom part, in y/x plot.

for $x_F = 10$ mol-%,

$$(1)(10) + (2.714)(5.72) = (0.945)(5) + (2.77)(7.5)$$

$$25.4 = 25.4$$

which is the proper balance equation for the low-boiling component.

Example 9 A liquid with $F = 54.1$ kmol/h, purity of $x_F = 99.7$ mol-% and contaminated by a high-boiling component is to be rectified to practically 100 mol-% = x_D. Along with the contaminations, $x_W = 54.8$ mol-% of the product is allowed to leave the column base. The relative volatility between pure product and contaminations is $\alpha = 13$. The molar heats of evaporation are equal. What are the reflux ratio and theoretical steps?

Answer From Eq. (2-4) the equilibrium curve can be calculated for $\alpha = 13$, and for the concentration range between 50 and 100 mol-% (Fig. 7.57). This separation will, obviously, require only a relatively low reflux ratio. With $r = 1$, five theoretical steps would yield a practically pure top product. At $x_D = 100$ mol-%, the runoff would contain all of the impurities:

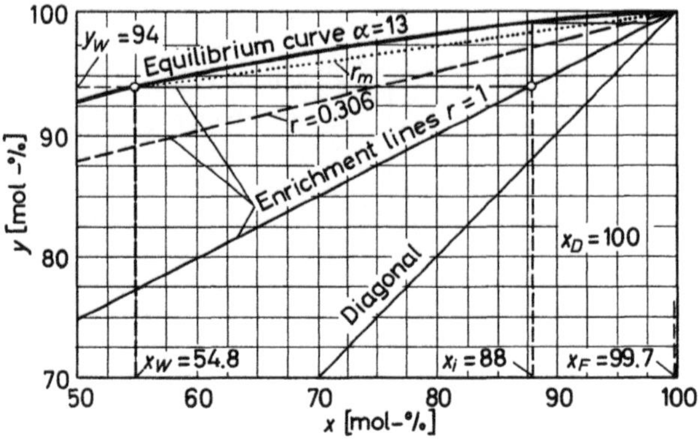

Fig. 7.57 Example 9. Concentrations in a straight enrichment column, in y/x plot.

$$F\frac{100 - x_F}{100} = (54.1)\frac{100 - 99.7}{100} = 0.1623\frac{\text{kmol}}{\text{h}}$$

which gives the quantity of

$$W = (0.1623)\frac{100}{100 - x_W} = 0.36 \text{ kmol/h runoff}$$

From

$$D = F - W \tag{7-108}$$

there follows the quantity of the top product

$$D = 54.1 - 0.36 = 53.74 \text{ kmol/h}$$

Eq. (3-40) gives, for equal molar heat of evaporation, the mass of vapor:

$$V = D(r + 1) \tag{7-109}$$

$$= (53.74)(2) = 107.84 \text{ kmol/h}$$

The reflux is calculated from the combined relations (3-48) and (3-50), or derived from Eq. (7-9):

$$L = V\frac{r}{r + 1} \tag{7-110}$$

and the reflux quantity is

$$L = 107.48 \frac{1}{2} = 53.74$$

From this and with $x_i = 88$ mol-%, $y_W = 94$ mol-%, and Eq. (7-105)

$$(54.1)(99.7) + (53.74)(88) = (0.36)(54.8) + (107.48)(94)$$

$$10{,}122 = 10{,}122$$

The assumed reflux ratio $r = 1$ amounts to about 6.5 times the minimum, which from Eq. (7-104) equals

$$r_m = \frac{100 - 94}{94 - 54.8} = 0.153$$

In practice, the separation could take place with a reflux ratio $r = 2\, r_m = 0.306$; the molar vapor load would equal only

$$\frac{0.306 + 1}{1 + 1} (100) = 6.5\%$$

of the vapor load at $r = 6.5\, r_m = 1$. This would mean more theoretical steps, and because of the low liquid load (only 30.6%) at $r = 1$, the separation effect of most packings would be slight and a taller column would be necessary.

Examples 8 and 9 demonstrate cases where the use of straight enriching columns causes high vapor rates. The question arises whether a flowsheet of Fig. 7.54 would not be advantageous for heat economy. Even a once-through evaporator at the column base might be preferable. That, or a thin-film evaporator would yield a vapor concentration superior to the phase equilibrium concentration correlated to x_W, as can be seen in Chapter 9. In that event the minimum reflux ratio, and therefore the actual reflux ratio, would be reduced.

D. Direct Heating of Columns

In the column arrangements of Figs. 7.40, 7.41, and 7.43, the reboilers were steam heated; that direct method of supplying heat is permissible provided the runoff is water or mostly water. In that event, the column calculation is the same as for indirect heating. Only the weight of the runoff is increased by that of the heating steam if the latter is saturated at the column base pressure. The quantity drained equals that of reflux at the base. If, however, the share of the low boiler in the runoff water cannot be neglected, the following facts should be considered.

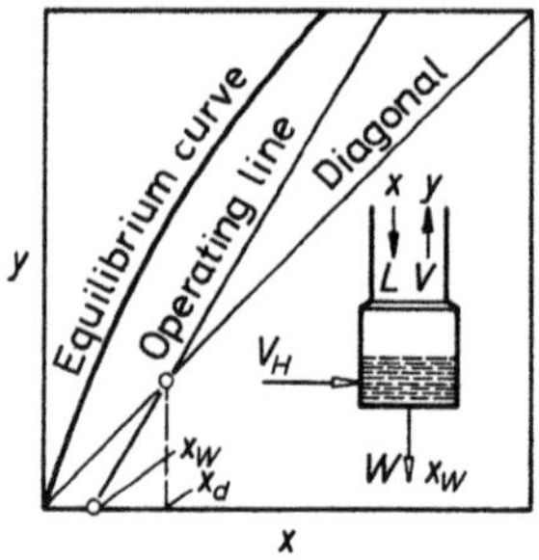

Fig. 7.58 Stripping line in y/x plot, for direct heating.

Starting from principles, consider the continuous rectification of a binary, with the molar heats of evaporation virtually independent of the composition. Let the heating steam equal V_H (kmol/h) (Fig. 7.58), then the mass balance for the stripping function is:

$$V_H + L = V + W \tag{7-111}$$

and

$$Lx = Vy + Wx_W \tag{7-112}$$

From this, the equation for the stripping line becomes

$$y = \frac{L}{V}x - \frac{W}{V}x_W \tag{7-113}$$

The molar vapor quantity $V = V_s$ in the stripper is equal to a constant

$$V_H = V = V_s = \text{constant} \tag{7-114}$$

In other words, the moles evaporated will equal those condensed (of heating steam.) Then Eq. (7-111) gives

$$L = W \tag{7-115}$$

and from Eqs. (7-113), (7-114), and (7-115):

$$y = \frac{W}{V_H}(x - x_W) \tag{7-116}$$

At the intersection of the stripping line and the diagonal,

$$y = x = x_d \tag{7-117}$$

From this, Eqs. (7-113), and (7-116)

$$x_d = \frac{W}{L - V} x_W \tag{7-118}$$

and

$$x_d = \frac{W}{W - V_H} x_W \tag{7-119}$$

which presupposes knowing vapor quantity $V = V_s$ and V_H and reflux quantity $L = L_s$ in the stripper or the runoff quantity W.

At the intersection of the stripping line with the abscissa

$$y = 0 \tag{7-120}$$

From this and Eq. (7-113) the abscissa value becomes

$$x = \frac{W}{L} x_W \tag{7-121}$$

and from Eq. (7-116)

$$x = x_W \tag{7-122}$$

The stripping line is fixed by x_W and x_d.

In a direct-heated rectifier having a stripping and an enriching function, the balancing equations

$$V_H + F = D + W \tag{7-123}$$

with

$$Fx_F = D\,x_D + W\,x_W \tag{7-124}$$

must be satisfied (Fig. 7.59). By adopting Eq. (7-114) the vapor quantity in the

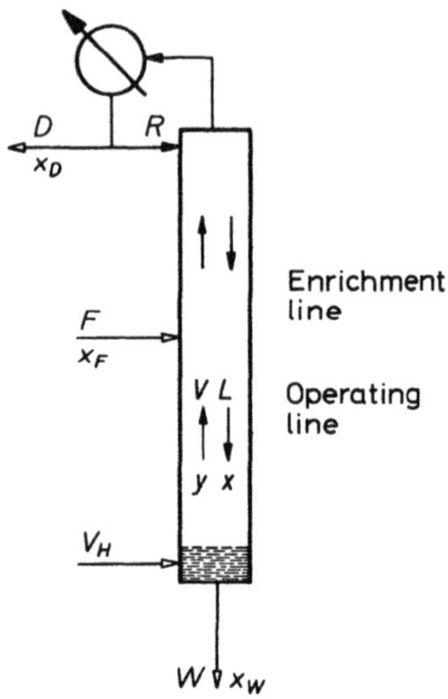

Fig. 7.59 Straight enrichment column with direct heat. Typical concentrations and motions of phases.

stripper becomes, according to Eq. (3-20)

$$V_H = D\,(r + 1) + F\,(f_i - 1) \tag{7-125}$$

where factor f_i characterizing the thermal condition of feed F follows from Eq. (3-33). The first term on the right-hand side of Eq. (7-125) is identical to the vapor quantity (in the enriching section)

$$V = V_e = D\,(r + 1) \tag{7-126}$$

For the top product, Eqs. (7-123), (7-124), and (7-125) will give

$$D = F\,\frac{x_F - x_W\,f_i}{x_D + x_W\,r} \tag{7-127}$$

which, combined with Eq. (7-125), establishes the heating steam requirement

$$V_H = F\left[(r+1)\frac{x_F - x_W f_i}{x_D + x_W r} + f_i - 1\right] \tag{7-128}$$

Note the presence of concentrations, of feed conditions, and of the reflux condition.

In a comparison of direct versus indirect heat for a given F and an equal x_F the condition where both operations have equal vapor loads V and equal top product D of equal composition x_D, is of paramount interest. This corresponds to case a) in Fig. 7.60, where the reflux ratio r is steady and

$$W_{\text{ind}}(x_W)_{\text{ind}} = W_{\text{dir}}(x_W)_{\text{dir}} \tag{7-129}$$

With indirect heating, the runoff must be less than with direct heat by the amount of the steam condensate V_H

$$W_{\text{ind}} = W_{\text{dir}} - V_H \tag{7-130}$$

After noting Eq. (7-119), combining Eqs. (7-129) and (7-130) will yield

$$(x_W)_{\text{ind}} = \frac{W_{\text{dir}}}{W_{\text{dir}} - V_H}(x_W)_{\text{dir}} = x_d \tag{7-131}$$

Thus both heating methods will give the same stripping line, which in the case of indirect heat will intersect the diagonal at the runoff concentration. However, for the directly-heated column the number of theoretical steps is increased by the amount of the stripping line extension between x_d and $(x_W)_{\text{dir}}$ in the y/x plot. In general that difference is slight; in Fig. 7.60 a) it roughly equals one theretical step.

In case a) the runoff concentration for direct heat may also be understood as that concentration resulting by mixing the runoff for indirect heating with the heating steam condensate.

To compare the two methods of heating for equal runoff concentrations x_W, the other variables F, x_F, x_D and r being equal, refer to case b) in Fig. 7.60. It would appear that the number of theoretical steps is less for direct than for indirect heat. Using the ratio of the molar top product quantities, note Eqs. (7-127) and (3-23),

$$\frac{D_{\text{dir}}}{D_{\text{ind}}} = \frac{x_F - x_W f_i}{x_D + x_W r}\,\frac{x_D - x_W}{x_F - x_W} \tag{7-132}$$

and from the ratio of the corresponding molar vapor quantities in the stripping device, which are about equal to the molar quantities of the heating steam (note Eqs. (7-128) and Chapter 3, Section 4),

$$\frac{V_{dir}}{V_{ind}} = \frac{(V_H)_{dir}}{(V_H)_{ind}} = \frac{(r+1)\dfrac{x_F - x_W f_i}{x_D + x_W r} + f_i - 1}{(r+1)\dfrac{x_F - x_W}{x_D - x_W} + f_i - 1} \tag{7-133}$$

combined with $f_i = 1$ yields

$$\frac{V_{dir}}{V_{ind}} = \frac{(V_H)_{dir}}{(V_H)_{ind}} = \frac{D_{dir}}{D_{ind}} = \frac{x_D - x_W}{x_D + x_W r} \tag{7-134}$$

the quantities of the top product and vapor are also less for direct heat. That, however, means that more of the low boiler is lost in the runoff. Only for $x_W = 0$ does Eq. (7-134) hold true and $D_{dir} = D_{ind}$.

If it is stipulated that for both methods of heating, identical vapor loads ($V = V_S$) and identical steam quantities are assumed, with values F, x_F, x_D, and x_W remaining unchanged, it follows from case c) in Fig. 7.60 that the directly heated column requires fewer steps in both the stripping and the enriching function. However, since it is desired that

$$(V_H)_{dir} = (V_H)_{ind} \tag{7-135}$$

it follows that for $f_i = 1$

$$D_{dir}\,(r_{dir} + 1) = D_{ind}\,(r_{ind} + 1) \tag{7-136}$$

and in view of Eq. (7-132)

$$D_{dir} = D_{ind}\,\frac{r_{ind} + 1}{r_{dir} + 1} = D_{ind}\,\frac{x_D - x_W}{x_D + x_W r_{dir}} \tag{7-137}$$

it appears again that direct heat will yield less top product, and hence lose more of the low boiler than will indirect heat.

The difference in the reflux ratios results from Eq. (7-133) for $(V_H)_{dir}/(V_H)_{ind} = 1$ and $f_i = 1$

$$r_{\text{ind}} = \frac{x_D - x_W}{x_D + x_W r_{\text{dir}}}(r_{\text{dir}} + 1) - 1 \tag{7-138}$$

It is presented in the lower part of Fig. 7.61 versus runoff concentration x_W for two different x_D values.

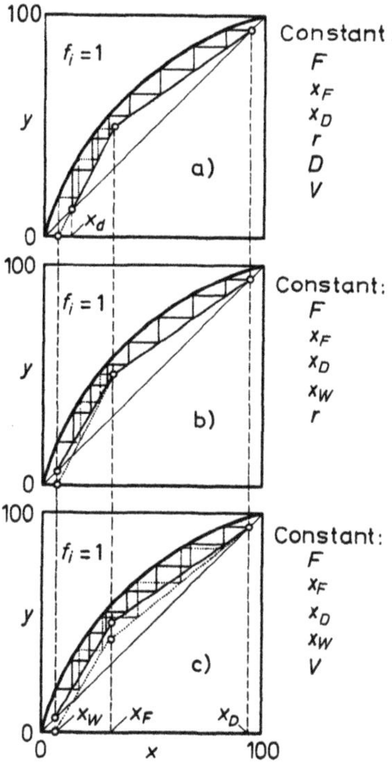

Fig. 7.60 Locations of operating lines, typical concentrations, and theoretical steps, for direct and for indirect heating. Three different operating conditions.

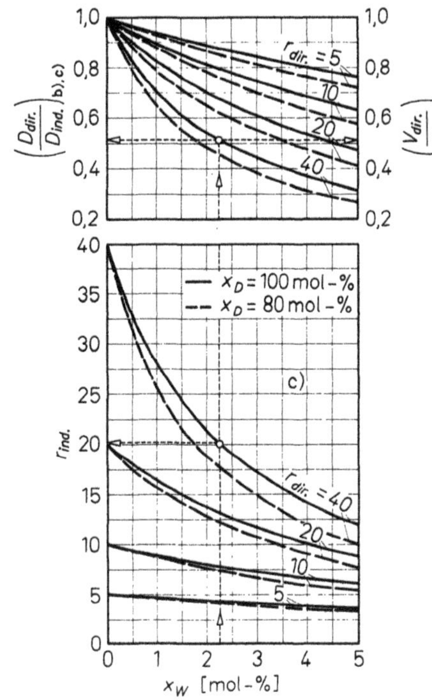

Fig. 7.61 Direct versus indirect heating.

The upper part of Fig. 7.61 shows the difference in the quantities of top product for cases b) and c). In case b) that difference is the same as that of the vapor.

If with direct heat the steam is not injected at the column base as in Figs. 7.58 and 7.59, but rather between the base and packing, the bottom concentration is the same as that in the liquid runoff of the lowest step. Then another theoretical step must be added. Evidently, in the planning of directly heated columns the runoff concentration must be clearly stated in order to obviate mistakes.

Fig. 7.62 shows a multicolumn flowsheet for making prime spirits from an alcohol water mash. Of the five columns, four are directly heated.

Other commercial applications of direct heat are in the separation of ammonia-water mixtures, and of mixtures of water with amines or acetone.

E. Saving Heat in Rectification

Depending on composition and boiling performance of raw materials, and on the mechanics of separation, the heat requirement will fluctuate, and together with the investment, control the processing cost for the project in hand. A high heat consumption also means a high coolant consumption for water, air, or brine. Frequently, the operational cost will surpass the investment, in which event heat economy is especially important.

If the mixture to be decomposed is liquid at delivery and at low temperature, the heat of condensation of the top product, the heat of the liquid bottom product, or both may serve to preheat the feed as shown in Fig. 7.62. As the enthalpy of the feed increases, the heat requirement diminishes [Eq. (3.62)].

In the discussion of combined rectifiers working at different pressures (Fig. 7.39) it had been pointed out that the condenser of the column of high pressure could serve as a reboiler to the column of lower pressure. That hookup is employed in the separation of air (Fig. 7.63). By superimposition of the columns with the condenser and evaporator a compact construction and minimum loss of cold are achieved.

If the runoff of the rectifier is water, vapor compression by ejector can effectively reduce the heat required by the rectifier.[25] In Fig. 7.64 the ejector partly recovers

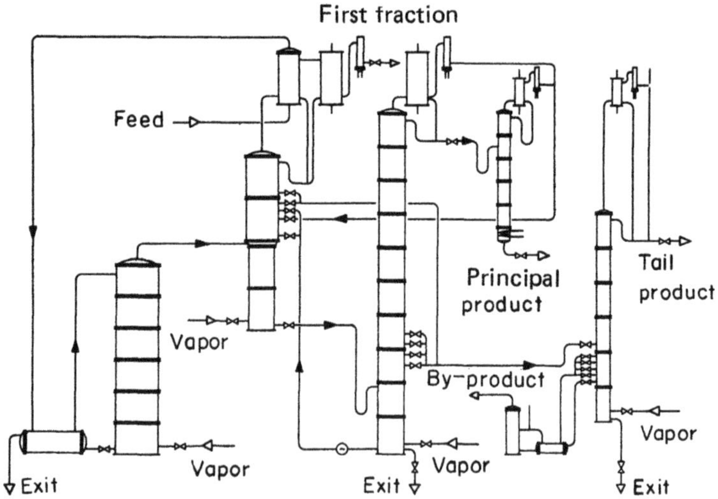

Fig. 7.62 Flowsheet of a plant for making prime spirits with most columns directly heated.

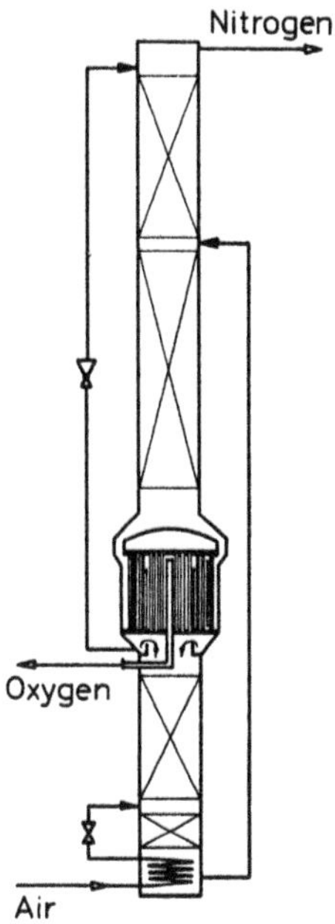

Fig. 7.63 Flowsheet in air separation.

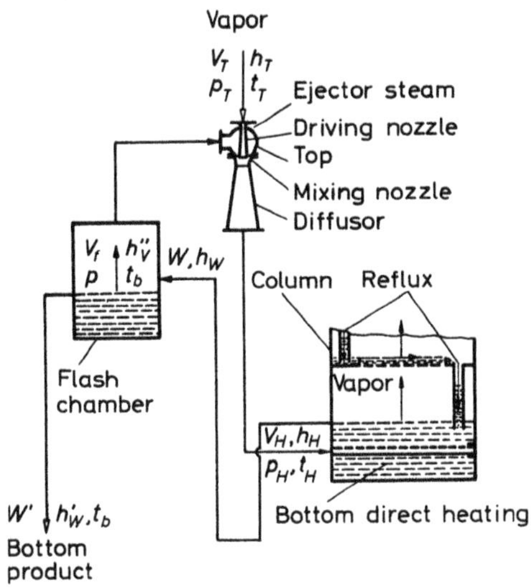

Fig. 7.64 Column heating by ejector steam.

the heat in the runoff of a stripper. The ejector consists of the head with the propelling nozzle, and the mixing nozzle with the diffuser. As there are no moving parts, the ejector is reliable in operation without being expensive to build. Moreover, its material of construction may be any of the common engineering materials, porcelain, or plastics.

In the flowsheet of Fig. 7.64, runoff W having enthalpy h_W is supplied to a flash chamber where the vapor liberated, V_f having an enthalpy h''_V, pressure p, and temperature t_b is aspired by the steam of enthalpy h_T, pressure p_T, and temperature t_T, while the runoff remaining after expansion

$$W' = W - V_I \tag{7-139}$$

having enthalpy h'_W, and temperature t_b is drained from the flash chamber. The vapor mixture from the ejector

$$V_H = V_T + V_f \tag{7-140}$$

having an enthalpy h_H, a pressure p_H, and a temperature t_H is fed directly to the reboiler of the stripper, where the pressure is p_H or less

$$p < p_H < p_T \tag{7-141}$$

To determine the steam requirement V_T of the ejector the characteristics of that piece of equipment must be known, such as the delivery factor

$$\varphi = V_f / V_T \tag{7-142}$$

which depends on temperature t_b for different pressures p_T, temperatures t_T and p_H, t_H. Then, with Eq. (7-140)

$$V_T = V_H \frac{1}{1+\varphi} \tag{7-143}$$

and

$$V_f = V_H \frac{\varphi}{1+\varphi} \tag{7-144}$$

Fig. 7.65 presents ejector characteristics.

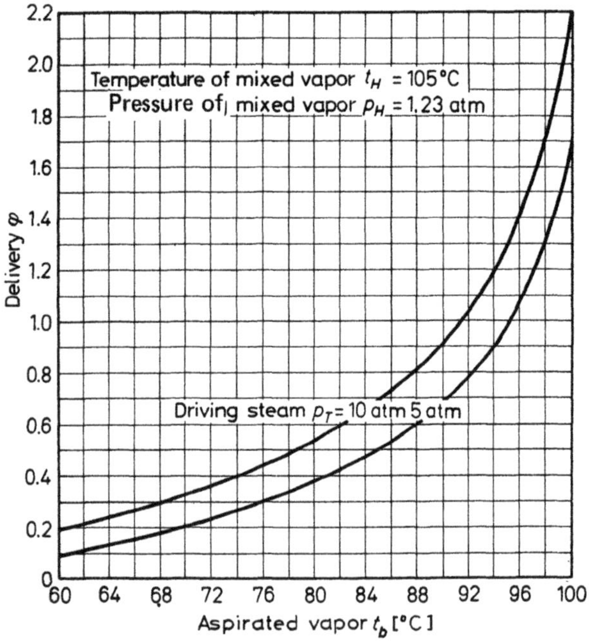

Fig. 7.65 Delivery of a steam ejector as a function of the temperature of the aspirated vapor. Ejector made by Wiegand of Karlsruhe, Germany.

A heat balance around the expansion chamber yields the vapor quantity produced (disregarding heat losses)

$$V_f = W\frac{h_W - h'_W}{h''_V - h'_W} \tag{7-145}$$

where the difference

$$\lambda = h''_V - h'_W \tag{7-146}$$

equals the heat of vaporization at pressure p and temperature t_b.

Given the condition of vapor mixture leaving the ejector, p and t_b must be such that both Eqs. (7-145) and (7-144) are satisfied. To that end, assume several temperatures of the expanded vapor t_b, and calculate V_f and from this V_T, using the equation

$$V_T = V_H - V_f \tag{7-147}$$

the W and V_H values being known from the column calculations and from Eq. 7.145 the φ values follow from Eq. (7-142) and are entered into the ejector plot $\varphi = f(t_b)$. Intersections i_1 and i_2 of the operating curves with the calculated curve yield the expanded vapor temperatures t_{b1} and t_{b2}; note the schematic plot in Fig. 7.66 for two steam pressures p_{T1} and p_{T2}.

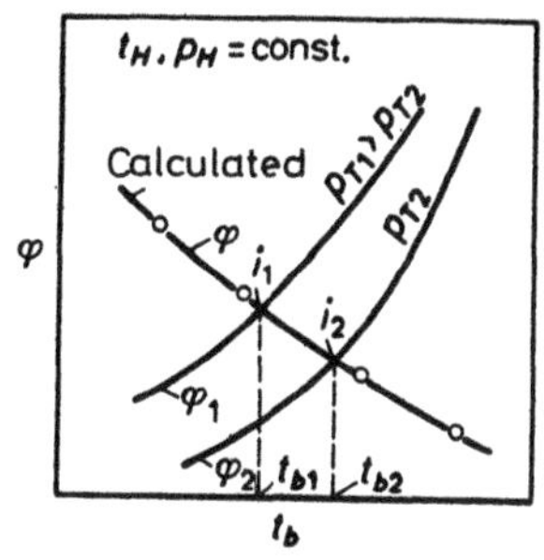

Fig. 7.66 Graph for finding the temperature of the expanded vapor, (diagrammatic,) for arrangement of Fig. 7.64.

Another application of vapor compression in rectification is presented in Fig. 7.67. The vapors leaving the top of a rectifying column with temperature t_D and pressure p_D are mechanically compressed to a point where they can heat the reboiler. That kind of heat recovery will pay off where the components involved have boiling points close to each other, and where the pressure loss of the column is low. The compression must be sufficient to overcome the temperature difference[25]

$$\Delta t = t_H - t_D = (t_W - t_D) + \vartheta \tag{7-148}$$

where

$$\vartheta = t_H - t_W \tag{7-149}$$

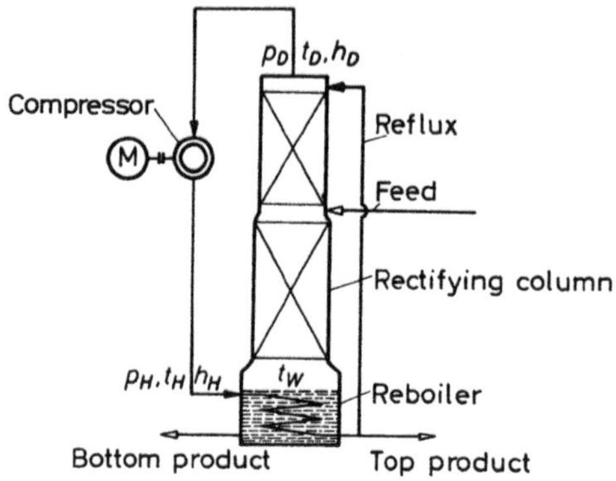

Fig. 7.67 Heating a rectifying column by the top product vapors compressed in an electric driven unit.

In weighing the economy of a separation involving the use of a compressor for the top product, the expense for purchase and operation of the electric driver must be considered. The compression of the vapor can be described by an enthalpy-entropy plot as shown in Fig. 7.68. It follows the polytrope 1-2 to enthalpy

$$h_H = h'_D + \frac{\Delta h_{ac}}{\eta_{ic}} \tag{7-150}$$

Δh_{ac} is the adiabatic compression difference and η_{ic} is the indicated compressor efficiency.

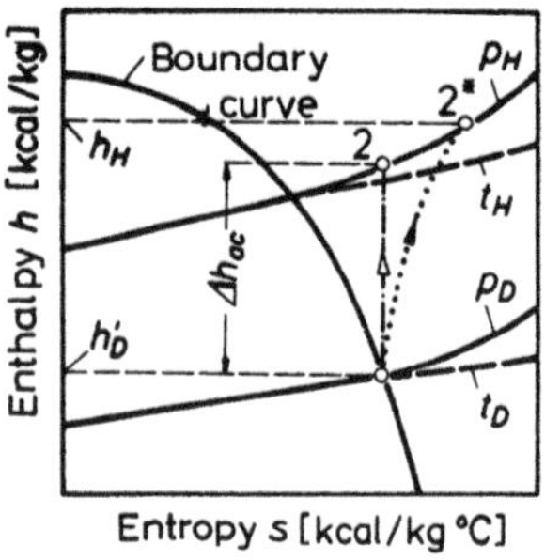

Fig. 7.68 Qualitative changes in state during compression by an electrically driven unit, in enthalpy-entropy plot.

Since the energy supplied by the electric motor must equal the energy absorbed by the compressor, the motor size (HP) becomes

$$N = \frac{427}{(3600)(75)} \frac{\Delta h_{ac} V_H}{\eta} \tag{7-151}$$

where η = overall efficiency of compression.

Example 10 Ammonia is to be recovered continuously by thermal decomposition of ammonium carbonate solution in a straight stripping column. According to Fig. 7.69, case a), this can be achieved by feeding the unpreheated solution to a stripper column the bottom of which is heated directly by steam.

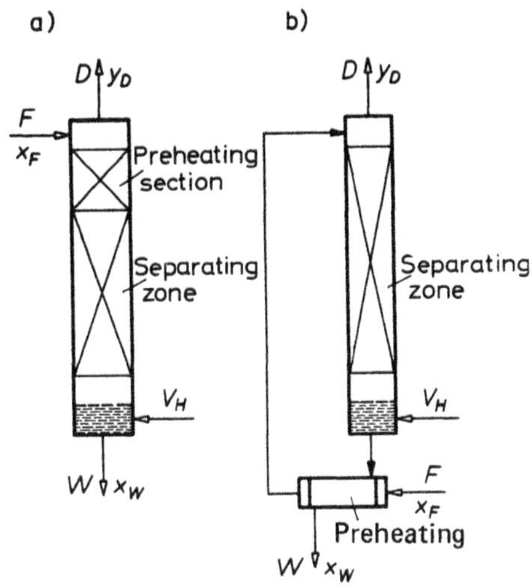

Fig. 7.69 Directly heated stripper; a) with. b) without feed preheated by waste heated for example 10.

By how much will the heat requirement be reduced if the feed is preheated by the runoff heat of the bottom product–case b).

The quantity of ammonium carbonate solution is 17,000 kg/h, the temperature is 20 C; upon thermal decomposition through dissociation it will give, at 68 C and normal pressure, 72% by weight of water, 15% by weight of ammonia, and 13% by weight carbon dioxide.

The top product is a vapor with 35% by volume of water, 48 volume % ammonia, and 17 volume % carbon dioxide. The waste is to contain less than 1% ammonia. The column operates at normal pressure.

Note that only about 2/3 of the ammonia is present as ammonium carbonate in the original mixture, while 1/3 is already dissolved.

Answer According to case a) of the flowsheet 7.69, the solution is first heated to dissociation temperature in the top part of the column. This forms carbon dioxide on one hand, and an ammonia-water mixture on the other, partly as a vapor and partly as a liquid phase of the above mentioned composition. The latter is split up in the lower column part, and the runoff is virtually free of ammonia.

Within the separation zone, only the ammonia–water system need be considered. Therefore, the heat consumption Q (kcal/h) may be derived from Eq. (7-81) which holds for straight stripper columns, if the thermal decomposition is allowed for as follows:

In an ordinary rectification, and for cold feed, the heat requirement would be composed, partly of preheat and partly of rectifying heat. The original solution dissociates at 68 C. To find the overall heat requirement

$$Q_T = Q_p + Q \tag{7-152}$$

where Q_p = sum of the preheat to bring the $(NH_4)_2CO_3$ solution from the initial to the dissociation temperature and the heat of dissociation, and Q = the heat for the rectification of the NH_3/H_2O mixture formed by dissociation, calculated by Eq. (7-81) with h_F = enthalpy at the dissociation temperature.
For the data specified, rectification of the ammonia–water mixture gives: $F = 8.30$ kmol/h, $x_F = 18.1$ mol-% ammonia, and $y_D = 58$ mol-% ammonia: the CO_2 formed is considered as an inert gas and disregarded.

The y/x plot for the ammonia–water mixture (Fig. 7.70) reveals that at higher inlet temperatures more theoretical steps are needed, because of the steeper rise in the stripping lines than at lower feed temperatures.

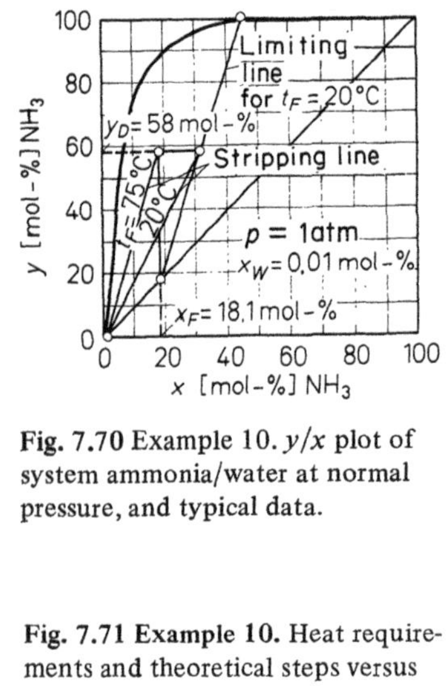

Fig. 7.70 Example 10. y/x plot of system ammonia/water at normal pressure, and typical data.

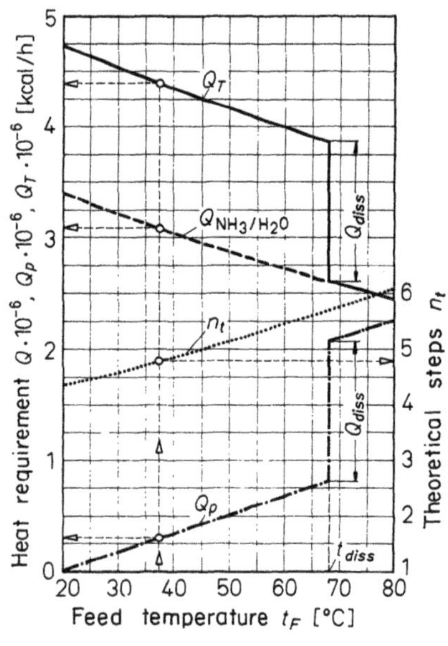

Fig. 7.71 Example 10. Heat requirements and theoretical steps versus feed temperature.

Fig. 7.71 sums up the results. It shows how the number of steps and the heat requirements change with temperature. The slight advantage of having fewer steps in the rectifying zone at lower feed temperature is neutralized by the increased building height for the preheating zone and the larger column diameter caused by the increased vapor load. The curve giving the number of steps refers to the separation of ammonia and water only. The corresponding change in the heat required is shown by the dotted line.

The lower vapor loads at feed temperatures above dissociation temperature (Fig. 7.72) entail an appreciable diameter reduction. Thus, advantages of hookup b) are

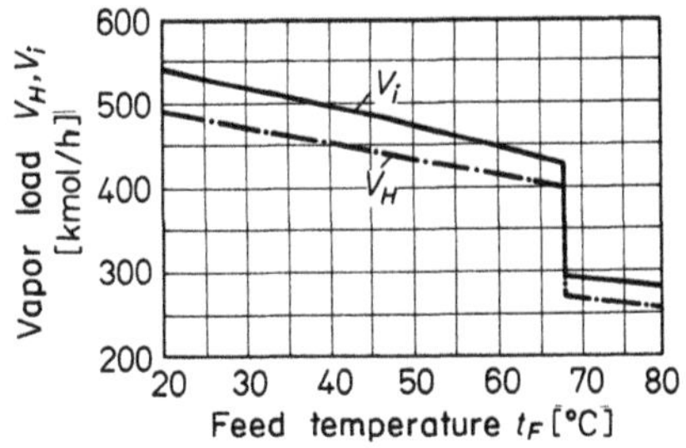

Fig. 7.72 Example 10. Vapor loads below feedpoint and at column bottom versus feed temperature.

the lower column cost and the lesser use of utilities. The juxtaposition of these savings and the expenses for the heat exchanger justify a preference for hookup b).

VIII Detail Planning of Separating Columns

Once the number of theoretical steps and the flows of vapor and liquid have been calculated for a given separation, and a choice of column internals has been made, the principal dimensions of the equipment can be arrived at. A searching analysis of the fluid motions in the column, or of the mechanisms of separation is not attempted; rather, means will be offered for establishing specification data sufficient for actual construction of process equipment for industrial use.

1. GENERAL VIEWPOINTS IN THE SELECTION OF COLUMN TYPES

Chapter 6 presents some angles for the comparison of separation devices; some additional viewpoints that may carry weight in the ultimate decision for a given situation should also be reviewed.

For historical reasons, packed columns predominated in the chemical industries of the past, and plate columns have been customary in petroleum refining. As one might expect, the petrochemical industries made use of both types. If the standard practice of the day did not avail itself to the optimum answer, it was due to insufficient experience or to the lack of design data. Factors that control a decision on use or nonuse of column internals, other than cost, are listed under five headings, as follows:[1]

A. Physical and Chemical Properties of Systems

Generally, a packed column is more suitable for use with foaming systems because the foam will collapse more readily in them. The tray column offers advantages where the systems contain sludges or fats. Thus, the moving discs of valve plates, especially the valve guides moving in their cages located near the vapor streams, have a self-cleaning action that prevents clogging of the openings.

Tray designs do not generally favor the use of nonmetallic construction materials; hence, ceramic parts are limited to packed columns handling corrosive systems. Examples include the manufacture of hydrochloric, sulfuric, and other acids, alkalies, and oxidants.

Systems liberating considerable heats of solution or reaction are preferably treated in plate columns in which pipes for cooling may be mounted on plates inside the body of liquid, which promotes effective heat transfer. Such devices are, however, infrequent in rectification.

Use of packed columns is recommended with heat-sensitive substances that might decompose inside column bases; moreover, packed columns permit heating by low-pressure steam and can be of lighter construction.

B. Characteristics of Fluid Phases

Generally, in packed columns and through the main load zone, liquids are in a dispersed condition, and vapors are in a state of turbulence. This promotes good material separation for systems whose resistance to mass transfer rests in the vapor phase. The reverse condition takes place in tray columns where the vapor phase is dispersed while the liquid is in turbulent flow; this allows for good material separation in systems whose resistance to mass transfer is in the liquid phase.

Packed columns are often thought to be unsuitable for low liquid loads, and tray columns with baffles near weirs and crenelated downspouts are recommended. However, the tests mentioned in Chapter 6 have shown that dumped Pall rings and webbed packing can operate at very low liquid loads. Aluminum Pall rings, for instance, as applied in tall-oil refining, can operate at liquid loads of only 1.5 m^3 per m^2 of the free column section per hour, with a separating effect of 1.6 theoretical steps per meter of depth. The pressure loss observed was 22 mm water per meter of packing depth. Based on the theoretical step, that is half the loss of a well designed tray column.

It had been thought that the maximum loadability of packed columns was less than that of plate types. This is true only for thick-walled packing elements of small free clearance volume. For thin-walled elements of large free clearance volume the loadability may exceed that of plate types.

C. Packed Columns versus Plate Columns

It used to be thought that large diameter packed columns were less efficient, and because of their need for redistributors, they were regarded as more expensive than plate columns. This may be true in some cases, but is not necessarily so. Operating conditions and materials of construction are the deciding factors.

Thus, in columns for diethanol amines, packing of ceramics, steel or polypropylene may be employed. The latter is increasingly preferred because of its low weight, easy handling, resistance to corrosion, and to incrustation—deposits on packing increase the pressure loss. Interim periods between two runs can thus be shortened considerably. Since the density of polypropylene is less than that of water, the column cleaning may be done by flotation and treatment with steam and water. Special polypropylene packing elements may stand temperatures to 130 C.[2]

In many separations the total weight of packed columns exceeds that of plate columns, and generally column internals are attached in the shop rather than on the site. Plate columns are often preferred to packed columns, and plate columns are chosen if sidestreams are to be produced (Fig. 7.44). The refining of crude oil is one example. Extraction points are easier to design for plate columns than for packed columns; the same holds for feedpoints.

Habitually, packed columns are considered to be more economical and more practical when their diameter does not exceed 400 to 600 mm, as it is difficult to fit small columns with trays, and their cost per unit column section is relatively high. On the other hand, tray columns are economical for sizes exceeding 900 mm because redistributors for packing are costly.

There are exceptions, as in tall-oil refining. While a plate column requires a stainless quality steel, the packed column may have aluminium rings, and only the support construction must be stainless. Use of aluminum for the plates would be impossible in view of the 273 C temperature for reasons of mechanical strength.

D. Operation

In a batch operation aiming at the recovery of a high percentage of volatile low boiler (-narrow cut-), the packed column is favorable because of its low liquid holdup. In batch rectifications it may be desirable to rain a larger quantity of liquid during intervals. In a conventional bell-plate column, the vapor stacks will prevent the liquid from raining through. Handholes in plates may permit the plates to run empty before the onset of the next run, however, the necessary liquid reserve is recovered more quickly than for any other fractionator. Depending on the task specified, in continuous service as well, the aforementioned factors and the ultimate column cost will decide on the final choice of column internals; note Chapter 6.

A seasonal market condition, or a gradual expansion in the demand for certain products may require that a rectifying column work dependably over a wide load range. Feed load fluctuations in excess of 2.5 to 1 generally favor plate columns, and especially valve-type plates, if the reflux ratio, and hence the specific steam demand, must be held to ensure the proper load for good separation. The valve plate stays efficient within wide load limitations. At low loads, part of the valves stay closed; that provides an adequate dynamic pressure loss to prevent raining through. As opposed to this situation the efficiency of a sieve plate would be lost by raining through (Figs. 6.54 and 6.55).

When no fluctuating feed load is expected, sieve plates or packed columns will perform satisfactorily, although sieve plates must be checked for economy if the phase level changes along the column depth. Sieve plates would then have to be designed for variable loads within the column, and the valve tray would seem attractive in view of its wide load band.

E. Availability

If a short delivery is required, packing elements can normally be had from stock, while plates entail design work and special fabrication. The availability of the packing wanted in a given case may become an important consideration. As a rule, however, deliveries will be more critical for components such as column shells, compressors, and heat exchangers, so that the procurement of the distilling gear in large plants is rarely momentous. Procurement does become significant if during a shutdown period certain internals or packing must be replaced promptly.

2. PACKED COLUMNS

When the admissible pressure loss is small, packed columns are preferred. This applies primarily in the working of high-molecular and high-boiling, temperature sensitive materials that tend to decompose and polymerize, thereby limiting the temperatures at the column bottom. Many organic materials can be rectified in vacuum only, and then the pressure loss may decide the choice of packing, as it fixes the absolute pressure at the column foot with constant column-head pressure.

Table 8.1. is a specification of some presently popular ceramic packings, shown pictorially in Fig. 8.1. In their ultimate evaluation, their cost and functional properties, as separating power, loadability, and pressure loss, decide, as discussed in Chapter 6, in a juxtaposition with metallic Pall rings and Interpack elements. It had been seen that

Table 8.1. Ceramic packing elements

Type	d (mm)	Size (kg/m³)	Number (Pieces/m³)	Surface (m²/m³)	Free Volume %	a/ϵ^3 (1/m)
Raschig Ring	25	620	46,000	200	73	520
	50	520	6 400	90	78	214
	80	535	1 530	60	77	122
	25	620	46,000	220	73	352
Pall Ring	50	520	6 400	120	78	142
	80	540	1 530	75	76	86
	25	700	75,000	260	69	360
Berl Saddle	50	600	8 000	120	73	147
	80	550	2 200	60	75	–
	25	610	84,000	255	74	321
Novalox Saddle	50	530	9 300	120	79	131
	80	595	1 870	92	80	72

the separating effect and pressure loss of a packing bed are the greater and the loadability is less the smaller the element dimensions, and small elements have a greater load dependence of separation than larger ones.

Even now the planning of an industrial packed column is not always straightforward. The bibliography, while listing many fundamental and statistical efforts in the field, is mostly specialized so that the results are subject to limitations.

An exact computation of packed columns on theoretical grounds is virtually unfeasible because of the complexity of the flow of the phases and the nonuniform mass transfer in the irregularly dumped packing elements. The only way to investigate the correlation between the physical data of the mixtures rectified and the operational conditions is offered by the experiment. The main disturbances are rivulet

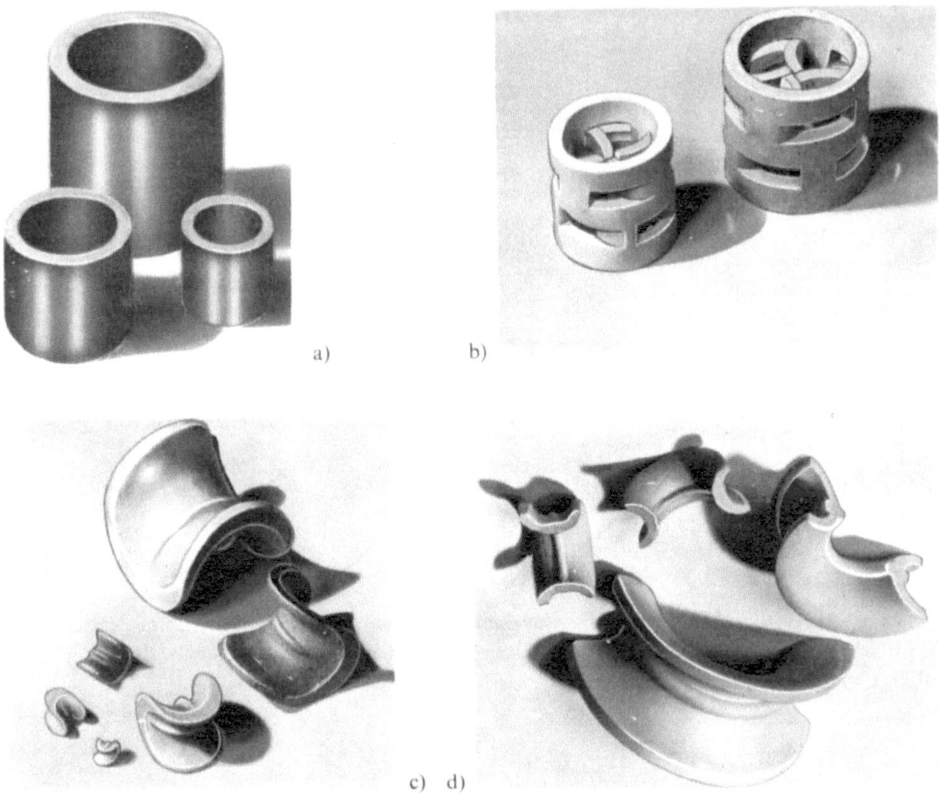

Fig. 8.1 Packings for distillation. a) Raschig rings, b) Pall rings, c) Berl saddles, d) Novalox saddles.

formation and blowby;[3] they are particularly noticeable with greater packing depth. Moreover, in large columns the probability of irregular flow of the phases is much greater than for small diameters. To avoid excessive blowby of the phases caused by an orientation of elements along the shell circumference, the ratio

$$\phi' = \frac{d_s}{d} \tag{8-1}$$

of the column diameter and the element diameter should be no less than 10.[4] For ϕ much above 10, separation will deteriorate more or less, depending on the degree of liquid mal-distribution u which may vary within the limits of

$$0 \lesssim u \lesssim 1 \tag{8-2}$$

The degree of separation η (%) of a packed column could be established as

$$\eta = \frac{\left.\dfrac{n_t}{H}\right|_{u>0}}{\left.\dfrac{n_t}{H}\right|_{u=0}}(100) \tag{8-3}$$

the ratio operation involving mal-distribution ($u > 0$) against an operation without irregularities ($u = 0$). Inclusion of Eq. (8-5) gives

$$\eta = \frac{100}{1 + \dfrac{r}{r+1}\dfrac{u}{m}\left(u + \dfrac{2.5}{r}\right)\dfrac{4}{15}} \tag{8-4}$$

where r = the reflux ratio, and m = mixing per theoretical step, approximately

$$m = 27\left(\frac{d}{d_s}\right)^2 = \frac{27}{\phi^2} \tag{8-5}$$

If the separation effect $(n_t/H)_{u=0}$ of a packed column is known, barring mal-distribution, Eq. (8-4) combined with Eq. (8-5) will supply the separation effect including mal-distribution, and the relation

$$\left(\frac{n_t}{H}\right)_{u>0} = \frac{\eta}{100}\left(\frac{n_t}{H}\right)_{u=0} \tag{8-6}$$

permits computing the separation. For total reflux, $r = \infty$ or $L/V = 1$, Eq. (8-4) becomes[5]

$$(\eta)_{r=\infty} = \frac{100}{1 + 0.266\dfrac{u^2}{m}} \tag{8-7}$$

Fig. 8.2 reveals that for reflux ratios $r = \infty$ and $r = 10$, ϕ values exceeding 30, and mal-distribution exceeding 0.3 the separation efficiency falls below 50%. The smaller the reflux ratio, the more the efficiency drops.

Hence, one prerequisite for good separation in a packed column is uniform liquid distribution at the feedpoint or column head, and proper redistributor design. A device, according to Fig. 6.21, with roughly 6 distributing points per dm^2 demonstrated that the u values definitely depend on the load. Specific liquid loads $L^* = 8$ to $12\ m^3/m^2\ h$ gave u maxima = 0.07, and $L^* = 4$ to $6\ m^3/m^2\ h$ gave $u = 0.25$.

According to Eq. (8-5), in columns exceeding 1 m in diameter good mixing, or in the

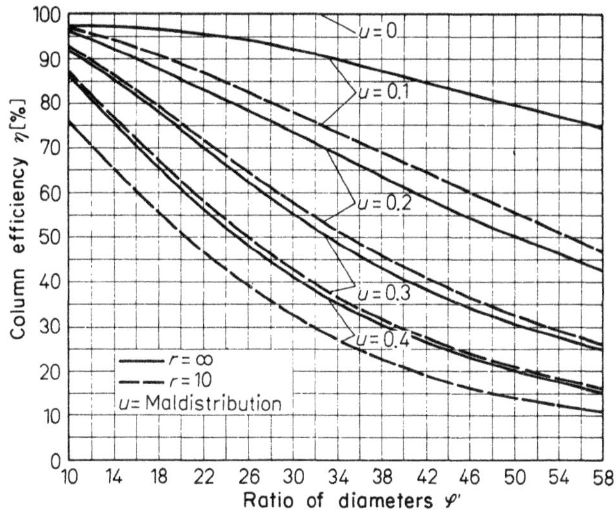

Fig. 8.2 Packed columns. Efficiencies versus ratio column size/packing size, for various degrees of mal-distribution. Total reflux.

event of mal-distribution a good separation, cannot be achieved. The reflux must then be gathered in devices for remixing and redistribution.

A good packing element therefore must compensate for any segregation taking place and help effective lateral mixing if it is to be used successfully in large columns.

Barring mal-distribution, a good packing element of nominal size d (mm) should yield in its main load range a mean theoretical number of steps (equivalent to the separating power of a depth H of 1 m) around

$$\left(\frac{n_t}{H}\right)_{u=0} \approx \frac{75}{d} \tag{8-8}$$

For testing columns determining n_t/H for $u = 0$, if possible the ratio $\phi' = 10$ should be held fairly closely in order to minimize the effect of an eventual mal-distribution. The latter affects the separation to a minor extent when mixing is intense and the ϕ' values are correspondingly low; however, it serves as a measure of the unevenness of the fluid motion across the column section.

Low ϕ' values (below 10) are undesirable because efficiency may be lost by blowby caused by a lineup of packing elements along the column shell. Mal-distribution appears especially at low liquid loads, and for packed columns with ϕ' larger than 10, the separation may be influenced by the loading. Rather, the liquid distribution would be improved by higher load, and indeed the experimental curves in Chapter 6 confirm this for columns of $d_S = 500$ mm and 800 mm.

The plot in Fig. 8.3 will aid in the estimation of a vapor load under flooding conditions in a packed column. It was derived from the general relationship[6]

$$\frac{(V\mu_V)^2 a v_L^m \Psi}{g\epsilon^3 \rho_V \rho_L} = f\left[\left(\frac{L\mu_L}{V\mu_V}\right)\left(\frac{\rho_V}{\rho_L}\right)^{1/2}\right] \tag{8-9}$$

Eq. (8-9) is evaluated by means of a flooding curve[7,8] plotted in a log-log coordinate system. The "packing factor" α/ϵ^3 (where α (m^2/m^3) is the specific packing surface, and ϵ is its relative clearance volume) is a variable primarily dependent on configuration and dimensions of elements, tightness of packing, and wetting at the flooding point. This must be found by experimentation, and some data are given in Table 8.1. Furthermore:

ρ_V = specific weight of vapor (kg/m^3)
ρ_L = specific weight of liquid (kg/m^3)
ν_L = its viscosity (cP)
Ψ = ratio of specific weights of water and liquid
g = acceleration of gravity (m/s^2)

The explicit form of Eq. (8-9) permits a calculation of the vapor velocity at flooding by the relation

$$(u_V)_{fl} = \left[(O_V)_{fl} \frac{g}{\frac{a}{\epsilon^3}\frac{\rho_V}{\rho_L}\nu_L^m \Psi}\right]^{1/2} \tag{8-10}$$

where $(O_V)_{fl}$ is the ordinate corresponding to the abscissa value

$$A = \frac{L\mu_L}{V\mu_V}\left(\frac{\rho_V}{\rho_L}\right)^{1/2} \tag{8-11}$$

(note Fig. 8.3).

The viscosity exponent is $m = 0.2$. Also, in Fig. 8.3. an upper and a lower load line are drawn. In between, operating conditions agree with the main load range of Fig. 6.1.

In the layout of a packed column, a vapor velocity should be chosen that offers a sufficient margin of safety, as far as vapor velocities at fluctuating loads are concerned. On the other hand, packed columns are not suited to low loads, certainly not in vacuum operation where the dynamic loadability of the packing will change appreciably along the column height. As a result, shell diameters have been increased for the upper column section, especially with tray columns since packed towers would demand the use of liquid redistributors.

Failures of packed columns in industrial use have been attributed to lack of design data. To date there is no generally useful theory for the calculations in a given application

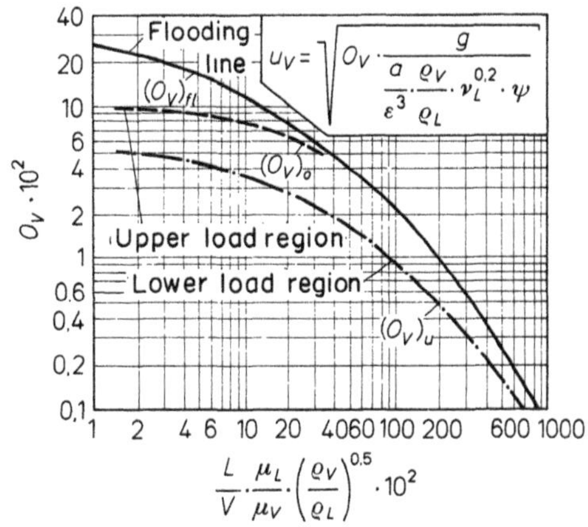

Fig. 8.3 Packed columns. Chart for finding the flooding point.

and experimentation continues to be a necessity. Still, some facts about diverse packings made of ceramics, plastics, and metals, have become established and are discussed in the following sections.

A. Ceramic Raschig Rings

Ceramic packing is popular in the separation of systems that would corrode metallic equipment at higher temperatures. One well-established packing element is the Raschig ring, named for its inventor; its diameter equals its height, and it is typically made of ceramic material. It frequently serves as a standard in comparisons with other elements (Fig. 8.1.a).

a. Separation, Pressure Loss, and Loadability

Best separation $(n_t/H)_o$ of an industrial packed column takes place at top loads; note the dotted line in Fig. 8.3. Table 8.2.[12] presents, for various systems, maximum separations achieved with the packing of dumped Raschig rings 1 m high, at top loads and for different ring sizes between 8 and 50 mm. The diameter ratios ϕ' are between 10 and 22.

The results apply to binary mixtures and were obtained at total reflux and with uniform liquid distribution, no marked mal-distribution, and at various pressures.

Fig 8.4[12] shows the marked dependence of the separation of the packing size; the theoretical steps of a column 1 m high $(n_t/H)_{o1}$ at top load $(u_V)_o$ and at a load of $u_V = 0.65(u_V)_o$ are plotted versus ring size d.

The points entered are those for ethanol/water at normal pressure in Table 8.2. Fig. 8.5 presents, for the same mixture and the same pressure, and for two packing sizes, the theoretical steps $(n_t H)$ and n_t/H versus height H of the packing.[12]

Table 8.2.

Dumped Raschig rings. Maximum separation per meter of depth at upper load limit, depending on ring sizes and systems.

Raschig Ring	φ'	Symbol	p (mm Hg)	$(n_t/H)_{o1}$ (1/m)	x_W (mol-%)	Mixture
8 × 8 × 1.4	12.5	○̍	760	11.0	14	Benzene/ Ethylene Chloride
			400	9.4		
			100	6.3		
		♀	760	12.4	50	
			400	9.8		
			100	8.6		
		-○	760	7.3	4	Benzene/ Toluene
			500	7.0	6	
			250	6.5	4	
			100	5.0	6	
		ˋ○	760	8.4	5	Ethylene Chloride/ Toluene
			100	6.6		
			50	5.8		
			15	4.9		
		○	760	10.8	3.8	Ethyl Alcohol/ Water
			500	7.8	3.0	
			250	7.3	3,0	
			100	7.3	3.7	
			50	6.2	2.7	
25 × 25 × 3	16	□	760	5.2	2.2	
			500	4.7	3.1	
			250	4.4	2.5	
			100	4.1	2.5	
			50	3.7	2.2	
35 × 35 × 4	21.4	Δ	760	4.0	4.2	
50 × 50 × 5	15	●	760	3.2		

The fact that the curves above $H = 1$ m level off may be explained by a coincidence of the blowby of the liquid on the one hand and of the vapor on the other.[3]

In the calculation of the vapor velocity $(u_V)_o$ (m/s) at top load, where optimum separation is expected, the vapor/liquid ratio may again be derived from

$$(u_V)_{o,L/V=1} = K\left(\frac{\rho_L}{\rho_V}\right)^{1/2} \tag{8-12}$$

where ρ_L and ρ_V are again the specific weights of the liquid and vapor. Factor K for Raschig rings of ceramic material follows from Fig. 8.6. If the liquid/vapor ratio L/V differs from unity, it may be corrected by

$$(u_V)_{o,L/V\gtrless 1} = \frac{(u_V)_{o,L/V=1}}{\left(\frac{L}{V}\right)^{0.3}} \tag{8-13}$$

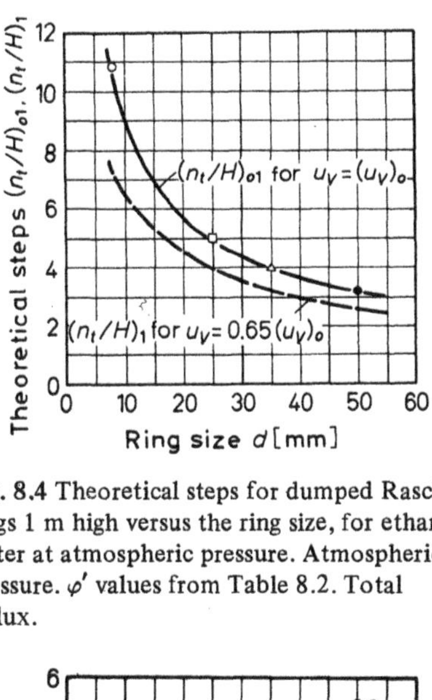

Fig. 8.4 Theoretical steps for dumped Raschig rings 1 m high versus the ring size, for ethanol/water at atmospheric pressure. Atmospheric pressure. φ' values from Table 8.2. Total reflux.

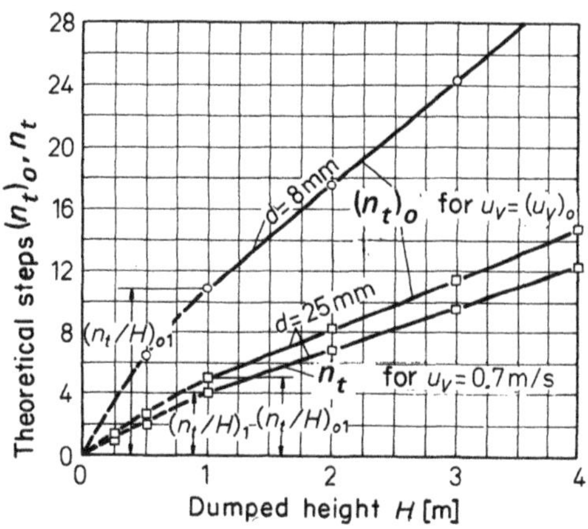

Fig. 8.5 Theoretical steps for dumped Raschig rings versus depth, for ethanol/water at atmospheric pressure. φ' values from Table 8.2. Total reflux.

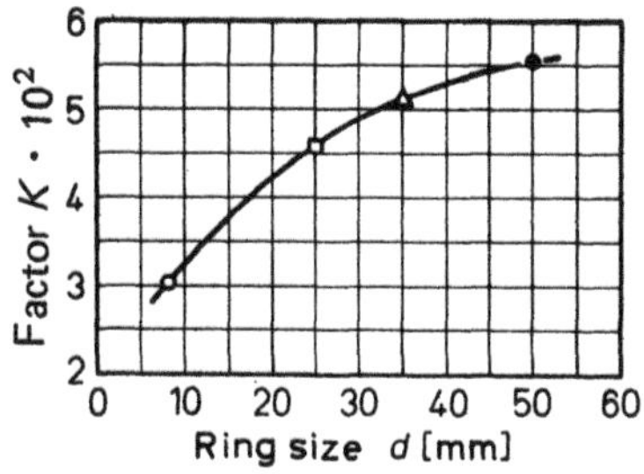

Fig. 8.6 Ceramic Raschig rings of Table 8.2. $L/V = 1$. Vapor velocity at top loads. Packing constant.

b. Dumped Height

To find the height H of a rectifying column expected to achieve a separation equivalent in n_t theoretical stages, first it is necessary to know the theoretical steps equivalent to 1 m of packing depth $(n_t/H)_1$. This is related to the condition of flow in the main velocity range of ceramic Raschig rings,

$$(u_V)_u < u_V < (u_V)_o \quad (8\text{-}14)$$

by a relation found from experiments on such rings[9-13]

$$\frac{(n_t/H)_1}{\xi_1} = 0.13 \quad (8\text{-}15)$$

The flow factor ξ_1 for 1 m of packing depth is

$$\xi_1 = \frac{2g}{u_V^2 \rho_V} \frac{\Delta p}{H} d\,(10^{-3}) \quad (8\text{-}16)$$

where $\Delta p/H$ (mm water) is the pressure loss of the vapor flowing through, having a specific weight of ρ_V (kg/m^3), and d (mm) is the ring size.

Fig. 8.7 shows the theoretical steps $(n_t/H)_{o1}$ at top load versus the correlated $(\xi_o)_1$ and 0.13 $(\xi_o)_1$ for the systems and conditions of Table 8.2.[12] This confirms the inter-relation of Eqs (8-15) and (8-16).

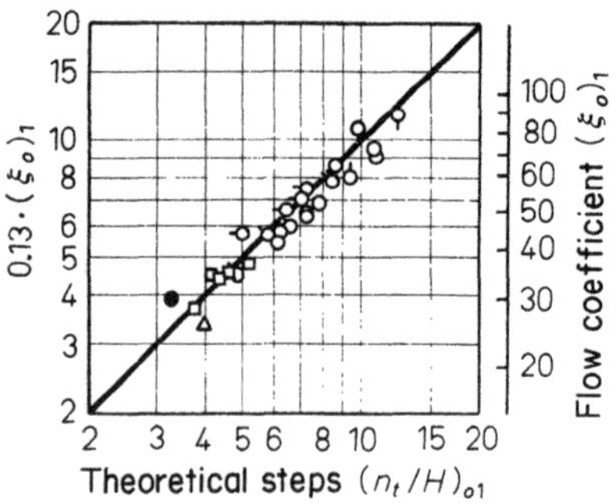

Fig. 8.7 Ceramic Raschig rings of Table 8.2. Flow coefficients for a packed column 1 m high, top load, versus the theoretical steps.

To calculate the height H (m) of a Raschig column intended to separate with n_t theoretical steps, an empirical relation (8-12) was developed from tests and from Eqs (8-15) and (8-16).

$$H = 1 + (n_t - 0.13\xi_1)\frac{10 + d}{120} \quad (8\text{-}17)$$

Eqs. (8-12) to (8-17) demonstrate the effect of packing dimensions upon load and the effectiveness of a packed column. For small packing elements the admissible vapor velocities are reduced, and the flow coefficients are increased because of high pressure losses compared with larger elements. As a rule, small packing sizes increase the column section required and reduce packing depth.

The minimum packing depth is achieved for the top load range, at which maximum separation is due. By Table 8.3, pressure losses may occur in a 1-m packing of ceramic rings that are between 70 and 300 mm water. From this, and with Eqs. (8-12) and (8-13), Eq. (8-16) will yield the flow coefficient $\xi_1 = (\xi_o)_1$, and from Eq. (8-17) the corresponding minimum packing depth will be found.

Table 8.3.

Ceramic Raschig rings. Pressure loss at top load, and total reflux.

Ring Dimension (mm)	$(\Delta p/H)_{o1}$ (mm H_2O/m)
8 X 8 X 1.5	230–330
25 X 25 X 2	100–140
35 X 35 X 4	75–90
50 X 50 X 5	70

In Chapter 6 it is stated that for a given dynamic vapor load the separation by packing is better the more liquid fed to the column. Eqs. (8-15) and (8-16) confirm this; the coefficient ξ_1 is increased because of the greater pressure loss accompanying the greater liquid load, and in consequence the specific separation $(n_t/H)_1$ is improved. An equation[14] that considers the influence of the liquid load closely enough may also serve to estimate the packing depth H (m) for ceramic Raschig rings over the main load range:

$$H = 4.2\, n_t \left(\frac{d v_L}{L'} \right)^{1/2} \tag{8-18}$$

where n_t = the theoretical steps for separation, and d (mm) = the ring size. L' (kg/m^2 h) is the specific liquid load, and v_L (cP) is its dynamic viscosity at operating temperature.

For Eqs. (8-17) and (8-18) to be applicable the phase distribution must be uniform. Mal-distribution can be corrected by applying Eq. (8-4) for the column efficiency.

Fig 8.8. shows the typical dependence of separation upon the load, as found earlier in Chapter 6. It was seen from Figs. 6.15 and 6.31 that the dependence of the separation efficiency on the load is less pronounced at lower pressures and for larger elements. Less separation at very low loads is typical for most packings. This is due firstly to flow conditions less conducive to good mass transfer, and further to incomplete and irregular wetting.

The lessening of the separation effect with lower pressure is caused by a combination of several influences; pressure reduction will change the physical properties of the vapors and liquids. The much lowered specific weight of the vapor in a vacuum means less

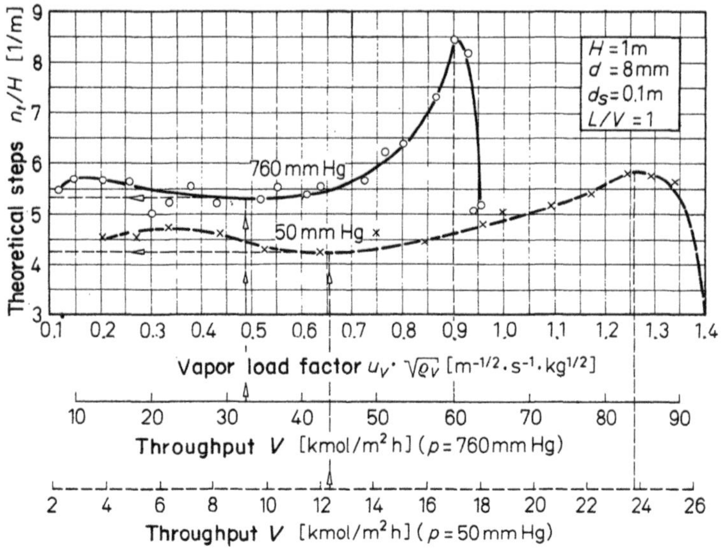

Fig. 8.8 Ceramic Raschig rings 8 X 8 X 1.5 mm in a packed column 1 m high, $\varphi' = 12.5$. Ethylene chloride/toluene mixture. Load curves.

throughput. The flow around the packing element is disturbed. The greater viscosity and surface tension may cause the liquid in the stack to contract into rivulets or at least seep along the individual element, which extends the paths of interchange. This affects separation unfavorably. Fig 8.9 explains how the throughput V_o, corresponding to the top load of a column packed with small ceramic Raschig rings, and the correlated separation $(n_t/H)_{o1}$ vary with the system pressure.

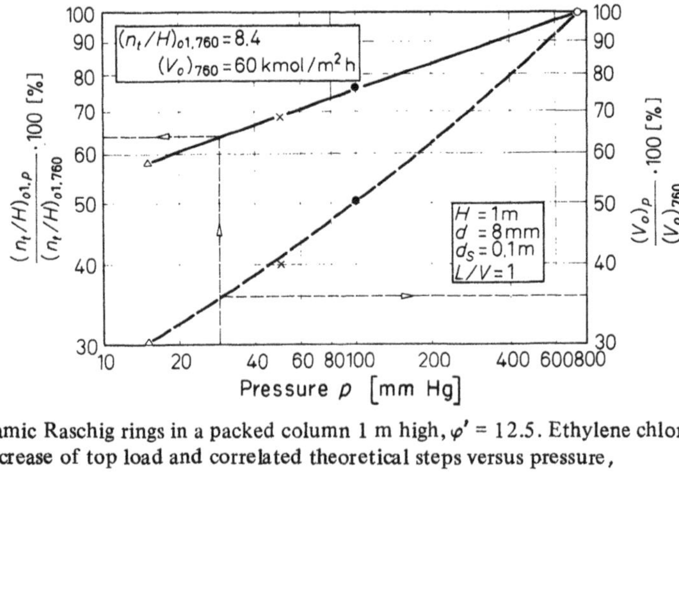

Fig. 8.9 Ceramic Raschig rings in a packed column 1 m high, $\varphi' = 12.5$. Ethylene chloride-toluene mixture. Decrease of top load and correlated theoretical steps versus pressure,

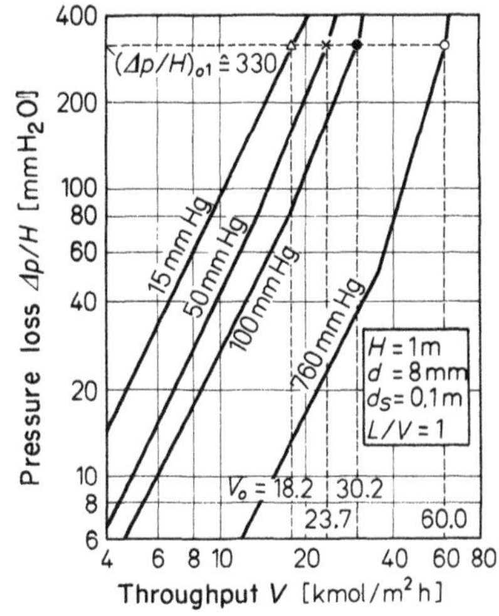

Fig. 8.10 Ceramic Raschig rings in a packed column 1 m high. Pressure loss at various pressures. Ethylene chloride/toluene mixture.

According to Fig. 8.10 the pressure loss undergone by the vapor in passing through packing 1 m deep is larger for a given throughput at smaller pressures than at higher pressures. In the log-log presentation the function is expressed by straight lines that have slopes around 2 in the lower load range. These increase more markedly at increasing pressure in the upper range. Another rise of the pressure loss line occurs above the top load where maximum separation has been exceeded and the column begins to flood. The less the pressure the less the differences in the slopes of the lines, which at 15 mm Hg merge into a single line. The reason is that as the pressure dwindles the liquid quantities become less and less, and hence the pressure loss dependence must come closer to that for unwetted packing, which would appear as a straight line in the log-log plot. The exponential rise of the pressure loss at increasing loads is explained by increased liquid volume, and the ensuing tightening of the remaining vapor passages in the packing.

Example 4 An ethanol-water mixture is to be rectified in a straight enrichment column at ordinary pressure, at total reflux. With a bottom concentration of 4.2 mol-% ethanol, a virtually azeotropic concentration is desired at the column top. Find the depth of packing for a column containing ceramic Raschig rings of 25 x 25 x 2 mm size.

Answer 4 By the method explained in Chapter 3, about $n_t = 15$ theoretical steps will be needed. The azeotropic composition at the column top corresponds to a specific weight of the vapor of $\rho_V = 1.43$ kg/m^3 and a specific weight of the liquid of ρ_L equal to 760 kg/m^3. For $d = 25$ mm, Fig. 8.6. gives a constant $K = 4.58 \cdot 10^{-2}$. Eq. (8-12) gives a top load velocity of

$$(u_V)_{o,L/V=1} = 4.58 \cdot 10^{-2} \left(\frac{760}{1.43} \right)^{1/2} = 1.07 \text{ m/s}$$

Eq. (8-16) yields, for the correlated flow coefficient, $(\Delta p/H)_{o1} = 125$ mm water as the top pressure loss.[11] (also note Table 8.3)

$$\xi_1 = (\xi_o)_1 = \frac{(2)(9.81)}{1.07^2(1.43)} \frac{(125)}{1}(25)(10^{-3}) = 37.4$$

From this, Eq. (8-17) gives a minimum packing depth of

$$H = 1 + [15 - (0.13)(37.4)] \frac{10 + 25}{120} = 3.96 \text{ m}$$

This result agrees with the one from Fig. 8.5, according to which $n_t = 15$ steps correspond to $H = 4$m.

For vapor load V' (kg/m² h):

$$V' = u_V \rho_V (3600)$$

Since $\frac{L}{V} = 1$

$$L' = V' = (1.07)(1.43)(3600) = 5460 \text{ kg/m}^2\text{ h}$$

and Eq. (8-18) will yield, with $\mu_L = 0.6$ cP

$$H = (4.2)(15)\left[\frac{(25)(0.6)}{5460}\right]^{1/2} = 3.3 \text{ m}$$

or 16% less packing depth than by Eq. (8-16).

B. Plastic Packings

Plastic packing may be applied for corrosive systems if metallic materials fail. Their use is limited, however, by the level of the operating temperature which must for polypropylene, for example, never exceed 130 C. An advantage is in the low packing weight, which means getting away with a simple understructure and relatively easy and quick installation. Experience shows that polypropylene shapes are installed in one-tenth the time required for ceramics. Thus higher investment for propylene packing may be recovered by savings in time.

a. Configurations

Some shapes familiar from metallic and ceramic productions will be satisfactory for plastic production as well. It has been observed that certain plastics are less readily

Fig. 8.11 Packing element made of plastic. Grooved to improve wetting. Internal ribs to improve rigidity. Windows to improve pressure loss. 1) Grooves, 2) Ribs, 3) Windows.

wetted than ceramic or metal objects, and therefore less of a separating effect was expected. This, resistance to chemicals, mechanical endurance, and thermal sensitivity, along with the manufacturing procedure, has been grounds for a limited application in the past.

From a process designer's viewpoint a plastic packing may be attractive; note the element made of Lupolen material presented in Fig. 8.11. The wetting has been improved by helical grooving, and the rigidity has been enhanced by the provision of internal ribbing.

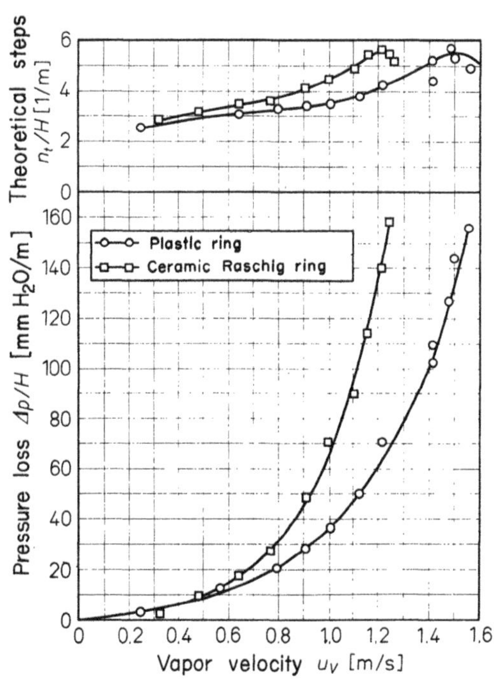

Fig. 8.12 Plastic packing ring of Fig. 8.11 and ceramic Raschig ring, both 25 × 25 mm size. Dumped height 1.1 m, column diameter 0.5 m. Methanol/ethanol system, 760 mm Hg. L/V = 1. Separations and pressure loss.

This also increases the specific surface. To reduce the pressure loss, the opposing points of the wall have cutouts.[16]

b. Plastic Ring versus Ceramic Ring

In the rectification of the methanol–ethanol system at ordinary pressure, and in the rectification of *trans*-decalin–cis-decalin at 10 mm Hg, comparable results were obtained (Figs. 8.12 and 8.13). The column packing was of the type shown in Fig. 8.11, and 25 X 25 mm ceramic Raschig rings were used in parallel tests. The maximum load carried was slightly greater for the plastic packing, and the pressure loss per theoretical step was slightly lower. This is a valuable advantage in vacuum operations.

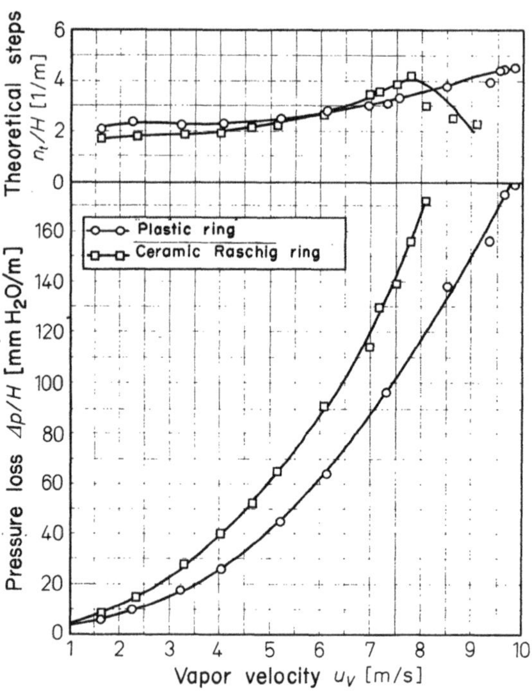

Fig. 8.13 Plastic packing ring of Fig. 8.11 and ceramic Raschig ring, 25 X 25 mm. Dumped height 1.1 m, column diameter 0.5 m. *Trans*-decaline/*cis*-decaline system 10 mm Hg. $L/V = 1$.

C. Metallic Pall Rings

It is seen in Chapter 6 that Pall-ring columns are suitable especially in vacuum rectifications; moreover, they are economical to acquire. Extended tests permit a number of conclusions.[17–22]

Tests to determine the separation n_t/H were run at $L/V = 1$; the corresponding liquid loads and hence the absolute quantities worked differed for the various systems for a given vapor load $u_V\rho_V^{1/2}$; note Fig. 8.14. The specific liquid loads $L^*(m^3/m^2h)$ are referred to a unit of column section and time, against the vapor load factor $u_V\rho_V^{1/2}$ (m^{-1} s^{-1} $kg^{-1/2}$).

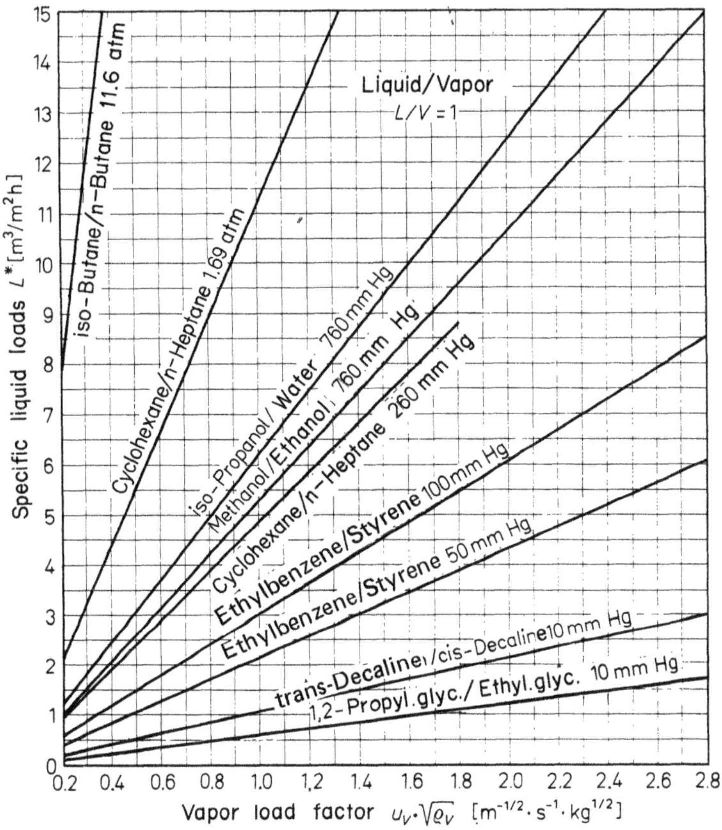

Fig. 8.14 Specific liquid loads at the column head versus the vapor load factor for various systems at $L/V = 1$.

a. Charging and Packing Density

The clearances in a packing layer will affect not only the separation and the pressure loss but also the maximum loadability. Depending on the way the packing elements are charged into the column, packing tightness and clearances vary, and the operational data scatter noticeably. Fig. 8.15 considers four different procedures for charging a column of 500-mm diameter and 2-m packing height:

1. The elements are dumped into the dry column and planed; there are 19,260 pieces per cubic meter of packing.
2. The elements are randomly dumped into the dry column; there are 18,460 pieces per cubic meter of packing.
3. The elements are dumped over a central cone into the dry column; there are 18,500 pieces per cubic meter of packing.

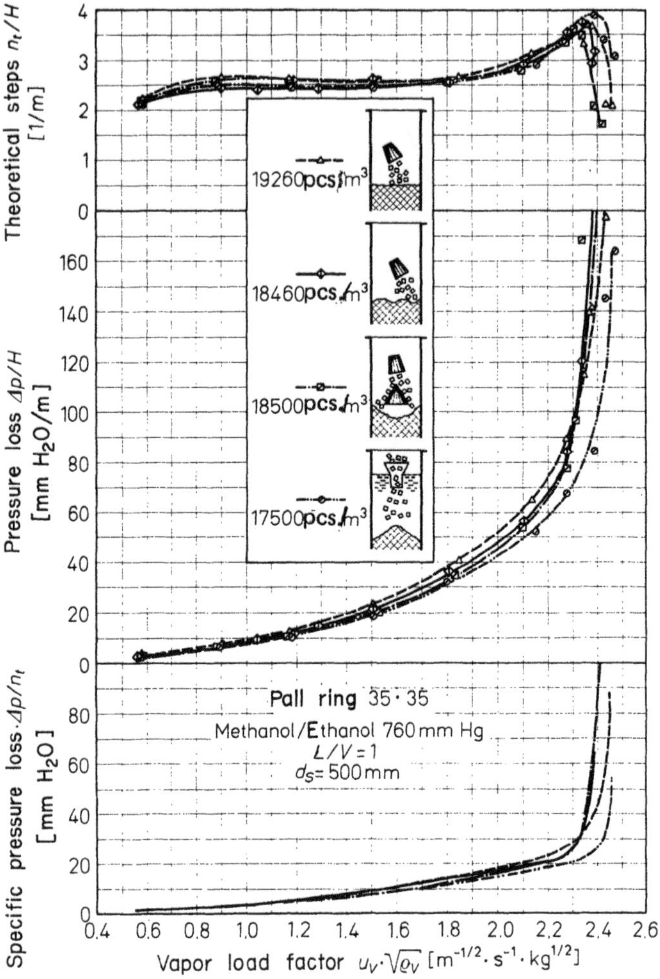

Fig. 8.15 Loading procedure. Influence on separation, pressure loss, and capacity of 35 mm metallic Pall rings.

4. The elements are dumped into the water-filled column via a central funnel; there are only 17,500 pieces per cubic meter of packing.

As might be expected, the more tightly packed column of case 1 gives more pressure loss but also better separation than the less tightly packed column of case 4. The packings of cases 2 and 3 have about equal numbers of pieces, and thus generate about the same pressure loss. The loadabilities differ only slightly. The essential point is that pressure losses for the different loading procedures are only slightly different over the entire load range up to about 90% of the upper limit. This is a significant property of the Pall ring; also note Chapter 6.

b. Effects of Packing Size

Fig 8.16 shows a marked effect of ring size upon separation, pressure loss, and loadability. It applies to the rectification of methanol–ethanol at normal pressure with total reflux in a 500-mm column with 2 m of packing depth. The curves confirm that separation and pressure loss are greater for smaller rings, and maximum loadability is less than for

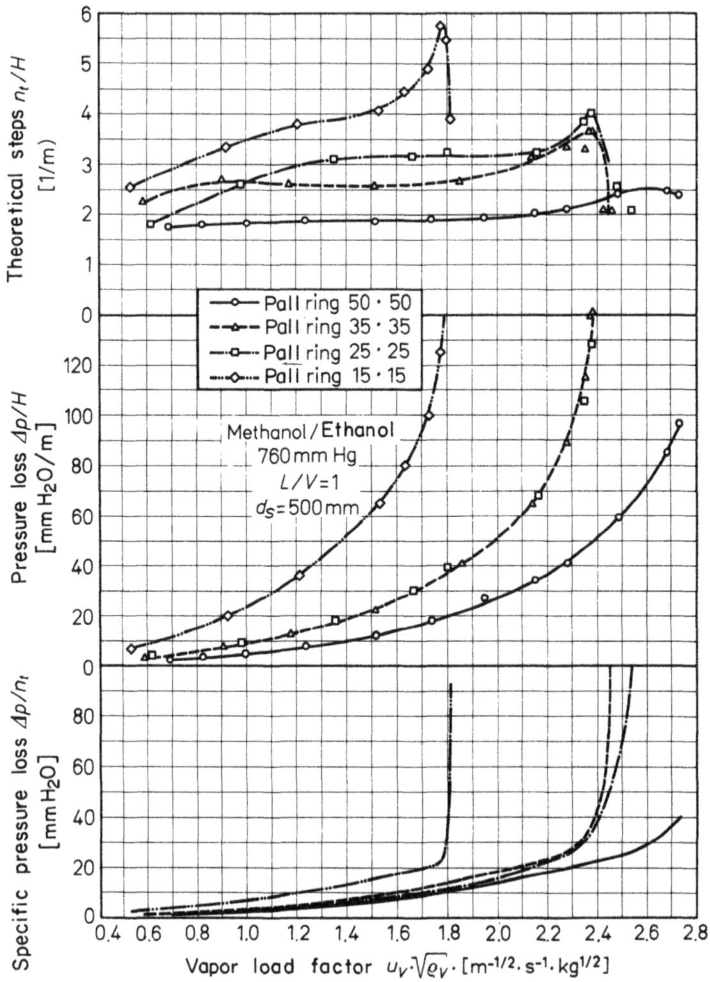

Fig. 8.16 Pall ring packing. Methanol/ethanol system at 760 mm Hg. Ring size versus separation, pressure loss, and capacity.

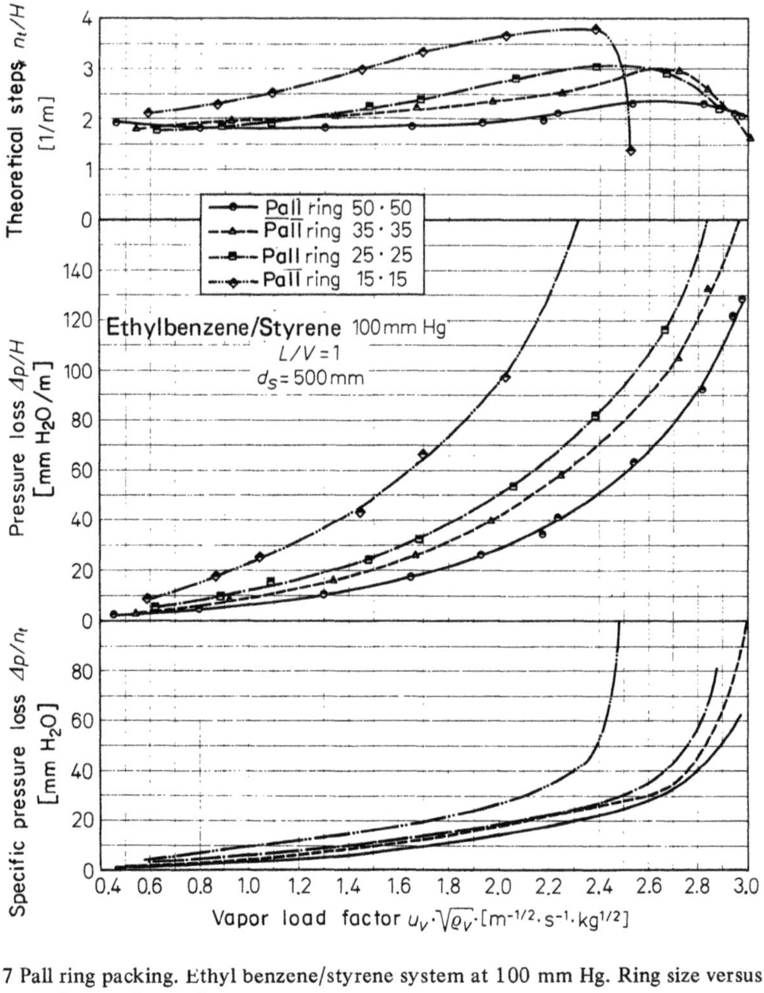

Fig. 8.17 Pall ring packing. Ethyl benzene/styrene system at 100 mm Hg. Ring size versus separation, pressure loss, and capacity.

large rings. The ethyl benzene–styrene system gave similar results with 100 mm Hg pressure at the column top (Fig 8.17).

The 50-mm Pall ring is recommended in vacuum work; it gives best results regarding the specific pressure loss based on the theoretical step.

c. Effects of System and Liquid Load

The higher the system pressure, and from Fig. 8.14 the specific liquid load, the higher, by Fig. 8.18 the separation and the pressure loss for any given vapor load factor. While at ordinary pressures the vapor load factor is less, the absolute quantity throughput is still greater than for vacuum systems.

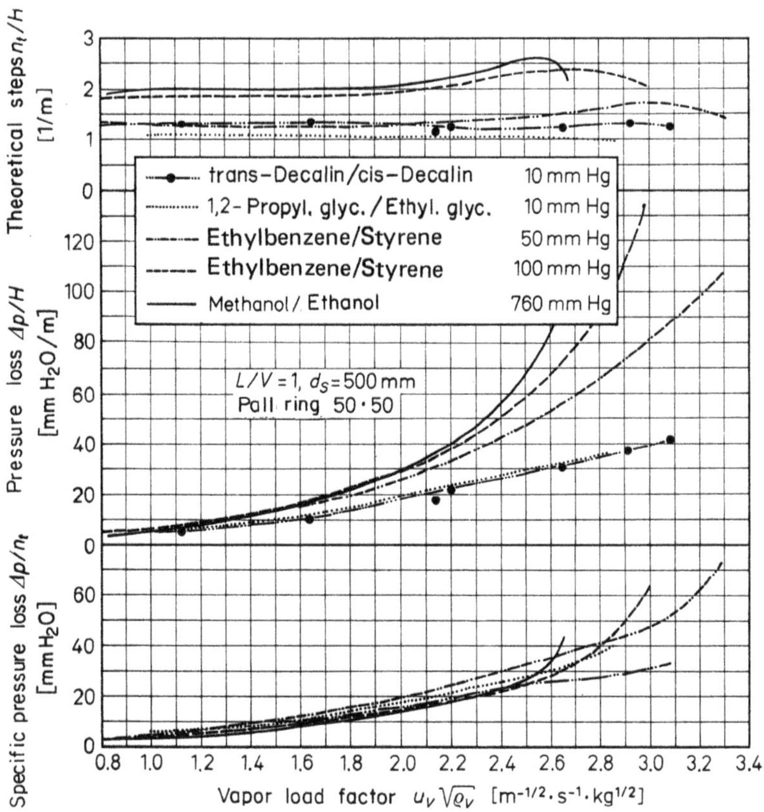

Fig. 8.18 Dumped 50-mm metallic Pall rings. Influence of the system on the separation, pressure loss, and capacity.

The flood loading for a stack of packing can be found from Fig. 8.3 and Eq. (8-10); at this time only systems with liquid viscosities below 20 cP have been investigated, although recently treatment by industrial distillation of much more viscous mixtures has been considered.

Simulated tests have been run with diverse pairs of air/liquid phases, with liquid viscosities to 90 cP, at different liquid loads and temperatures, (Fig. 8.19). The increase in pressure loss as a function of the load factor was the more marked the higher the liquid viscosity.[23]

The data in Fig. 8.19 were obtained in a glass-shell column (Fig. 8.20). Thus flooding could be observed. This began at around the loads corresponding to the pressure loss lines of Fig. 8.19. Note the summary evaluation of flooding loads for metallic packing in Sec. 2.F. of this chapter. The points labeled F are for loads at which the entire packing is submerged completely.

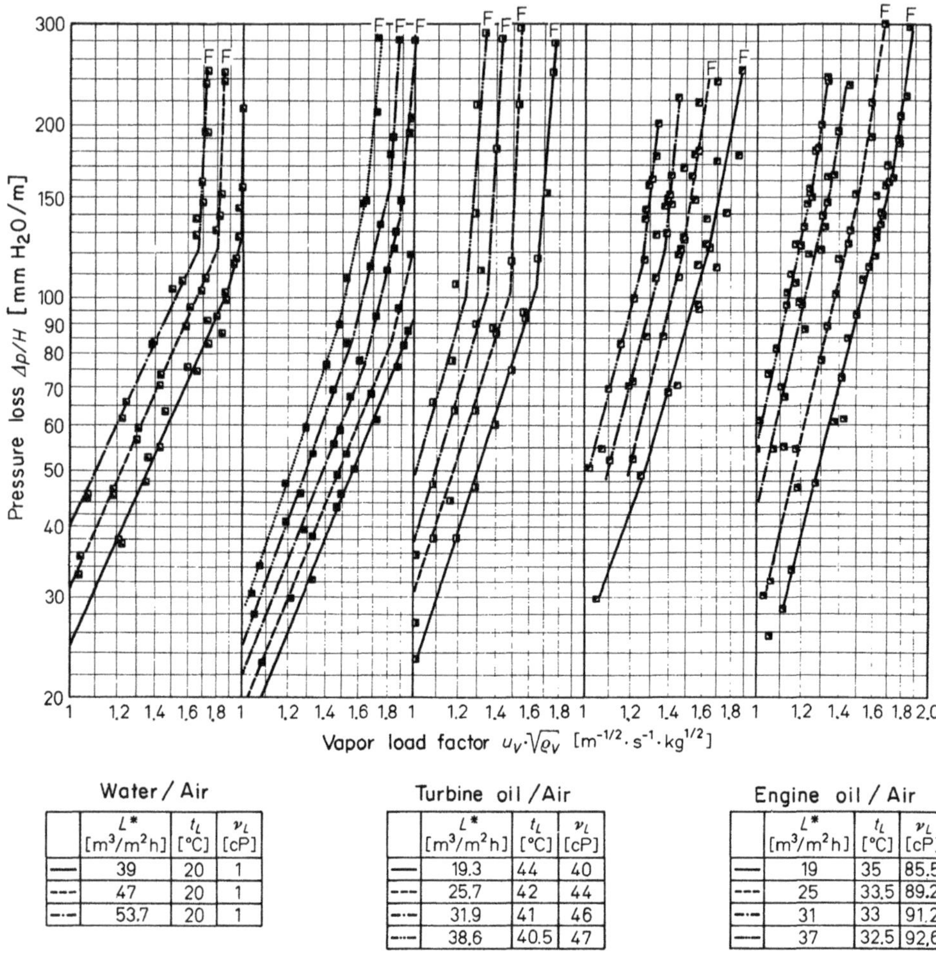

Water / Air

	L^* [m³/m²h]	t_L [°C]	ν_L [cP]
——	39	20	1
---	47	20	1
-·-	53.7	20	1

Turbine oil / Air

	L^* [m³/m²h]	t_L [°C]	ν_L [cP]
——	19.3	44	40
---	25.7	42	44
-·-	31.9	41	46
-··-	38.6	40.5	47

Engine oil / Air

	L^* [m³/m²h]	t_L [°C]	ν_L [cP]
——	19	35	85.5
---	25	33.5	89.2
-·-	31	33	91.2
——	37	32.5	92.6

Glycol / Air

	L^* [m³/m²h]	t_L [°C]	ν_L [cP]
——	20.7	33.5	17
---	27	33	17.2
-·-	33.4	32	18
-··-	39.9	30.5	18 8
······	46.5	31	18 6

Engine oil / Air

	L^* [m³/m²h]	t_L [°C]	ν_L [cP]
——	25	48	51.1
---	32.3	47	52.3
-·-	39.2	48	51.1
-··-	45.75	47.5	51.8

Fig. 8.19 Dumped 25-mm Pall rings. Pressure loss at various liquid loads and viscosities versus the vapor load factor. Vapor used is air.

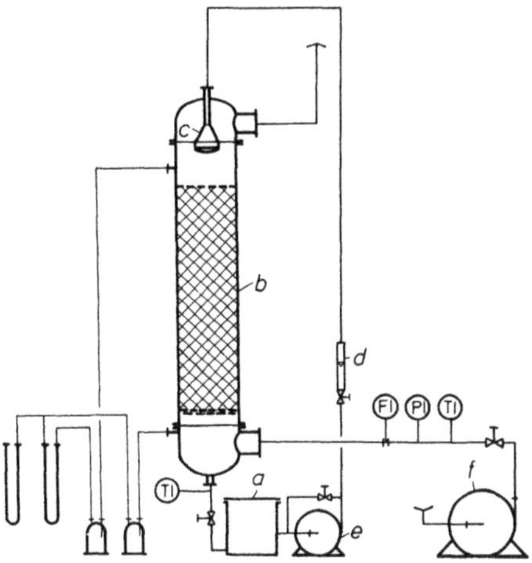

Fig. 8.20 Test setup for measuring pressure losses and flooding points of packed columns: a) liquid tank b) column c) distributor d) rotameter e) pump f) blower.

d. Effects of Packing Depth

It had been found earlier[24] that the separation by ceramic Raschig rings per unit of packing depth diminishes when going down because of the effects of blowby and rivulet formation along the column. Their specific separation n_t/H is less for greater packing depths than for small ones. This characteristic of ceramic Raschig rings is also evident from the curves of Fig. 8.5.

For metallic Pall rings, the effect of the packing depth upon the specific separation appears less pronounced, (Fig. 8.21). Differences in evaluation of different depths rest on different distributions within the available space. The data apply to a 500-mm column filled with 50-mm Pall rings and for the methanol–ethanol system at 760 mm Hg combined with a total reflux; the liquid/vapor ratio is 1.

The curves of Fig. 8.22 lead to the same conclusion. They apply to the rectification of ethyl benzene–styrene in a 500-mm column. The body of 35-mm Pall rings has, for about the same specific density of 19,200 pieces per m^3 and 2-m depth, about the same specific separation and the same maximum load as for 3.95-m depth.

Likewise, for an 800-mm column (Fig. 8.23) containing 35-mm Pall rings, differences in specific separations for various packing depths must be explained by different packing densities. For higher densities, pressure loss and separation are greater than at lower densities. For 50-mm metallic Pall rings the separations are virtually identical.

Figs. 8.21, 8.22, 8.23 all demonstrate that the specific pressure loss per theoretical step $\Delta p/n_t$ across the main load range is hardly influenced by the packing density. This is pointed out in section 2.C.1. of this chapter. The lower plots of these figures indicate the percentages of liquid entrainment by the vapor flow at the column head that becomes notable at flooding.

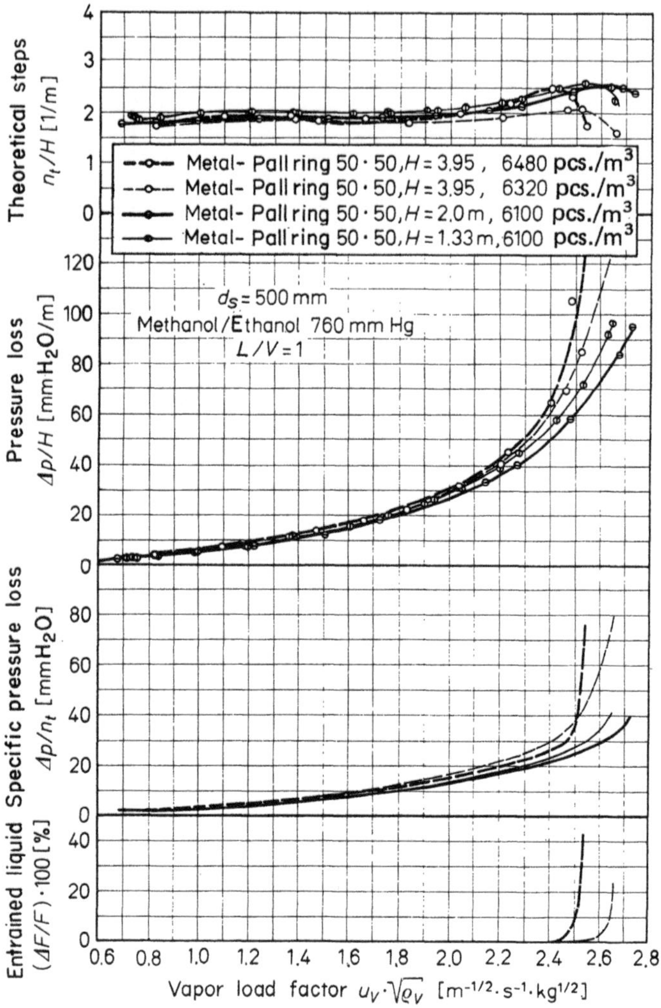

Fig. 8.21 500-mm Pall ring column. Influence of packing depth. Methanol/ethanol system. 760 mm Hg.

e. Effects of Column Diameter

It is a fact that the specific separation of a packed column decreases in the presence of mal-distribution as the column diameter increases. This is why theories for preestimating packing efficiencies have frequently failed to materialize. Usually it had been assumed that the flow densities of vapor and liquid across the column section were constant, and that the theoretical steps n_t/H were therefore independent of column size.

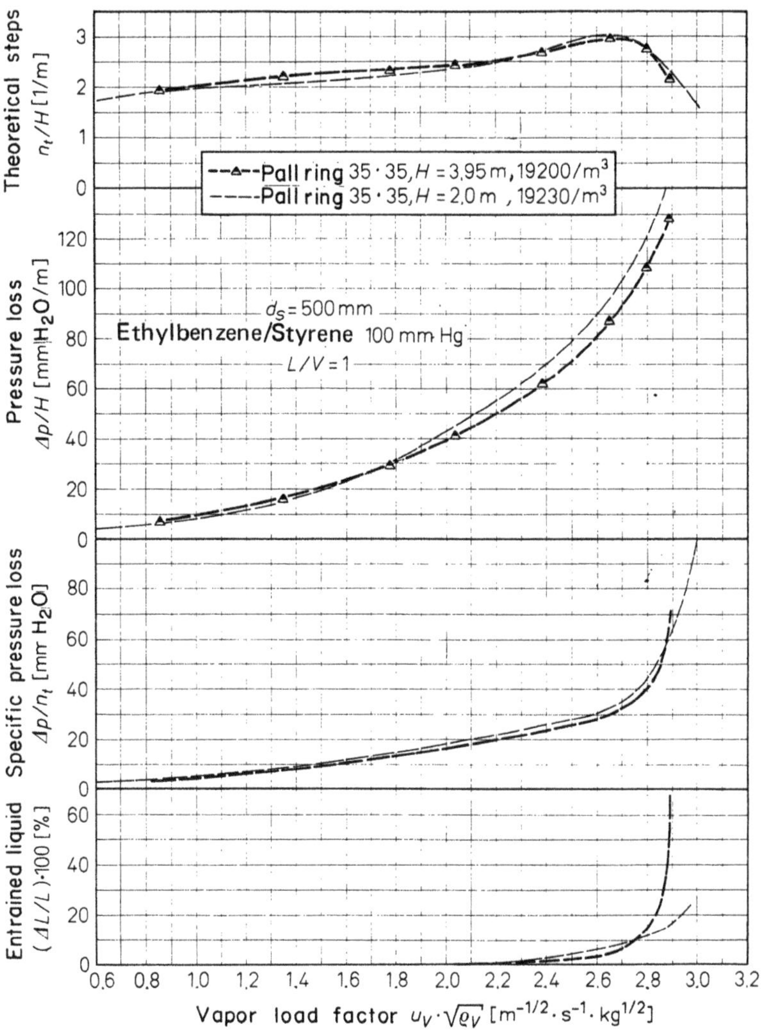

Fig. 8.22 500-mm Pall ring column. Packing depth versus efficiency. System ethyl benzene–styrene. 100 mm Hg.

Fig. 8.24 demonstrates the error involved, for a rectification of methanol–ethanol at normal pressure in columns with 15-mm metallic Pall rings, having sizes of 150, 500, and 800 mm in diameter, and equipped with equivalent liquid distributors with 6 supply points per dm^2. Only at maximum load, when the liquid load of the column is relatively high, is the separation with the 500-mm column the same as with the 150-mm column. Note that the curves for 500- and for 800-mm diameter show about the same load dependence of the separation, although, the sectional areas are 1 to 2.5.

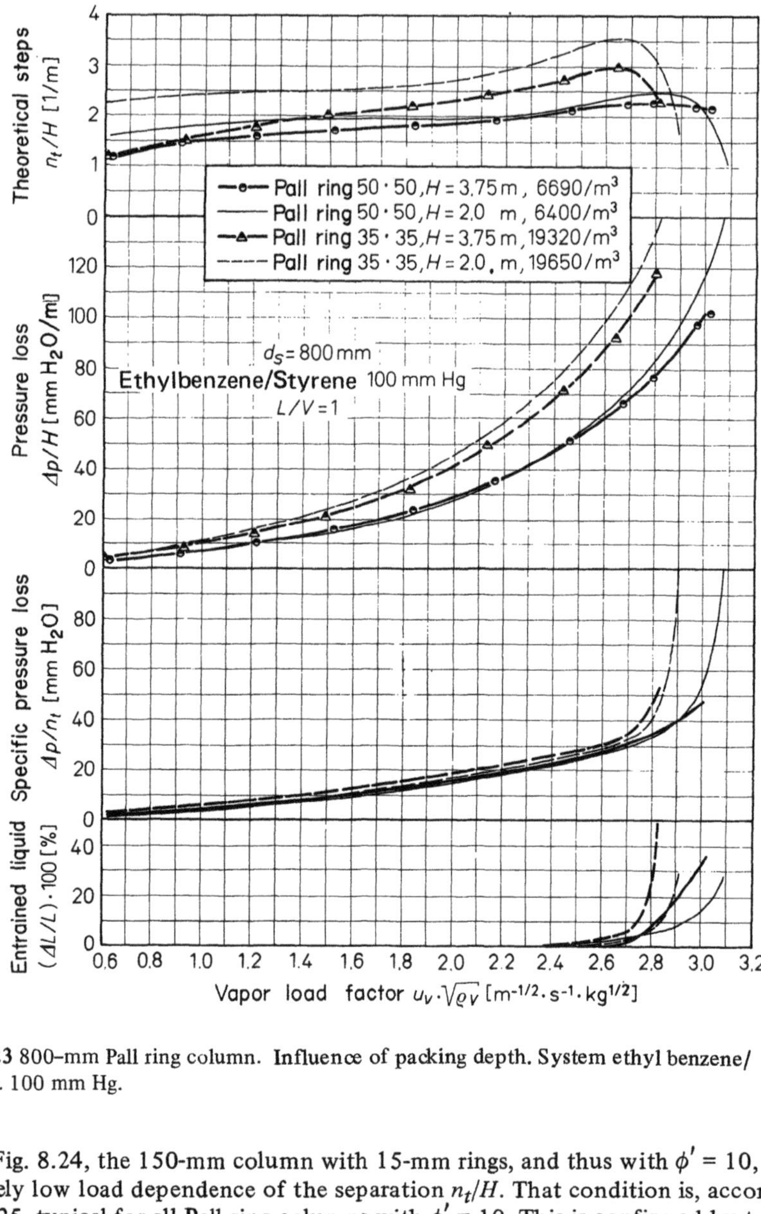

Fig. 8.23 800-mm Pall ring column. Influence of packing depth. System ethyl benzene/ styrene. 100 mm Hg.

In Fig. 8.24, the 150-mm column with 15-mm rings, and thus with $\phi' = 10$, has a relatively low load dependence of the separation n_t/H. That condition is, according to Fig. 8.25, typical for all Pall-ring columns with $\phi' = 10$. This is confirmed by tests involving other ring dimensions and column sizes, and for different systems.

This effect of column diameter, and therefore of ϕ' may be explained by Eqs. (8-4) and (8-5) and from the fact that for small liquid loads the mal-distribution is more marked than for larger ones. Fig. 8.22. reveals that for relatively small ϕ' values, and by

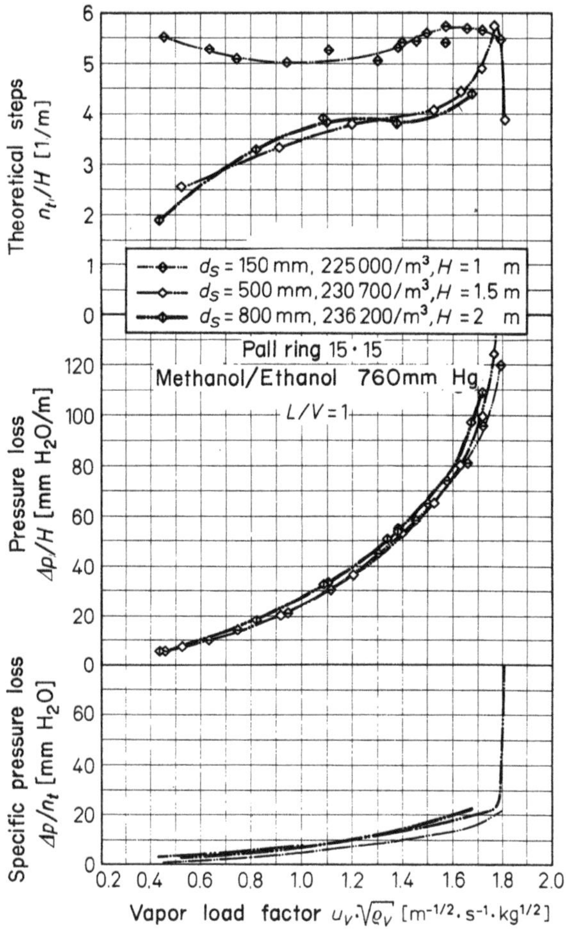

Fig. 8.24 Pall ring columns, constant ring sizes. Influence of column diameter.

Eq. (8-5) for large mixing coefficients m, a large mal-distribution u (a measure of the irregularity of the liquid densities across the column section) will affect the separation only slightly. This holds for $\phi' = 10$. For $\phi' = 20$, the separation deteriorates when the falling load increases mal-distribution.

The liquid redistributor in Fig. 8.26 has only one liquid feedpoint for 2.5 dm^2 of column area, the column having $d_S = 1200$ mm in diameter and a packing of 50-mm Pall rings to a height of $H = 5.5$ m. Under the circumstances no high specific separation n_t/H can be expected. From Fig. 8.27, it can be seen that it equals something between 1 and 1.8 theoretical steps per meter of packing height depending on the system and load factor.[25]

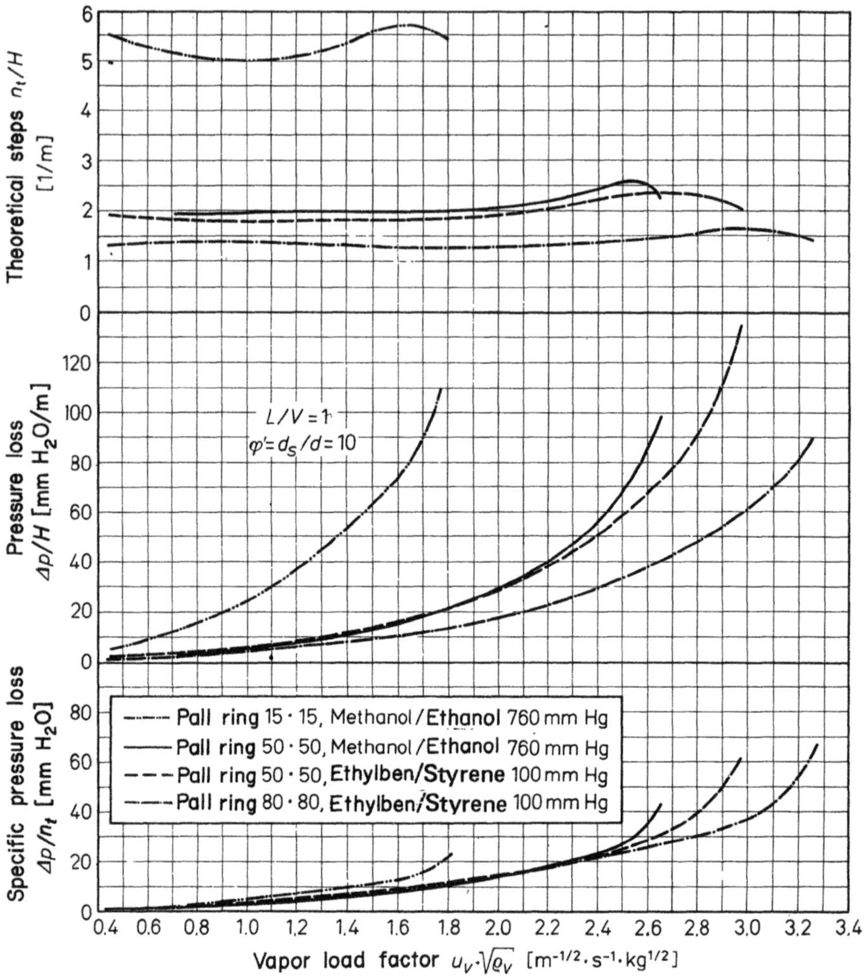

Fig. 8.25 Pall ring columns, packing size = 1/10 of column size. Influence of column diameter.

Measurements on concentrations along the column height indicate that, because of irregular liquid distribution at the column top, the specific separation was at times less in the upper column than in the lower zones where the distribution of the liquid improved gradually. The data of Fig. 8.27 were obtained from a setup outlined in Fig. 8.28.

Fig. 8.29 shows the influence of the column diameters upon column efficiency for various ring sizes. The system is ethyl benzene/styrene at 100 mm Hg. Within the main load range, the theoretical stages per unit of packing height n_t/H are higher for the 800-mm column than for the 500-mm column; the same goes for the pressure loss $\Delta p/H$,

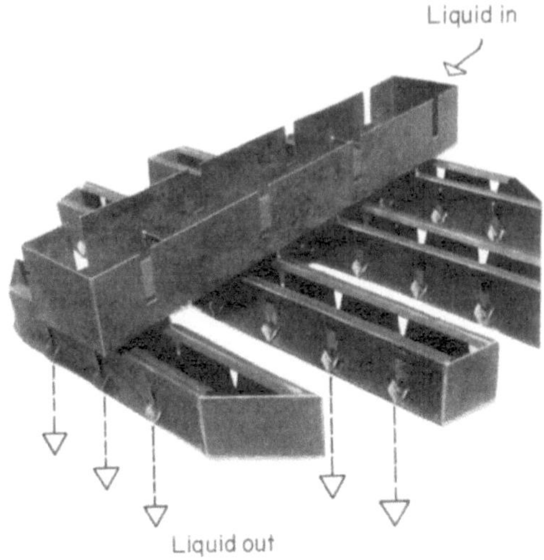

Fig. 8.26 Liquid redistributor for packed columns.

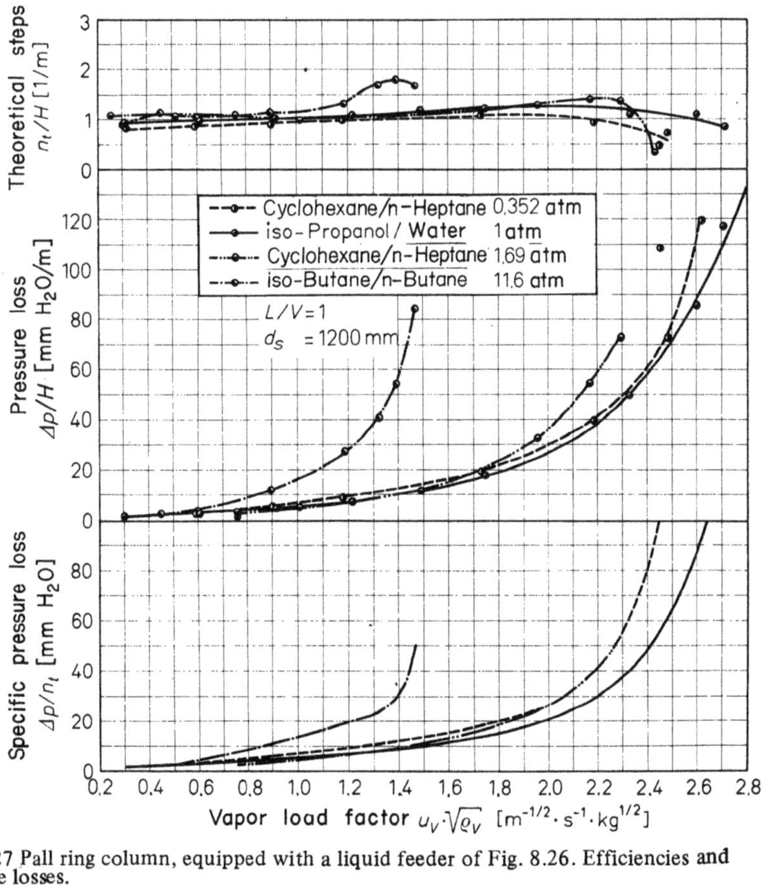

Fig. 8.27 Pall ring column, equipped with a liquid feeder of Fig. 8.26. Efficiencies and pressure losses.

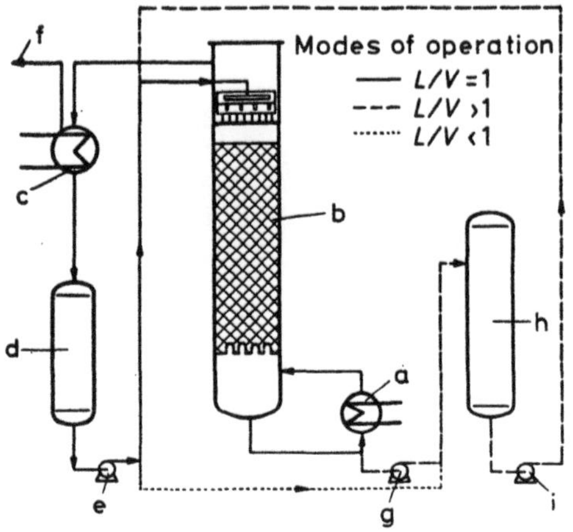

Fig. 8.28 Flowsheet for rectification tests. a) reboiler, b) column, c) condenser, d), h) holdover tanks, e), g), i), pumps, f) vacuum connection.

ring sizes being 15 to 35 mm. The probable reason is the greater packing density in the 800-mm column. For the 50-mm Pall rings the differences in the n_t/H values are negligible.

f. Specific Pressure Loss and Packing Depth

For a given packing element size, the separation and the pressure loss per unit of packing depth may differ greatly; note Figs. 8.15 to 8.18 and 8.21 to 8.25. However, for a constant load factor within the main load range, the pressure losses per theoretical stage differ up to ±20%, although the liquid loads of the systems in question differed up to a factor of 10. This may be due to an interrelation of flow condition and material exchange, but the fact is significant for the preestimation of the packing depth H, provided the pressure loss $\Delta p/H$ can be determined by calculation or by simulated tests, with sufficient accuracy. In that event, with a theoretical number of steps n_t

$$H = \frac{\dfrac{\Delta p}{n_t}}{\dfrac{\Delta p}{H}} \tag{8-19}$$

The mean specific pressure losses $\Delta p/n_t$ may be taken from Table 8.4.

For pall rings of the sizes 25, 35, and 50 mm, the mean specific pressure losses $\Delta p/n_t$ are almost the same at a given vapor load $u_V\rho_V^{1/2}$. A tendency toward higher values is shown by the 15-mm ring for $\phi' > 10$. This may be due to less effective separation because of the relatively large ratio of column diameter and packing element size (Fig. 8.2).

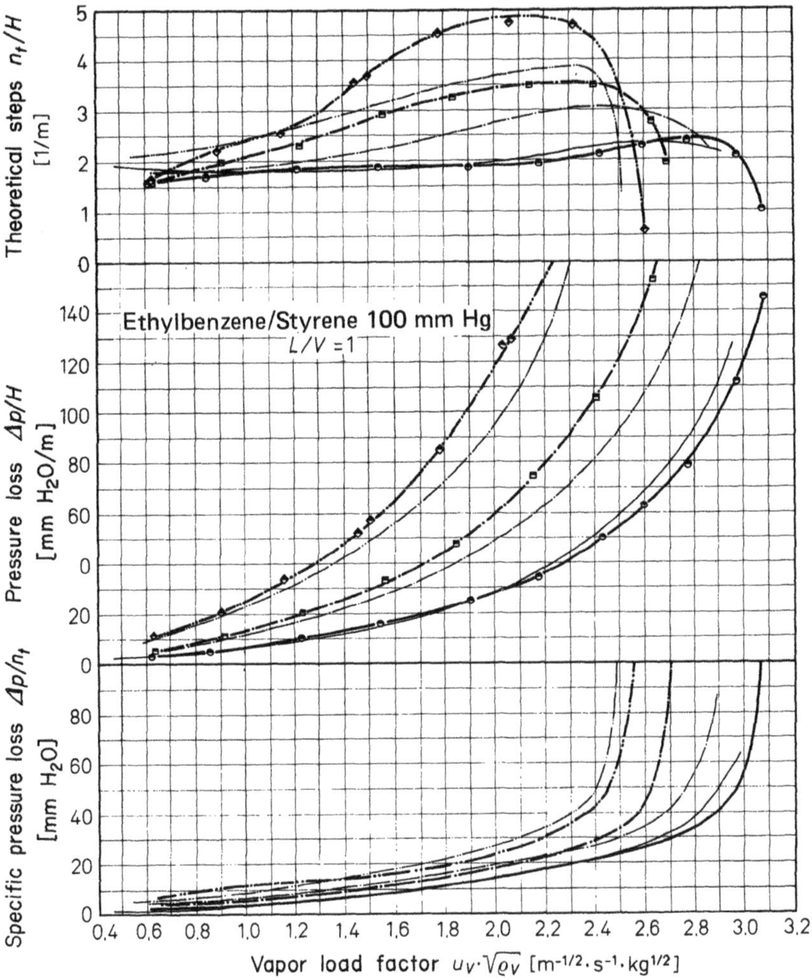

Curve	d_S	Packing	Number of pcs./m^3	Depth of dumped packing
——	0,5	Pall ring 50	6100	1,33
—o—	0,8		6400	2,0
—·—	0,5	Pall ring 25	47500	2,0
—·□·—	0,8		51500	2,0
—··—	0,5	Pall ring 15	224000	1,5
··◇··	0,8		236200	2,0

Fig. 8.29 Pall ring columns, various ring sizes. Influence of column diameter on efficiency and pressure loss.

Use of expanded metal in Pall rings (Fig. 8.30) as a means for reducing the specific pressure loss, was successful for the 25-mm size only (Fig. 8.31). For the 50-mm Pall ring the specific pressure loss in the main load range was unchanged. However, the maximum loadability was improved in both cases.

Table 8.4.

Metallic Pall rings.
Mean specific pressure loss at $L/V = 1$, (mm water).

$u_V \rho_V^{1/2}$ $(m^{-1/2} s^{-1} kg^{1/2})$	1.0	1.5	2.0	2.5	φ'	d (mm)
Mean Specific	4.0	8.0	14.0	24.0	10	80
Pressure Loss	4.0	8.5	16.5	27.5	10–16	50
Δp	4.0	8.5	16.5	–	14–23	35
――	4.5	10	18.5	–	20–32	25
n_t	4.5	11	-	–	10	15
(mm H_2O)	8	14	25	–	33–53	

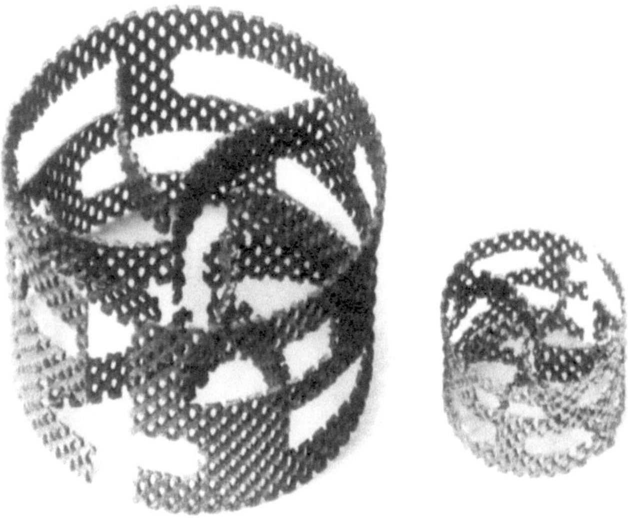

Fig. 8.30 Pall rings made of expanded metal.

g. Pressure Losses at Various Liquid Loads

The results mentioned in Section 2.C.1. though 2.C.6. of this chapter all referred to load conditions where the liquid/vapor ratio $L/V = 1$. For this, Fig. 8.14 presents the specific liquid loads L^* for various load factors $u_V \rho_V^{1/2}$. These are all below 15 m^3/m^2 h. Fig. 8.32 reveals that a marked increase in pressure loss must be expected if the liquid load becomes a multiple of that at $L/V = 1$.

One limitation is in the "dry" pressure loss Δp_{tr} that occurs when the liquid load $L^* = 0$. Fig. 8.33 presents this for several systems, ring sizes, column heights, and diameters, as relative values $\Delta p_{tr}/H$ in log-log coordinates versus the comparable air velocity u_A [note Eq. (6-15)]. It is seen that the dry pressure loss increases by an average of the 1.9th power of the velocity.

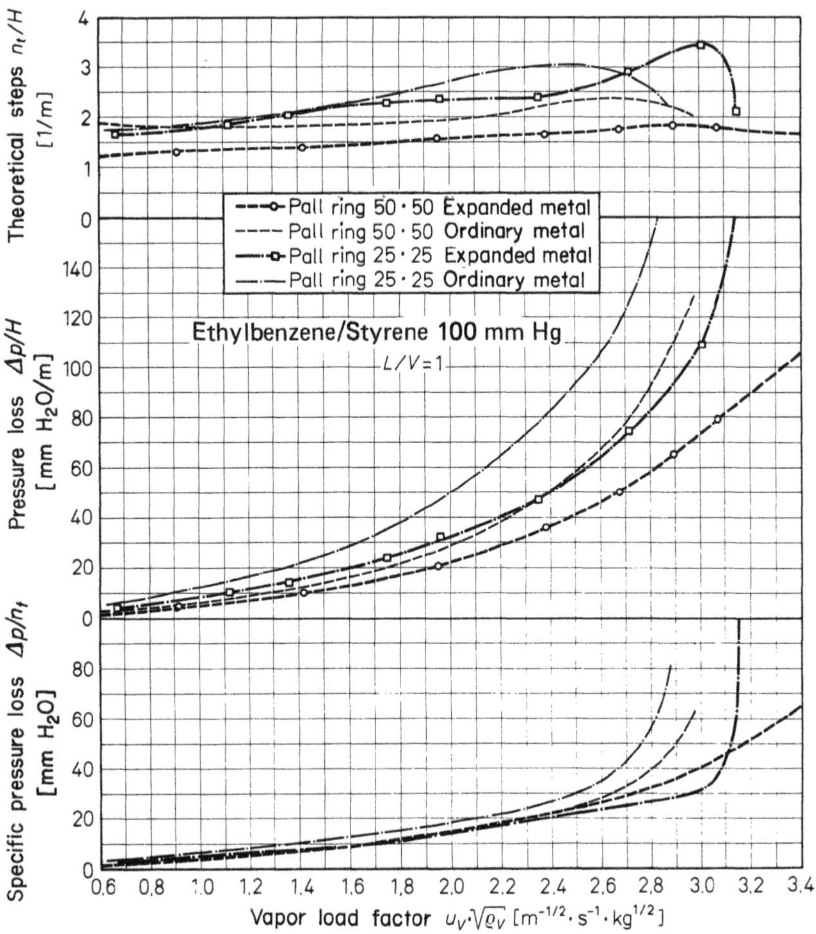

Fig. 8.31 Pall rings, expanded metal versus flat sheets. Efficiencies and pressure losses.

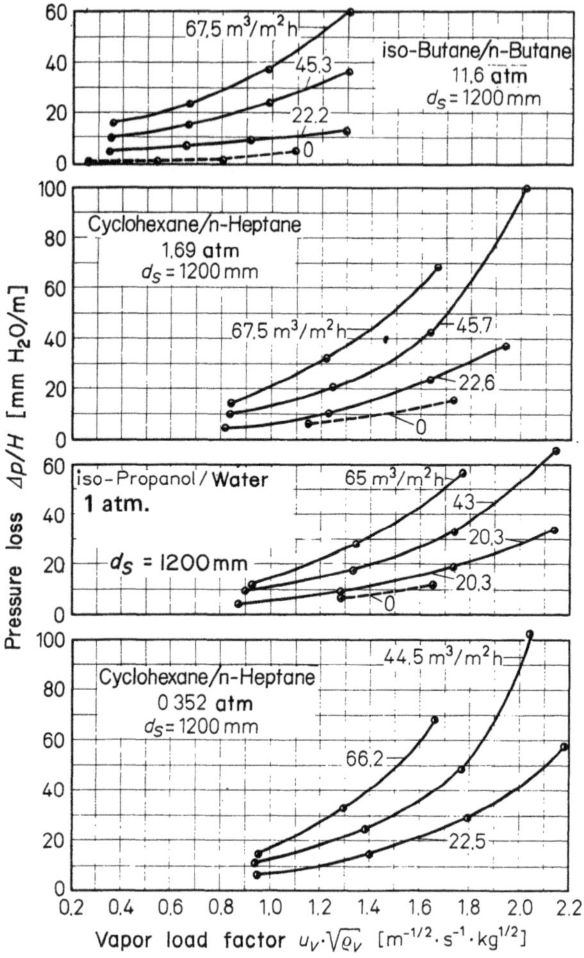

Fig. 8.32 Dumped Pall rings. Influence of liquid load on pressure loss.

D. Metallic Raschig Rings

Raschig rings are among the earliest packing elements used in separation work. Its simple configuration, with the diameter equal to the height, permits easy manufacture. The function of the thin-walled metallic type will be explained in some detail.

a. Dimensional Effects

The effects of ring size upon separation, load and pressure loss are shown in Fig. 8.34. For descending ring size, the separation and pressure loss increase, and the maximum admissible load becomes less. Again the curves are based on total reflux. They were ob-

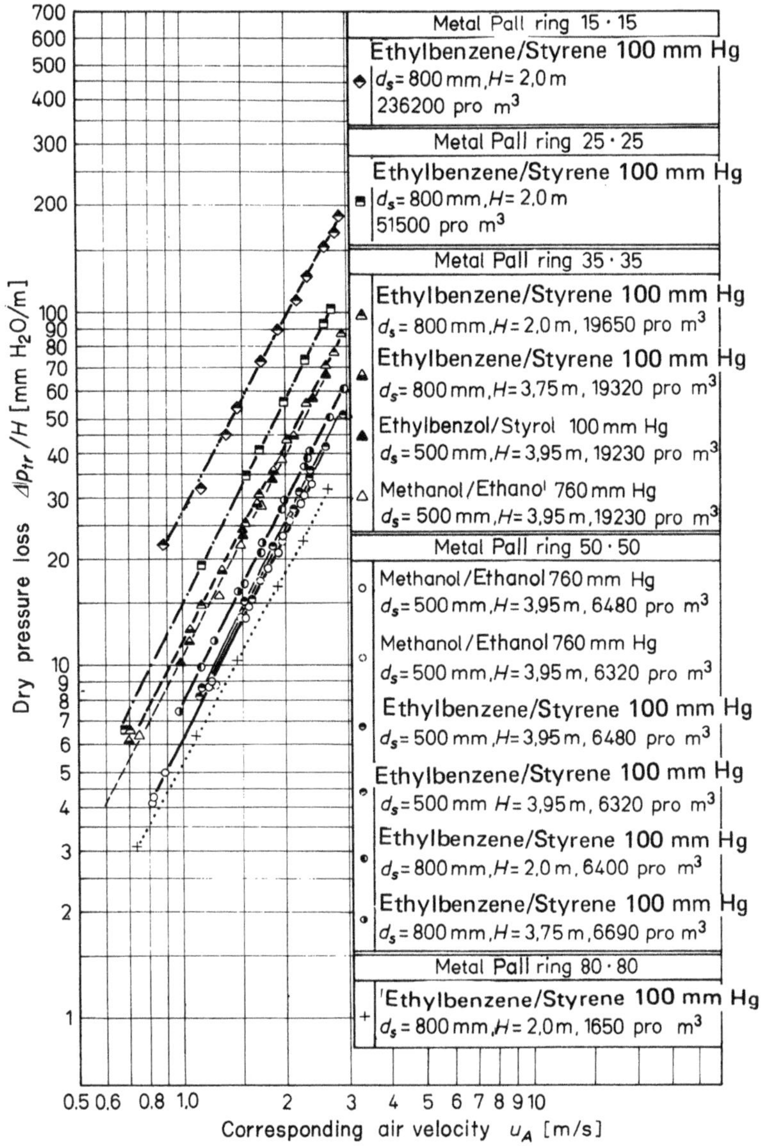

Fig. 8.33 Dry pressure loss as function of corresponding air velocity.

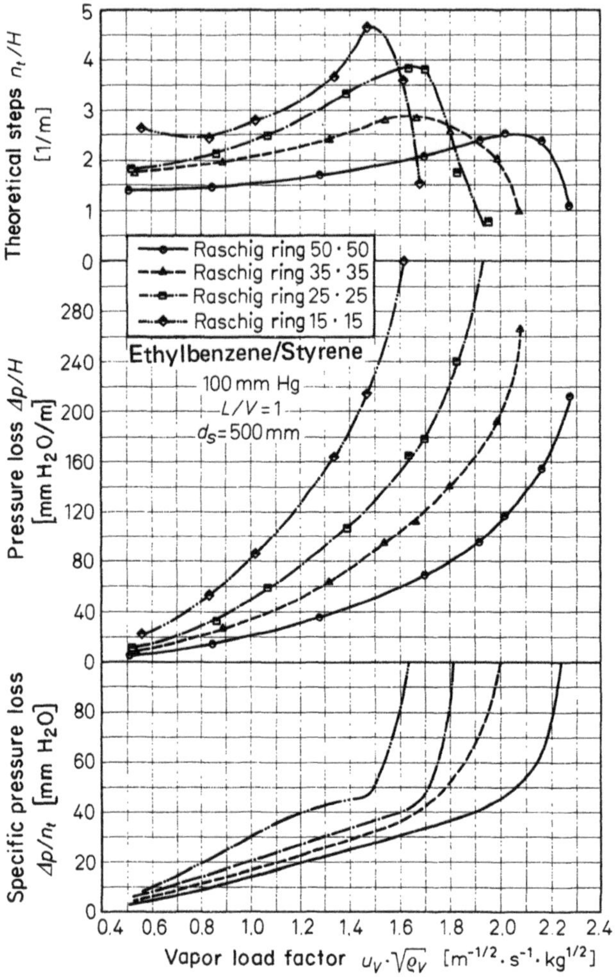

Fig. 8.34 Dumped Raschig rings. Sizes versus separation, pressure loss, capacity. Ethyl benzene–styrene system. 100 m Hg.

tained in a 500-mm test column with a 2-m packing depth. The liquid distributor used was the same as for the Pall ring tests. Thus Fig. 8.34 may serve for comparison.

b. Effects of Systems

In a test involving a 50-mm Raschig ring packing of 2-m depth, there was evidence of the tendency shown by Pall ring tests (Fig. 8.35). Systems at higher pressure and therefor higher liquid load gave the better separation, but the pressure loss was greater, and the loading less admissible. The latter could not be reached in tests with the 1,2-propylene glycol/ethylene glycol system at 10 mm Hg, due to a limitation in the capacity of the reboiler.

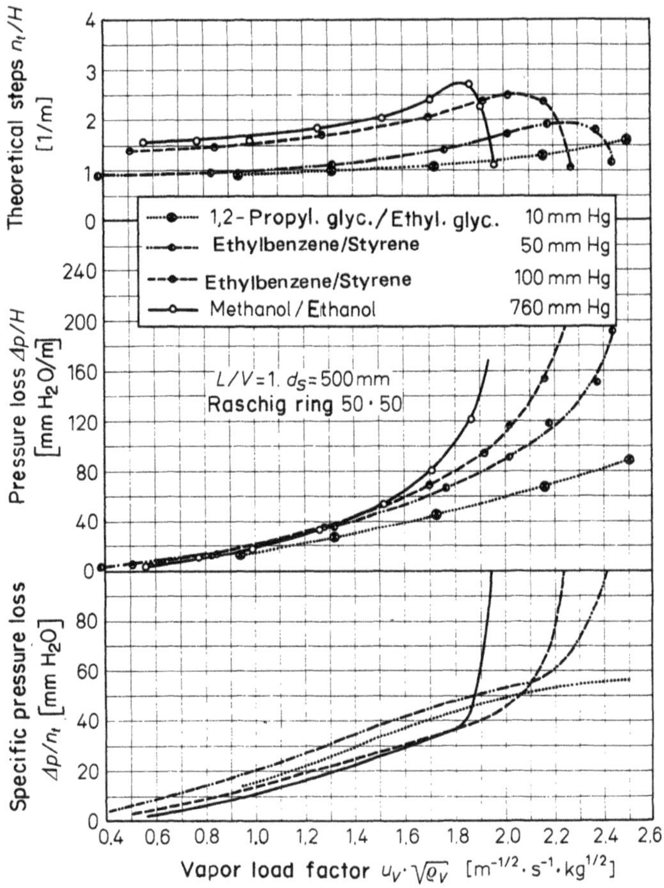

Fig. 8.35 Dumped Raschig rings. System versus separation, pressure loss, capacity.

E. Metallic Pall Rings versus Raschig Ring Columns

While it would appear that the separation is practically equivalent for metallic Pall rings and Raschig packing, differences exist as to pressure loss and limit loadings. The specific pressure loss, referred to as the theoretical separation step, is especially important because it is a yardstick for the usefulness of a packed column in the rectification of thermally unstable systems; this specific pressure loss must be small in view of the column-base temperature limitations.

Fig. 8.36 gives the specific pressure loss of Raschig-ring stacks, depending on packing element size and loading of the systems investigated. It is superior by 250 to 450 % to those of Pall-ring stacks, across the main load range.

The flooding point of metallic Raschig-ring stacks is at only 65 to 75% of the Pall-ring level.

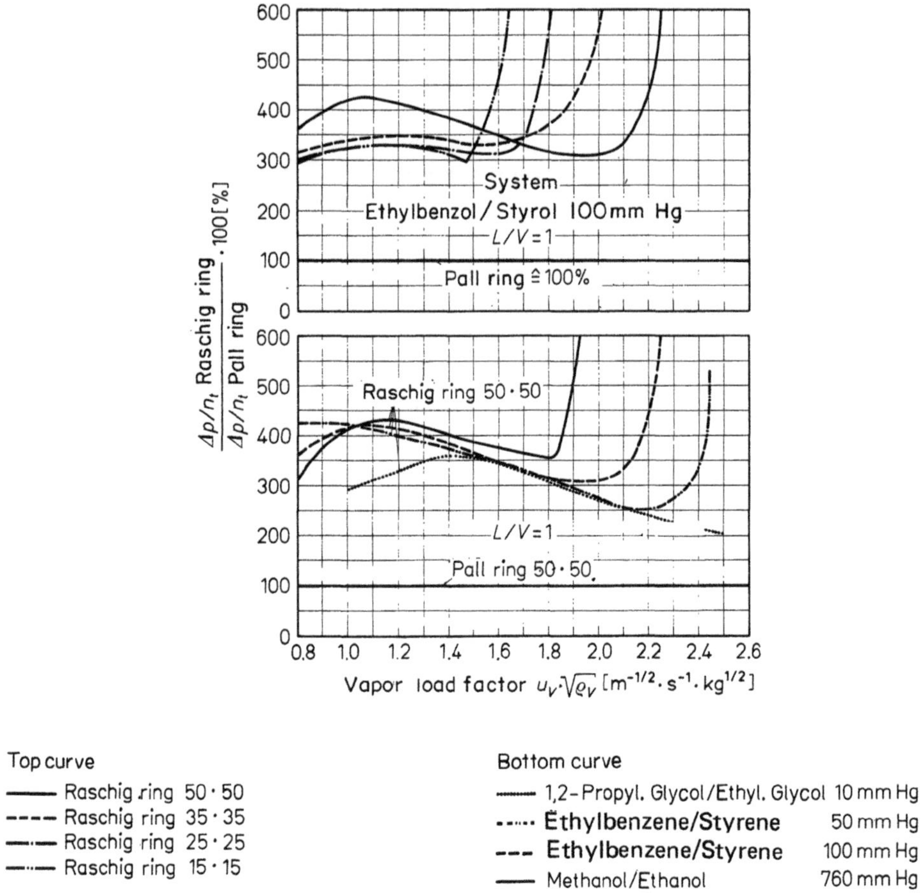

Fig. 8.36 Raschig rings and Pall rings, various sizes and systems. Specific pressure losses.

A cost analysis reveals that, depending on system and ring size, a column containing metallic Raschig rings may be anywhere between 30 and 100% more expensive than a Pall-ring column, if specified pressure losses are below 30 mm water per theoretical step, and for carbon steel construction of packing and shell. For stainless steel the cost picture may be in favor of the Raschig rings.

F. Maximum Loads for Metallic Pall-Ring and Raschig-Ring Columns

To test Eq. (8-9), all maximum loads reported in the figures of Section 2.C. of this chapter and Fig. 8.24 on packed columns were reviewed, and likewise the flood loads obtained for various systems and liquid loads on industrial columns as presented in

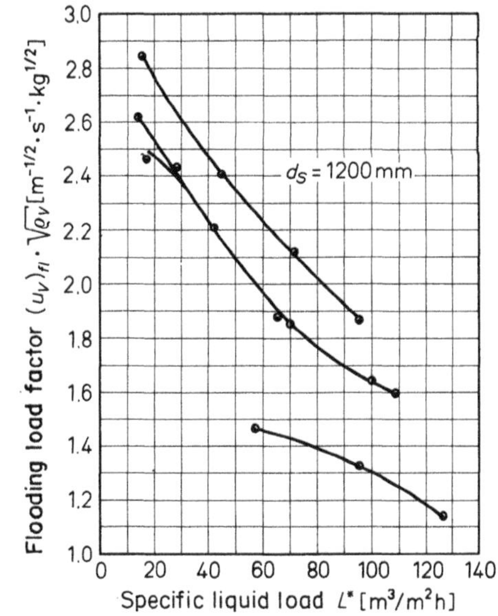

Fig. 8.37 Vapor load at flooding versus liquid load. Various sytems.

Fig. 8.37. The flood load is that point where the column becomes submerged, the separation drops off, and the pressure loss rises markedly.

According to Fig. 8.38, the best rendition of the measured points, allowing for 15% scatter, is obtained by making the viscosity exponent $m = 0.12$ instead of $m = 0.2$ as found in the bibliography, and by using the packing coefficients of Fig. 8.39 for flooded columns.

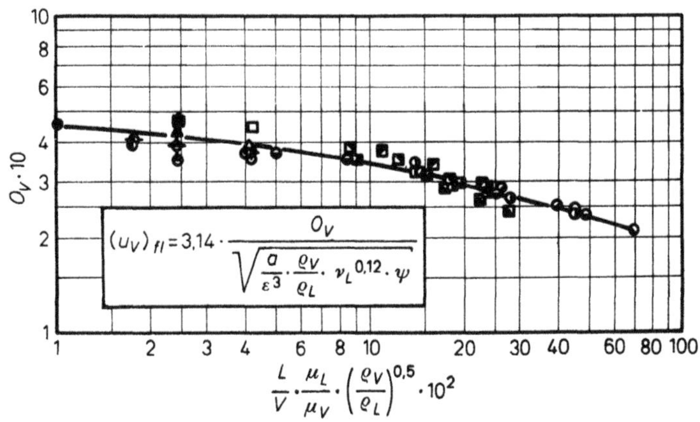

Fig. 8.38 Metallic ring packing (Pall, Raschig). Vapor velocities at flooding point. $(O_V) = (O_V)_{fl}$.

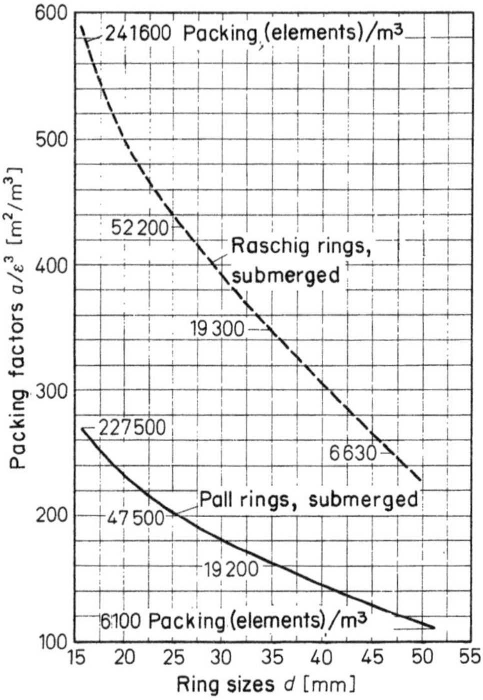

Fig. 8.39 Metallic ring packing (Pall, Raschig). Packing factors versus ring sizes. Dimensionless figures give pieces per cubic meter of column volume.

Then, the vapor velocity at the flooding point is calculated from:

$$(u_V)_{fl} = 3.14 \frac{(\sigma_V)_{fl}}{\left(\frac{a}{\epsilon^3} \frac{\rho_V}{\rho_L} \nu_L^{0.12} \Psi\right)^{1/2}} \tag{8-20}$$

The velocity to be used in finding the column diameter may amount to 70–80% of the figure thus calculated.

Other equations for this application are given in the bibliography.[26,27]

3. SPECIAL PACKINGS

Thermal instability of certain mixtures for separation by distillation or rectification may pose serious problems in engineering. Thus, thermal sensitivity of high boilers may cause chemical decomposition or polymerization once limiting temperatures are overstepped. For proper column operation those temperature boundaries must be considered, along with those for a suitable condensation and those imposed by the provision of a reasonable vacuum. This yields the specification data for pressures and temperatures at

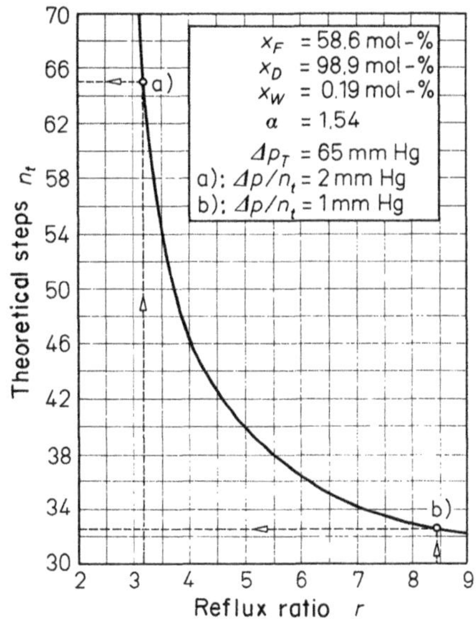

Fig. 8.40 Separation of a binary. Theoretical steps versus reflux ratio. a) and b) are operating points inside permissible pressure loss for two different column internals.

the top and at the bottom of the column, and therefore for the maximum admissible pressure loss across the column. The economy of a separation depends on the investment and on the running expenses. Consequently, in the case of a vacuum rectification the number of theoretical steps feasible with a given pressure loss is decisive.

Note the diagrammatic picture in Fig. 8.40. The theoretical steps n_t for the separation of a binary have been plotted against the reflux ratio r. Obviously, packing of low pressure loss p/n_t per theoretical step [case (a)] will allow accommodation of more theoretical steps at a lower reflux ratio inside the specified total pressure loss p_t, and therefore requires less separation energy than columns with packing of high specific pressure loss [case b)].

Separations falling within the pressure loss range of case a) are best handled by means of a web-type column (Fig. 6.23). When a coarse grid column is selected that, for vapor load coefficients below $u_V \rho_V^{1/2} = 2.5\ \mathrm{m}^{-1/2}\ 5^{-1}\ \mathrm{kg}^{1/2}$, also will yield pressure losses under $p/n_t = 1$ mm Hg, a large construction height might result which could be problematic. Therefore, it could only be considered for applications involving a moderate number of theoretical steps.

A. Web-Type Packing

Generally, web-type packing can be useful in all fields of the chemical and allied industries. Of the many productions where web-type columns have been applied, only menthol isomers, neraniol-geraniol, di-, and triethanol amines, di-, tri-, and tetraethylene glycols, higher fatty alcohols, and heavy water should be named.

Fig 8.41 gives various views of a type BX web packing popular in vacuum work.

Fig. 8.41 Type BX wire cloth packing made by Sulzer of Winterthur, Switzerland. a) Plan view, b) side view, c) looking through open passages.

The reflux distributor is the critical component of a web-type column. If it becomes plugged, the liquid distribution may become irregular and separation will suffer. Therefore, the reflux distributor must be accessible for cleaning or it must be removable.

Fig. 6.22 showed a successful reflux distributor for web-type columns. To counteract plugging of the outlet pipes, a filter of about 10μ mesh is sometimes inserted which must be frequently cleaned during the starting period and less frequently later on. The outlet pipes are customarily so arranged that there is an average of 10 openings per square decimeter of column area.

The thesis established in Chapter 6., to the effect that minimum equipment cost is not necessarily tied to a column laid out for maximum throughput, is brought back to mind. Rather, because of the drop in specific separation at increasing load, the optimum throughput will correspond to a vapor-load factor that for vacuum systems is less than 60–70% of the applicable flooding load and has been found[28–31] to equal an average of

$$(u_V \rho_V^{1/2})_{opt} = 2.4 m^{-1/2}\ s^{-1}\ kg^{1/2} \tag{8-21}$$

Using Eq. (6-23), the optimum column diameter (m) can be found from:

$$(d_S)_{opt} = 0.034 \left[V \left(\mu \frac{T}{p} \right)^{1/2} \right]^{1/2} \tag{8-22}$$

The corresponding specific separation of the packing is

$$\frac{n_t}{H} = 5 \text{ theoretical steps per meter} \tag{8-23}$$

independently of the diameter if properly installed. The characteristic load by Eq. (8-21) can be based on a pressure loss per theoretical step of only

$$(\Delta p/n_t)_{opt} = 5 \text{ mm } H_2O \tag{8-24}$$

which causes the column cost to be less than for all others in the same category.

Use of web-type columns in the separation of thermally sensitive materials at very low pressures saves money and is recommended. The small specific volume and therefore reduced dwell time in processing will be found favorable when compared to conventional equipment; note Table 8.5.

The arrangement is by order of increases in specific cost factors C_S. As explained for Eq. (6-37) in Chapter 6, the specific values are obtained from the data applying for a separation by an equivalent column filled with 50 × 50 mm Pall rings. Thus they indicate the cost multipliers for a column, compared with a Pall-ring design.

Table 8.5.

Packed stainless steel columns. Specific cost factors, established for the ethyl benzene/styrene system at 100 mm Hg. $L/V = 1$, $\Delta p/n_t = 5$ mm water, and d_S (coarse grid) = 0.8 m reference diameter.

$\Delta p/n_t = 5$ mm H_2O

Column Type	C_S	$u_V \rho_V^{1/2}$ ($m^{-1/2} s^{-1} kg^{1/2}$)	ν_V ($m^3/m^3/s$)
Wire Web Packing	0.77	2.40	0.057
Pallring 50 × 50 × 1	1.00	1.15	0.324
Pallring 80 × 80 × 1.5	1.25	1.25	0.406
Fine Grid	1.25	1.30	0.429
Coarse Grid	1.40	1.60	0.578
Square Grid	2.00	1.30	0.428

The figures for the load factors $u_V \rho_V^{1/2}$ and those for the specific column volumes v_V reveal that, for the given pressure loss, web packing permits maximum throughput and is remarkably compact. Type BX packing is eminently suited to rectifications of minimum pressure loss per theoretical stage. That comprises the separation of thermally unstable systems and applications where the column-base pressure may not exceed a maximum set by the available evaporator surface, and the heating steam pressure for the reboiler is limited.

Even when a separation demands many theoretical steps, the use of a web packing may be advantageous in view of the much reduced construction height, compared with conventional columns. Then, the CY type offers desirable features. Table 8.6 presents the processing and design data for types BX and CY.

Table 8.6.

Sulzer BX and CY type wire web packing. Specifications. 100 mm Hg nominal pressure. a = specific surface (m^2/m^3), l_z = peak to peak (m/m), α_z = inclination with column centerline (°), $u_V \rho_V^{1/2}$ = load factor ($m^{-1/2}\ s^{-1}\ kg^{1/2}$) n_t/H = theoretical steps (1/m), and $\Delta p/n_t$ = specific pressure loss (mm H_2O).

Wire-Web Packing	BX	CY
$a(m^2/m^3)$	500	700
l_z(mm)	10	6
α_z(°)	30	45
$u_V \rho_V^{1/2}(m^{-1/2} s^{-1} kg^{1/2})$	2.4	1.8
n_t/H (1/m)	5	10
$\Delta p/n_t$ (mmH_2O)	5	5

Finally, Fig. 8.42 reveals that, particularly at low loads, the initial liquid distribution does influence the separation by a type BX web packing appreciably. The column diameter is 500 mm, the packing height is 2 m. The reflux was supplied by the distributors of Fig. 6.14 (lower curve) and of Fig. 6.22 (upper curve).

Fig. 8.43 shows an industrial installation consisting of several web-type columns, base heaters, and overhead condensers.

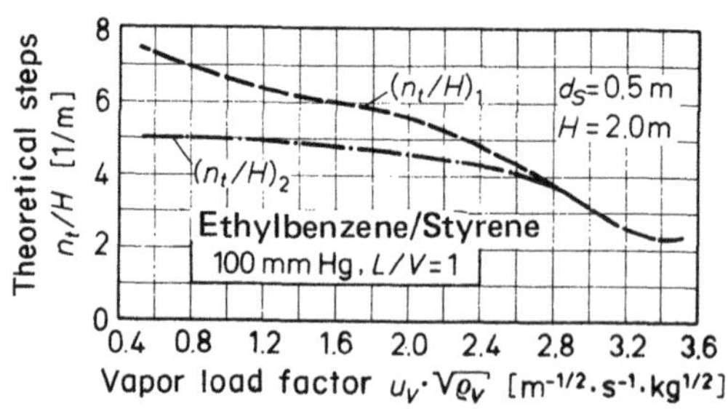

Fig. 8.42 Liquid distribution and separation by wire cloth packing. 1) Feeder of Fig. 6.22, 2) Feeder of Fig. 6.14.

Fig. 8.43 Multi-column separation plant fitted with Sulzer type wire web packing.

B. Grid Packing

Its relatively ample clearances provide this type of packing with high loadability and low pressure loss per theoretical step. This is another type of column internal that lends itself to applications involving temperature sensitive materials to be split up in a vacuum, such as fatty acids and tall oil. Of the grids in Chapter 6, coarse grids have the best loadability. They are especially suitable where the capacity of a production column is to be enhanced by exchanging packing elements.

The diameter of a grid-type column should be so established that the throughput will not exceed 80% of the flood load. The latter can be found from Fig. 8.44 in which the

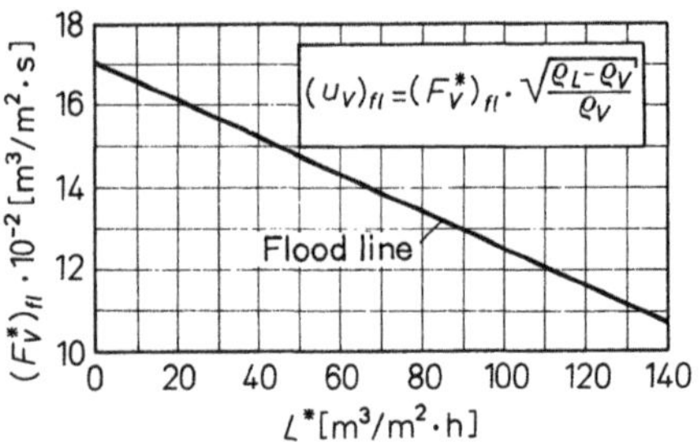

Fig. 8.44 Coarse grid packed columns by Glitsch of Dallas, Texas, and GHH of Sterkrade, Germany. Determination of flood point.

vapor load factor F_V^* (m/s) of the flood condition defined by Eq. (6-10) is plotted over the specific liquid load L^*(m³/m²h). The vapor velocity at the flood point then becomes

$$(u_V)_{fl} = (F_V^*)_{fl} \left(\frac{\rho_L - \rho_V}{\rho_L} \right)^{1/2} \tag{8-25}$$

At the optimum load of web-type packing $u_V \rho_V^{1/2} = 2.4 \text{ m}^{-1/2} \text{s}^{-1} \text{ kg}^{1/2}$, the specific pressure loss of the grid packing equals, from Fig. 6.23, around $\Delta p/n_t = 10$ mm water, about twice as high as for the web type. However, the optimum load of the grid packing, correlated to the minimum specific cost, is given by Fig. 6.28 as

$$(u_V \rho_V^{1/2})_{\text{opt}} = 3.8 \text{ m}^{-1/2} \text{ s}^{-1} \text{ kg}^{1/2} \tag{8-26}$$

which is much higher than that of the web type. However, the corresponding separation is only

$$(n_t/H)_{\text{opt}} = 0.75 \text{ 1/m} \tag{8-27}$$

The specific pressure loss is

$$(\Delta p/n_t)_{\text{opt}} = 43 \text{ mm water} \tag{8-28}$$

and the column diameter

$$(d_S)_{\text{opt}} = 0.027 \left[V \left(\mu \frac{T}{P} \right)^{1/2} \right]^{1/2} \tag{8-29}$$

Example 2. 46 kmol/h of ethyl benzene-styrene vapors with a top-of-column concentration around 60 mol-% are to be processed at a pressure of 100 mm Hg at 78 C. Find the diameter of a coarse-grid column for this duty; the liquid/vapor ratio being unity.

Answer In order to apply Eq. (8-25) in conjunction with Fig. 8.44, first assume a specific liquid load L^*. Then, find from F ig. 8.44 the load factor $(F_V^*)_{fl}$ and from Eq. (8-25) calculate vapor velocity $u_V = (u_V)_{fl}$. The diameter is then given by Eq. (3-39) or by Eq. (6-23). Then, the assumption of the specific liquid load

$$L^* = \frac{L \mu_L}{d_S^2 \frac{\pi}{4} \rho_L} \tag{8-30}$$

can be verified. In case of nonagreement, obtain a new $(F_V^*)_{fl}$ from Fig. 8.44 for the L^* of Eq. (8-30), and repeat.

In the present instance, the choice of $L^* = 14\ \text{m}^3/\text{m}^2\ \text{h}$ is confirmed by Eq. (8-30). From Fig. 8.44, this gives an $(F_V^*)_{fl} = 0.164$ m/s. From Eq. (7-7), $\rho_V = 0.481$ kg/m^3. With $\rho_L = 846$ kg/m^3, Eq. (8-25) eventually gives

$$u_V = (u_V)_{fl} = 0.164 \left(\frac{846 - 0.481}{0.481} \right)^{1/2} = 6.88 \text{ m/s}$$

and Eq. 3-39 gives the diameter

$$d_S = 0.149 \left(46 \frac{351}{100} \frac{1}{6.88} \right)^{1/2} = 0.72 \text{ m}$$

Since $L = V = 46$ kmol/h had been specified. from Eq. (8-30)

$$L^* = \frac{(46)(105)}{(0.72^2)\left(\frac{\pi}{4}\right)(846)} = 14 \text{ m}^3/\text{m}^2\text{h}$$

which confirms the assumption made. If the operating load is to amount to only 80% of the flood load, the column requires a diameter of

$$d_S = \left(\frac{0.72^2}{0.8} \right)^{1/2} = 0.8 \text{ m}$$

This is identical with the optimum diameter of Eq. (8-29)

$$(d_S)_{opt} = 0.027 \left\{ 46 \left[\frac{(105)(351)}{100} \right]^{1/2} \right\}^{1/2} = 0.8 \text{ m}$$

4. PLATE-TYPE COLUMNS

When no extreme demands are made regarding the pressure loss, columns with tray-type equipment are preferred to columns containing trickling or wetting internals, such as packing or grids. Most interchange trays achieve, by considered design, a more or less compulsory routing of the liquid and the vapor phase, which controls the surface contact, and produces specific results as to separation, loadability, and pressure loss, as illustrated by Fig. 8.45.[32]

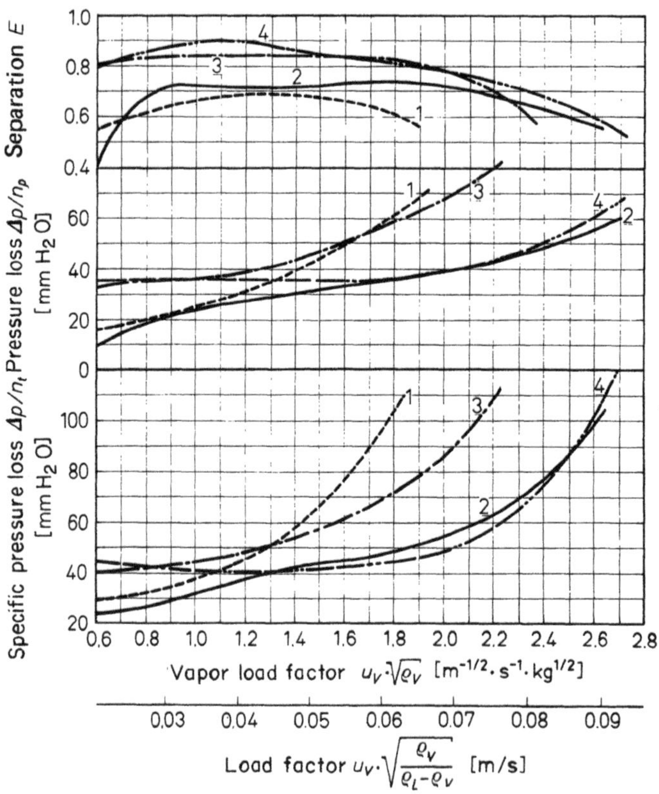

Fig. 8.45 Separation and pressure loss of various trays versus loads. Ethyl benzene/styrene system. 100 Hg, total reflux. Curves 1–4 are for designs to Figs. 6A, C, F, I. 1) Bell cap tray, 2) Sieve tray, 3) V-1 valve tray, 4) T-9-F sieve-and-valve tray; Tray diameter 800 m, submergence 38 mm.

The engineering aspects of tray design and layout have been dealt with in the existing bibliography,[33] however, the general strategy of tray layout will be discussed to facilitate quick estimations of capacities and dimensions.

A. Layout, Efficiency, and Capacity

All trays feature horizontal plates with perforations for the passage of vapor and sometimes attachments for guiding the vapor and providing a more intimate contact with the liquid usually passed crosswise over the plate. Occasionally radial flow is applied, as on centrifugal plates (Fig. 6.6D). Independently of the vapor, flow goes from the plate higher up through downcomers to the lower level. A downflow weir maintains a certain liquid level on the plate. A retaining weir ensures a seal on the downcomer and distributes the feed. In the column layout, the downflow weir, downspout, and seal pot must be considered as an assembly (Fig. 8.46). The simplest and cheapest arrangement is the operational seal, item 1, which dispenses with an inlet weir. Here, the distance between the edge of the down weir and the floor of the seal pot should be at least 13 mm. The downflow may also be sealed by the design in item 2, shown in Fig. 8.46 using an inlet weir. The most dependable seal is as shown by deep submergence (item 3).

Fig. 8.47 Downcomer designs.

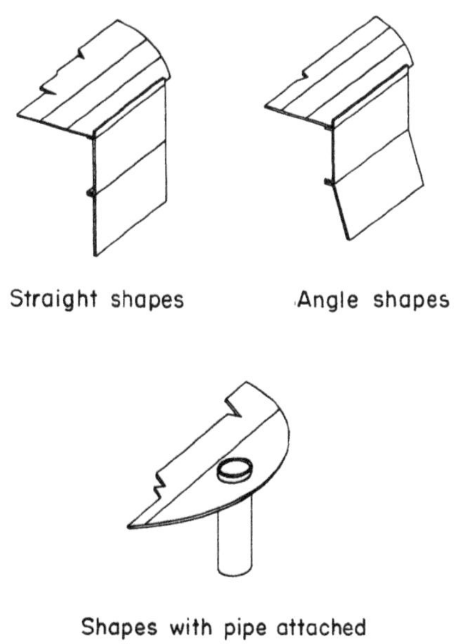

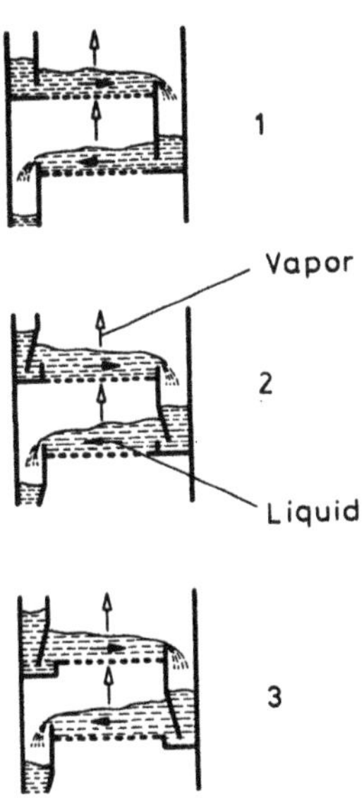

Fig. 8.46 Downcomers in crossflow tray stacks. 1) Levels in operation, 2) Holdback by inlet weirs, 3) Recessed pool.

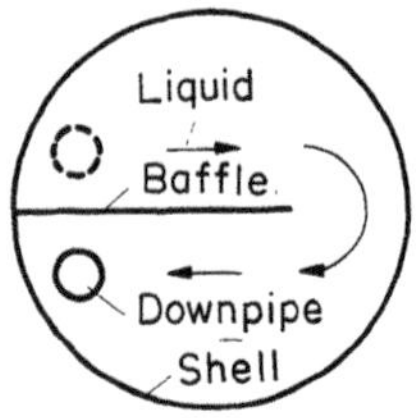

Fig. 8.48 Pattern of flow on plate.

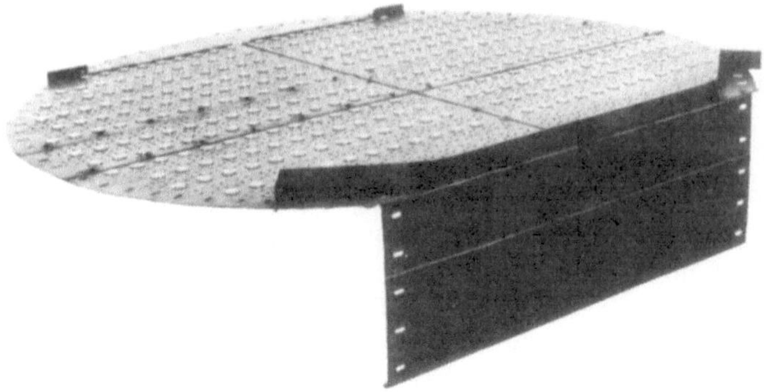

Fig. 8.49 Sieve plate equipped with type A valves, recessed weir, and straight skirt. Made by Koch Eng. Co., Wichita, Kansas.

The downcomer may have a straight or an inclined skirt, or there may be downtubes (Fig. 8.47). The straight skirt is the standard, but the bent shape may be economical in terms of using less floor space. Downtubes are customary for smaller columns, especially for low liquid loads. It is sometimes used in connection with reverse flow on plates (Fig. 8.48).

Fig. 8.49 shows a combined sieve and valve plate with a straight skirt. The valves are of type A of Fig. 6.10. Valve plate type T in Fig. 8.50 has a bent skirt. In Fig. 8.49 the

Fig. 8.50 Tray equipped with type T valves, segment weir, and bent skirt. Made by Koch Eng. Co., Wichita, Kansas.

plate has a recessed weir, and in Fig. 8.50, it has a segment weir. Recessed weirs have greater length for given downcomer areas, while the active surface is reduced; they are applied for greater specific weir loadings. For the operation of a plate, and its maximum loadability, the specific weir load determines the weir height.

Let: L = quantity of liquid (kmol/h)
μ_L = molecular weight of the liquid (kg/kmol)
L = specific weight of liquid (kg/m^3)
l_W = length of downflow weir (m)
h_{oW} = height of liquid above weir (m)

then

$$\frac{L}{l_W} = 6600\,(h_{oW}^3)^{1/2}\frac{\rho_L}{\mu_L} \tag{8-31}$$

Actually, the specific liquid load above the outlet weir

$$l' = \frac{L}{l_W}\,\frac{\mu_L}{\rho_L}\,\frac{1000}{3600} = 23\,\frac{1}{\text{s m}} \tag{8-32}$$

is considered the limit for conventional crossflow trays.[34] Higher values make a multiflow design mandatory unless special measures are devised; single-flow operation would result in raining through, or for large columns in liquid backup due to the liquid gradient.

According to Fig. 8.51, multiflow designs may be regarded as assemblies of single-flow plates.

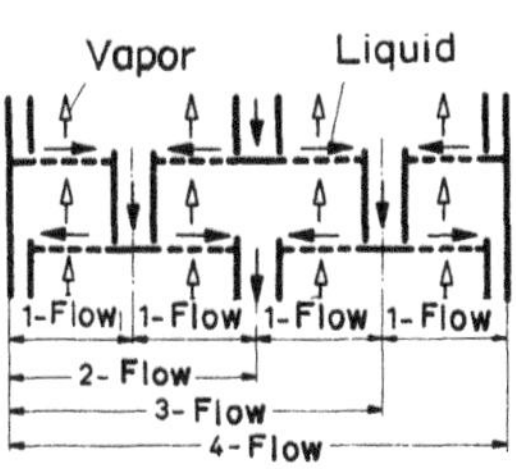

Fig. 8.51 Diagram of four-path flow cross-stream tray arrangement.

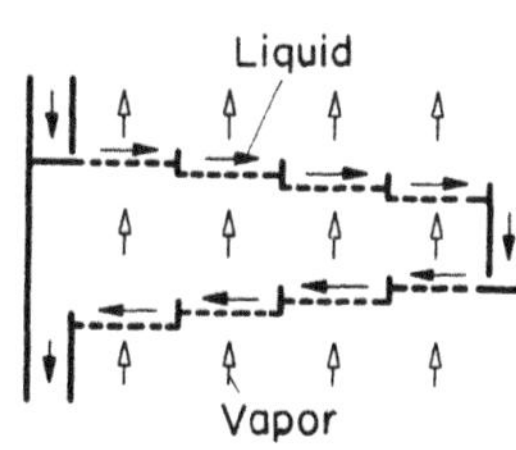

Fig. 8.52 Diagram of single-flow trays, cascade arrangement.

If with large plate diameters high liquid gradients are expected along the flow path, it is recommended to use crossflow plates in a cascade arrangement as shown in Fig. 8.52 for single flow. Multiflow plates may also employ the cascade principle.

Today mainly two basic constructions are in use, the package design and the sectional buildup; the former is encountered in columns up to 600-mm diameter. Fig 8.53 shows how five to seven plates are made into packages by spacers and through bolts, then slipped into place.

In sectional construction, the shell is fitted with manholes for the installation and dismantling of plate elements. The latter are joined by overlapping, by flanging together,

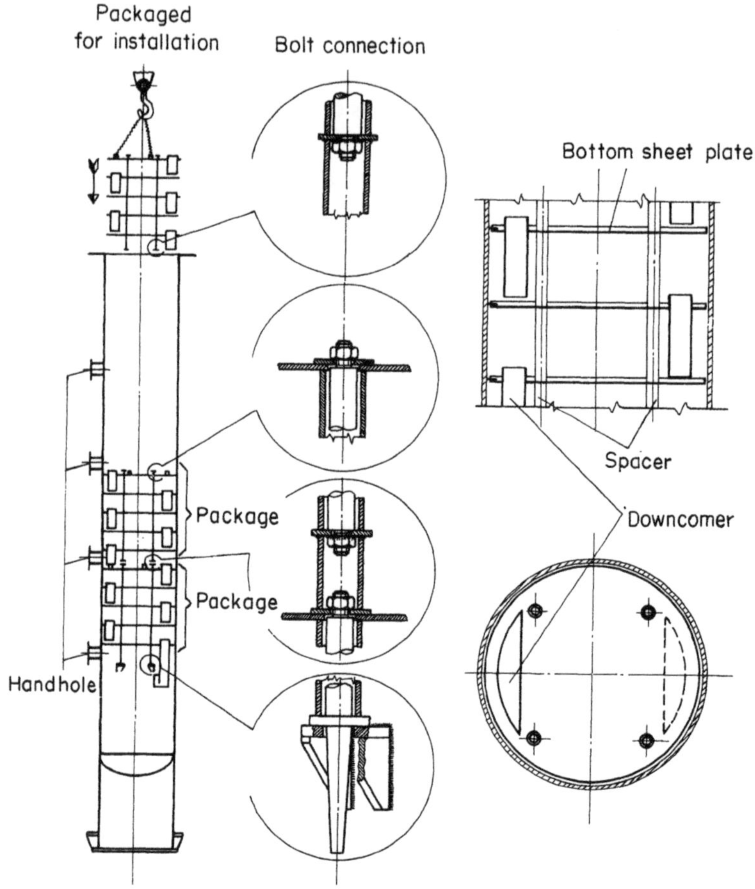

Fig. 8.53 Assembly of stacked trays in small-diameter columns.

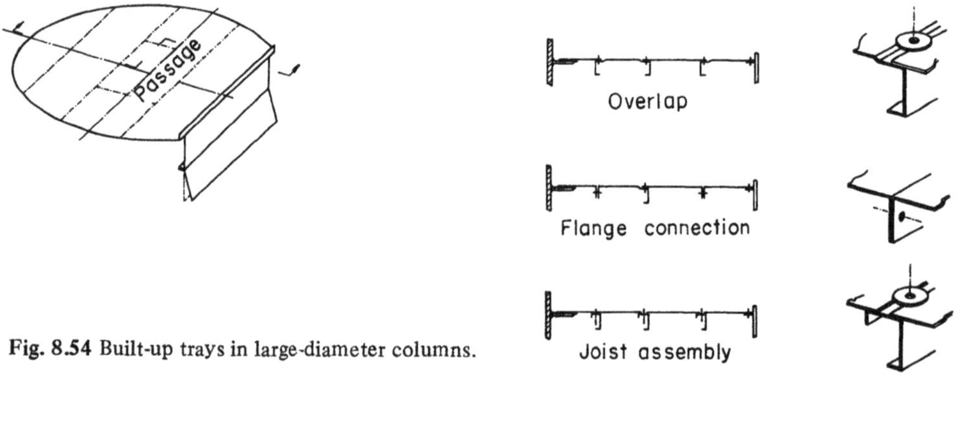

Fig. 8.54 Built-up trays in large-diameter columns.

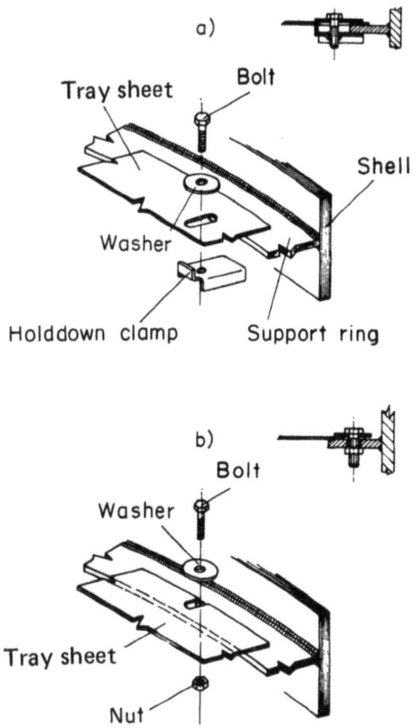

Fig. 8.55 Fastening of the trays to the shell a) Clamping, b) Bolting, Designs by Koch International of Bergamo, Italy.

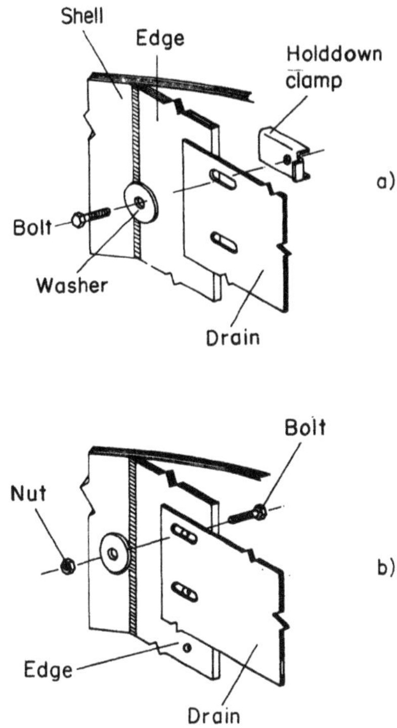

Fig. 8.56 Attaching baffles to shell. Also note Fig. 8.55.

or by support joists that are bolted or preferably welded for tightness. Again, each plate has a manhole that can be opened from below or from above.

The shell may have internal support rings (Fig. 8.55) or brackets for holding the components, either by direct attachment or by clamp plates applied from two sides (Fig. 8.56).

Special fastenings hold the passageways in place; Fig. 8.57 shows two designs for the inner area of a tray, and Fig. 8.58 offers two details for securing a cover to an internal ring flange; detail a) is for top entrance, and detail b) for entrance from below.

For a good fit, plates must be reasonably circular. Certain tolerances have been established for out-of-roundness (Fig. 8.59). Table 8.7 tabulates dimensions for support rings.

Large masses of liquid can be handled by radial flow trays that have been discussed in Chapter 6 together with other designs. The sectional view in Fig. 8.60[35] explains the function of this plate; the liquid enters centrally and flows radially toward a circular drain. Proper orientation of the slots in the expanded metal segments guides the vapor in such a way that tangential vapor flow is superposed to radial liquid flow,

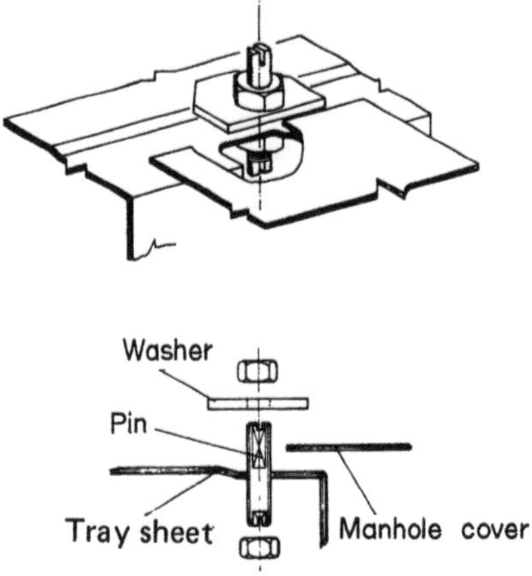

Fig. 8.57 Details of covers on passageways of plates inside columns.

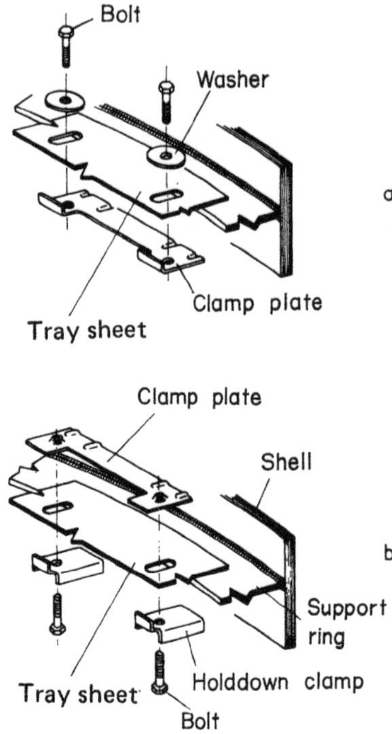

Fig. 8.58 Passageways inside shells. a) Access from above, b) Access from below. Designs by Koch International of Bergamo, Italy.

Eccentricity: max. 1%

$$1\% = \frac{\text{max. diameter.- min. diameter}}{\text{theoretical diameter}} \cdot 100$$

Column diameters [mm]	Tolerance [mm]			
	A	B	C	D
1600 and under	±2	±3	±2	±3
1600 - 3000	±3	±4	±3,5	±5
Over 3000	±4	±5	±5	±5

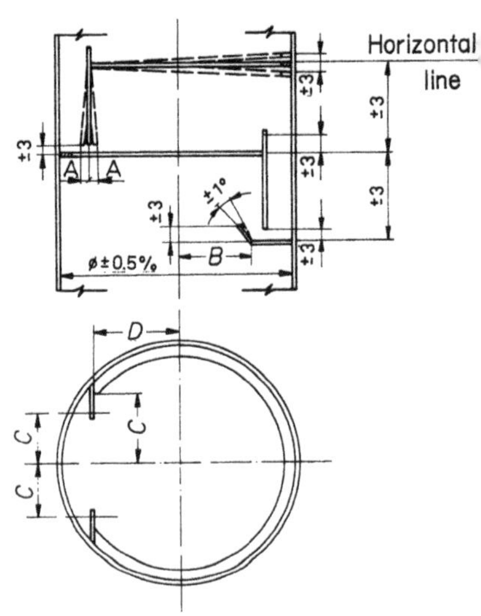

Fig. 8.59 Construction tolerances in assembling trays for various column sizes.

Table 8.7.

Support rings for rectifying trays, mm dimensions.
Column diameters—Width—Thickness, carbon steel
—Thickness, stainless steel.

Column Diameters	Width	Carbon Steel Thickness	Stainless Steel Thickness
1600 and under	40	10	6
1601 to 2300	50	10	6
2301 to 3000	60	10	6
3001 to 3500	60	13	6
3501 to 4600	80	13	6
4601 and larger	90	13	6

achieving a helical phase motion of extended length of passage.

The relatively long exit edges of the radial flow tray are especially favorable for high liquid loads. Double-flow types[36] of this tray (Fig. 8.61) have even longer weirs than quadruple-flow crossflow trays of conventional design. A cascade arrangement is likewise feasible (Fig. 8.62).

Trays having no controlled liquid flow, with both phases in simple countercurrent, are less frequent. This includes sieve plates without downcomers, also called dual flow

Fig. 8.60 Radial-flow tray, single-flow centrifugal type, using expanded metal plates. Made by Stahl Co. of Viernheim, Germany.

plates because the apertures will pass vapor as well as liquid; fig. 8.63. They can be designed for high throughputs and for small pressure loss per theoretical step, and are useful in cases where the capacity of a column is to be enhanced by exchanging internals for more effective ones. Being of simple design, such plates are inexpensive. However,

Fig. 8.61 Radial-flow tray and a dual flow type using expanded metal elements. Made by Stahl Co., Viernheim, Germany.

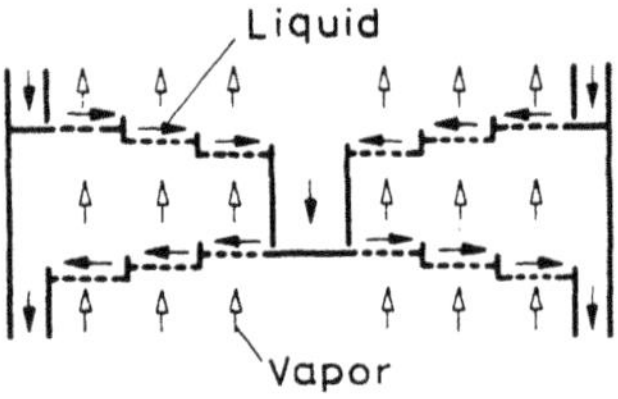

Fig. 8.62 Radial-flow trays, cascade arrangement.

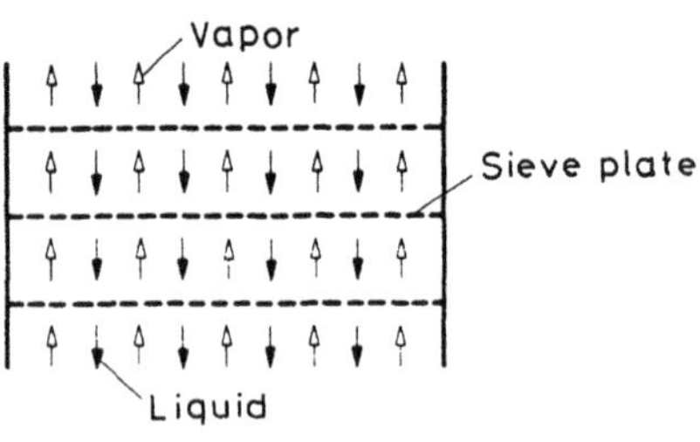

Fig. 8.63 Phase flow on Dual Flow plates. Vapor up, liquid going down.

they should not be proposed for cases of fluctuating loads, because of their lack of flexibility under that condition.

Generally, the efficiency E_T of the interchange plates is expressed as the ratio of the actually attainable separation effect and that of a theoretical step. While Eq. (6-7) deals with the enriching effect in the vapor phase, in principle an interchange ratio for liquids could also be defined.[37-39]

In Chapter 6 it had been demonstrated that liquid entrainment by vapor will affect separation. If

$$r_e = \frac{\Delta L}{L} \tag{8-33}$$

is the ratio of the liquid ΔL (kmol/h) and the reflux quantity L (kmol/h), the separation effect per plate E_T drops to E_{Te} according to[40]

$$E_{Te} = E_T \frac{1}{1 + r_e E_T} \tag{8-34}$$

For any given load, there are various r_e values, depending on plate design.

At any given point of an separation plate, the local separation for a specific phase flow is not identical with the overall separation for the plate. In principle, the overall separation is better the less backmixing of phases occurs. Furthermore, parallel liquid flow favors separation.[4] The positive influence of long liquid paths in large diameter columns cannot become dominating because of the undesirable effect of the liquid gradient.

An attempt based on the local separating effect E_l to determine its total separating E_T, according to relationship[41]

$$E_T = E_l f\left(E_l \frac{V}{L} mPe\right) \tag{8-35}$$

gave, in an evaluation of bell caps, results that were no better than 30% true. In Eq. 8.35, m is the slope of the equilibrium curve within a given concentration range, and Pe is a modified Péclet number which allows for the characteristic load and configuration. The definition of the local separation

$$E = \frac{\Delta y}{\Delta y_t} \tag{8-36}$$

is formally identical with that of the overall separation of the plate according to Eq. (6-7); Δy_l (mol-%) is the spotwise enrichment of the vapor actually achieved at a certain location of the plate, and y_t (mol-%) is theoretically feasible, assuming that the separation of the plate element in question equals that of one theoretical step.

Actually, for most plates $E_l \leqslant E_T$, and the E_T values found on small test columns are mostly less than the corresponding E_l values. Therefore, in column design it is safe to write

$$E_T = E_l \tag{8-37}$$

Basically, the separation depends on the physical properties of the mixture and therefore on the concentration and eventually operational factors such as distillation pressure, vapor/liquid ratio,[42] column load, and plate design. Recent research reports on the effects of phase flow and the distribution of resistances.[43]

For estimating, the data of Table 8.8 on separations, which apply to conventional trays, will suffice.[44] According to Chapter 6, recent plate concepts yield better results.

Table 8.8.

Conventional rectifying trays . Average efficiencies.

Systems	E_T
Alcohol/Water Mixture	0.6–0.8
Hydrocarbon Mixtures and Organic Liquids	0.5–0.6
High-boiling Liquids, Minerals, etc.	0.35–0.5

With the separation E_T known for a plate, the overall separation effect can be found; for a column

$$E_C = \frac{\log\left[1 + \left(\frac{m}{L/V} - 1\right) E_T\right]}{\log \frac{m}{L/V}} \tag{8-38}$$

and the actual number of plates becomes:

$$n_p = \frac{n_t}{E_C} \tag{8-39}$$

for a separation with n_t theoretical steps.

The upper load limit of a separating column is set by a variety of factors. If the pressure losses per plate are high, or if the dwell times in the liquid seal are short, flooding can take place through liquid backup. Another case of flooding may result from liquid entrainment by the vapor stream. Design details of exchange elements, and spacing of trays can increase such flood loads. However, there exists an upper limitation exclusively dependent on the system properties; this cannot be exceeded in the layout; it is characterized by that vapor velocity at which the disperse liquid phase is carried upward by the impetus of the flowing vapor. That vapor velocity depends on the system properties rather than on column details, and it is the deciding barrier to the column output.

The flow mechanics of a column depend on more or less accidental circumstances and cannot readily be described by an equation with any degree of accuracy. Mutual influences of a number of forces, gravity, buoyancy, and dynamic forces, depend on unstable properties of the continuous phases and prevent any successful use of idealized analytical concepts. The variability of the phase interfaces of vapor and liquid is due to pulsations of vapor passing through liquid. The turbulence of the rising vapor causes a formation of foam whose stability and distribution in the liquid spray are unpredictable, except for arriving at the ultimate fact that the limitations imposed by the nature of the system revert to the mass of the liquid, to the stability of the continuous and dispersed phases, from the stability, motion, deformation, and entrainment of droplets and their distribution, again influenced by the evolving vapor phase.

As to an interrelation between the factors characterizing the load maximum and the system properties, basic mechanisms are tied to flooding: the flow resistance of the dispersed liquid phase by the rising vapor, the splitting up of liquid droplets caused by their own lack of stability, and the turbulence in the vapor phase. If these forces cooperate, and if an adequate number of droplets are divided into sufficiently minute particles, the column must flood.

An accepted criterion of drop stability is the Weber number. It equals the ratio of the dynamic pressure generated by the vapor flow outside the drop and the inner excess pressure caused by the surface tension along the interface of the drop and the vapor.

If the vapor flows at velocity u_V and its specific weight is ρ_V, using g, the acceleration of gravity for the dynamic pressure p_s

$$p_s = \frac{u_V^2}{2g} \rho_V \tag{8-40}$$

For the internal excess pressure of a droplet of diameter d_T and surface tension σ

$$p_i = \frac{4\sigma}{d_T} \tag{8-41}$$

Hence the Weber number

$$We = \frac{p_s}{p_i} = \frac{1}{8} \frac{d_T u_V^2}{\sigma g} \rho_V \tag{8-42}$$

The resulting ratio of the Weber number and the droplet diameter

$$8 \frac{We}{d_T} = \frac{\rho_V}{g} \frac{u_V^2}{\sigma} \tag{8-43}$$

may be interpreted as the ratio of the flow energy per unit volume of rising vapor and the surface energy per unit surface of the entrained droplet. Within the range of the system load limit it will approach a constant value.

Depending upon their size, the droplets will move by the dynamic force resulting from the rising vapor, the force of gravity, and the buoyancy in an upward or downward direction at different absolute velocities. Very minute droplets will move at about vapor velocity, larger droplets will move upward at reduced velocity, and very large droplets will settle by gravity. In between there is a droplet size controlled by the metastable equilibrium and it will appear stationary to an observer outside the column; this is the size of all droplets of maximum diameter carried upward. Most entrained droplets are of that size; it is the entrainment that moves the bulk of the liquid upward, according to the third power of the diameter of the particles. On the other hand, in a turbulent environment some droplets may become unstable and disintegrate into smaller bodies, or else they may remain of stable size but merely become deformed and oscillate. Minute droplets may collide and grow by cohesion, eventually becoming unstable and disintegrating once more.

For the limiting load of the system, the ratio

$$\frac{We}{d_T} = \text{constant}$$

is almost axiomatic and independent of the system; the correlated vapor velocity becomes, from Eq. (8-43)

$$(u_V)_{fl} = K \left(\frac{\sigma}{\rho_V} \right)^{1/2}$$

By insertion of the surface tension value (dyn/cm), and the specific weight of the vapor ρ_V (kg/m^3), the constant assumes for low liquid loads a value of $K = 1.22$, and velocity $(u_V)_{fl}$ comes out in meters per second. However, in the vicinity of the critical point Eq. (8-45) will supply results that are low.

Around the center of gravity of a droplet impinged from below, with a specific weight ρ_L there exists an equilibrium between the dynamic force of the vapor and the difference between the force of gravity of the liquid droplet and the force of buoyancy according to the equation

$$\xi \, d_T^2 \frac{\pi}{4} \frac{u_V^2}{2g} \rho_V = \frac{4}{3} d_T^3 \frac{\pi}{8} (\rho_L - \rho_V) \qquad (8\text{-}44)$$

from which follows

$$u_V = \left(\frac{4 g d_T}{3 \xi} \right)^{1/2} \left(\frac{\rho_V}{\rho_L - \rho_V} \right)^{-1/2} \qquad (8\text{-}45)$$

In Eqs. (8-44) and (8-45) ξ is the coefficient of resistance. Eq. (8-45) demonstrates that the limit velocity of the vapor is inversely proportional to the square root of $\rho_V/(\rho_L - \rho_V)$, and that the relation

$$F_V^* = u_V \left(\frac{\rho_V}{\rho_L - \rho_V} \right)^{1/2} \tag{8-46}$$

is a useful correlation factor for the flood load of a column. It is often presented in this form, or a modification of it, as a function of the flow parameter (as defined by Eq. (8-11),) or as a function of the liquid load. On that basis, plots can be devised that permit ready answers to column size problems via separations involving different vapor and liquid rates.

In practice, Eq. (8-46) was proven to offer a satisfactory correlation; other correction coefficients relate to the plate spacing and sometimes to the vapor density, especially where the latter is considerable. Also, for certain systems deviating in the distillation of saturated hydrocarbon systems empirical factors have been employed.

Eq. (8-46) was derived from the equilibrium of forces on liquid droplets in suspension above the boiling or foaming liquid phase. The mean diameter of the drops had been assumed to be alike for all systems, which is unrealistic because the droplet size depends on physical properties such as surface tension and vapor density. Therefore, the formation of droplets above the boiling or foaming layer as a consequence of the energy display upon penetration of the liquid layer on a plate should be referred to a dimensionless group satisfying both system properties. The Weber number may fit this requirement; it could be related to coefficients that make Eq. (8-46) adaptable to systems deviating from the performance of saturated hydrocarbons, as far as most data indicate. That hypothesis awaits confirmation. Should the Weber number be confirmed, empirical correlations might explain occasional irregularities.

When a droplet having a relative velocity of $(u_V)_r$ remains stationary in a moving vapor phase, its size may be determined in view of the equilibrium of forces which are gravitational, buoyant, and dynamic. Following Eq. (8-44) one may write

$$d_T = \frac{3}{4} \xi \frac{\rho_V}{\rho_L - \rho_V} \frac{(u_V)_r^2}{g} \tag{8-47}$$

Evaluation of Eq. (8-47) requires knowledge of the resistance coefficient ξ, which is controlled by the Reynolds number of the vapor

$$(Re)_V = \frac{(u_V)_r \, d_T}{\nu_V} \tag{8-48}$$

with kinematic viscosity ν_V and surface tension σ; see Fig. 8.64.[46] Depending on system and load, the maximum diameter of entrained droplets is around $d_T = 0.5$ mm. Eqs. (8.47) and (8-48) together with Fig. 8.64 may be evaluated graphically.

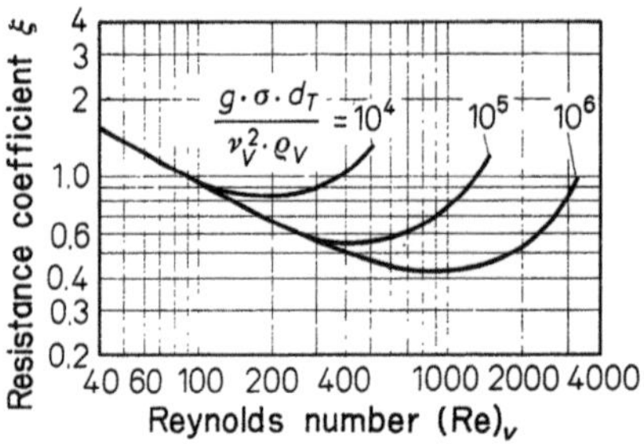

Fig. 8.64 Resistance coefficient of vapor versus Reynolds number and surface tension.

Another type of flooding which is liquid backup in downcomers may be due to insufficient liquid in the seal cup, or to high pressure loss on the plate, or by both. To counteract this the layout should be such that the level of the clear liquid does not exceed half the plate spacing; this is the case for opening ratios over 10% and adequate dwell times t_R (s) for the liquid in the downcomer. That is true for foaming systems such as most hydrocarbons (t_R = around 3.5 s), average foaming systems such as crudes ($t_R = 4.5$ s), and intensely foaming systems such as amines in vacuum columns ($t_R = 7$ s).

Depending upon plate spacing T_S (m) a calculated flow velocity u_d (m/s) of the liquid in the downcomer results from

$$u_d = \frac{T_S}{t_R} \tag{8-49}$$

The dwell time of the clear liquid becomes, for a liquid head of h_d (m)

$$(t_R)_d = \frac{h_d}{u_d} \tag{8-50}$$

If L (kmol/h) of liquid of a molecular weight of u_L (kg/mol) and a specific weight of ρ_L (kg/m^3) pass through a downcomer, its cross section A_d (m^2) must equal no less than

$$A_d = \frac{t_R}{T_s} \frac{L\,\mu_L}{3600\,\rho_L} \tag{8-51}$$

Because of the relatively low flow velocity of the liquid through a properly designed downcomer the main head loss h_u (mm of clear liquid) occurs on leaving the downcomer. It may be largely responsible for the liquid head h_d along with the pressure loss Δp_T of the vapor flowing onto the plate. It may be estimated at

$$h_u = 1.2\cdot 10^{-5} \left(\frac{L\mu_L}{\rho_L} \frac{1}{(A_d)_m} \right)^2 \tag{8-52}$$

where $(A_d)_m$ (m^2) is the smallest exit section for the liquid below the downcomer.

Distillation equipment is expected to function dependably over an extended time. Nevertheless it appears that some areas are left where further experimentation and experience would be of help. Thus, in some instances columns working at elevated pressures have flooded prematurely, as in demethanizing, where the operating pressure is near critical. This took place on plates of all kinds. Intense foaming in the downcomers was observed. The first thought was to reduce liquid levels in downcomers by reducing the pressure losses of the dry plates and those occurring when the liquid passes through the bottleneck between the downcomer and the next plate. However, at elevated pressures the vapor is distributed in the liquid in the form of tiny bubbles. A certain dwell time may be required for the vapor-liquid mix in the downcomer to agglomerate and decompose. Too brief a time and inadequate separation of vapor and liquid cause the foam in the downcomer to be too thin, and flooding will result. Calculation based on dwell time[49,50] applicable to low-pressure systems may fail for high pressures, especially if the foam density in the downcomer makes itself felt. Also, the "cascade effect" produced by the falling liquid, may affect foam formation.

Foam stability is presumably controlled by increased surface tension. Therefore, for binary mixtures, the most intense foaming may be expected at those points of the column where [51]

$$S = \frac{x\,(100 - x)}{(100 - x) + \dfrac{V_1}{V_2}\,\dfrac{x}{100}} \left[\frac{d\sigma}{d\left(\dfrac{x}{100}\right)}\right]^2 \tag{8-53}$$

has a maximum; x is the concentration of the component of less surface tension, V_1 and V_2 = molar volumes of the component having the least and most surface tension, and σ = surface tension. Mixtures tending to foam markedly do so mostly for $x < 5$ mol-%.

B. Bell-Cap Trays

The discussion of cost optimization in Chapter 6 shows definitely that the bell-cap plate, as built conventionally, is one of the most expensive column internals. Nevertheless it remains popular especially where a large load range with good separation is required, or where a variable vapor/liquid ratio is expected in operation, as in batch rectification at constant overhead concentration.

In the typical design of Fig. 8.65 the vapor stacks and bells are comparatively slender; recent designs feature flat bells and squat vapor stacks drawn from the plate material (note lower part of Fig. 8.65). Pressure loss and load conditions are thus improved.

a. Conventional Design

The admissible top vapor velocity, based on the free column surface, for plates fitted with conventional bell caps may be found from Eq. (8-46), $u_V = (u_V)_o$ (m/s).

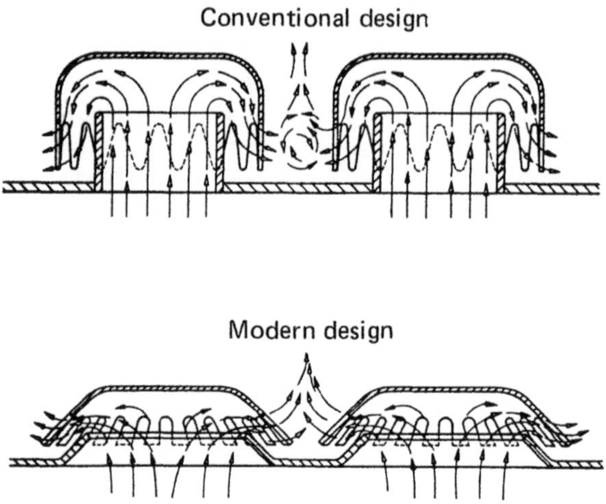

Fig. 8.65 Typical features of bell caps.

Tests have confirmed the validity[52,53] of the term

$$F_V^* = (F_V^*)_o = 5\cdot 10^{-2} \frac{(T_S - h_c)^{1/2}}{(d_c^2)^{1/3}} \tag{8-54}$$

where T_S = tray spacing (mm), h_c = height of caps (mm) (note Fig. 8.66), and d_c = diameter of caps (mm). Although the validity of Eq. (8-54) has been established only for liquid/vapor ratios of $0.6 \leqslant L/V \leqslant 1$, it is useful for other ratios as well, provided

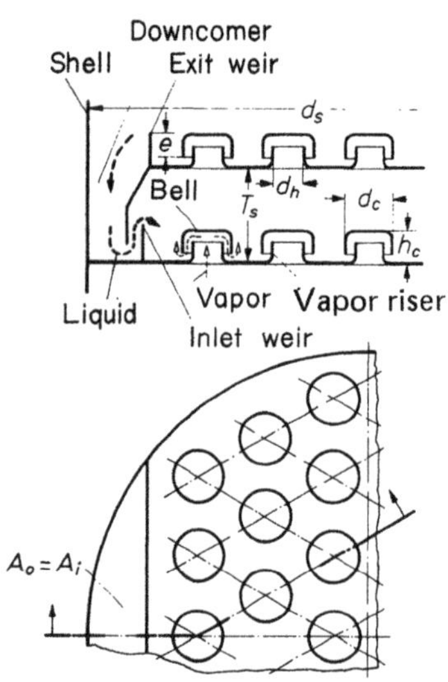

Fig. 8.66 Bell cap tray details, diagrammatic.

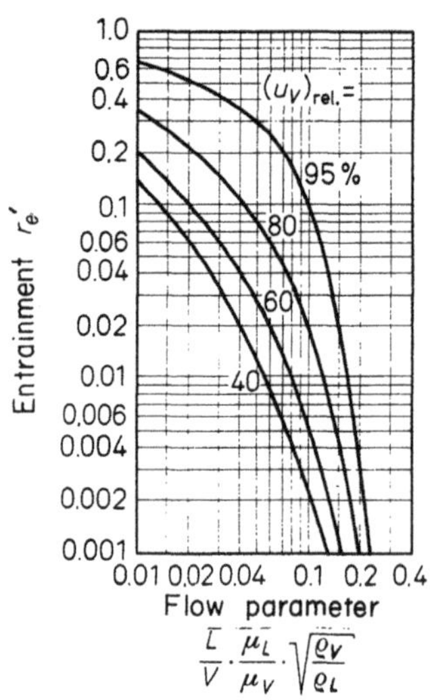

Fig. 8.67 Liquid entrainment on bell cap trays (54).

the rules explained in Chapter 4. for the dimensioning of the downcomers and on entrainment are observed. For the latter, Fig. 8.67 will supply a flow parameter, previously defined by Eq. (8-11) which is

$$r'_e = \frac{\Delta L}{L + \Delta L} \tag{8-55}$$

for the vapor load in question

$$(u_V)_{\text{rel}} = \frac{u_V}{(u_V)_{\text{fl}}} 100 \tag{8-56}$$

which with Eq. (8-33) yields

$$r_e = \frac{r'_e}{1 - r'_e} \tag{8-57}$$

The reduced separation can now be found from Eq. (8-34).

In standard designs, segments A_i and A_o defined by the inlet and exit weirs, will take about 10% of the column area. For best separation, the ratio of column area and working area outside of the caps will roughly amount to

$$\varphi_a = \frac{d_s^2 \frac{\pi}{4}}{d_s^2 \frac{\pi}{4} - n_c d_c^2 \frac{\pi}{4} - 2A_o} = 1.82 \tag{8-58}$$

where n_c = number of caps and d_s = column diameter. In that connection note Figs. 8.68 and 8.69.

When establishing the minimum weir length, refer to Eq. (8-32); also[44,55] the extra weir height should not exceed

$$h_{oW} = 3 \text{ to } 5 \cdot 10^{-2} \text{ m} \tag{8-59}$$

With L (kmol/h) of liquid overflowing the weir, the weir length becomes

$$l_W = 1.52 \cdot 10^{-4} \frac{L\, \mu_L}{\rho_L\, (h_{oW}^3)^{1/2}} \tag{8-60}$$

where μ_L = molecular weight of the liquid (kg/kmol) and ρ_L = specific weight of the liquid (kg/m^3).

Submergence e, which is generally between 25 and 30 mm, may be much less in vacuum columns to obtain low pressure loss. For the effect of e upon efficiency and loadability of plates, note the bibliography.[56] It appears that for low e values the sep-

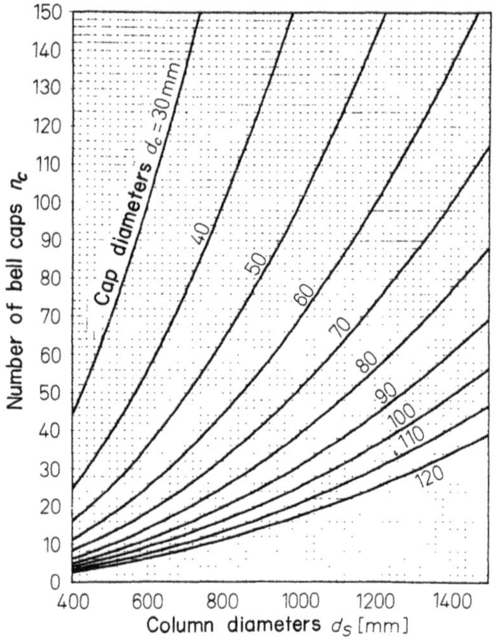

Fig. 8.68 Standard bell cap plates, φ_a = 1. 82. Number of caps, cap diameters and column diameters.

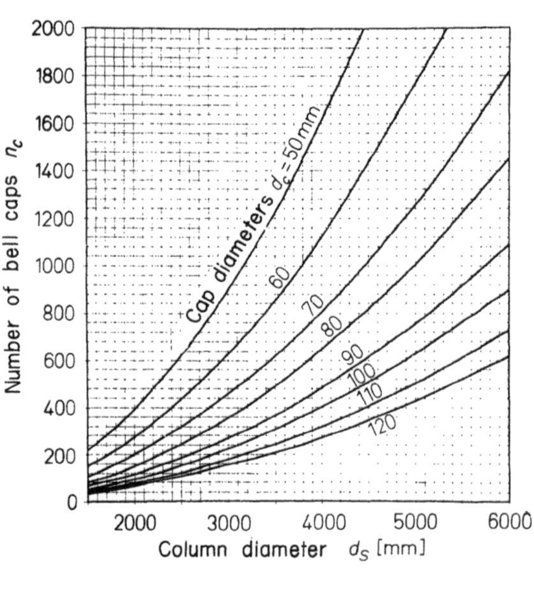

Fig. 8.69 Continuation of Fig. 8.68.

aration is virtually unchanged while the pressure loss of the vapor flowing through the plate is much less than for large e values. Practical results agree inasmuch as in vacuum columns submergence is frequently nil with large exchange ratios. It is also suggested that exchange occurs immediately upon entrance of the vapor into the liquid on the plate.[57,58]

Popular bell sizes d_c are between 50 and 200 mm; the large sizes are preferred in working encrusting liquors. Eq. (8-54) gives the effect of the bell diameter upon the load; plates with small caps carry higher loads than plates with larger caps. The preferred size is around d_c = 80 mm. From Fig. 8.70 the separation is better the smaller the cap size.

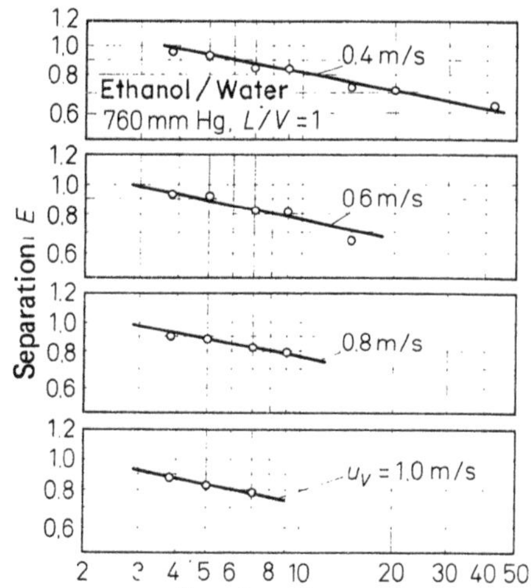

Fig. 8.70 Separation versus the cap size for various vapor velocities. Ethanol/water system 760 mm Hg, L/V = 1. Runoff concentration of 50 mol-%, d_S = 750 mm, φ_a = 1.82, T_S = 200 mm, e = 30 mm.

The minimum distance between caps is 50 mm. The distance from caps to weirs should be somewhat larger. The distance between the plate and the lower rim of the caps should be no less than 10 mm. Because of the pressure loss, the assembly of the vapor stack and cap should be designed for equal vapor sections; preferably the free areas between the stack and annular space should be 1:1. 1:1. 4. A standard bell-cap plate at ordinary loads will have a Δp from 30 to 60 mm water; the lower figures apply in vacuum work.

Regardless of design, the admissible vapor velocity must always be established for the highest loaded plate, and the variations in pressure, in temperature, and in vapor weight along the column must all be noted. This is especially true in vacuum work where the minimum load is usually at the column top. Sometimes, vacuum columns must have staggered diameters, or column internals of different load characteristics must be selected to allow for changes in load in columns of constant diameter.

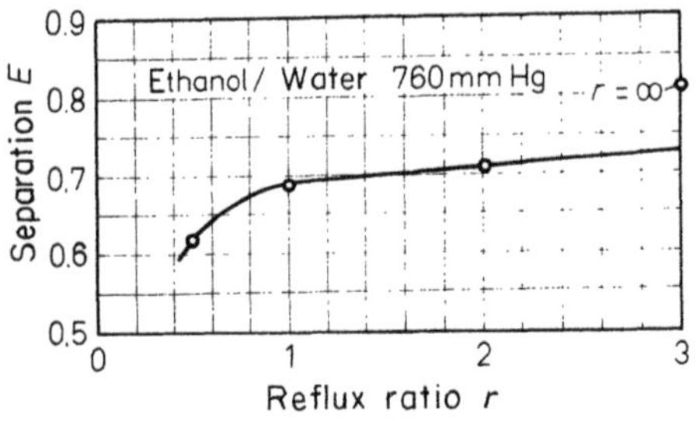

Fig. 8.71 Separation versus the reflux ratio for bell cap plates. Ethanol/water system, 760 mm Hg, $L/V = 1$. Runoff concentration 50 mol-%, $d_S = 750$ mm, $d_c = 90$ mm, $\varphi_a = 1.82$, $T_S = 200$ mm, $e = 30$ mm.

Another fact is that the separation becomes less for sinking liquid/vapor ratios. In the range of higher vapor loads, for low L/V ratios, the proportion of liquid entrained by vapor is greater than for higher—at any given vapor throughput, which by Fig. 8.67 and Eqs. (8-34) and (8-55) to (8-57) causes impaired separation. Even at average loads, the drop in separation is not negligible for L/V ratios below unity (Fig. 8.71).[42] Separation E for a bell-cap plate is plotted against the reflux ratio

$$r = \frac{L/V}{1 - (L/V)} \tag{8-61}$$

The separation drops from $E = 0.8$ for $r = \infty$ at $L/V = 1$, to $E = 0.6$ for $r = 0.5$ at $L/V \doteq 0$. That is important for enrichment columns in vacuum work because of the large relative volatility of the mixture resulting in a low reflux ratio.

Eq. (8-54) reveals the effect of plate spacing T_S upon the loadability of a column. The admissible vapor velocity increases and the required column size decreases as the plate spacing increases. A spacing $T_S = 350$ mm is considered average. For installation inside buildings T_S is reduced; in open-air installations it may be much greater. Petroleum refineries use T_S exceeding 600 mm.

Fig 8.72 shows details of a conventional design.

Example 3 For the conditions of Example (7.6) calculate the upper vapor velocity of the butanol column equipped with bell-cap plates. Dimensions of caps are $h_c = 75$ mm, $d_c = 75$ mm. Plate spacing $T_S = 500$ mm.

Answer Eq. (8-54) gives a F_V^* factor

$$F_V^* = (F_V^*)_o = (5 \cdot 10^{-2}) \frac{(500 - 75)^{1/2}}{(75^2)^{1/3}} = 5.8 \cdot 10^{-2}$$

This along with $\rho_V = 1.43$ kg/m^3 and $\rho_L = 790$ kg/m^3, combined with Eq. (8-46)

$$u_V = (u_V)_o = (F_V^*)_o \left(\frac{\rho_L - \rho_V}{\rho_V}\right)^{1/2} = (5.8 \cdot 10^{-2}) \left(\frac{790 - 1.43}{1.43}\right)^{1/2} \tag{8-62}$$

$$= 1.35 \text{ m/s}$$

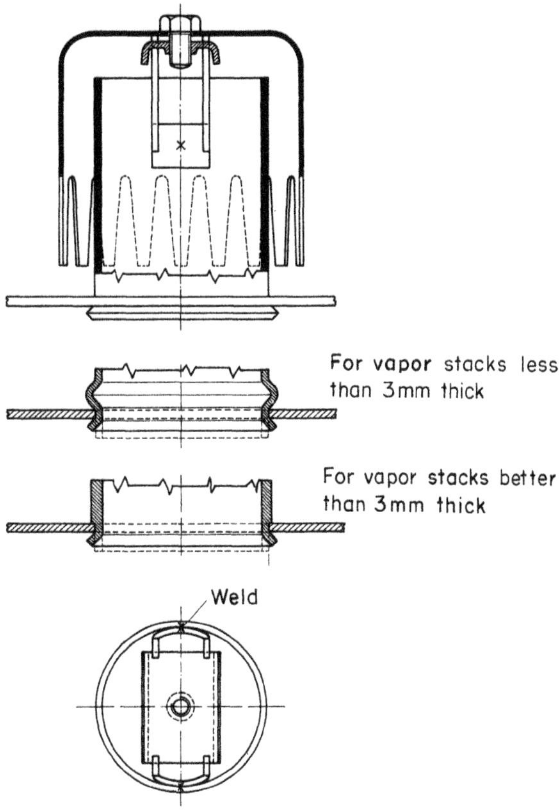

Fig. 8.72 Conventional bubble cap, made by Koch International of Bergamo, Italy.

From the data of Example 7.6, $V = V_1 = 64.3$ kmol/h, $p = 760$ mm Hg, $T =$ 366 K, and the column diameter becomes

$$d_S = 0.149 \left[\left(64.3\right) \left(\frac{366}{760}\right) \left(\frac{1}{1.35}\right) \right]^{1/2} = 0.72 \text{ mm}$$

and the number of caps from Fig. 8.68 becomes $n_c = 24$.

b. Recent Designs

Inspection of Figs. (6.42) and (8.45) will reveal that the bell-cap plate in its conventional design does not rate favorably from an operational viewpoint. Inherently more appealing is the modern variant featuring flat caps and low vapor stacks presented in the lower portion of Fig. 8.65, where the mutual interference of neighboring caps is reduced to a minimum. The pressure loss is low, the loadability is high because the cap spacing may be less than 10 mm, the openings are more liberal, and the liquid head is relatively low because of less cap height.

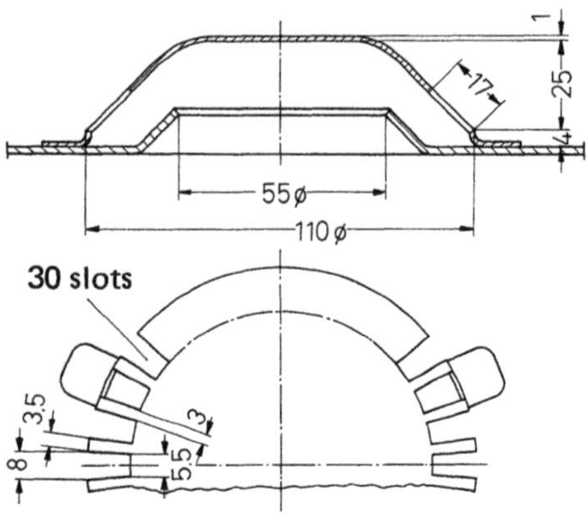

Fig. 8.73 Squat bubble cap made by ACV Stage-Schmidding of Cologne, Germany.

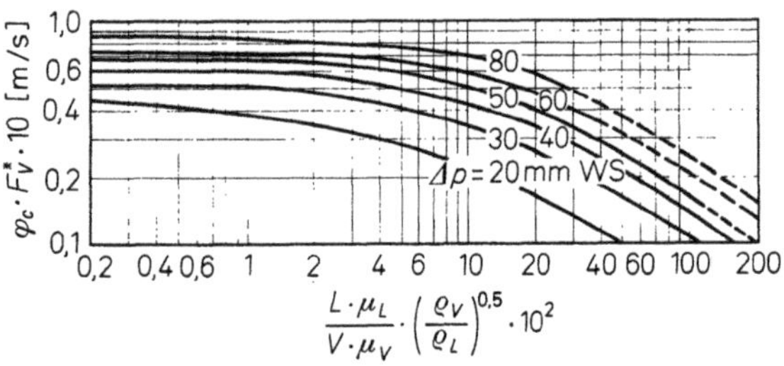

Fig. 8.74 Squat bubble caps. Vapor velocities versus pressure loss.[59]

Fig. 8.73 gives the principal dimensions of a modern flat cap.[59] It can be arranged on the tray at 120 mm spacing. For these dimensions Fig. 8.74[59] allows determination of the superficial vapor velocity u_V (m/s). The plot was drawn for the load factor from Eq. (8-46) and the flow parameter of Eq. (8-11). For the separation data given, the abscissa on the curve for the admissible pressure loss Δp (mm water) can be calculated, and the correlated ordinate $\varphi_c F_V^*$ can be read off. In this connection

$$\varphi_c = \frac{d_S^2 \frac{\pi}{4} - A_i - A_o}{d_S^2 \frac{\pi}{4}} \tag{8-63}$$

is the ratio of the working area of the cap and the superficial column area. A_i and A_o are again the sectors defined by the inlet and outlet weir on the tray.

For thin-walled caps, of conventional design, the value of $\varphi_a = 1.82$ corresponds, by Eq. (8-58), to an aperture ratio φ_h, defined by

$$\varphi_h = n_c \left(\frac{d_h}{d_S}\right)^2 \tag{8-64}$$

the sum of all vapor stack areas, by the column area, of around 10%. Plates with flat caps may be designed for greater φ_h which gives more vapor throughput and less pressure loss.

Regarding details of available cap types, such as crenelated shapes, slotted caps, flat rims, stackless construction, etc., reference is made to the bibliography.[4,33,54,60,61]

C. Sieve Trays

Like the bell-cap plate, this is one of the earliest developments, having been used for many years in the chemical industries and in petroleum refining. For the performance of this design note the bibliography.[4,62–66]

Early designs featured 2 to 4 mm perforations, which sometimes caused plugging or incrustation. Nowadays, holes between 10 and 25 mm are customary.

The sieve plate is cheapest as a rule. Dual flow plates of suitable design are even less costly with respect to unit column area, but they cannot be used indiscriminately because of their wide load range.

a. Standard Sieve Plates

Generally, perforated sheets are used with hole areas between 8 and 15% with reference to the working area between the weirs. Maximum loadability is obtained for 15% openings; above that the loadability improves little. Plates with small holes have excellent separation over a relatively wide range. If d_h is the hole diameter and t_h the hole spacing, the aperture ratio becomes

$$\varphi_h' = 0.906\left(\frac{d_h}{t_h}\right)^2 \tag{8-65}$$

From this, and allowing for the spaces of inlet and exit weirs and downcomers A_i and A_o and lost areas A_l (fasteners and stiffeners,) the overall aperture ratio results, eventually

$$\varphi_h = \varphi_h' \frac{d_S^2 \dfrac{\pi}{4} - (A_i + A_o + A_l)}{d_S^2 \dfrac{\pi}{4}} \tag{8-66}$$

Based on tests,[4] the surficial column area may be computed to obtain the admissible vapor velocity $u_V = (u_V)_o$ (m/s). For $\varphi_h = 8\%$

$$F_V^* = (F_V^*)_o = 4.5 \cdot 10^{-3} \frac{T_S^{1/2}}{d_h^{1/5}} \tag{8-67}$$

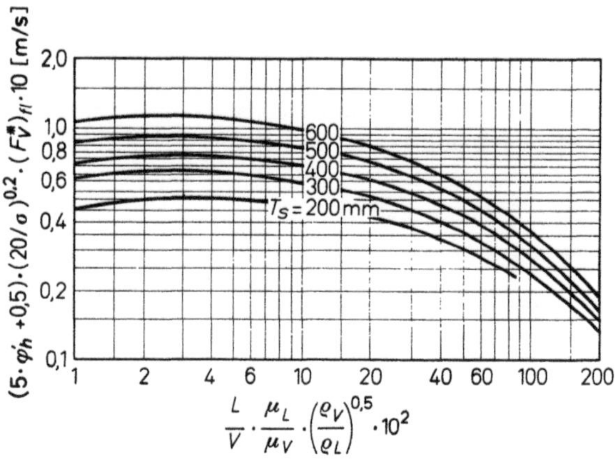

Fig. 8.75 Sieve plate columns. Vapor velocities at flooding point.[54]

where plate spacing T_S and hole diameter d_h must be written in millimeters. By means of relation (8-62) derived from Eq. (8-46), $(u_V)_o$ (m/s) is thus determined.

On the other hand, one can start from the vapor velocity $u_V = (u_V)_{fl}$ at flooding, making use of a plot as shown in Fig. 8.75.[54] For the abscissa specified, read the ordinate and compute from this the surface tension σ(dyn/cm), and from aperture ratio φ'_H the load factor $F_V^* = (F_V^*)_{fl}$, which in turn yields the flooding velocity by Eq. (8-46). At low loads the liquid will rain through the sieve holes. The range of the greatest efficiency for plates with $\varphi'_H = 10\%$ of open area is around 3–1 to 4–1.

In the layout of a sieve plate one may also start from the minimum vapor velocity $(u_h)_m$ (m/s) in the hole at which the liquid L (kmol/h) will not yet rain through, and the operation will remain stable.[64] Tests[63] have revealed that for sieve plates of customary design, a vapor load factor (referred to the hole) of

$$(u_h)_m \rho_V^{1/2} = 10 \text{ m}^{-1/2} \text{ s}^{-1} \text{ kg}^{1/2} \tag{8-68}$$

will correspond to that velocity. This, combined with a relation based on equivalent diameters, and a relation[64,67] applied to the plate and downcomer areas

$$d_S = 3\,(d_h)_H + (d_{Ao})_H \tag{8-69}$$

will yield, with V_m (kmol/h) minimum vapor quantity, for the equivalent diameter of the total hole area

$$(d_h)_H = \left(\frac{4}{\pi}\,\frac{V_m}{3600}\,\frac{M_V}{\rho_V}\,\frac{1}{(u_h)_m}\right)^{1/2} \tag{8-70}$$

while the equivalent diameter of the segmental area claimed by the downcomer, $A_o = A_i$, is found from

$$(d_{Ao})_H = \left(\frac{4}{\pi}\,\frac{L}{3600}\,\frac{\mu_L}{\rho_L}\,\frac{1}{u_d}\right)^{1/2} \tag{8-71}$$

Eqs. (8-69), (8-70), and (8-71) will now supply the minimum vapor quantity

$$V_m = \frac{\rho_V\,(u_h)_m}{\mu_V}\left[\frac{d_S\left(\frac{\pi}{4}\,3600\right)^{1/2} - \left(L\,\frac{\mu_L}{\rho_L}\,\frac{1}{u_d}\right)^{1/2}}{3}\right]^2 \tag{8-72}$$

which must be maintained to prevent liquid from raining through the sieve holes. Liquid velocity u_d inside the downcomer follows from Eq. (8-49). The dependence of the entrained liquid mass from the load of a sieve plate is seen in Fig. 8.76. By use of Eq. (8-55), (8-56), and (8-57), eq. (8-34) will supply the reduced separation effect.

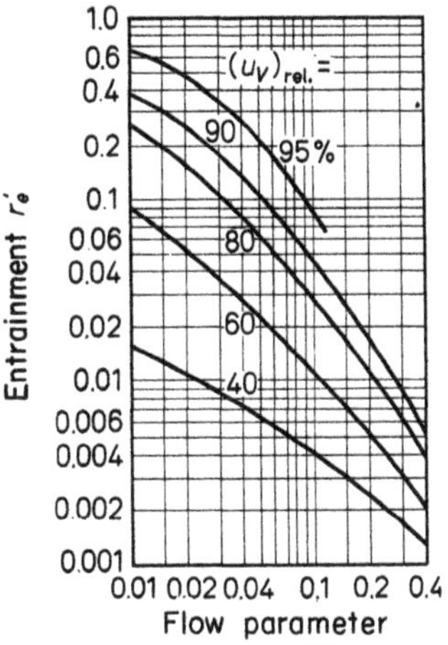

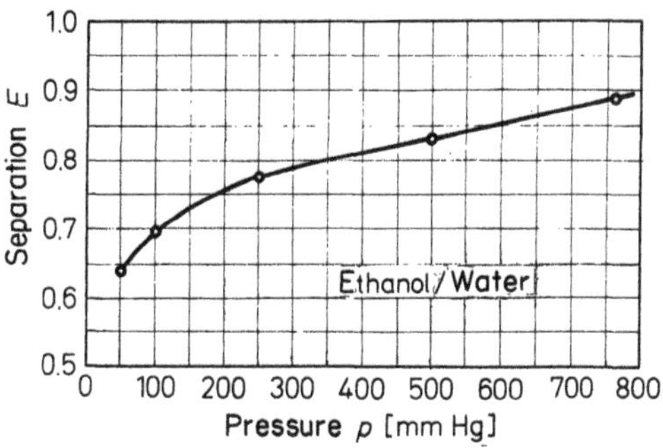

Fig. 8.77 Sieve plates. Separation versus system pressure. Ethanol/water system. $V = L = 30$ kmol/m² h. $d_S = 400$ mm, $d_h = 2.5$ mm, $t_h = 7$ mm, $\varphi_h = 7.8\%$, $T_S = 135$ mm, $h_w = 28$ mm.

◄ **Fig. 8.76** Sieve plates. Entrainment versus flow parameter[54].

Separation by properly designed sieve plates is not inferior to that of bell-cap plates; within the useful range of elevated loads it may even be markedly superior to that of conventional bell caps (Fig. 8.45). As for other designs, a lowered system pressure means less effective separation[4,41] (note Fig. 8.77); this fact should be kept in mind when establishing plate numbers.

In columns for normal and higher pressures, the height of runoff weirs may be up to

$$h_W = 50 \text{ to } 100 \text{ mm} \tag{8-73}$$

while for low-pressure columns that dimension should not exceed

$$h_W = 15 \text{ to } 40 \text{ mm} \tag{8-74}$$

The center distance should be around

$$t_h = (2.5 \text{ to } 4)\, d_h \tag{8-75}$$

up to a maximum of 75 mm. Within the operational range, pressure losses should amount to Δp = 30–50 mm water per tray.

Example 4 A column for separating a multiple mixture is to be fitted with sieve plates. Project data are as follows:

V = 35 kmol/h vapor at the column head
L = 15 kmol/h liquid at the column head
T_S = 400 mm tray spacing
d_h = 13 mm hole size
ρ_V = 0.046 kg/m^3 specific weight of the vapor
ρ_L = 990 kg/m^3 specific weight of the liquid
σ = 20 dyn/cm surface tension
$\mu_V = \mu_L$ = 100 kg/kmol molecular weight

Answer The hole spacing will be, from Eq. (8-75):

$$t_h = (3)(13) = 39 \text{ mm}$$

The aperture ratio for the perforated area

$$\varphi_h' = 0.906\left(\frac{13}{39}\right)^2 = 0.1$$

For the abscissa of Fig. 8.75

$$\left(\frac{15}{35}\right)\left(\frac{0.046}{990}\right)^{0.5} = 3 \cdot 10^{-3}$$

the ordinate is, at T_S = 400 mm

$$[(5)(0.1) + 0.5]\left(\frac{20}{20}\right)^{1/5} (F_V^*)_{fl}\,(10) = 0.7$$

for

$$(F_V^*)_{fl} = 0.07$$

and with this, from Eq. (8-46), a vapor velocity at flooding

$$(u_V)_{fl} = 0.07 \left(\frac{990 - 0.046}{0.046}\right)^{1/2} = 10.3 \text{ m/s},$$

corresponding to a vapor load factor of

$$(u_V)_{fl}\, \rho_V^{1/2} = 10.3\,(0.046)^{1/2} = 2.2 \text{ m}^{-1/2} \text{ s}^{-1} \text{ kg}^{1/2}$$

For a slightly greater value, a sieve plate with $d_h = 12.5$ mm and $\varphi_h = 0.123$ has been tested (Figs. 6.54 and 6.55).

By Eq. (8-67), the load factor at the top load of

$$(F_V^*)_o = 4.5 \cdot 10^{-3} \frac{400^{1/2}}{13^{1/5}} = 54 \cdot 10^{-3}$$

for which Eq. (8-62) will supply the corresponding vapor velocity

$$u_V = (u_V)_o = 54 \cdot 10^{-3} \left(\frac{990 - 0.046}{0.046}\right)^{1/2} = 7.9 \text{ m/s},$$

which is approximately 70% of the flooding load. Since in Eq. (3-39) the relation

$$22.4 \frac{T}{273} \frac{760}{p} = \frac{\mu_V}{\rho_V} \tag{8-76}$$

is identical, the column diameter becomes

$$d_S = \left(\frac{4}{3600\,\pi}\right)^{1/2} \left(V \frac{\mu_V}{\rho_V} \frac{1}{u_V}\right)^{1/2} \tag{8-77}$$

or

$$d_S = 0.0188 \left(35 \frac{100}{0.046} \frac{1}{7.9}\right)^{1/2} = 1.85 \text{ m}.$$

From Eq. (8-68) the minimum vapor velocity in the perforations becomes

$$(u_h)_m = \frac{10}{(0.046)^{1/2}} = 46.6 \text{ m/s}.$$

In view of the relatively low liquid load, a downcomer area of 6.5% of the total plate area may be selected, which is not too much over the recommended minimum of 5%. The surface of the downcomer becomes

$$A_o = 0.065\,\frac{1.85^2\,\pi}{4} = 0.174\ \text{m}^2$$

and the flow velocity of the liquid is only

$$u_d = \frac{L\,\mu_L}{\rho_L\,A_o\,3600} = \frac{(15)(100)}{(990)(0.174)(3600)} = 0.0024\ \text{m/s}, \qquad (8\text{-}78)$$

Use of Eq. (8-49) would give an admissible dwell time of 7 s, and the velocity would be

$$u_d = \frac{0.4}{7} = 0.057\ \text{m/s}$$

The minimum vapor quantity would, by Eq. (8-72) become, if $L = Lm = (15/35)V_m$

$$V_m = \frac{(0.046)(46.6)}{100}\left[\frac{1.85\left(\dfrac{\pi}{4}\,3600\right)^{1/2} - \left(15\,\dfrac{100}{990}\,\dfrac{1}{0.0024}\right)^{1/2}}{3}\right]^2 = 16\ \text{kmol/h},$$

which by Eq. (8-77) corresponds to a minimum vapor velocity, in reference to the free column area, of

$$(u_V)_m = \frac{4}{\pi\,3600}\,\frac{V_m}{d_S^2}\,\frac{\mu_V}{\rho_V} \qquad (8\text{-}79)$$

$$= 3.53\cdot 10^{-4}\,\frac{16}{1.85^2}\,\frac{100}{0.046} = 3.6\ \text{m/s}$$

This gives a correlation of vapor velocities

$$(u_V)_{fl} : (u_V)_o : (u_V)_m = 2.9 : 2.2 : 1.$$

By Eq. (8-70) the equivalent diameter of the perforation area is

$$(d_h)_H = \left(\frac{4}{\pi}\,\frac{16}{3600}\,\frac{100}{0.046}\,\frac{1}{46.6}\right)^{1/2} = 0.511\ \text{m},$$

and the number of holes follows from

$$n_h = \left(\frac{(d_h)_H}{d_h}\right)^2$$

$$= \left(\frac{0.511}{0.013}\right)^2 = 1550 \tag{8-80}$$

This means an aperture ratio referred to the perforation area of

$$\varphi_h' = \frac{n_h\, d_h^2 \dfrac{\pi}{4}}{d_S^2 \dfrac{\pi}{4} - (A_o + A_i + A_1)} = 0.1 \tag{8-81}$$

$$= \frac{1550\,(0.013)^2 \dfrac{\pi}{4}}{1.85^2 \dfrac{\pi}{4} - \dfrac{(6.5 + 6.5 + 10)\,1.85^2}{100}\dfrac{\pi}{4}} = 0.1$$

if the areas of the inlet and exit weirs, A_i and A_o, are allowed 6.5% each, and the loss area A_l is allowed 10% of the column surface. Eq. (8-65) also gave $\varphi_h' = 0.1$.
For the aperture ratio referred to the free column area

$$\varphi_h = n_h\left(\frac{d_h}{d_S}\right)^2 \tag{8-82}$$

there results

$$\varphi_h = 1550\left(\frac{0.013}{1.85}\right)^2 = 0.077$$

which is close to the result given by Eq. (8-67)

If the perforation area were retained, and a hole diameter of $d_h = 9.5$ mm were adopted, the number of drillings would become

$$n_h = 1550\,\frac{13^2}{9.5^2} = 2906$$

and, for a hole spacing of $t_h = (3)(9.5) = 28.5$ mm, the subdivision of Fig. 8.78 would be obtained. Note that the "h" central row of holes, angular to the perforation

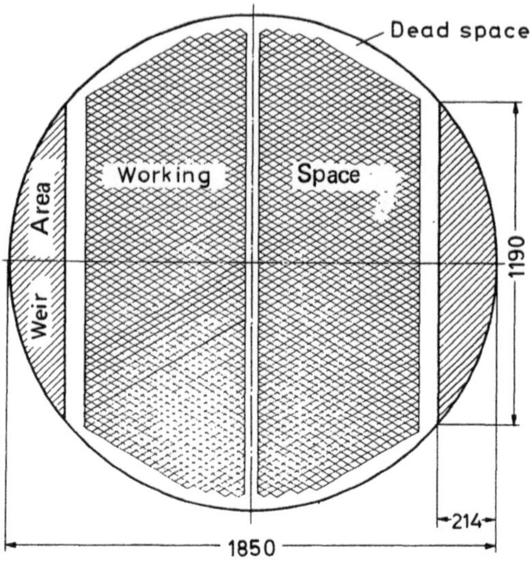

Fig. 8.78 Layout and principal dimensions for a sieve plate; Example 4.

pattern, is omitted to accommodate a reinforcing joist. The two rows affected will be at least 70 mm apart, and the distance between the outer holes and the shell will be at least 50 mm.

b. Sieve Plates without Downcomers (Dual-Flow)

In plates labeled "dual-flow" the holes have two different functions–passing the vapor and passing the liquid. At low vapor loads the liquid will rain through the holes, and there is no visible head of liquid. Separation and vapor loss are low. As the vapor load and/or the liquid load increase, the liquid volume grows. With an adequate load, a pulsating liquid bed with undulating motions will form, and periodically liquid will flow from the wave crests through the perforations. The points of downflow are transient. The load at which this state occurs depends on the system and is around 40 and 70% of the flooding load. Under this condition, the liquid quantity slowly increases with the vapor load, and the separation reaches a maximum immediately before reaching the flood stage. For loads below the range, separation drops rapidly.

To estimate the vapor velocity $u_V = (u_V)_o$ (m/s) at the [illegible] load, from Eq. (8-46) the load factor may be employed[68]

$$F_V^* = (F_V^*)_o = 0.4\,\varphi_h' \left\{1 - \frac{1}{1 + \left[0.7\left(\dfrac{V\,\mu_V}{L\,\mu_L}\right)^2 \dfrac{\rho_L}{\rho_V}\right]^{1/3}}\right\} \tag{8-83}$$

At minimum load, approximately

$$(F_V^*)_m = 0.55\,(F_V^*)_o \tag{8-84}$$

The larger the bore area, the higher the admissible vapor velocity. Mostly, the optimum aperture ratios φ_h' are presumably 20 to 25% of the working area. For higher figures separation gets worse. Figures of $\varphi_h' = 30\%$ should be chosen only when the capacity of available columns is to be enhanced through the exchange of plates. Normally it is recommended that

$$\varphi_h' = \frac{n_h\, d_h^2 \frac{\pi}{4}}{d_S^2 \frac{\pi}{4} - A_l} = 0.2 \qquad (8\text{–}85)$$

A_l is lost area due to fasteners and stiffeners.

Holes with d_h = 10 to 25 mm are best suited to pressure and to vacuum operations. Hole spacing exceeding 70 mm are not recommended. When installing plates, the side with the sharp hole edges should face the vapor flow.

Mostly, plate spacings between 450 and 600 mm appear to be preferred. For smaller spacings the loadability of the plates drops off. Vacuum columns carry moderate vapor loads for least pressure loss, and smaller plate spacings are more economical than the ones pertaining to pressure columns.

In laying out vacuum columns, it may become necessary to increase the perforated zone toward the column top in order to make up for increased vapor load factors, in view of changes in vapor density along the tower. Dual-flow plates are suitable for systems of high vapor density. In the vapor load range of 40 to 98% of flooding, their separation is hardly less than for any other well-conceived tray. Also, their maximum loads are comparable to those of other high-capacity trays. In vacuum operation they have low pressure loss per theoretical step and excellent loadability. Owing to their low liquid holdup and easy cleaning, dual-flow plates will advantageously handle mixtures tending to polymerize and materials with much solid content.

D. Valve Trays

At one time, bell-cap trays used to be the preferred ones. Nowadays, valve trays are favored for engineering and economic reasons. They have better liquid and vapor capacities. In general, their separation is better. Their liquid flow gradient and pressure loss are less. They are lighter and cheaper, and more easily installed or removed. Therefore, and because of easier cleaning, they require less maintenance

When the liquid and vapor profiles along the column vary greatly, sieve plate construction along the column must be modulated, and this may render a sieve plate more costly than a valve plate column. The discussion of cost optimization in Chapter 6 revealed that a combination of sieve and valve trays with venturi openings offer slight cost advantages compared to ordinary sieve trays on the basis of specific pressure loss (Fig. 6.57).

Since the advent of valve trays, it has been noted that valves are less susceptible to acquire deposits of coke, polymers, and other solids likely to cause plugging than are bells. While the space between cap and vapor stack may become plugged, valves moving up and down are self-cleaning, and their better performance seems to be due to the direct flow of the vapor toward the liquid.

Of late, the valve type T, in Fig. 8.79 (top), is increasingly employed for systems that promote fouling. Presumably, an incrustation becomes less likely to occur when the valve is guided by narrow rather than by wide tracks, as for type A of Fig: 8.79. Finally, cokes and polymers tend to settle in spots protected from the flow, primarily behind projections, causing higher pressure loss and lower plate efficiency.

Low pressure loss is important not only in vacuum rectifications but also in fractionation work that may require 100 or more plates. Frequently, the temperature and hence the pressure at the bottom set limits in design when only low-pressure steam is available for column heat.

Typ T

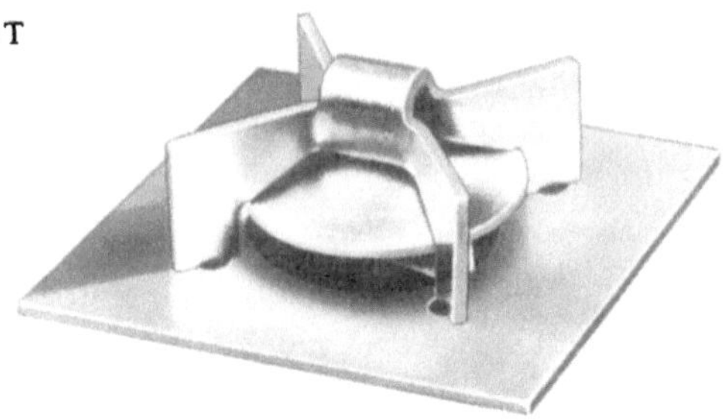

Typ A

Fig. 8.79 Type A and type T valves, made by Koch of Wichita, Kansas, and Bergamo, Italy.

Valve type T9 with venturi openings (Fig. 8.80) offers good chances for reducing the dry pressure loss at the bottom and thus the overall pressure loss (Figs. 6.46 and 6.47).

Experiences with crude oil topping columns fitted with specified Monel valves indicated that type A valves were corroded to paper thickness after one year, and the plate openings had been enlarged to such an extent that the lower trays had to be replaced. This may be explained in part by the friction of the type A legs in the holes which destroyed the passivated oxide film supposedly preventing corrosion. Indeed valves do move up and down even at high vapor loads. This does not apply to stainless steel valves that have no passivating films. Laboratory tests indicated slight corrosion and a probable life of 5 to 6 years. Actually, most A type valves, distributed

Fig. 8.80 Type T-9, made by Koch.

over the entire area, were corroded to paper thickness within one year. Also, the guide legs were attacked. It is thought that valve rotation was responsible, but the Coriolis forces seem insufficient to make the valves revolve, and leverages are inadequate. It may be that rotation results because of changing liquid flow conditions on one side of the valves.

Another explanation may be derived from occurrences outside of the guide legs. In Fig. 8.79 these have been press formed from the body of type A, and bent back to a diameter below that of the valve plate. The latter acquire openings that allow liquid to enter into the recess behind the legs. The intense wear on the legs may be changed to erosion by the vapor–liquid mixture and the vortices generated. As compared to type A, in type T erosions are less likely to occur as the vertical portions of the guides are streamlined and properly distribute the vapor in the liquid.

In the main load range of Figs. 6.46 and 6.47, the separation is fairly steady. For rectifications, in which the specific pressure loss is immaterial, the best layout for such plates is expected for the high load region. For that condition, Figs. 8.81 A and B allow finding the optimum column diameter d_S (mm) from the reduced vapor load V_R (m^3/s) and the reduced liquid load L_R (m^3/h). The figure for the column diameter is read off at the intersection of the parallel to the abscissa through V_R with the parallel to the ordinate through L_R on the d_S line for single-flow and double-flow operation.

The equation for the reduced vapor load is

$$V_R = \frac{V}{3600} \frac{\mu_V}{\rho_V} \frac{1}{f_s f_{T_S}} \frac{\rho_V}{\rho_L - \rho_V} \tag{8-86a}$$

and that for the reduced liquid load

$$L_R = L \frac{\mu_L}{\rho_L} \frac{1}{f_s f_{T_S}} \tag{8-86b}$$

where f_S = a system coefficient that measures the foaming tendency of a mixture, f_{TS} = a coefficient considering plate spacing, ρ_V = specific weight of the vapor (kmol/h) and ρ_L = specific weight of the liquid (kmol/h).

Figs. 8.81 A and B also reveal the difference in column diameters for single-flow and for double-flow plates, respectively, and for relatively large loads. Thus, with single-

Fig. 8.81 Value plates. Estimating the optimum column diameter.[69]

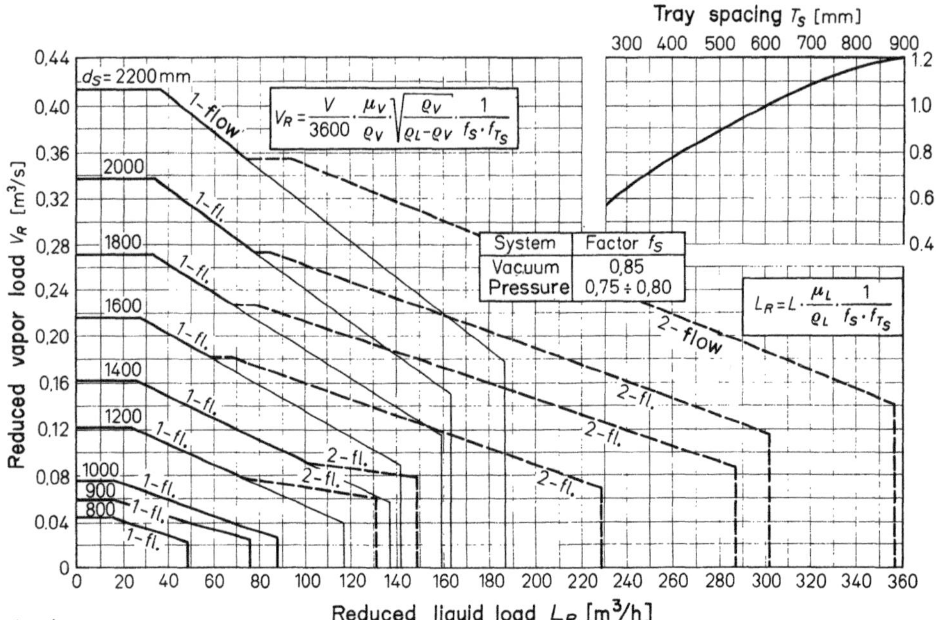

A ▲

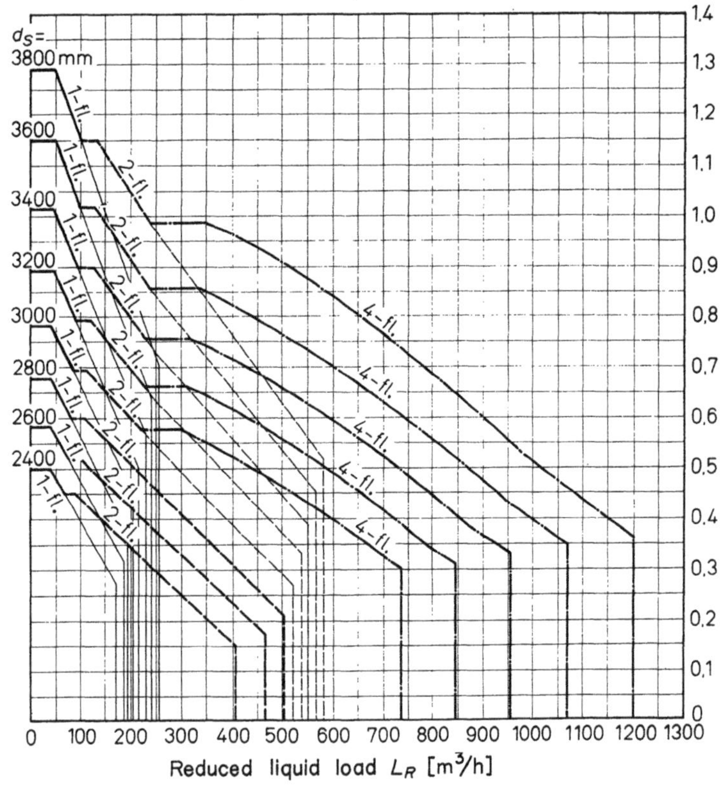

B ▶

flow plates, L_R = 160 m^3/h and V_R = 0.22 m^3/s will call for a diameter of d_S = 2200 mm; double-flow plates will require d_S = 2000 mm.
Based on Eq. (8-46) and the relation

$$\frac{V}{3600}\frac{\mu_V}{\rho_V} = u_V \frac{d_S^2 \pi}{4} \tag{8-87}$$

V_R and the load factor $F_V^* = (F_V^*)_o$ are connected by

$$(F_V^*)_o = V_R \frac{f_s f_{T_S}}{d_S^2 \dfrac{\pi}{4}} \tag{8-88}$$

Figs 8.81 A and B become inapplicable when the liquid load on the plates is less than 3.5 m^3/m^{2}h. Then, the vapor load must be assumed lower and hence the column diameter must be made larger than that of the above figures, in order to avoid a blowing off of the limited liquid quantity, and a dropoff in separation.

Example 5 A column is to separate a mixture of ethyl benzene and styrene under vacuum. It is to be fitted with valve-type plates spaced at 500 mm. The column is to operate at test conditions, at total reflux, with a vapor throughput of 23.7 kmol/h.

Answer The specific weights are for the vapor ρ_V = 0.485 kg/m^3 for the liquid ρ_L = 850 kg/m^3. Molecular weight is approximately $\mu_V = \mu_L$ = 106 kmol/kg. From Eq. (8–86a)

$$V_R = \frac{23.7}{3600}\frac{105}{0.485}\frac{1}{(0.85)(0.9)}\left(\frac{0.485}{850 - 0.485}\right)^{1/2} = 0.0445\ \frac{\text{m}^3}{\text{s}}$$

From Eq.(8–86b)with $L = V$ = 23.7 kmol/h

$$L_R = 23.7\ \frac{105}{850}\frac{1}{(0.85)(0.9)} = 3.9\ \text{m}^3/\text{h}.$$

From this, Fig. 8.81 A yields a column diameter of d_S = 800 mm. Therefore, by Eq. (8-88)

$$(F_V^*)_o = 0.0445\ \frac{(0.85)(0.9)}{0.8^2\ \dfrac{\pi}{4}} = 0.068\ \text{m/s}$$

Fig. 8.81 Valve plates. Estimating the optimum column diameter (69).

and by Eq. (8-62) the vapor velocity

$$(u_V)_o = 0.068\left(\frac{850 - 0.485}{0.485}\right)^{1/2} = 2.84\,\text{m/s}$$

and the vapor load factor

$$(u_V)_o\, \rho_V^{1/2} = 2\ \text{m}^{-1/2}\ \text{s}^{-1}\ \text{kg}^{1/2}$$

This last value corresponds, by Figs. 6.46 and 6.47 to a load where the separation by valve plates begins to deteriorate.

It had been noted in Chapter 6 that a combination of the operational properties of valve plates with the economical sieve plate produces an optimum design T9F, for which part of the plate is fitted with T9 type valves (with venturi nozzles), and another part with normal sieve-type perforations.

The performance of that plate is excellent (Fig. 8.45). Its admissible maximum vapor load, corresponding to 60% separation, is about 50% superior to the bell cap type. Across the entire range of applications its separation is much better than for the bell cap by an average of 30%. Another favorable characteristic of the T9F is its low pressure loss per theoretical step. The venturi action of the vapor stacks is remarkable. When joined to a favorable column design the T9F plate offers cost advantages. The maximum loads of standard valve plates may be represented graphically[70] by plotting the load factor of Eq. (8-46) over the flow parameter of Eq. (8-11).

E. Slotted Sieve Trays

The friction losses caused by a liquid flowing over a crossflow plate create a head differential in the bubbling layer between inlet and outlet. At high liquid loads, that is, at high flow velocities of the bubbling layer, the hydrodynamic gradients on sieve plates may be up to 15 mm per meter of flow length. At considerable flow lengths, such as 5 m or more, for the level difference between inlet and outlet weir, an undesirable result ensues which is an irregular distribution of the vapor phase across the column area. However, the local vapor velocities on a plate are in reverse proportion to the local resistances, and consequently on a plate where a gradient exists, the zone near the exit weir will pass a larger portion of the vapor than the zone near the inlet weir, as far as active plates go. For the turbulence on a sieve plate, the vapor load near the inlet weir is important. If that part of a large sieve plate is subjected to a much lowered vapor load, it tends to rain through. However, very probably that part will pass no vapor, and this impedes the mass transfer. To avoid such a condition, one must allow more dry pressure loss. When the overall pressure loss is specified, fewer conventional sieve plates can be accommodated. To achieve the desired separation, the column must be run at a high reflux rate, which means higher operating expense.

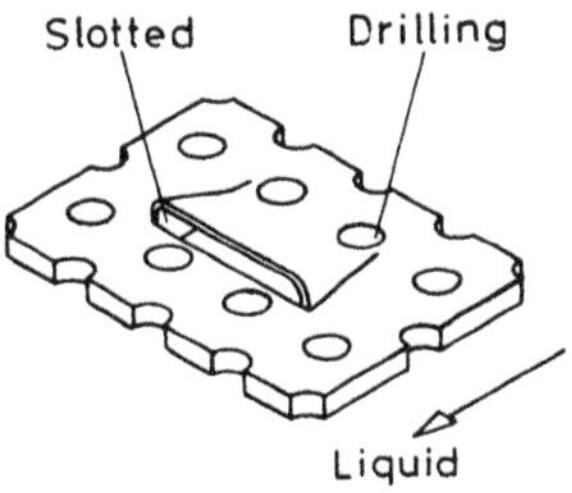

Fig. 8.82 Slotted sieve plate, designed by Union Carbide Corp., Tonawanda, N.Y.

Fig. 8.83 Bubbling promotor at inlet side of slotted sieve plate designed by Union Carbide Corp., Tonawanda, N.Y.

One may readily visualize a sieve tray for which the gradient is eliminated, so that the dry pressure loss becomes very slight and a turbulent layer can be maintained between inlet and outlet weir. The type of plate under consideration has press-formed slots that deflect part of the vapor stream in the direction of the liquid flow. This will impart additional momentum to the liquid. The proportion of vapor deflected by the slots is so controlled by dimensions and numbers of slots that the impulse given suffices to overcome friction, and the turbulent layer moves without appreciable gradient. Fig. 8.82 shows these slots and their arrangement on a perforated plate.

The liquid moving onto the perforation area of the plate is normally clear and quiescent. The static head of that clear liquid is greater than the head of liquid intermixed with vapor. Hence, for conventional plates the dry pressure loss and the overall pressure loss must be sufficiently high to allow the vapor to penetrate the clear liquid at the plate inlet. By suitable design[71,72] (note Fig. 8.83) an increase in the dry pressure loss can be waived. The perforated sheet at the plate inlet has been inclined to lower the static head of clear liquid, and the vapor phase can readily penetrate the liquid layer. Bubbling starts right at the beginning of the flow path; accordingly this particular device has been labeled "bubbling promoter."[71] A slotted sieve plate equipped with a bubbling promoter guarantees turbulence across the entire active area with least pressure loss.

The improvement in the overall tray efficiency as compared to its local efficiency[73] depends on the relative slope of the equilibrium curve as compared to the operating line (the liquid/vapor ratio being constant), and on the intermixing of the liquid on the plate. The liquid/vapor ratio on a conventional sieve plate is, however, not constant, and intermixing is unavoidable. In that event, a large portion of the vapor will pass through a restricted plate area. These facts will impair mass transfer so much that the theoretical improvement in separation will not serve adequately. However, the slots will provide a fairly uniform motion of the bubbling layer across the full plate width, which largely precludes liquid intermixing; also the liquid/vapor ratio is almost constant over the entire flow path of the turbulent layer. For those two reasons, slotted sieve plates of large diameter can offer good separation per plate; plates with unidirectional liquid flow and extended gradient-free liquid paths yield virtually 100% separation in the separation of ethyl benzene and styrene.

As the thickness of the bubbling layer increases, so does the pressure loss of the actual plate and the separation due to the extended duration of contact between vapor and liquid phases[72] (Fig. 8.84). The pressure loss per actual plate increases

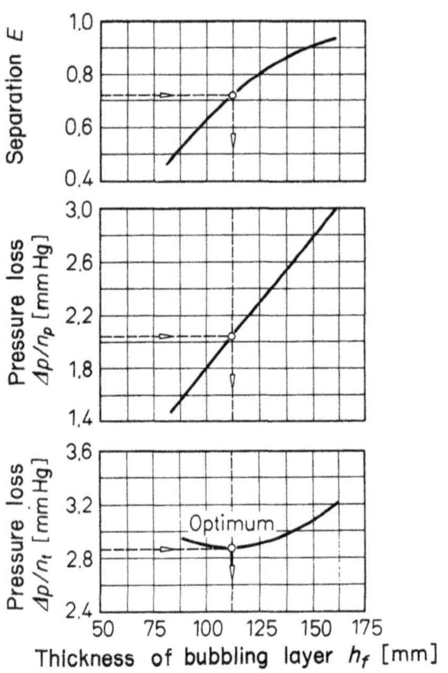

Fig. 8.84 Slotted sieve plate. Separation and pressure loss versus thickness of bubbling layer.

in proportion to the head of turbulent layer, while the increase in separation becomes less as the layer grows. Plotting the pressure loss per theoretical step (that is the ratio of the pressure loss of an actual plate and its separation) over the head of the turbulent layer yields a curve that has a minimum for a specific head. That head should be made a criterion in the layout of the trays.

Fig 8.84 refers to an industrial-scale separation of the system ethyl benzene–styrene, and it applies to that part of the enriching column where the pressure is

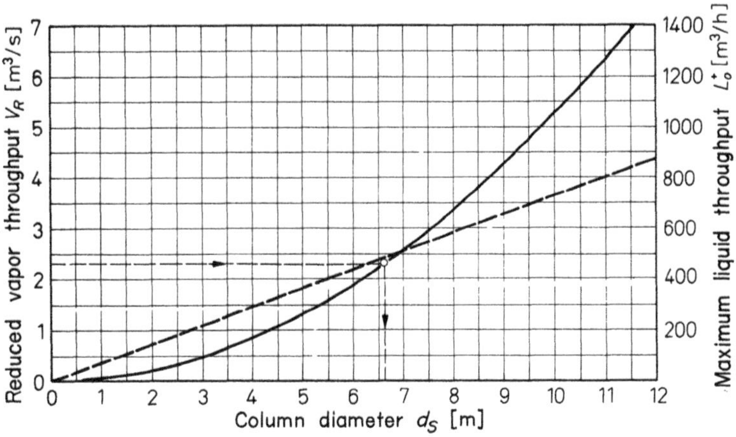

Fig. 8.85 Estimating the diameter of slotted sieve plates.[74]

around 80 mm Hg and the styrene concentration is 30 to 35% by weight. Because of their low pressure loss per theoretical step, slotted sieve trays are advantageous in vacuum rectifications. Within a specified overall pressure loss, such trays will accommodate more theoretical steps than will conventional types; slotted sieve plates may save considerable operational expenses.

Figs 8.85 and 8.86 (A to C) will aid in estimating diameters and plate spacings for that type of plate.[71] Fig. 8.85 gives the interrelation of the reduced vapor load V_R (m^3/s) and the maximum liquid throughput L_o^+(m^3/h) corresponding to L_o (kmol/h)

$$V_R = \frac{V}{3600} \frac{\mu_V}{\rho_V} \left(\frac{\rho_V}{\rho_L - \rho_V} \right)^{1/2} \tag{8-89}$$

and

$$L_o^+ = L_o \frac{\mu_L}{\rho_L} \tag{8-90}$$

versus the plate diameter d_S (m), where μ (kg/kmol) and ρ (kg/m^3) are again the molecular weights and specific weights, suffixes V and L referring to the vapor quantity V (kmol/h) and liquid quantity L (kmol/h).

To determine the required column size d_S (mm) for a given vapor throughput, first calculate the reduced vapor throughput V_R. Then, verify that the given liquid throughput is less than L_o^+ of Fig. 8.85. If the desired liquid load exceeds the maximum allowed, either increase the diameter or select a tray type of superior liquid capacity.

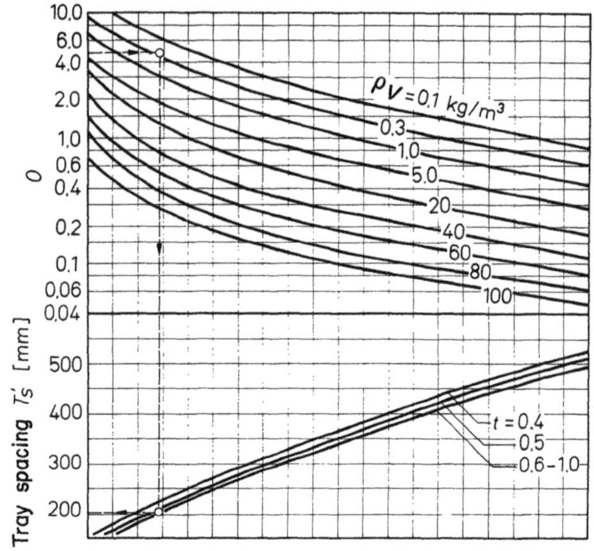

Fig. 8.86 A Slotted sieve plates. Estimating the spacing in view of the backup in a downspout.[74]

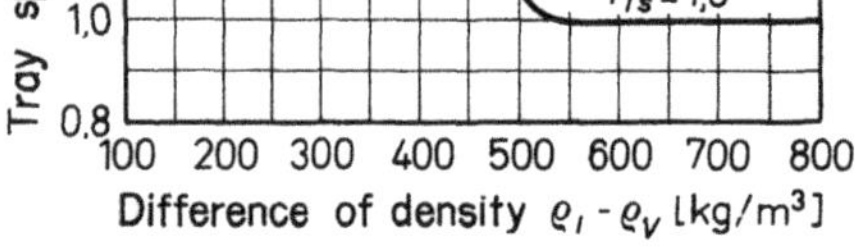

Fig. 8.86 B Slotted sieve plates. Estimating plate spacing factor as a function of the specific weights of the vapor and liquid.[74]

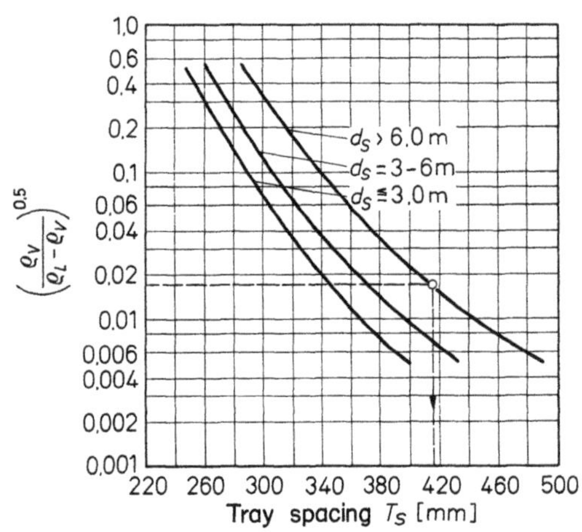

Fig. 8.86 C Slotted sieve plates. Plate spacings in view of entrainment.[74]

Tray spacing T_S required for proper operation of a crossflow-type tray must satisfy two criteria; a minimum plate spacing is needed to avoid flooding, which may be due to excessive backup in the downcomer or to excessive entrainment of a liquid by a vapor.

Figs. 8.86 (A and B) indicate the minimum plate spacing to avoid flooding by liquid backup in the downcomer. First read, from Fig. 8.85, the V_R value for a given d_S. From this, and the given vapor throughput V^+ (m³/s) corresponding to V (kmol/h)

$$V^+ = \frac{V}{3600} \frac{\mu_V}{\rho_V} \tag{8-91}$$

ordinate O in Fig. 8.86(A) defined by

$$O = 8.38 \cdot 10^{-2} \frac{V^+}{V_R} \tag{8-92}$$

can be computed, and the required plate spacing can be read off $(T'S)$ (mm) as a function of the specific weight of the vapor ρ_V and the desired flexibility t of the plate. In the case of foaming systems, T'_S must be modified by a coefficient given in Fig. 8.86 (B). The actual plate spacing T_S then becomes:

$$T_S = T'_S f_{TS} \tag{8-93}$$

Fig. 8.86C presents the plate spacing T_S (mm) required to avoid an excessive proportion of liquid in the vapor, depending on column diameter d_S (m) and density ratio $\rho_V/(\rho_L-\rho_V)^{1/2}$. For the actual construction select the larger of the two T_S values.

Example 6 For the separation of the ethyl benzene–styrene mixture, the top plates have been specified as follows:

$V^+ = 131$ m³/s
$L^+ = 127$ m³/h
$\rho_V = 0.256$ kg/m³
$\rho_L = 833$ kg/m³
$t = 0.7$ tray flexibility

Find the plate diameter and spacing.

Answer Combining Eq. (8-89) with Eq. (8-91) yields

$$V_R = 131 \left(\frac{0.256}{833 - 0.256}\right)^{1/2} = 2.293 \text{ m}^3/\text{s}$$

For this, Fig. 8.85 will supply $d_S = 6.65$ m; selected size 6.75 m. For this, Fig. 8.85 gives $V_R = 2.4$ m³/s.
From this, by Eq. (8-92):

$$0 = 8.38 \cdot 10^{-2} \frac{1.31}{2.4} = 4.574$$

from which, by Fig. 8.86, at $t = 0.7$

$$T'_S = 205 \text{ mm}$$

Likewise, from Eq. (8-93), with $f_{TS} = 1$, by Fig. 8.86 (B), again

$$T_S = 205 \text{ mm.}$$

By Fig. 8.86(C), for $\rho_V/(\rho_L - \rho_V)^{1/2} = 0.0175$

$$T_S = 414 \text{ mm.}$$

For the construction, $T_S = 450$ mm is suggested.

F. Multi-Downcomer Trays

The basic components of this type are presented in Fig. 8.87. Perforated sheets are located between a number of downcomers terminating in the vapor space of the plate bel Two superimposed plates are arranged with longitudinal axes at right angles.[75]

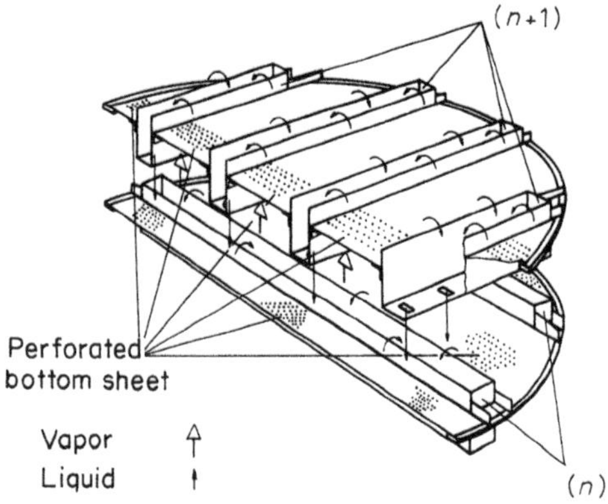

Fig. 8.87 Buildup of a MD (Multi-Downcomer) plate. Union Carbide Corp., Tonawanda, N. Y.

The downcomers have circular or rectangular openings that allow the liquid to escape downward. These are so arranged that the jet emerging from the downtube encounters the turbulent zone of the plate below. The width of the downcomers is up to 125 mm. The relative areas of the perforation and of the downcomers depend on the number of downcomers, which is in relation to the ratio of vapor and liquid loads. For low liquid loads, a single central downcomer may suffice. Even for large column diameters, the moment of resistance of the downcomer is sufficient to impart to the plate the necessary strength and stiffness without requiring additional braces or brackets as discussed in (Section 4.A. of this Chapter). Thus the mechanical and the functional requirements of this plate can readily be met by suitable design. Multi-downcomer plates are attached to the column by a continuous clamp ring. No auxiliary joists or supports are needed. The number of downcomers does not affect the construction of the plate above or below, and can be chosen to suit the liquid load; in this way the different loads of the enriching and the stripping sections can be considered. Multi-downcomer plates may be laid out for assembly and removal from above, from below, or sidewise. The spacing of the downcomers is a function of the liquid load. However, for convenient construction it should be no less than twice the width. Multi-downcomer trays are known in industry as MD-trays.

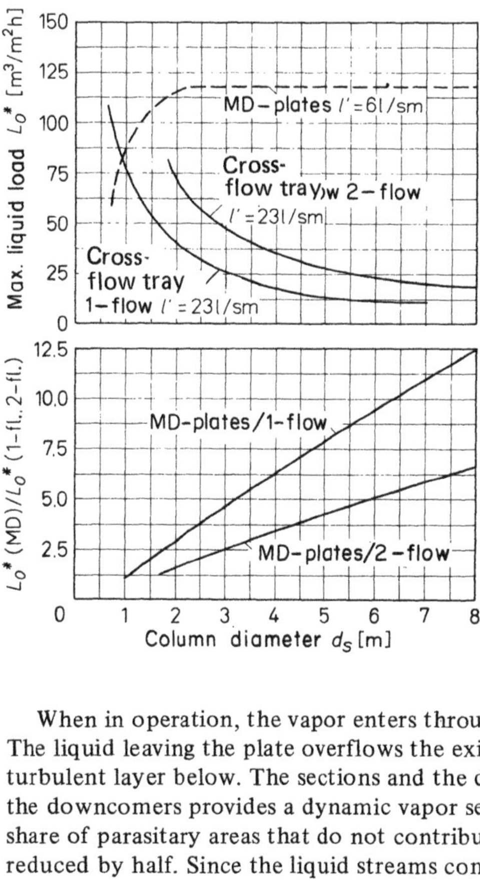

Fig. 8.88 MD plates versus conventional single and two pass flow trays for columns of varied size. Maximum liquid loads.

When in operation, the vapor enters through the perforations into the turbulent layer. The liquid leaving the plate overflows the exit weir, flows downward and enters the turbulent layer below. The sections and the depth are so dimensioned that the liquid in the downcomers provides a dynamic vapor seal. This makes seal cups unnecessary. The share of parasitary areas that do not contribute to the materials exchange is thereby reduced by half. Since the liquid streams coming in from the sides of the downways, the specific liquid load of the weir is limited to

$$l' = \frac{L}{l_W} \; \frac{\mu_L}{\rho_L} \left(\frac{1000}{3600}\right) = 6 \, \frac{1}{\text{s m}} \tag{8-94}$$

where l_W = weir length (m). The minimum of downcomers for a liquid quantity L (kmol/h) of molecular weight μ_L (kg/kmol) and specific weight ρ_L (kg/m^3) follows from this. Because of the low load on the weir, the thickness of the turbulent layer is essentially controlled by the height of the weir; therefore separation and pressure loss will vary little for changing loads. The quarter-turn displacement of the downcomers causes intermixing of the part streams, and an irregular liquid supply is soon corrected. On conventional trays, however, the high weir loads [Eq. (8-32)], will cause heavy turbulent layers that demand large plate spacings and cause high pressure losses, but will be hard to justify by the modest increase in separation achieved.

The downcomers of this type of tray terminate in the vapor space, and are not pulled low as in the conventional designs. Therefore, a multi-downcomer plate of a given diameter has a relatively large share of surface available for vapor passages. A quadruple-flow conventional plate, for instance, would require half its area for downcomers and seal cups. As the mass transfer takes a certain active area, that particular layout demands a large increase in diameter. Now, high specific liquid loads occur mostly in high-pressure processing, the first cost of column shells becomes appreciable because of the large diameters. For multi-downcomer plates, much of the shell cost is saved through reduction of useless plate area. The upper portion of Fig. 8.88 compares the maximum liquid loads for a multi-downcomer plate and a conventional single-or double-flow tray of the crossflow type. It is seen that the maximum liquid load for the customary crossflow trays diminishes hyperbolically as the diameter increases, while it remains constant independently of the column diameter for multi-downcomer plates.

The lower part of Fig. 8.88 shows the relative maximum liquid loads per square meter of column area, for a multi-downcomer plate and for a single or double flow conventional

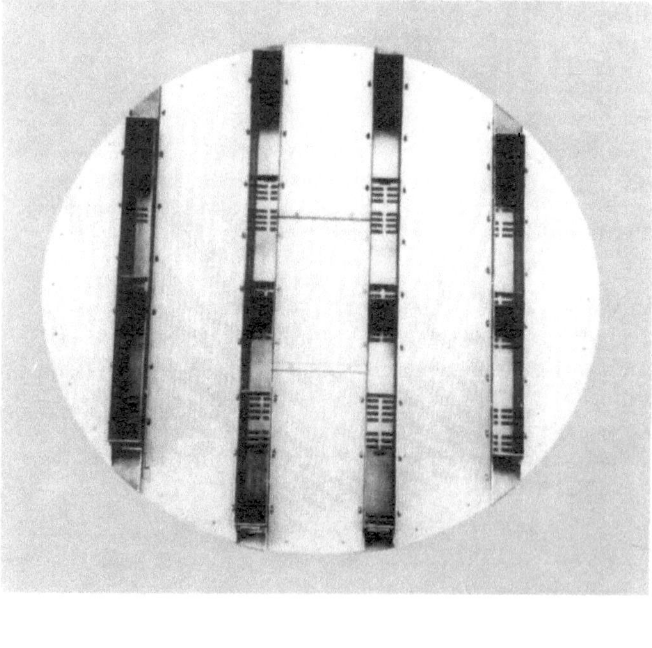

Fig. 8.89 Plan view of MD sieve plate. Union Carbide Corp., Tonawanda, N.Y.

plate. For instance, for a 6-m diameter column the liquid capacity of a multi-downcomer tray is nine times that of a single-flow crossflow tray, and five times that of a double-flow tray.

The maximum vapor load of a multi-downcomer plate is achieved through the use of a single central downcomer. The active area of a single-flow crossflow tray is mostly around 80% of the column area. The active surface of a multi-downcomer plate with a central downcomer 125 mm wide, as used for a low liquid load, is 93% of the entire area of a 2000-mm column. In other words, the vapor throughput alone would be 14% larger for a multi-downcomer tray than for a single-flow conventional tray.

On a multi-downcomer tray, the vapor and liquid phase move almost countercurrently. As the horizontal flowpaths are short, the local separation effect is not increased much. However, for multiflow plates of conventional design the horizontal flowpaths are not much longer; furthermore there is the risk of irregular liquid distribution which reduces the efficiency. Therefore the separation by multi-downcomer plates is equivalent to, or superior to, a multiflow standard plate.

Thus, the multi-downcomer design is basically suited to all rectifications at very high liquid loads. Wherever it is desired to avoid the drawbacks of the multi-flow design characterized by high liquid load, the multi-downcomer plate offers economical advantages through reduced diameter and plate spacing. When columns are rebuilt it allows higher throughput and better product purity because of more plates per unit height. Because of its high liquid capacity, this plate is used in high-pressure work, as in ethylene–ethane and propylene–propane separation, demethanizing, butadiene purification, and xylol separation.

Figs 8.89 and 8.90 show two examples of industrial scale multi-downcomer sieve plates. Figs. 8.91 (A,B) will assist in finding the plate diameter d_S (mm) and number n_d

Fig. 8.90 MD sieve plate assembly. Union Carbide Corp., Tonawanda, N.Y.

of the downcomers. Figs. 8.92 (A, B) will establish the tray spacing T_S (mm).

The reduced vapor throughput (m^3/s) becomes

$$V_R = V^+\left(\frac{\rho_V}{\rho_L - \rho_V}\right)^{1/2} \tag{8-95}$$

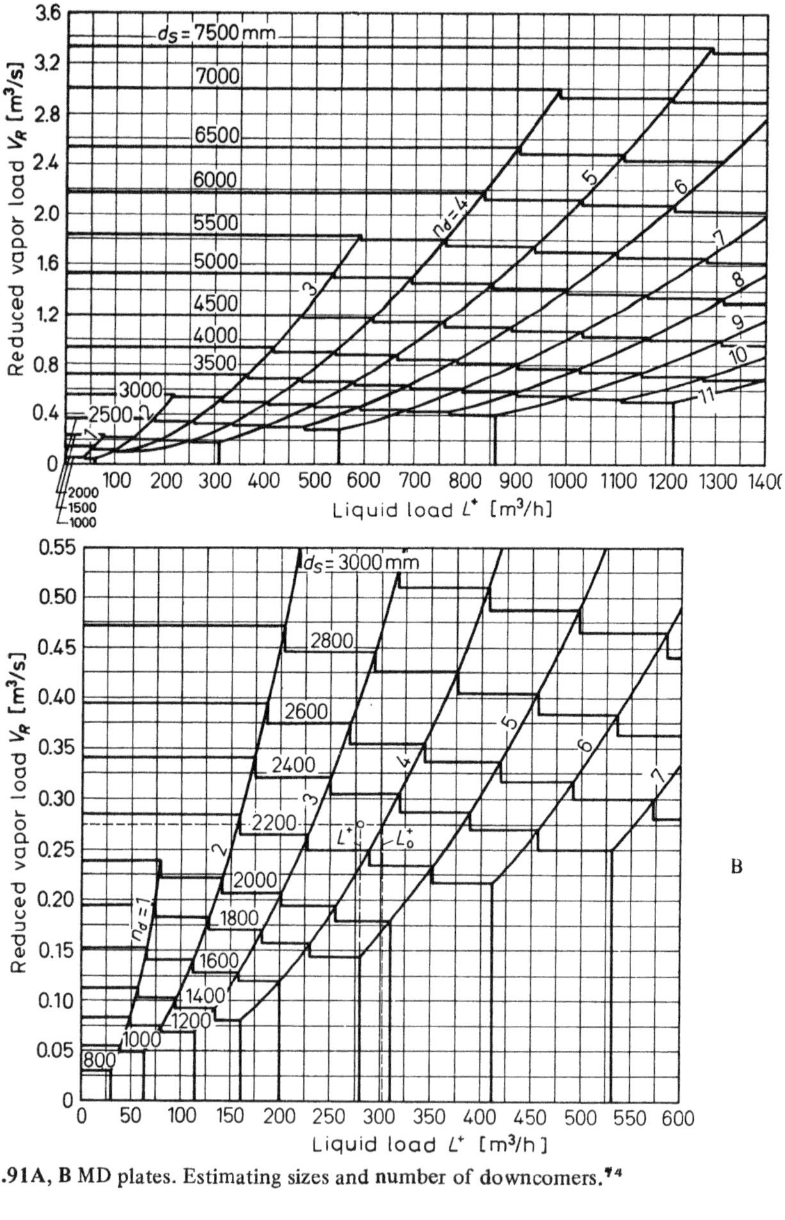

Fig. 8.91A, B MD plates. Estimating sizes and number of downcomers.[74]

where V^+ (m^3/s) is the actual vapor quantity from Eq. (8-91). The liquid throughput

$$L^+ = L \frac{\mu_L}{\rho_L} \tag{8-96}$$

The throughputs of liquid and vapor yield the operating point in Fig. 8.91 (A) or (B). Departing from that point, one may either go upward or horizontally to the right until the nearest n_d line is encountered, to find the number of downcomers n_d and diameter d_S. Then, select the optimum combination of d_S and n_d. In the event that the operating point is located close to an n_d line, use the lower n_d and higher d_S value.

To find the plate spacing, first find O from Eqs. (8-91) and (8-92). Fig. 8.92 (A) will then give plate spacing T_S' (mm) for the given vapor density ρ_V (kg/m³) and plate flexibility. T_S' will be correct under certain conditions, such as maximum specific weir load of the downspout, for which the n_d lines of Fig. 8.91 (A, B) will hold, and ρ_L/

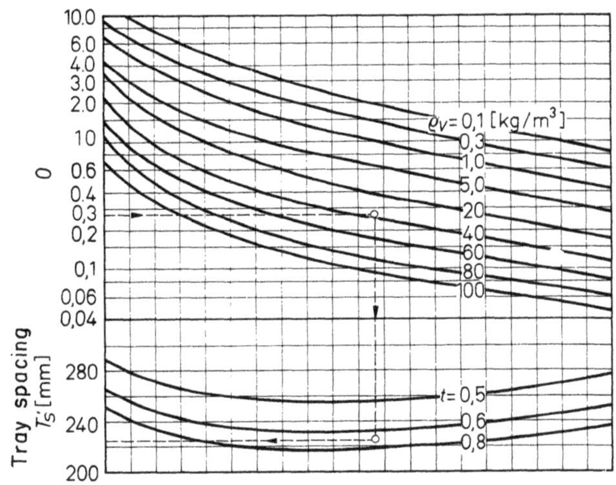

Fig. 8.92 A MD plates. Finding the tray spacing.[74]

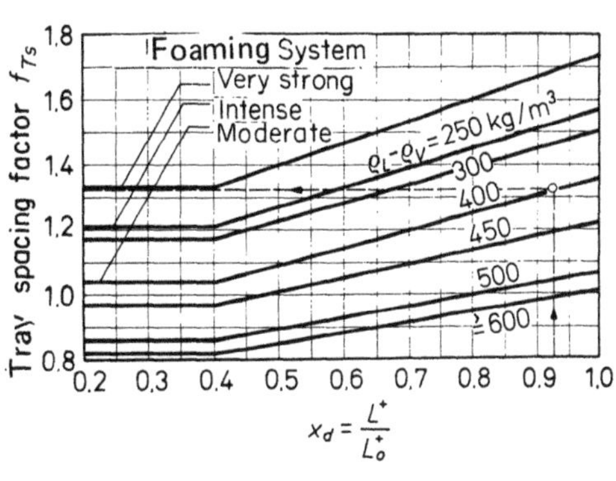

Fig. 8.92 B Tray spacing factor versus densities of vapor and liquid, and liquid load.[74]

$(\rho_L - \rho_V) = 1$. If the intersection of the diameter selected with the liquid throughput L^+ is not a n_d line, the proportional load on the downspout

$$x_d = \frac{L^+}{L_o^+} \tag{8-97}$$

must be found, where L_o^+ is the maximum liquid load for the selected d_S and n_d values. Fig. 8.92 (B) will now give the plate spacing coefficient f_{TS} as a function of x_d and $(\rho_L - \rho_V)$ even for foaming systems.

The required plate spacing for proper operation then follows from equation

$$T_S = T_S \, f_{TS} \, \frac{\rho_L}{\rho_L - \rho_V} \tag{8-98}$$

Example 7 The specification for an ethane/ethylene column reads:

$V^+ = 0.9 \text{ m}^3/\text{s}$
$L^+ = 280 \text{ m}^3/\text{h}$
$\rho_V = 37.2 \text{ kg/m}^3$
$\rho_L = 432 \text{ kg/m}^3$
$t = 0.7$ tray flexibility

Find the plate diameter, number of downcomers, and plate spacing.

Answer From Eq. (8-95)

$$V_R = 0.9 \left(\frac{37.2}{432 - 37.2}\right)^{1/2} = 0.276$$

With this, from Fig. 8.91 (A), for

$d_S = 2300$ mm
$n_d = 4$
$L_o^+ = 302 \text{ m}^3/\text{h}$

Furthermore, from Eq. (8-92)

$$0 = 8.38 \cdot 10^{-2} \left(\frac{0.9}{0.276}\right) = 0.273$$

and then from Fig. 8.92 (A) the plate spacing

$$T'_S = 225 \text{ mm}$$

Eq. (8-97) will yield

$$x_d = \frac{280}{302} = 0.927$$

For a difference in densities of

$$\rho_L - \rho_V = 432 - 37.2 = 394.8 \text{ kg/m}^3$$

Fig. 8.92 (B) will supply

$$f_{TS} = 1.32$$

and Eq. (8-98) will yield the corrected plate spacing

$$T_S = (225)\,(1.32)\left(\frac{432}{394.8}\right) = 325 \text{ mm}$$

G. Multiflow Trays with Concentric Downcomers

For separations involving large liquid quantities, multiflow plates with concentric downcomersare recommended along with the aforementioned multiple downcomer plates[76] (Fig. 8.93). The active plate area may be fitted with the elements presented in Fig. 6.6 or,

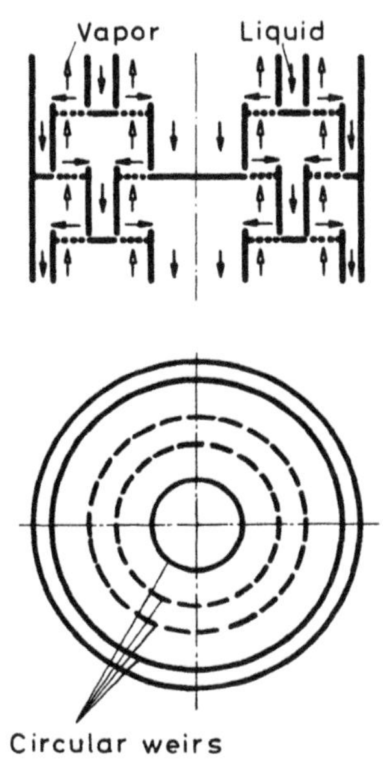

Fig. 8.93 Tow pass flow tray with concentric circular weirs, made by Stahl of Viernheim, Germany

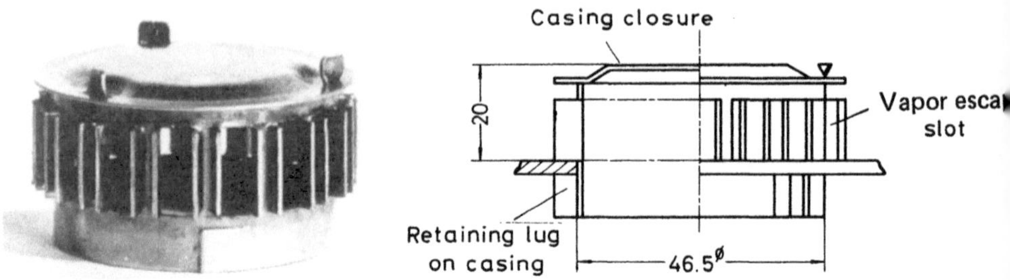

Fig. 8.94 Varioflex cap, made by Stahl of Viernheim, Germany.

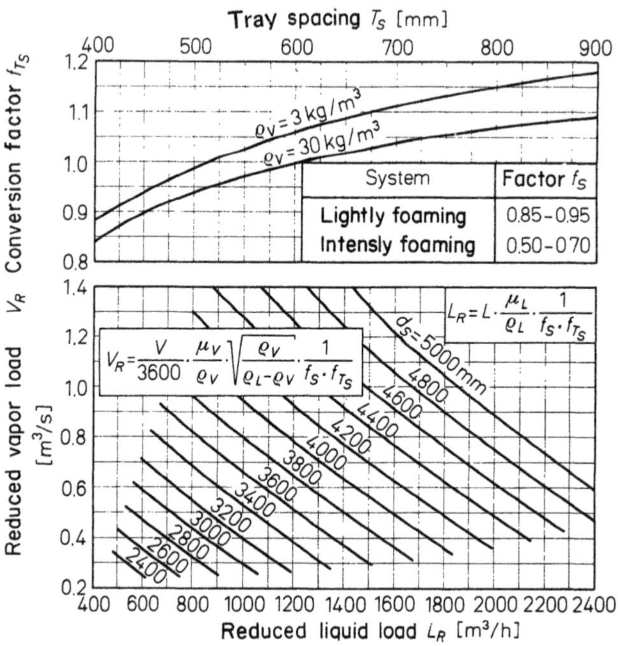

Fig. 8.95 Dual-flow plates[77] with concentric weirs and expanded metal sheets. Estimating column diameters.

for instance, the Varioflex cap of Fig. 8.94. In the latter, the rim below the plate reduces the pressure loss of the vapor at the inlet, while the exit loss is reduced by vanes around the periphery.

The plots of Fig. 8.95 permit estimating the diameters of columns fitted with concentric double-flow downcomers if the active plate area consists of expanded metal as in the centrifugal tray of Fig. 6.6. The procedure in the use of the figure is the same as for Fig. 8.81, explained in connection with Eqs. (8-85) and (8-86).

H. Multiflow Tray with Central Liquid Redistribution

This design can handle a large liquid load. In Fig. 8.96 several collectors placed under the tray will pass the liquid to a central downcomers for intimate intermixing, and dis-

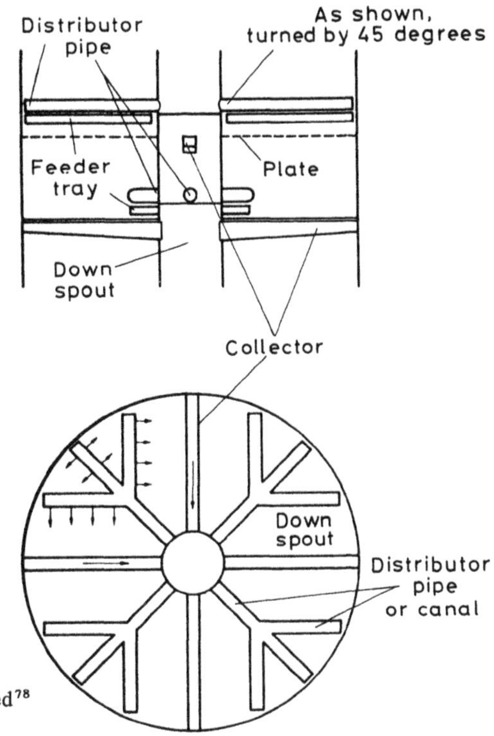

Fig. 8.96 Multiflow tray with central liquid feed[78] made by Schmidding of Koeln, Germany.

tributing pipes and ducts will move it to the next plate underneath.[78] The result is a liquid flow that is unidirectional and favors separation. The continuous central floor supports the plates and allows the proper design clearances to be maintained even for relatively large column diameters.

I. Turbogrid Tray

This design has gained popularity in petrochemical plants because of its simplicity. It consists of parallel strips of metal or other materials. Like the dual-flow plate of Section 4.C.b of this chapter, It has no return spouts (Fig. 8.97). Its separation versus its load characteristic is similar to that of the dual flow design. No particularly defined motion is imparted to the liquid; it descends randomly through the free spaces between the strips whose width is generally between 6 and 12 mm.

The opening ratio may be up to 20% of the column area. In working[79] nonfoaming systems a fairly low plate spacing (100 to 150 mm) may be selected, but it mostly amounts to 200–450 mm.[80]

Turbogrid plates have relatively low pressure loss, but their range of optimum separation is limited. Their pressure loss per theoretical step is about the same as that of

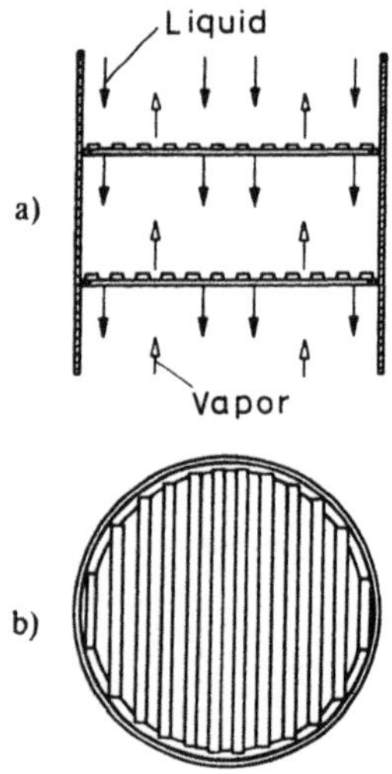

Fig. 8.97 Turbogrid tray design by Shell.
a) Section and flow scheme, and plan view, b).

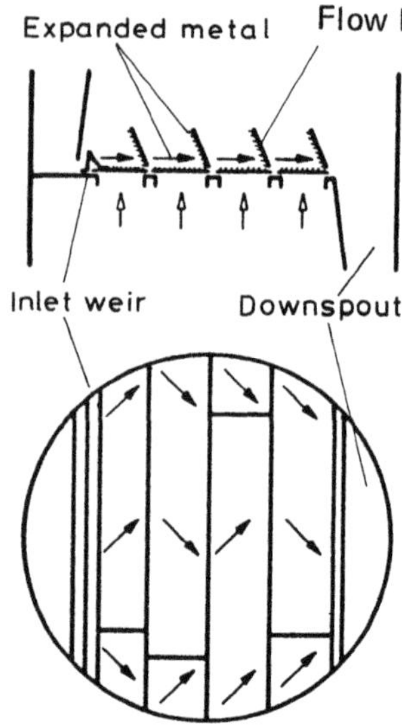

Fig. 8.98 Perform tray, made by Germania of Marxstadt, Germany.

conventional designs, but their throughput per unit of column area is greater.[81] The simple construction means low cost. It is inherently more resistant to fouling than conventional plates and therefore applied to the working of residues and given large openings and rounded strips.

J. The Perform Tray

Like the centrifugal tray of Fig. 6.6, this design uses expanded metal for the plate elements (Fig. 8.98). This is a crossflow type subdivided into segments. The extended flowpath for the liquid, and the phase intermixing caused by the arrangement of the segments and flow breakers cause an improved separation and reduce entrainment; the pressure loss is lessened and the throughput is improved.

K. The Leva Film Tray

Fig 8.99 is a diagram of this construction that was conceived for vacuum separations with minimum pressure loss per theoretical step. It consists of flat sheets interlinked at about 40 mm spacings and in groups of up to 18 elements. Short downspouts for the liquid and vapor phases are about 50 mm wide and reminiscent of the dual-flow design.[84],

Although the separation per tray, depending on the load, amounts to only 10–20%, the low plate spacings permit many steps per unit of column height. The pressure loss per step

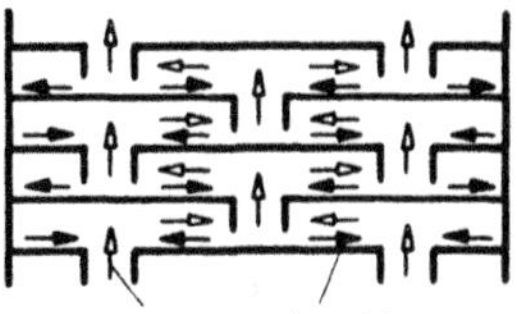

Fig. 8.99 The leva film tray.

less than 10 mm water, is extremely low at average load factors and cannot be excelled by any other tray. The load range approaches those of conventional designs. Even at relatively low liquid loads, adequate separation is obtained.

5. PRESSURE LOSSES IN RECTIFICATION

An effective design of a rectification column and its reboiler demands knowledge of the pressure loss caused by the column internals. It controls, for a given column head pressure, the boiling point of the bottom product, and from this and the temperature of the heating medium, the effective temperature difference for operation. Separations requiring many theoretical steps may impose the use of column internals of low pressure loss even though the product is thermally stable, if that temperature difference becomes unduly low for transferring the process heat.

Available equations for finding the pressure loss generated by a given packing are often of the type

$$\frac{\Delta p}{H} = \xi(d, L', \nu_L)\, u_V^n\, \rho_V \tag{8-99}$$

where

H = unit packing height (m)
Δp = pressure loss (mm water)
u_V = vapor velocity based on free column area (m/s)
ρ_V = specific weight of the vapor (kg/m^3)
d = size of packing element (mm)
L' = specific liquid load (kg/m^2 h)
ν_L = viscosity of the liquid (cP)
ξ = flow coefficient of the wetted packing
n = exponent of vapor velocity

However, use of that type of relationship will often give answers inconsistent with measurements, and in Chapters 6 and 8 reference has been made to Eq. (8-99).

For estimates, Fig. 8.100 combined with Table 1 will serve. The process specification will supply the abscissa and ordinate data, and the corresponding pressure loss is given by the plot.

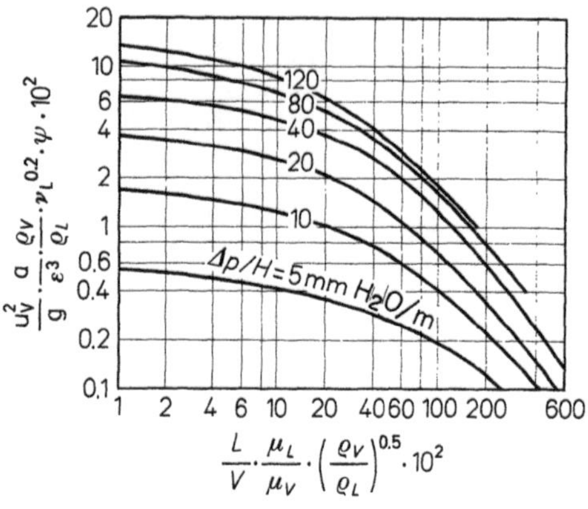

Fig. 8.100 Packed columns. Finding the pressure loss.

Likewise, an accurate calculation of the pressure loss for plate-type columns is not genera feasible. Thus even for bell-cap plates, the deviations of the pressure loss differ from a median value according to various authors by more than 50%.[86] Therefore, reference should be made to Chapters 6 and 8, giving for sieve and valve type plates

$$\Delta p = \xi(\varphi_h, d_h, s_T) u_V^n \rho_V + h_c' \qquad \text{(8-10}$$

The results will, within limits, line up and may be extrapolated for given conditions. In Eq. (8-100):

ξ = flow coefficient
φ_h = aperture ratio
d_h = diameter of openings or vapor stacks
S_T = plate thickness
n = exponent of velocity
u_V = vapor velocity (m/s)
ρ_V = specific weight of vapor (kg/m^3)
h_c' = height of clear liquid on plate (mm H_2O)

Fig. 8.101 applies to valve plates and is based upon Eq. (8-100).[69] In analogy to the corresponding load of atmospheric air per valve, which is equivalent to the vapor through-put V^+ (m^3/s) of a plate with n_h valves

$$\frac{V^+}{n_h}\left(\frac{\rho_V}{\rho_A}\right)^{1/2} = 0.92\, \frac{V^+}{\varphi_h}\left(\frac{d_h}{d_S}\right)^2 (\rho_V)^{1/2} \qquad \text{(8-101)}$$

the top portion of the plot represents the dry pressure loss Δp_{dr}(mm water) according to the first term of the Eq. (8–100). The central portion gives the head $(h_c')_{38}$ (mm) of clear liquid for a weir height of 38 mm. The lower portion serves to convert weir heights h_W (mm water).

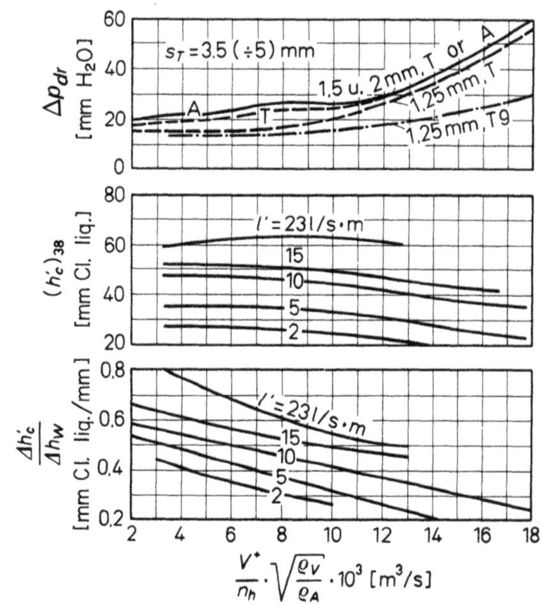

Fig. 8.101 Valve plates (69). Finding the pressure loss.

For the second term of Eq. (8-100):

$$h'_c = \left[(h'_c)_{38} + \frac{\Delta h'_c}{\Delta h_W} (h_W - 38) \right] \frac{\rho_L}{1000} \tag{8-102}$$

Given the aperture ratio φ_h, the hole diameter d_h of the bottom sheet, column diameter d_S and vapor throughput V^+ and specific weight ρ_V Eq. (8-101) determines the abscissa value for which Fig. 8.101 will yield the pressure losses at different specific weir loads l' (1/sm)

$$l' = \frac{L}{l_W} \frac{\mu_L}{\rho_L} \frac{1000}{3600} \tag{8-103}$$

With a single-flow design, the calculation should yield, for a liquid quantity L (kmol/h) of μ_L (kg/kmol) molecular weight and ρ_L (kg/m^3) specific weight that overflows a weir l_W (m) long, a l' value below 23 1/s m; note also Eq. (8-32). For greater l' values, a multi-flow design must be chosen.

Fig 8.101 has been plotted from tests with the air/water system, and its results have been tested for loads up to 85% of flooding.

IX Partial Distillation

The mixed vapors worked in rectifying columns are produced in vaporizers of the kind typically employed in concentrating operations.[1] There are: first, the still kettles heated in miscellaneous ways, the forced circulation vaporizers, and the once-through vaporizers; second, the falling-film and thin-layer vaporizers with revolving internals; and third, the flashing vaporizers.

In the preconstruction engineering of a distilling plant, it is generally assumed that the vapor generated is at phase equilibrium with the liquid or the runoff. For the first and the third group of vaporizers this will be true enough, but the second group behaves differently.

Other than servicing rectifying columns, vaporizers may be applied for pure distillations proceeding either continuously or intermittently.

1. DECOMPOSITION OF LIQUID BY CONTINUOUS PARTIAL DISTILLATION

The liquid mixture to be distilled is fed continuously to the vaporizer that is being heated at a constant rate. The calculation of the distilling operation is treated separately for the three groups of equipment outlined above insofar as differences exist in performance and results achieved.

A. Continuous Distillation in Still Kettles. Recirculating Vaporizers, and Once-Through Vaporizers

These three alternatives are presented diagrammatically in Fig. 9.1. Arrangement a) features a still kettle; to improve heat transfer at the wall and to prevent incrustation an agitator may be provided. Arrangement b) illustrates the principle of having a recirculating heater ("reboiler") located externally. Arrangement c) shows distillation by a single passage through a heater, frequently applied in tube stills in oil refining, for example. In any event, it will be assumed that the vapor leaving the heater, D (kmol/h) of concentration y_D (mol-%) of a low boiler, is at best at phase equilibrium with the liquid W (kmol/h) leaving heater with concentration x_W (mol-%) of a low boiler. The quantities are correlated with the feed F (kmol/h) by the mass balance

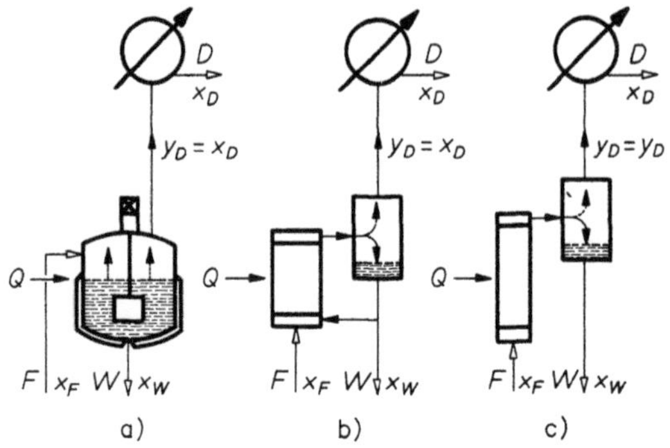

Fig. 9.1 Diagrams of continuous partial distillation processes: a) still kettle., b) recirculation type, c) once-through type.

$$D = F \frac{x_F - x_W}{y_D - x_W} \tag{9-1}$$

where y_D is considered to be the equilibrium concentration of the vapor for liquid concentration x_W

$$y_D = f(x_W) \tag{9-2}$$

For an ideal binary mixture, with x_W given and relative volatilities known,

$$y_D = \frac{\alpha\, x_W}{100 + x_W\,(\alpha - 1)}\, 100 \tag{9-3}$$

For an ideal multicomponent mixture, a relation may be derived from Eq. (2-31). The vapor distilled off will follow from Eq. (9-1), and the runoff quantity is

$$W = F - D \tag{9-4}$$

Partial distillations according to scheme (9-1) can be run continuously only if the relative volatility of the system, (as between the mixture and the components separated) is very large, and no extreme purity is specified, as in residual distillation. This is one area of usefulness for the thin-layer distillation to be discussed later.

Example 1 $F = 10$ kmol/h of a liquid containing $x_F = 70$ mol-% of low boiler are to be decomposed so as to have no more than $x_W = 10$ mol-% of a low boiler in the runoff. The relative volatility is $\alpha = 50$. What quantity must be evaporated, and what is the composition?

Answer For $x_W = 10$ mol-%, and $\alpha = 50$, Eq. (9-3) will give a vapor concentration

$$y_D = \frac{(50)(10)}{[100 + 10(50 - 1)]}\,(100) = 85 \text{ mol-\%}$$

and the quantity of distillate follows from Eq. (9-1), with $F = 10$ kmol/h

$$D = (10)\ \frac{70 - 10}{85 - 10} = 8 \text{ kmol/h},$$

meaning that 80% of the feed must be vaporized. The runoff, Eq. (9-4),

$$W = 10 - 8 = 2 \text{ kmol/h}$$

contains (2)(0.1) = 0.2 kmol/h, or [0.2/(0.7)(10)] (100) = 2.85% of low boiler which are lost, while the distillate is still contaminated with 15 mol-% of high boiler.

It is seen that even for systems of large relative volatilities the separation by simple partial distillation is imperfect. Only for α – values exceeding 100, as in cases of residual distillation, can a closer separation be achieved. Of evaporators of the vertical-tube, recirculation type, for large capacities, the Robert evaporator (named for its inventor) in Fig. 9.2, and the evaporator with an external heater, Fig. 9.3, have been effective. They may be regarded as standard designs; however, there are others. Standard designs of recirculation evaporators always feature return tubes for the untreated liquid, either within the heater tube assembly itself, or externally; this latter arrangement permits convenient cleaning.

In Fig. 9.2 the evaporator consists of a vertical cylindrical shell with attached tube sheets, vertical risers and a central downpipe. The heating medium flows around the outside of the tubes; in most instances steam is used for that purpose. The vapor bubbles generated entrain the boiling liquid upward, employing the principle of the air lift, so that a liquid-vapor mixture leaves at the upper tube ends. The remaining

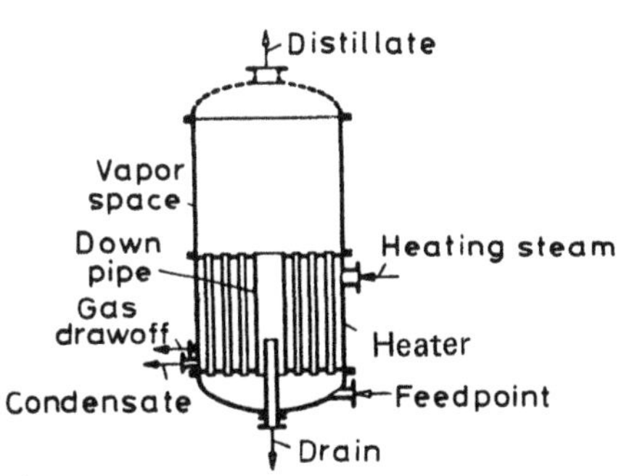

▲ **Fig. 9.2** Robert type evaporator.

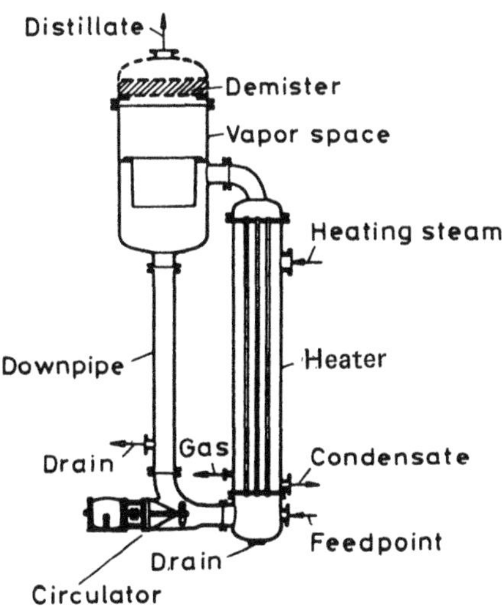

Fig. 9.3 Evaporator with external heater. ▶

liquid returns by the downpipe. The liberated vapor is eventually passed through a demister and liquified again in a condenser. If the heating steam entering the tube bundle region is dry and saturated, or only slightly superheated, and if the slight increase in boiling point caused by the hydrostatic head in the lower part of the evaporating tubes is disregarded, the temperature difference between steam and liquid is virtually constant. The heat interchange in the vaporizer is thus characterized by the material yielding its heat and the material absorbing heat have essentially constant temperatures; one condenses as the other is vaporized.

Fig 9.3 presents distillation apparatus fitted with an external heater and a circulator permitting elevated liquid velocities and heat transfer coefficients. This forced circulation evaporator is especially suitable where the liquid is viscous, the material of construction for the heat exchanger is costly, or where the temperature differential available is small. The circulator shown is frequently omitted from the assembly.

In distillation engineering, proper sizing of the heat transfer area.is difficult, but critical from the performance viewpoint. The required heat transfer area, F (m^2) for a vaporizer transmitting the heat Q (kcal/h) with a temperature difference $(t_H - t_L)$ (C) between steam temperature t_H and boiling point t_L prevailing in the vapor space equals

$$F = \frac{Q}{k\,(t_H - t_L)} \tag{9-5}$$

The overall coefficient k (kcal/m^2 h C) results from various thermal resistances

$$k = \frac{1}{\dfrac{1}{\alpha_H} + \dfrac{s}{\lambda} + \dfrac{1}{\alpha_L}} \tag{9-6}$$

where:

α_H = heat transfer coefficient of the condensing vapor (kcal/m^2 hC)
s = thickness of transmitting wall (m)
λ = heat conducivity coefficient of wall (kcal/mhC)
α_F = heat transfer coefficient of the boiling liquid (kcal/m^2 hC)

Eq. (9-6) is primarily applicable to flat walls, but may be used with thin-walled vaporizing tubes.

If the transmitting wall is multi-layered, $s_{\mathrm{I}}, s_{\mathrm{II}}, \ldots s_N$, having heat conducivity coef $\lambda_{\mathrm{I}}, \lambda_{\mathrm{II}}, \ldots \lambda_N$, s/λ of Eq. (9-6) becomes

$$\sum \frac{s}{\lambda} = \frac{s_{\mathrm{I}}}{\lambda_{\mathrm{I}}} + \frac{s_{\mathrm{II}}}{\lambda_{\mathrm{II}}} + \ldots \frac{s_N}{\lambda_N} \tag{9-7}$$

If, for instance, intense fouling is expected, this must be allowed for in calculating the heat transfer

If quiescent saturated steam condenses on a vertical wall, forming a liquid film that flows downward on an interchanging surface of height H (m), the film heat transfer coefficient becomes for steady and laminar flow

$$\alpha_H = 0.943\left(\frac{r_C \rho_C^2 \lambda_C^3}{\eta_C \vartheta_H H}\right)^{1/4} \tag{9-8}$$

where r_C = heat of condensation of the heating steam at the prevailing pressure (kcal/kg) ρ_C = the specific weight (kg/m^3), λ_C = the heat conductivity coefficient (kcal/mhC), and η_C = the dynamic viscosity of the condensate film (kg/hm^2). ρ_C, λ_C, and η_C are material constants of the condensate and related to the average condensate film temperature t_C which follows from

$$t_C = t_{WH} + \frac{\vartheta_H}{2} \tag{9-9}$$

where t_{WH} = wall temperature on the steam side and $\vartheta_H = t_H - t_{WH}$ is the temperature difference between the steam and the wall.

Frequently the steam will condense on the wall in the form of drops, while other points are not wetted at all. This drip condensation will occur when the steam or the exchanging surface are contaminated by oil or grease. Then heat transfer coefficients are about 6 to 20 times larger than those for pure film condensation[3]. Mostly one will be on the safe side by figuring from Eq. (9-8). Disregarding low values for the temperature difference ϑ_H it will supply the smallest α_H.[4]

The heat transfer coefficient for the liquid mixture boiling in the tubes, for normal circulation[5]

$$\alpha_L = (\alpha'_L)_{q,p}\,\frac{\rho_L}{\rho'_L}\left(\frac{\dfrac{\lambda_L}{\lambda'_L}\cdot\dfrac{c_L}{c'_L}}{\dfrac{\sigma_L}{\sigma'_L}\;\dfrac{r_L}{r'_L}}\right)^{1/2}\left(\frac{\rho'_V}{\rho_V}\;\frac{\eta'_L}{\eta_L}\right)^{1/4} \tag{9-10}$$

where $(\alpha'_L)_{q,p}$ = known heat transfer coefficient of a reference substance, e.g., water, at the same heat load q (kcal/m^2 h) and the boiling pressure p (mm Hg). Subscripts L and V refer to the liquid and vapor phase, and $'$ points to the reference substance. If this is water, $(\alpha'_L)_{q,p}$ becomes [6]

$$(\alpha'_L)_{q,p} = 1.85\left(\frac{p}{760}\right)^{0.4} q^{0.75} \tag{9-11}$$

for heat loads of

$$q = 15{,}000 \text{ to } 200{,}000 \text{ kcal/m}^2\,\text{h} \tag{9-12}$$

Eq. (9-10) supplies an apparent film coefficient between the tube wall and the boiling liquid, referred to the apparent temperature difference $\vartheta_L = t_{WL} - t_L$ (C) on the liquid side with t_{WL} as the mean wall temperature on the liquid side, and t_L as boiling temperature belonging to pressure p (mm Hg) in the vapor space.

In Eq. (9-10): σ_L = surface tension of the boiling liquid against the vapor (kg/m), λ_L = the conductivity (kcal/mhC), c_L = the specific heat, r_L = the heat of vaporization of the liquid, ρ_L = the specific heat of the liquid (kg/m^3), ρ_V = the specific heat of its vapor (kg/m^3) and η_L = the viscosity of the liquid as present (kg s/m^2). All material constants in Eq. (9-10) are based on the arithmetic mean t_L^+ between the average temperature of the tube wall t_{WL} on the liquid side and the boiling temperature t_L in the vapor space.

$$t_L^+ = t_{WL} - \frac{\vartheta_L}{2} \tag{9-13}$$

When natural circulation cannot be depended upon, or is known to be inadequate, heat transfer can be improved by forcing the liquid through the boiling tubes by means of a circulator (Fig. 9.3). The ensuing saving in materials is desirable especially where their cost is high.

To find the heat transfer coefficients in forced circulation, the Nusselt-Kraussol equation[7]

$$Nu = 0.032\, Re^{0.8}\, Pr^{0.37} \left(\frac{H}{d_i}\right)^{-0.054} \tag{9-14}$$

may be used if the temperature difference on the liquid side $\vartheta_L = t_{WL} - t_L$ is below 6 C and Re exceeds 10,000. With smaller temperature differences, little vapor is generated at the heating surface; heat transfer should follow the same laws as in the case of heating a liquid flowing through a pipe.

Forced circulation will essentially permit heating up of the liquid in a vaporizer tube if the slight evaporation at the top end of the tube is disregarded. When the liquid leaves the evaporator tube it will be superheated and eventually vaporize in the vapor space.

In Eq. (9-14)

$$Nu = \frac{\alpha_L\, d_i}{\lambda_L} \tag{9-15}$$

is the Nusselt number with d_i (m) as the inner tube diameter.

$$Re = \frac{u_L\, d_i}{\nu_L} \tag{9-16}$$

is the Reynolds number with u_L (m/s) as the flow velocity of the liquid in the vaporizer tubes, and ν_L (m^2/s) is the kinematic viscosity of the liquid.

$$Pr = \frac{\nu_L \, c_L \, \rho_L}{\lambda_L} \tag{9-17}$$

is the Prandtl number. The material constants λ_L, c_L, and ρ_L have the same significance as in Eq. (9-10). Boiling point t_L is the reference temperature.

To evaluate Eqs. (9-8) and (9-10), the values for $\vartheta_H = t_H - t_{WH}$ and $\vartheta_L = t_{WL} - t_L$, and wall temperatures t_{WH} and t_{WL} must be known. The specific heat loads can be calculated from the assumption that the heat transferred from the steam to the tube wall equals that passed through the tube wall and transmitted to the circulating liquid. Thus disregarding the effect of the difference of the outer and inner tube diameters, the heat transferred per unit time and area becomes the yardstick of the specific heat load q (kcal/m^2 h)

$$q = \alpha_H (t_H - t_{WH}) = \frac{\lambda}{s} (t_{WH} - t_{WL}) = \alpha_L (t_{WL} - t_L) \tag{9-18}$$

and by eliminating t_{WL} and solving for t_{WH}[1]

$$t_{WH} = \frac{\dfrac{\alpha_H}{\alpha_L} t_H + \alpha_H \dfrac{s}{\lambda} t_H + t_L}{1 + \alpha_H \dfrac{s}{\lambda} + \dfrac{\alpha_H}{\alpha_L}} \tag{9-19}$$

Thus, wall temperature t_{WH} must be selected to satisfy Eqs. (9-18) and (9-19). First compute the α_H and α_L values for various assumed wall temperatures t_{WH} from the $t_H > t_{WH} > t_L$ range, and with these find the results from Eq. (9-19). Plot these points over the assumed wall temperatures t_{WH} and connect them by a line. The intersection of that curve with the diagonal gives the wall temperature t_{WH} that satisfies Eq. (9-19).

Knowing heat transfer coefficients α_H and α_L, the overall coefficient, can be determined by Eq. (9–6).

The heat requirement Q (kcal/h) for partial distillation is found from a heat balance, disregarding losses

$$Q = F \left[\frac{x_F - x_W}{y_D - x_W} (\lambda_D + h'_D - h'_W) + h'_W - h_F \right] \tag{9-20}$$

where x_D= molar heat of evaporation of the distillate (kcal/kmol), h'_D = molar enthalpy of the distillate (kcal/kmol), h'_W = molar enthalpy of the runoff (kcal/kmol), and h_F = molar enthalpy of the feed (kcal/kmol). Now the area of heating surface can be found by Eq. (9-5).

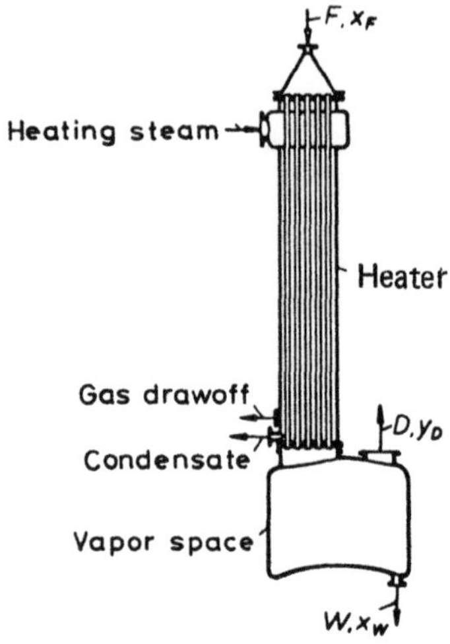

Fig. 9.4 Falling-film evaporator made by Wiegand of Karlsruhe, Germany.

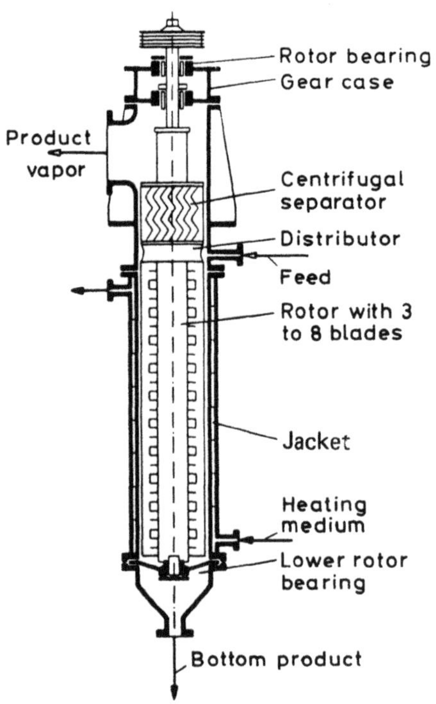

Fig. 9.5 Thin-layer evaporator with rigid wiper blades, made by Luwa of Zurich, Switzerland.

B. Continuous Distillation in Falling Film Evaporators and Thin Film Evaporators.

Falling film evaporators, and thin-film evaporators with revolving internals are especially useful in working thermally unstable mixtures that permit no extended holdover. Either device requires only small effective liquid volumes since the heated tube wall is wetted immediately. As a result the holdover period is 1 minute or less.

Fig 9.4 is a diagram of a falling film evaporator. As compared to the recirculating evaporator the heating device is mounted on top of the vapor space. The liquid is fed to the top tube plate by a suitable distributor, and enters the externally heated tubes, flowing downward along the insides of the tubes and vaporizing on their way. This motion is supported by the vapor formation. The residual liquid and the vapor are separated in the vapor space; special baffling is sometimes employed. The residual liquid forms the runoff.

Thin-layer evaporators achieve brief holdover times for the liquid by mechanical means. The best known makes are the Luwa and the Sambay. The liquid to be evaporated runs down the inside of a vertical heated cylinder and is kept in motion by wipers attached to a revolving shaft. In the Luwa unit (Fig. 9.5) the connection between shaft and wipers is rigid and there is a 1- to 2-mm clearance between the

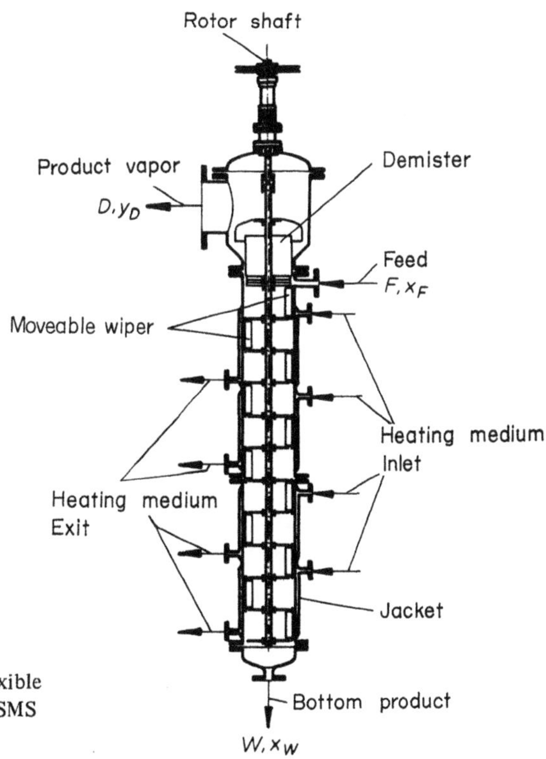

Fig. 9.6 Thin-layer evaporator with flexible wipers, design Sambay made by Luwa-SMS of Butzbach Germany.

wiper and the shell that controls the film thickness. In the Sambay design (Fig. 9.6) the wipers are hung onto the shaft; the thickness of the liquid film adjusts itself depending on the wiper pressure and the viscosity of the boiling liquid. Fig. 9.6 also reveals how subdivision of the jacket can grade the boiling performance of the liquid by variation of the temperatures at different levels.[8]

Continuous partial vaporization by thin-film treatment in falling-film and thin-layer devices is basically different from the treatment in recirculating evaporators. This will be shown for a binary, again using molar units, and giving quantities in terms of moles, and compositions as mole percentages of low boiler in the mixture.

Note Fig. 9.7; across a heated cylindrical wall there moves a thin layer of the liquid mixture to be vaporized, which arrives at point $h = H$ at boiling temperature.

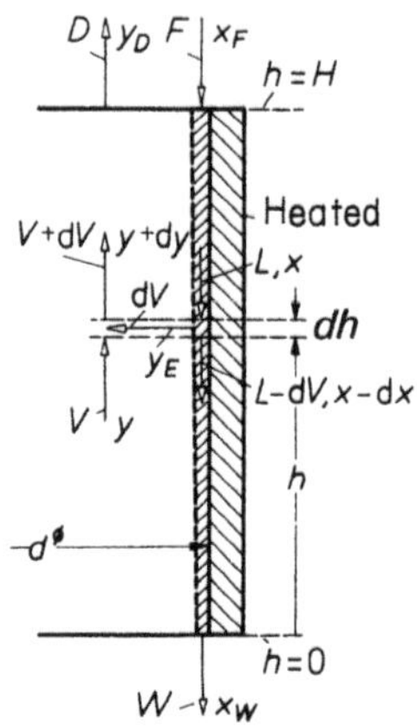

Fig. 9.7 Quantities and concentrations in falling-film and thin-layer vaporizers.

Let F = the charge in kmol/h. The mixed vapor formed at the liquid layer is assumed to escape immediately. With d_i (m) = the internal diameter of the heating jacket, the heat transfer area per infinitesimal length dh becomes

$$df = d_i \pi dh \tag{9-21}$$

with k (kcal/m² hC) = the overall coefficient from the heating medium through the wall t to the liquid, the heat passing hourly per length dh, dQ (kcal/h) becomes

$$dQ = kdf\vartheta = kd_i\pi dh\vartheta \tag{9-22}$$

where ϑ (C) = the temperature difference between the heating medium and the liquid feed. From the latter, at point h the infinitesimal amount dV (kmol/h) is vaporized, which by means of the molar heat of vaporization λ_D (kcal/kmol) at boiling pressure p becomes

$$dV = \frac{kd_i \pi dh \vartheta}{\lambda_D} \tag{9-23}$$

At a point $h + dh$ there moves a liquid quantity L (kmol/h) containing x (mol-%) of a low boiler. Along the infinitesimal path dh there evolves, by Eq. (9-23), the mass of vapor dV (mol/h) of a concentration y_E (mol-%) at the interface of mixed vapor and liquid. At point h there remains a quantity of liquid $(L-dV)$ (kmol/h) of a low-boiler content reduced by dx (mol-%), while the quantity of mixed vapor V (kmol/h) with y (mol-%) of a low boiler will be the infinitesimal amount dV (kmol/h) along dh, with a concentration increase of dy (mol-%) of low boiler. The process is described by

$$Lx = dVy_E + (L - dV)(x - dx) \tag{9-24}$$

Disregarding the second-order infinitesimal $dDdx$, one obtains the differential equation

$$dV = L \frac{dx}{y_E - x} \tag{9-25}$$

Equating this relationship with Eq. (9-23) yields the differential equation

$$L \frac{dx}{y_E - x} = \frac{k\, d_i\, \pi\, dh\, \vartheta}{\lambda_D} \tag{9-26}$$

which is the mathematical formulation of the vaporizing process at any given point. In it, the variable liquid quantity L may be replaced by

$$L = F \exp\left(-\int_x^{x_F} \frac{dx}{y_E - x}\right) \tag{9-27}$$

which results from the integration of differential Eq. (9-25) considering that the infinitesimal increase of the mass of mixed vapor equals the infinitesimal reduction dL of the mass of liquid. With Eq. (9-27) the quantity ratio for runoff W and feed F becomes

$$m = \frac{W}{F} = \exp\left(-\int_{x_W}^{x_F} \frac{dx}{y_E - x}\right) \tag{9-28}$$

Graphic integration of Eq. (9-26) within the limits of $h = 0$ and $h = H$, and of $x = x_W$, the runoff concentration, to $x = x_F$, the feed concentration, will give the height H of the evaporator.[9]

The approach is simplified by adopting constant averages for λ_D, ϑ, and k. In that event, Eq. (9-26) and (9-28) will supply

$$H = F\,(1 - m)\,\frac{\lambda_D}{k d_i \pi \vartheta} \qquad (9\text{-}29)$$

For a graphic solution of Eq. (9-28) the correlation between y_E and x would have to be known; this can be obtained from the vaporization equilibrium. In Fig. 9.8 y is the low boiler concentration of vapor dV generated at point h along the infinitesimal path dh from liquid of concentration x. In the thin-layer evaporator with rotating wipers, vapor formed of liquid is likely to be removed at once. In a boiling process, at the boiling temperature the liquid absorbs sufficient heat for prompt vaporization.

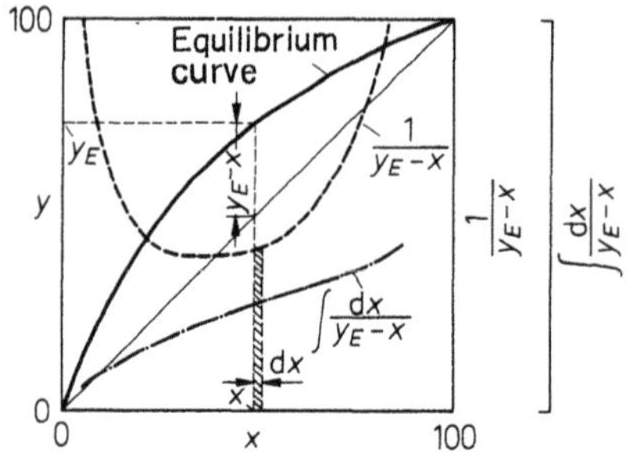

Fig. 9.8 y/x plot applied to partial distillation. Falling-film and thin-layer evaporators.

At the vapor-liquid interface phase equilibrium supposedly exists. The above integrals can be solved graphically if there is essentially no concentratior gradient in the liquid film in the radial direction, the difference $(y_E - x)$ is represented by the ordinate length between the diagonal and equilibrium curve in the y/x plot.

Measurements on thin-film evaporators with rotating elements[10] have confirmed the theory; obviously the wipers cause intimate mixing of particles of various coaxial liquid layers of the film.[9]

For a given feed concentration x_F and for a given quantity ratio m the concentration x_W of the runoff can be found from Eq. (9-28), using the transformation

$$\ln\frac{1}{m} = \int_{x=x_W}^{x=x_F} \frac{dx}{y_E - x} = J_F - J_W \qquad (9\text{-}30)$$

where J_F = integral value correlated to the initial concentration x_F and J_W = integral value correlated to the final concentration x_W of the runoff.

First plot the reciprocals of the differences $y_E - x$ given by the ordinate lengths between the diagonal and equilibrium curve over x. That gives the $1/(y_E - x)$ curve (Fig. 9.8). For any desired x_W assume a numerical value J_W. Read a numerical value for the area below the curve $1/(y_E - x)$, from x_W to a chosen x_F, and plot the sum as an integral value J_F over x_F. Thus, any number of J_F's can be computed for the

J_W assumed at point x_W. Integral curve $\int dx/(y_E - x)$ is determined in this fashion. From this, and from Eq. (9-30), the runoff concentration can be found for a given feed concentration x_F and for a given quantity ratio m.

For a given $x = x_F$, read a J_F from the integral curve in Fig. 9.8. For a chosen ratio m, integral value J_W can be computed from Eq. (9-30), and the desired end concentration $x = x_W$ may be found from the integral curve. With ratio m established, the x_W values can be found for any x_F assumed. In Fig. 9.9 these are plotted against x_F. The

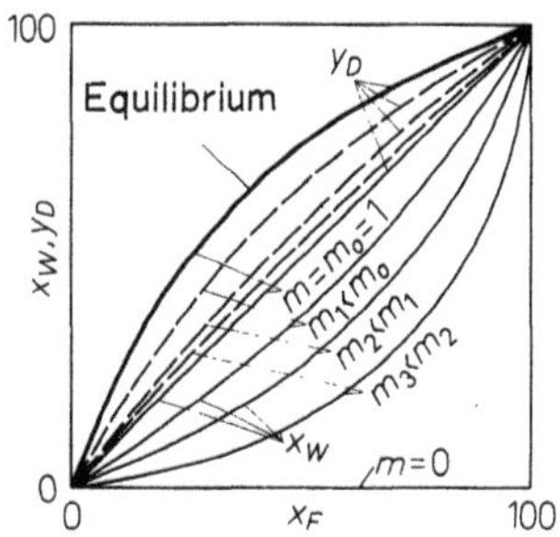

Fig. 9.9 Partial distillation in falling-film and thin-layer evaporators. General plot.

curve thus obtained gives concentration x_W to which the liquid has been reduced from an initial concentration x_F, and for a quantity ratio m. These ratios can be varied, and new curves result; x_W is a functional of x_F with m as the parameter. For $m = 1$, the x_W curve merges with the diagonal; for $m = 0$ it merges with the abscissa. By means of curves $x_F = f(x_W)$ for m = a constant, and a mass balance, curves can be plotted that give the concentration y_D of the mixed vapor leaving the evaporator for a given feed concentration x_F and a ratio m. From the mass balance equations

$$Dy_D + Wx_W = Fx_F \tag{9-31}$$

and

$$D + W = F \tag{9-32}$$

it follows that

$$y_D = \frac{Fx_F - Wx_W}{F - W} \tag{9-33}$$

By substitution of the mass ratio $m = W/F$,

$$y_D = \frac{x_F - mx_W}{1 - m} \tag{9-34}$$

The mixed vapor leaving the thin-film evaporator is

$$D = F\,(1 - m) \tag{9-35}$$

Fig 9.9 presents the general outline for the $y_D = f(x_F)$ curves; the concentration changes caused by different quantity ratios can readily be appreciated. When, for instance, the mass ratio is varied in a thin-film evaporator, such a plot illustrates the concentration changes. At the limit $m \to 1$, or $D \to 0$, the y_D line merges with the equilibrium curve, and for $m \to 0$, or $D \to F$, with the diagonal.

If within the concentration range in question, the relative volatility α of the mixture is virtually independent of the concentration, and hence constant at a constant boiling point, the integral in Eq. (9-30) may also be solved analytically if the right-hand end of Eqs. (9.2 - 9.4) replaces y_E in which connection $\alpha_{12} = \alpha$, $x_1 = x$, and $y_1 = y$, and

$$m = \frac{1}{\left(\dfrac{x_F}{x_W}\right)^{\frac{1}{\alpha-1}} \left(\dfrac{100 - x_W}{100 - x_F}\right)^{\frac{\alpha}{\alpha-1}}} = \frac{W}{F} \tag{9-36}$$

and Eq. (9-34) becomes

$$y_D = x_F + \frac{x_F - x_W}{\left(\dfrac{x_F}{x_W}\right)^{\frac{1}{\alpha-1}} \left(\dfrac{100 - x_W}{100 - x_F}\right)^{\frac{\alpha}{\alpha-1}} - 1} \tag{9-37}$$

Film coefficient α_L of the liquid boiling in the falling-film evaporator can be estimated from Nusselt's data.[2,11] From these[12] the mean film coefficient in a falling-film vaporizer is

$$\alpha_L = \frac{r_L\,\rho_L^2}{3\,\eta_L\,\vartheta_L\,H}\,(s_1^3 - s_2^3) \tag{9-38}$$

where film thickness s_1 at the feedpoint, at the beginning of vaporization, and at laminar velocity distribution, is given by the Hagen-Poíseuille Law as

$$s_1 = \left(\frac{3\eta_L F \mu_F}{\rho_L^2 \pi d_i}\right)^{1/3} \tag{9-39}$$

where F (kmol/h) = the feed quantity of molecular weight μ_F (kg/kmol), and d_i = the internal diameter of the heating tube.

Film thickness s_2 of the residual liquid leaving the vaporizer is

$$s_2 = \left(s_1^4 - \frac{4\lambda_L \eta_L \vartheta_L H}{r_L \rho_L^2}\right)^{1/4} \tag{9-40}$$

The variables in these equations had been discussed in Section 1.A. of this chapter. Their evaluation requires knowledge of the effective vaporizer height H (m) and the interior diameter of the heating tubes d_i (m) that must be found by trial and error.

With saturated steam as the heating medium, the film coefficient α_H on the steam side is computed by Eq. (9-8), and eventually the overall coefficient k is yielded by Eq. (9-6).

Figs. 9.10 and 9.11 present the correlations of measured coefficients in the distillation of the mixtures *trans*-decaline/ cis- -decaline, and 1,2-propylene glycol/ ethylene glycol for different temperature differentials

$$\vartheta = t_H - t_L \tag{9-41}$$

the temperature of the heating steam t_H, and the mean boiling point t_L of the liquid at the vaporizer outlet. The evaporation ratio is

$$\nu = \frac{D}{F} 100 \tag{9-42}$$

and the specific liquid load l_R (m^3/hm) is

$$l_R = \frac{F \mu_F}{\rho_F d_i \pi n_R} \tag{9-43}$$

that feed quantity (m^3) supplied per meter of internal tube circumference per hour in a falling-film evaporator of n_R tubes.

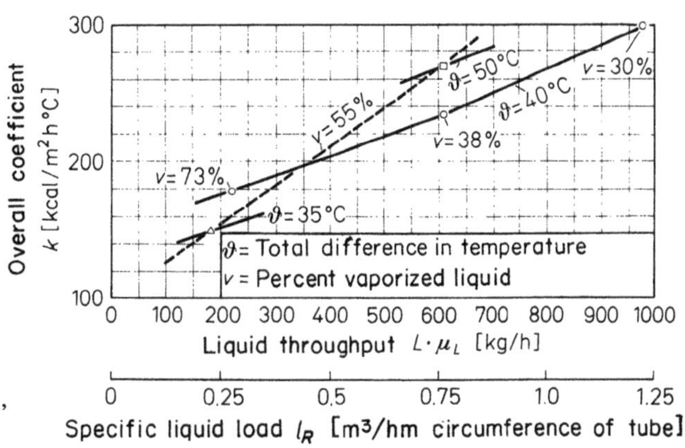

Fig. 9.10 Falling-film evaporator. Overall coefficient versus liquid load, liquid throughput, and temperature difference between heating steam and liquid. Boiling pressure 10 mm Hg. *Trans-,cis*-decaline with 40 mol-% *trans*-decaline. Boiling point 75° C. Heater tubes V2A, 48 X 2 X 2000.

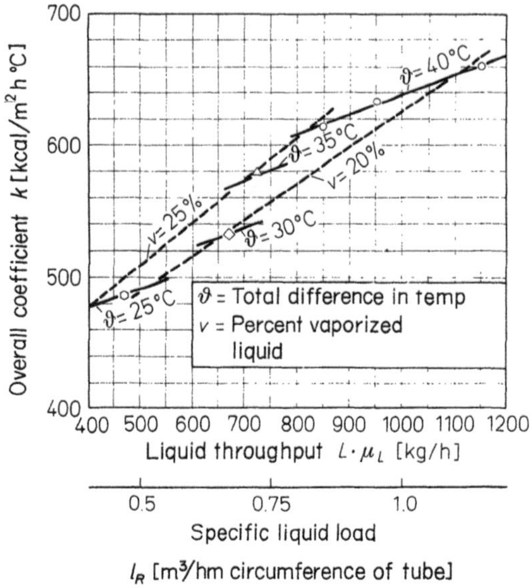

Fig. 9.11 Falling-film evaporator. Overall coefficient versus liquid load, liquid throug put, vaporized liquid, and temperature difference between heating steam and liquid. B ing pressure 10 mm Hg. 1,2-propylene glycol–ethylene glycol with 22 mol-% 1,2-propylene glycol. Boiling point 99°C. 7 heater tubes V2A, 48 × 2 × 2000.

Eq. (9-29) applies to a single evaporator tube. Hence, in the case of a falling-film evaporator of n_R tubes, for d_i the equivalent diameter should be used

$$d_J = n_R d_i \tag{9-44}$$

Then heat transfer coefficient α_L (kcal/m^{2}h C) of the vaporizing film of a thin-film evaporator with flexible wipers may be determined from measurements[10]

$$\alpha_L = 23.7\lambda_L \left(\frac{d_i\, n_r}{\eta_L}\right)^{1/3} \tag{9-45}$$

where n_r (1/m) = turns of rotor (rpm) and d_i = internal diameter of the evaporator. Conductivity λ_L (kcal/m h C) and dynamic viscosity η_L (kg s/m^2) of the liquid film must refer to the arithmetic mean of the wall temperature on the liquid side and its boiling point. With α_L and α_H known, Eq. (9-6) again yields the overall coefficient k.

Fig. 9.12 offers overall coefficients for various applications, plotted against the specific distillate quantity d (kg/m^2 h)

$$d = \frac{D\mu_D}{d_i \pi H} \tag{9-46}$$

Now, given Eq. (9-29) and (9-35)

$$\frac{D}{H} = \frac{k d_i \pi \vartheta}{\lambda_D} \tag{9-47}$$

combining Eqs. (9-46) and (9-47) gives the correlation

$$k = d\,\frac{\lambda_D}{\mu_D}\,\frac{1}{\vartheta} \tag{9-48}$$

only to be used however, in connection with Fig. 9.12.

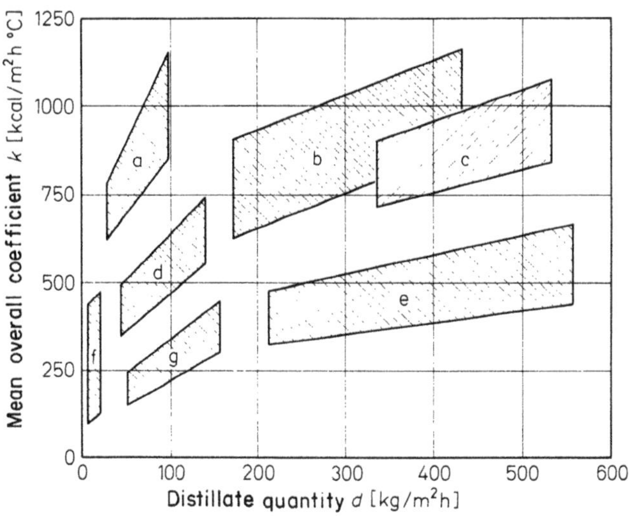

Fig. 9.12 Thin-layer vaporizer with flexible blades, (Luwa of Zurich). Overall coefficients. (a) Dehydration of organics, (b) Distillation of organics, (c) Solvent recovery by distillation, (d) Stripping of organic components, (e) Distillation of high boilers, (f) Deodorizing, (g) Stripping of organics.

Temperature difference ϑ follows from the trend of the concentration along the vaporizer surface, as far as the boiling liquid is concerned. It can be obtained from Eq. (9-9) for α = constant, by means of Eq. (9-26), assuming a constant heating load

$$h = H\,\frac{\left(\dfrac{x}{x_W}\right)^{\frac{1}{\alpha-1}}\left(\dfrac{100-x_W}{100-x}\right)^{\frac{\alpha}{\alpha-1}} - 1}{\left(\dfrac{x_F}{x_W}\right)^{\frac{1}{\alpha-1}}\left(\dfrac{100-x_W}{100-x_F}\right)^{\frac{\alpha}{\alpha-1}} - 1} \tag{9-49}$$

Point h of the evaporator can thus be computed for a given liquid concentration x of the liquid film. Then, by means of the boiling line the temperature curve $t_L = f(h)$ may also be found. By replacing in Eq. (9-37) concentrations y_D and x_F by the y and x values belonging to a given vaporizer level one obtains

$$y = x + \frac{x - x_W}{\left(\frac{x}{x_W}\right)^{\frac{1}{\alpha-1}} \left(\frac{100 - x_W}{100 - x}\right)^{\frac{\alpha}{\alpha-1}} - 1} \tag{9-50}$$

and the mean vapor concentration y prevailing at a given evaporator level can be stated as a function of the corresponding liquid concentration x. Again, the heat requirement can be found from Eq. (9-20).

Example 2 A thin-layer evaporator is to decompose, at atmospheric pressure, $F = 4$ kmol/h of a benzene–toluene mixture. The feed concentration is 20 mol-% of benzene, the runoff W is allowed $x_W = 10$ mol-% of benzene. The feed enters at the boiling point. What is the effective evaporator height if the heating medium is saturated steam of $t_H = 116$ C, and a mean overall transfer coefficient of $k = 800$ kcal/m^2 h C is assumed?The evaporator diameter is $d_i = 0.3$ m, and the rotor speed is to be adjusted to the thermal requirement.

Answer First find the quantity ratio $m = W/F$ by Eqs. (9-30) and (9-36) from feed and runoff concentrations. For graphic evaluation note Fig. 9.13. The $1/(y_E - x)$ curve was computed from the equilibrium curve and graphic integration. For $x_F = 20$ mol-% Eq. (9-30) supplies

$$\ln \frac{1}{m} = \int_{x_W=10}^{x_F=20} \frac{dx}{y_E - x} = 0.715$$

and a quantity ratio of

$$m = 0.49$$

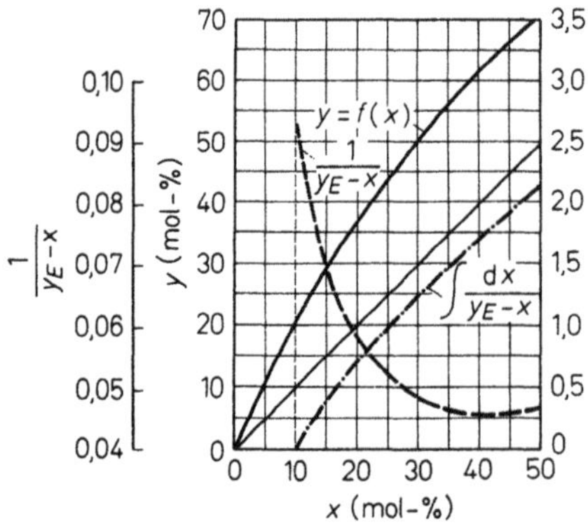

Fig. 9.13 Example 2. Graph for finding lines defining the quantity ratios for a thin-layer vaporizer.

From Eq. (9-34) the concentration of the mixed vapor at the exit y_D is computed

$$y_D = \frac{20 - (0.49)(10)}{1 - 0.49} = 29.6 \text{ mol-\%}$$

The relative volatility of the benzene–toluene mixture at atmospheric pressure, $\alpha = 2.4$, is practically independent of the concentration. Therefore, y_D may also be found analytically

$$y_D = 20 + \frac{20 - 10}{\left(\frac{20}{10}\right)^{\frac{1}{2.4-1}} \left(\frac{100 - 10}{100 - 20}\right)^{\frac{2.4}{2.4-1}} - 1} = 29.9 \text{ mol-\%}$$

Thus by Eq. (9-36)

$$m = \frac{1}{\left(\frac{20}{10}\right)^{\frac{1}{2.4-1}} \left(\frac{100 - 10}{100 - 20}\right)^{\frac{2.4}{2.4-1}}} = 0.496$$

Eqs. (9-49) and (9-50) reveal the changes in liquid concentration x and vapor concentration y in the vapor nucleus along the evaporator wall. Note Fig. 9.14 with the outline of the liquid temperature t_L according to the boiling curve in the concentration range considered.

For completeness, the phase equilibrium concentration y_E of the vapor escaping at the liquid surface has been plotted in Fig. 9.14.

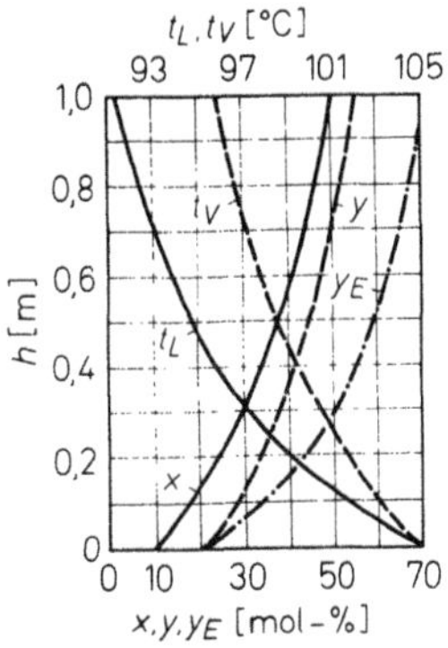

Fig. 9.14 Example 2. Concentrations and temperatures along a thin-layer vaporizer. x = liquid concentrations in film, y = concentration in nucleus of vapor, y_E = concentration of equilibrium vapor, t_L = boiling point of liquid, t_V = temperature of vapor mixture in nucleus.

If it is assumed that the dewpoint of the mixed vapor concentration y controls the mixed vapor temperature t_V, the t_V line in Fig. 9.14 shows the change in mixed vapor temperature in the vapor nucleus along the vaporizer wall. The temperature difference against the liquid temperature t_L averages at 3 C.

In practice, the concentration difference is usually less than in this example, and the difference $t_V - t_L$ should likewise be smaller. It follows that because of the brief holding time the effect of a possible rectification superimposed on to the thin-layer distillation would be negligible.

From the $t_L = f(h)$ line in Fig. 9.14 the mean liquid temperature becomes $t_L = 95$ C and the difference $\vartheta = t_H - t_L = 116 - 95 = 21$ C. To $y_D = 29.9$ mol-% belongs a molar heat of evaporation $\vartheta_D = 7810$ kcal/kmol.[9] With these data, Eq. (9-29) yields an evaporator height of

$$H = \frac{4(1 - 0.49)(7810)}{(800)(0.3)\pi(21)} = 1 \text{ m}$$

From Eq. (9-35) the distillate quantity becomes

$$D = 4\,(1 - 0.49) = 2.04 \text{ kmol/h}$$

and Eq. (9-46) gives, with $\mu_D = 88$ kg/kmol the specific value

$$d = \frac{(2.04)(88)}{(0.3)\pi(1)} = 190 \text{ kg/m}^2\text{h.}$$

Eq. (9-48) yields

$$k = 190\left(\frac{7810}{88}\right)\left(\frac{1}{21}\right) = 800 \text{ kcal/m}^2\text{ h C}$$

as specified. According to area b in Fig. 9.12 d and k supply an operating point of the evaporator.

Fig 9.15 is analogous to Fig. 9.9 and will indicate the liquid and vapor exit concentrations, for specific quantity ratios at various liquid inlet concentrations, for the benzene–toluene mixture in the feed concentration range $0 \leqslant x_F \leqslant 50$ mol-%. For small m values, that is, high evaporation ratios, the higher boiler may be increased in the runoff at will, but the actual recovery in the runoff is

$$W(100 - x_W) = mF\,(100 - x_W) \tag{9-51}$$

However, the lower boiler cannot be enriched at will in the vapor phase, as for high m values, or small evaporation ratios, the vapor concentration maximum is the phase equilibrium concentration for the feed concentration. To improve separation, part of

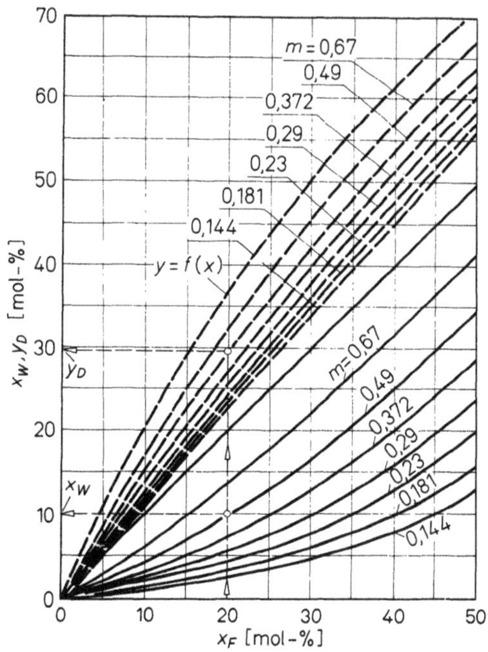

Fig. 9.15 Mixture of benzene and toluene (760 mm Hg). Vapor exit concentration y_D and runoff concentration x_W versus feed concentration x_F for various quantity ratios m.

the distillate may be recycled to the evaporator; see Fig. 9.16. That, however, increases the required evaporating surface and heat consumption.

When a binary mixture is separated by means of a thin-layer evaporator, a number of theoretical steps can be saved. Noting Eqs. (3-28), (9-34), and (9-35), the characteristics of a stripping column on one hand and a thin-layer evaporator on the other will be compared

$$\underbrace{\tan\beta = \frac{L}{V} = \frac{1}{1 - \dfrac{W}{L}}}_{\text{Stripping Column}} = \underbrace{\frac{1}{1-m} = \frac{y_D - x_W}{x_F - x_W} = \frac{F}{D}}_{\text{Thin-Layer Evaporator}} \tag{9-52}$$

and therefore

$$m = 1 - \frac{1}{\dfrac{L}{V}}. \tag{9-53}$$

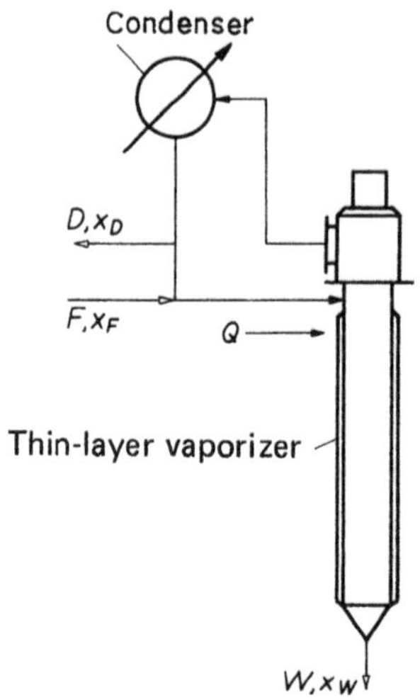

Fig. 9.16 Hookup for partial thin-layer distillation, total condensation of vapors, and partial recycling of the distillate.

In other words, to maintain the liquid/vapor ratio L/V in the stripping column, a thin-layer evaporator must be operated at the quantity ratio given by relation (9-53). Since, in view of Eq. (9-34), the correlation of vapor concentration y and liquid concentration x at any level of the thin-layer evaporator must be

$$y = \frac{x - mx_W}{1 - m} \tag{9-54}$$

the intersection of this line with the stripping line, for which the coordinates are $x = x_F$, $y = y_D$ according to Eq. (9-34), fixes the limit below which the sequence of steps between the stripping line and equilibrium curve in the y/x presentation, indicating the number of theoretical steps saved. Relation (9-54) is the material balance equation for the thin-layer evaporator. m can be determined with Eqs. (9-28) or (9-36) if x_F is replaced by x and consequently F by L.

Example 3 A continuous rectifying column separates benzene and toluene. In the stripper part the liquid/vapor ratio $L/V = 1.13$, and the runoff concentration $x_W = 10$ mol-% benzol. How many stripping steps are saved through the use of a thin-layer vaporizer instead of a reboiler? The plant works at standard pressure.

Answer From $L/V = 1.13$ follows by Eq. (9-53)

$$m = 1 - \frac{1}{1.13} = 0.1142$$

From Fig. 9.17, corner points 1 and 2 indicate the collocation of the equilibrium concentrations on the theoretical steps, and points 1* and 2* give the

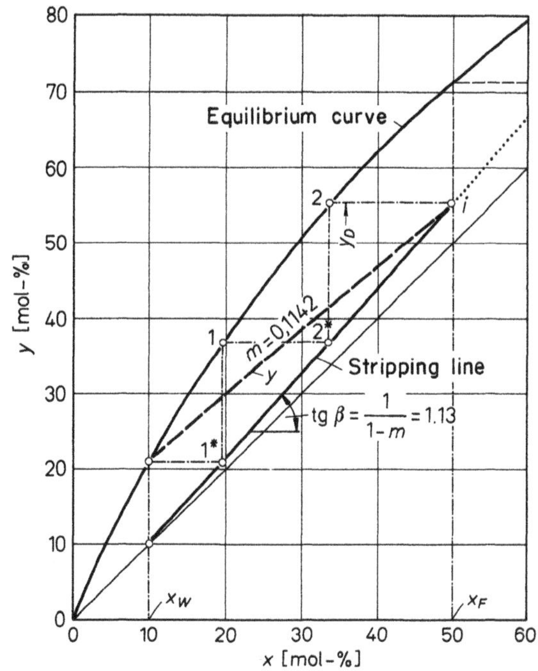

Fig. 9.17 Example 13. Benzene/toluene, 760 mm Hg. Number of steps for a stripper column, with a thin-layer vaporizer as reboiler.

concentrations of liquid and vapor at some level below a plate. In this case the number of theoretical steps saved is $n_t = 2$. The thin-layer vaporizer would then perform like a reboiler and two theoretical stripping steps for which the mol ratio of the vapor and the liquid is fixed by the value (1/(1-m). In Fig. 9.17 the dotted line gives the correlation of vapor and liquid concentrations at

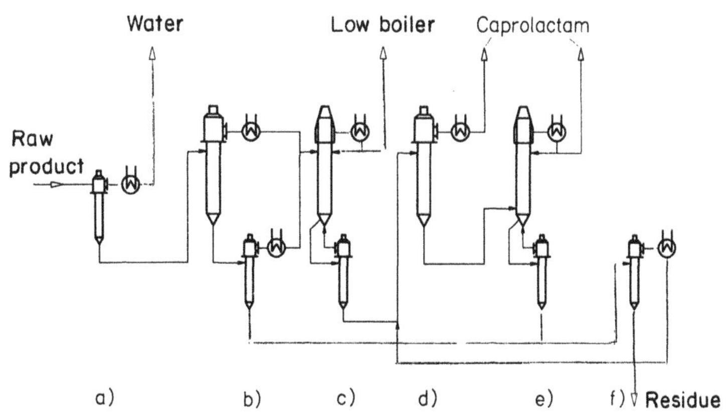

Fig. 9.18 Recovery of pure caprolactam by the Luwa method: (a) dehydration, (b) basic distillation in compound evaporators, (c) rectification of first fraction, (d) distillation, (e) basic rectification, (f) distillation of residue.

different vaporizer levels, and the solid stripping line gives vapor and liquid concentrations in the column of identical separation.

Fig. 9.18 shows how thin-layer vaporizers with rotors for continuous partial distillation can be employed as straight vaporization devices or as processors when combined with rectifiers as discussed in Chapter 6. The rectifiers then function as strippers with partial recycling of the condensates formed in the exchangers.

C. Continuous Distillation in Flashing Evaporators

In many chemical production processes it may be advantageous to first subject a reaction mixture under elevated pressure to a flash distillation, before feeding it to the separation equipment for further treatment. In that way, lower boiling components may be partly recovered. Moreover, the flashing technique may become interesting when the liquid mixture to be distilled contains solids or high boiling components likely to undergo cracking or cause fouling or incrustation of tubular surfaces at points of vapor formation. In such cases, vaporization by forced circulation is rarely an effective remedy. Fig. 9.19, scheme a), shows a flash distillation of a mixture F (kmol/h) under elevated pressure p_F, and at elevated temperature, with a low boiler content of x_F (mol-%) and an enthalpy of h_F (kcal/kmol). Upon flashing to lower pressure p vapor D (kmol/h) is liberated, with enthalpy h''_D (kcal/kmol) and a content of y_D (mol-%) low boiler reduced to W (kmol/h) with enthalpy h'_W (kcal/kmol) and x_W mol-% in the runoff. Assume y_D and x_W to be corresponding phase equilibrium concentrations as described by Eq. (9-3) for the case of a binary.

From heat balance

$$Fh_F = Dh''_D + Wh'_W \tag{9-55}$$

there follows, with the molar heat of vaporization λ_D (kcal/kmol) and the molar liquid enthalpy at boiling h'_D (kcal/kmol)

$$h''_D = \lambda_D + h'_D \tag{9-56}$$

and by substitution:

$$F - D = W \tag{9-57}$$

the relation for the vapor formed by flashing is

$$D = F\,\frac{h_F - h'_W}{\lambda_D + h'_D - h'_W} \tag{9-58}$$

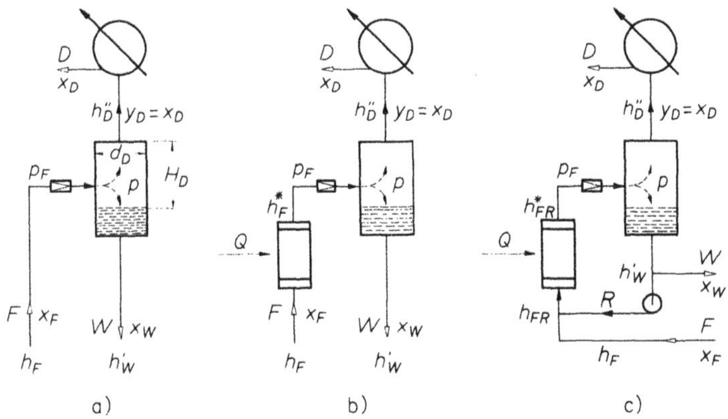

Fig. 9.19 Flash distillation: (a) simple flashing, (b) flashing after preheat, (c) flashing with recirculation type preheater.

It states that in general an appreciable quantity of distillate can be obtained through simple flash distillation only for large feed volumes F.

Scheme b) in Fig. 9.19 offers another way to obtain more vapor by preheating the feed to a high enthalpy h^*_F, but economic limits are usually set by the heating medium and operating pressure. When, however, a specific distillate quantity must be secured through flashing, scheme c) in Fig. 9.19 may be considered. By continuous circulation of a mass of liquid R and intermixing with feed F prior to entering the preheater, the mass of vapor

$$D = (F + R)\frac{h^*_{FR} - h'_W}{\lambda_D + h'_D - h'_W} \tag{9-59}$$

can be generated. In that case, the preheater must preheat the quantity $F + R$ of enthalpy

$$h_{FR} = \frac{Fh_F + Rh'_W}{F + R} \tag{9-60}$$

up to enthalpy h^*_{FR}. The corresponding temperature should be judiciously selected to avoid bubble formation on the product side of the tube wall in order to sidestep incrustation. To preheat the mixture containing the quantities F and R, whose share of low boiler is

$$x_{FR} = \frac{Fx_F + Rx_W}{F + R} \tag{9-61}$$

the heat required becomes

$$Q = (F + R)\left[\frac{x_{FR} - x_W}{y_D - x_W}(\lambda_D + h'_D - h'_W) + h'_W - h_{FR}\right] \tag{9-62}$$

This equation can also be derived analytically from thermal and mass balance considerations. The vapor quantity follows from Eq. (9-59) and from the balance

$$D = (F + R)\,\frac{x_{FR} - x_W}{y_D - x_W} = F\,\frac{x_F - x_W}{y_D - x_W} \tag{9-63}$$

Diameter d_D is given by

$$d_D = \left(\frac{4}{3600\,\pi}\,\frac{D}{\rho_D}\,\frac{\mu_D}{u_D}\right)^{\frac{1}{2}} \tag{9-64}$$

and may be based on the admissible vapor velocity u_D (m/s)[13]

$$u_D = 0.036\left(\frac{\rho_W - \rho_D}{\rho_D}\right)^{1/2} \tag{9-65}$$

or on the necessary minimum height H_D (m) of the vapor space (note Fig. 9.19 a))

$$H_D = \frac{4}{\pi\, d_D^2}\,\frac{D\,\mu_D}{\rho_L\, v_Z} \tag{9-66}$$

The admissible vapor load is[1]

$$v_Z = 3600\ \text{m}^3/\text{m}^3\text{h} \tag{9-67}$$

This is that vapor volume (m^3) that may be charged per cubic meter of vapor space per hour. The specific weight of the vapor ρ_D (kg/m^3) can be found from

$$\rho_D = \frac{\mu_D}{62.3}\,\frac{p}{T} \tag{9-68}$$

with known molecular weight μ_D (kg/kmol) and given pressure p (mmHg) and temperature T (K). ρ_W (kg/m^3) is the specific weight of the expanded liquid, and thus it is identical with that of the runoff having concentration x_W. Eqs. (9-64) to (9-68) will serve in calculations of the vapor spaces of evaporators of the circulation, once-through, and falling-film types.

2. SEPARATION OF LIQUIDS BY DISCONTINUOUS PARTIAL DISTILLATION

As opposed to the thin-layer distillation, in this method of thermal separation the masses and concentrations change with time. Even a batch still distillation can, on the whole, achieve only a partial separation of components. Basically the procedure consists of subjecting the original mixture B_1 (kmol) containing x_{B_1} (mol-%) low boiler, to a single distillation in a kettle, and condensing the vapor liberated from it, V (kmol) with y (mol-%) of low boiler immediately after generating it, and passing the condensate formed, without branching off any reflux, toward the product receiver (Fig. 9.20). Often, especially in the batch distillation of multiple mixtures,

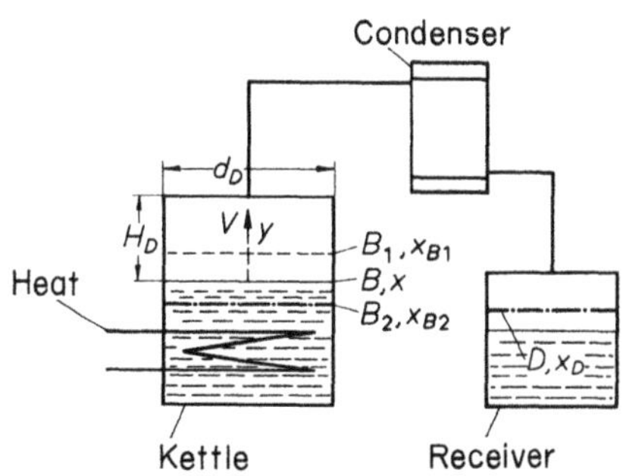

Fig. 9.20 Batch still.

several distillate fractions (cuts) are secured. Let the kettle contents at the end of the run equal B_2 (kmol) and their share of low boiler equal x_{B_2} (mol-%). Then the mass balance equals

$$\ln \frac{B_1}{B_2} = \int_{x_{B_2}}^{x_{B_1}} \frac{dx}{y - x} \tag{9-69}$$

where y and x are correlated concentrations of vapor and liquid. Assuming x and y to be correlated phase equilibrium concentrations, the integral in this equation can be solved by graphic integration through use of the equilibrium curve for a binary, as in the case of Eq. (9-30).

The entire quantity of distillate

$$D = B_1 - B_2 \tag{9-70}$$

has the mean concentration

$$x_D = \frac{B_1\, x_{B_1} - B_2\, x_{B_2}}{D} \tag{9-71}$$

If the relative volatility within the concentration range considered is virtually constant D and x_D may be found directly, avoiding the graphical evaluation of Eq. (9-69). Analogously to the procedure of Eqs. (9-36) and (9-37), the distillate quantity becomes[15]

$$D = B_1\left[1 - \frac{1}{\left(\dfrac{x_{B_1}}{x_{B_2}}\right)^{\frac{1}{\alpha-1}}\left(\dfrac{100 - x_{B_2}}{100 - x_{B_1}}\right)^{\frac{\alpha}{\alpha-1}}}\right] \tag{9-72}$$

and

$$x_D = x_{B_1} + \frac{x_{B_1} - x_{B_2}}{\left(\dfrac{x_{B_1}}{x_{B_2}}\right)^{\frac{1}{\alpha-1}}\left(\dfrac{100 - x_{B_2}}{100 - x_{B_1}}\right)^{\frac{\alpha}{\alpha-1}} - 1} \tag{9-73}$$

Example 4 A batch still is to split up $B_1 = 10$ kmol of a binary with an initial concentration of $x_{B_1} = 10$ mol-% of low boiler. The residue B_2 is allowed $x_{B_2} = 1$ mol-% of low boiler. In the concentration range between x_{B_1} and x_{B_2} the relative volatility at the pressure existing is virtually constant and $\alpha = 10$. Find distillate quantity D (kmol) and its concentration x_D (mol-%), and the still residue B_2 (mol).

Answer Eq. (9-72) gives distillate quantity D

$$D = 10\left[1 - \frac{1}{\left(\dfrac{10}{1}\right)^{\frac{1}{10-1}}\left(\dfrac{100 - 1}{100 - 10}\right)^{\frac{10}{10-1}}}\right] = 3 \text{ kmol}$$

Eq. (9-73) gives the mean distillate concentration

$$x_D = 10 + \frac{10 - 1}{\left(\dfrac{10}{1}\right)^{\frac{1}{10-1}}\left(\dfrac{100 - 1}{100 - 10}\right)^{\frac{10}{10-1}} - 1} = 31 \text{ mol-\%}$$

By Eq. (9-70) the still kettle will retain

$$B_2 = 10 - 3 = 7 \text{ kmol}$$

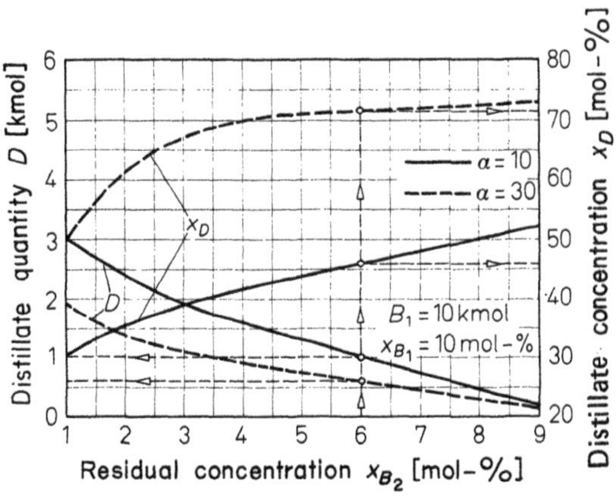

Fig. 9.21 Example 4. Batch distillation. Distillate quantity and concentration versus residual concentration.

Fig. 9.21 presents the changing distillate quantity D, its composition x_D, and still concentration x_{B_2}. For comparison, the lines for a mixture of $\alpha = 30$ are likewise shown.

Dimensions d_D (m) and H_D (m) of the vapor space of the still may be determined from Eqs. (9-64) to (9-66). However, the spatial load v_Z ($m^3/m^3 h$) should be checked from the equation

$$v_z = 500 + \frac{40}{10^{-2} + (H_D)^{1.5}} \tag{9-74}$$

This will ensure that there will be a minimum of entrainment in the vapor.

X Partial Condensation

Partial condensation might be understood as partial distillation in reverse. It is used infrequently. However, its use sometimes offers advantages, especially when the mixture yielded is a vapor as a result of the nature of the process.

1. PARTIAL CONDENSATION IN DEPHLEGMATORS

Partial distillation can partially decompose a liquid mixture, and the same end can be reached through the partial condensation of a vapor mixture. The reflux condenser of a rectifying column functions as a dephlegmator when it liquefies only that part $L_T = V_T - D$ (kmol/h) of the incoming vapor V_T (kmol/h) that contains the low boiler y_T (mol-%), while the remaining vapor D (kmol/h) is liquefied in the condenser-cooler and removed at reduced temperature (Fig. 10.1).

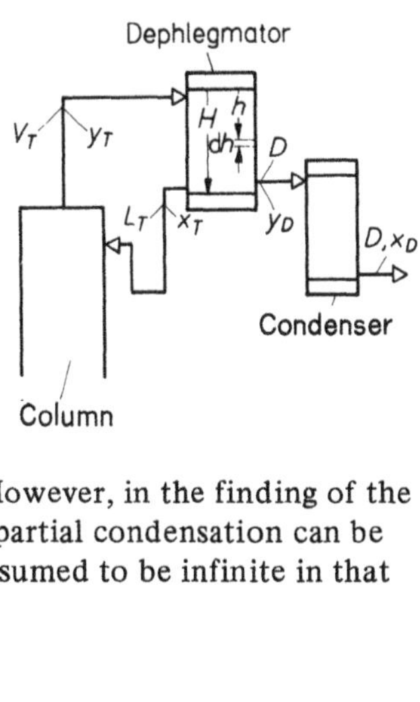

Fig. 10.1 Dephlegmator hookup at column head.

Then, the rectifier itself is computed as in Chapter 3. However, in the finding of the theoretical steps, the effect of vapor enrichment through partial condensation can be taken into account. The vapor side film coefficients are assumed to be infinite in that calculation.

A. Separating Effect

If V (kmol/h) = the mass of mixed vapor condensed at point h (m) of the dephlegmator, condensate dV (kmol/h) is formed at dh (m) because of the mass balance[1] of the differential equation

$$dV = \frac{dy}{y - x_E} (V_T - V) \tag{10-1}$$

from which, by integration within limits $V = 0$ and $V = V_T - D = L_T$ and $y = y_T$ and $y = y_D = x_D$ and substitution of the reflux ratio

$$r = r_D = \frac{L_T}{D} = \frac{V_T - D}{D} \tag{10-2}$$

there follows

$$\ln (r_D + 1) = \int_{y=y_T}^{y=x_D} \frac{dy}{y - x_E} \tag{10-3}$$

where x_E equals the concentration of the condensate formed at its interface with the vapor at a given condenser level, and y equals the correlated vapor concentration. For infinite mass transfer on the vapor side, this may be regarded as the phase equilibrium concentration of x_E. This allows a graphic solution of the integral of Eq. (10-3) by using the equilibrium curve for binaries.

The results of such an evaluation are the theoretical enrichment curves $y_T = f(y_D)$, (note Fig. 10.2). These correlate the vapor inlet concentration y_T and the vapor exit concentration $x_D = y_D$ for various reflux ratios, $r = r_D$, as the parameter. For a given product concentration y_T of the low boiler in the vapor entering the dephlegmator can be read off from the proper enrichment line. The parallel to the abscissa passing through y_T in the y/x plot intersects the enrichment line at a point to which concentration x_T of the reflux from the dephlegmator is correlated. From a mass balance

$$x_T = y_T - \frac{1}{r_D} (x_D - y_T) \tag{10-4}$$

The pair of values y_T, x_T gives the origin of the enrichment line and of the sequence of steps in the y/x plot. Since $y_T < y_D = x_D$, the dephlegmator itself represents a number of rectifying steps. This is indicated in Fig. 10.2 for two different reflux ratios $r_D = r_1$ and $r_D = r_2 > r_1$.

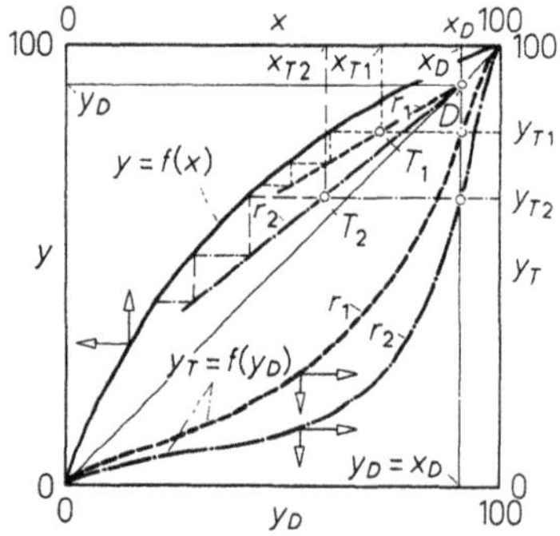

Fig. 10.2 Theoretical enrichment in dephlegmator, and its application in the determination of number of steps in rectifier.

The deviations from the theoretical enrichment and dephlegmator calculations are dealt with in publications.[2–10] In practice, enrichments are less than expected by theory, especially for large reflux ratios. The larger the ratio of the mass transfer coefficient and the condenser load, the better is the condenser function. When in a dephlegmator equipped with vertical tubes the vapor flows at low-cooling surface load, meaning low vapor velocities, and at low reflux ratios, countercurrently upward, with the cooling water moving downward, separation may be high.

The separation was determined by tests on a condenser fitted with turbulators[3] with optimum conditions maintained, and results were close to theoretical. On the other hand, high vapor velocities and reflux ratios gave separation much below the theoretical.

In the development of Eq. (10–3) it had been assumed that no rectification effect would be superimposed onto the partial condensation. This situation has been discussed by several authors.[11,12] Also the separation by dephlegmator was found to be better the less the condensation load is per unit area. Some tests gave results superior to those for straight partial condensation.

B. The Reflux Ratio

If the relative volatility within the range of concentrations in question is considered constant, the integral of Eq. (10-3) can also be solved analytically. The interrelation between x_E and y at infinite material transfer on the vapor side is given by the equilibrium curve, and hence by Eq. (2-4), the latter will supply for $x_1 = x_E, y_1 = y$, and $\alpha_1 = \alpha$

$$x_E = \frac{y}{\alpha 100 - y(\alpha - 1)} 100 \qquad (10\text{-}5)$$

By inserting the term of Eq. (10-5) for x_E into Eq. (10-3), integration will yield[10]

$$r_D = \left(\frac{y_D}{y_T}\right)^{\frac{\alpha}{\alpha - 1}} \left(\frac{100 - y_T}{100 - y_D}\right)^{\frac{1}{\alpha - 1}} - 1 \qquad (10\text{-}6)$$

Example 1 A dephlegmator must enrich a mixed vapor entering at $y_T = 60$ mol-% of low boiler content to $y_D = 75$ mol-% by partial condensation. What should the theoretical reflux ratio be? What will be the concentration of the condensate? Let the relative volatility of the binary in question be constant through the concentration range $y_T \leqslant y \leqslant y_D$ i.e. $\alpha = 1.5$.

Answer By Eq. (10-6) r_D becomes

$$r_D = \left(\frac{75}{60}\right)^{\frac{1.5}{1.5 - 1}} \left(\frac{100 - 60}{100 - 75}\right)^{\frac{1}{1.5 - 1}} - 1 = 4$$

and by Eq. (10-4)

$$x_T = 60 - \tfrac{1}{4}(75 - 60) = 56.25 \text{ mol-\%}$$

C. Dephlegmator Dimensions

Let k = the overall coefficient of heat transfer between the condensing vapor and the coolant in the dephlegmator (kcal/m^{2}hC), ϑ = the corresponding total temperature difference (C), λ_V = the molar heat of condensation (kcal/kmol), and U = total circumference of tubes exchanging heat (m).

The vapor condensing across an infinitesimal level difference dh (m) may be expressed by (in terms of kmol/h)

$$dV = \frac{k\, U\, dh\, \vartheta}{\lambda_V} \tag{10-7}$$

Differential Eq. (10-1) gives, for the vapor condensing at level h, K (mol/h), by integration

$$V = V_T \left[1 - \exp\left(-\int_{y_T}^{y} \frac{dy}{y - x_E}\right)\right] \tag{10-8}$$

By combining Eqs. (10-1), (10-7), and (10-8) there results the differential equation

$$dh = \frac{1}{k\,U} \cdot \frac{\lambda_V}{\vartheta} \cdot V_T \exp\left(-\int_{y_T}^{y} \frac{dy}{y - x_E}\right) \frac{dy}{y - x_E} \tag{10-9}$$

The theoretical height of the dephlegmator can be found from this, by integration within limits $h = 0$ and $h = H$, up to $y = y_T$ and $y = y_D = x_D$. That would lead to the question of the actual validity of coefficient k and of the temperature difference. Moreover, the remarks covering optimum separation of Section 1.A of this chapter should be remembered.

The number of tubes and thus their total circumference must be chosen to yield some tube length such that the specific condensate load at the actual total temperature difference and the mass transfer surface does not become excessive.

However, Eq. (10-9) does not yield an accurate answer because of the relationship of the heat transfer coefficient α(kcal/m^{2}hC) on the condensation side where

$$\alpha_V = 0.943 \left(\frac{r_L\, \rho_L^2\, \lambda_L^3}{\eta_L\, \vartheta_V\, H}\right)^{1/4} \tag{10-10}$$

subscript L refers to the mass of condensate formed L_T, and there is the dependence on the tube length H (m) and on the temperature difference

$$\vartheta_V = t_V - t_{WV} \tag{10-11}$$

where t_V equals the condensation temperature of vapor (C) and t_{WV} = corresponding temperature of tube wall (C). Instead, it is recommended that an average be used for the overall coefficient,

$$k = \frac{1}{\frac{1}{\alpha_V} + \frac{s}{\lambda} + \frac{1}{\alpha_K}} \tag{10-12}$$

where α_V refers to the mean condensate temperature of Eq. (10-10)

$$t_L = t_{WV} + \frac{\vartheta_V}{2} \tag{10-13}$$

Its determination requires, by Eq. (10-11), knowledge of the mean condensation temperature t_V. According to theory this would be the phase equilibrium temperature for the mean surface concentrations of the condensation film produced, which is figured from Eq. (10-4).

For a liquid coolant flowing through tubes, film coefficient α_K (kcal/m^2 hC) is found from Eqs. (10-9)–(10-14), if the proper material coefficients for the coolant are used in Eqs. (9-15)–(9-17).

With dephlegmator tube thickness s (m) conductivity λ, Eq. (10-12) yields the mean overall coefficient k (kcal/m^{2}hC). With this Eq. (10-9) supplies the relation (based on $\lambda_V = \lambda_T$)

$$H = (V_T - D)\frac{\lambda_T}{kU\vartheta} = r_D D\,\frac{\lambda_T}{kU\vartheta} \tag{10-14}$$

where the average of the condensate concentration λ_T (kcal/kmol) must be used for molar heat of condensation, according to Eq. (10-4), and for

$$\vartheta = t_T - t_K \tag{10-15}$$

the difference between the equilibrium vapor temperature t_T and the mean temperature t_K of the coolant is written.

The correctness of the assumed wall temperature t_{WV} can be checked by analogy through the use of Eq. (9-18).

D. Applications

A regeneration of reflux and simultaneous enrichment of the overhead vapors does not constitute the only use of dephlegmators. Fig. 10.3 reveals various applications of that component other than that shown by scheme a). In scheme b), a separation by continuous thin-layer distillation is improved through partial condensation of the vapor phase and the recycling of the condensate to the vaporizer.

The chemical industry makes frequent use of hot cooling, in which vapor formed by a reaction is liquefied in a condenser in order to fully or partly recover the heat of reaction. The condensate may return to the reactor, and some low boiler may be removed, as in scheme c) which avoids intermediate storage. Likewise, high boilers may be isolated, as in scheme d). Again, recovery of some low or high-boiling component through partial condensation of the vapor may occur at any other point in a process; note scheme e) in Fig. 10.3.

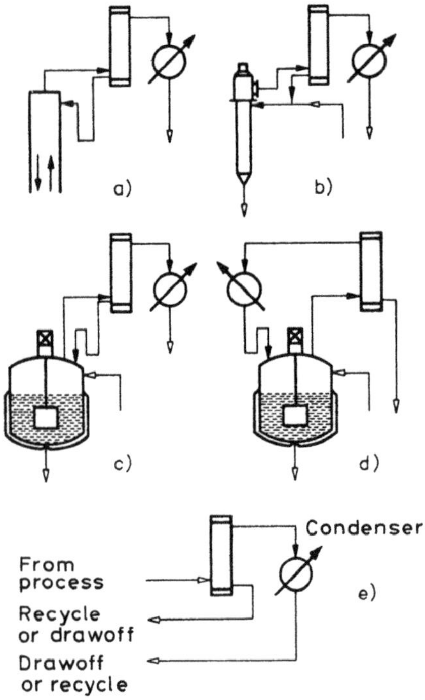

Fig. 10.3 Dephlegmator applications. a) column rectification, b) thin-layer evaporators, c) and d) reactors with hot cooling, e) processing, general.

The efficiency of the separation of some low or high-boiling material is controlled by its percentual share as a vapor phase, and also by the difference in boiling points between T_2 (K) of the higher boiler or mixture, and T_1 (K) of the lower boiler or mixture.

As a rule in such conditions, one estimates the relative volatility α from

$$\log \alpha = 8.9 \frac{T_2 - T_1}{T_2 + T_1} \tag{10-16}$$

which relation was worked out by Trouton and Clapeyron;[13] Eq. (10-6) can then be employed.

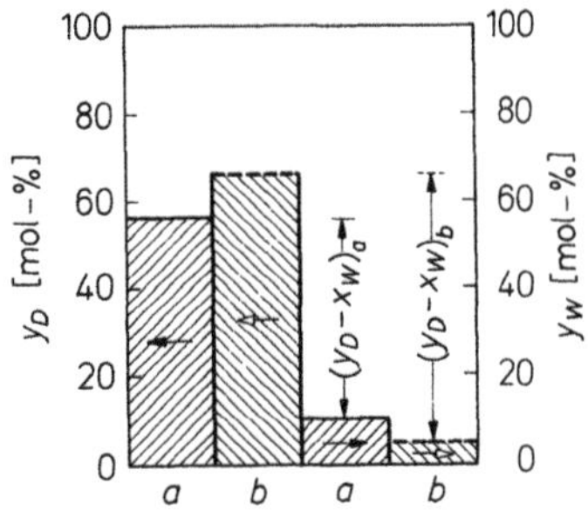

Fig. 10.4 Separation by straight thin-layer distillation versus distillation plus dephlegmation. System of $\alpha = 4.5$ and $x_F = 36$ mol-%.

Fig. 10.4 illustrates the fact that separation efficiency in thin-layer distillation can be much improved by partial condensation of the vapor and recycling the condensate to the vaporizer as reflux. The ensuing greater vaporizer load requires, either increased vaporizing surface or a greater effective total temperature difference between the heating medium and the boiling liquid.

Example 2 In a plastics production plant several ways exist to recover oligomers from the solvent cycle. Some of these involve the reworking of vapors from the degassing stage by partial condensation in a dephlegmator, with the residual vapors being totally liquefied in a condenser. Investigate the performance of that setup for oligomer concentrations of 1, 2, and 4 mol-% in the inlet vapor. Condensation is to take place at 8 mm Hg.

Answer First find the relative volatility from Eq. (10-16) with the boiling points at 8 mm Hg, $T_1 = 303$ K for the rotating solvent, and $T_2 = 393$ K for the oligomers:

$$\log \alpha = (8.9) \frac{393 - 303}{393 + 303}$$

$$\alpha = 14.1$$

By eq. (10-6) the exit vapor concentrations can be found for the various inlet concentrations of 1, 2, and 4 mol-%, assuming several r_D values. There is the interrelation

$$\frac{D}{V_T} = \frac{1}{r_D + 1} \tag{10-17}$$

where V_T = vapor entering and D equals the vapor leaving. The percentage of partially condensed vapor out of the inlet vapor is

$$\frac{L_T}{V_T} 100 = \frac{r_D}{r_D + 1} 100 \tag{10-18}$$

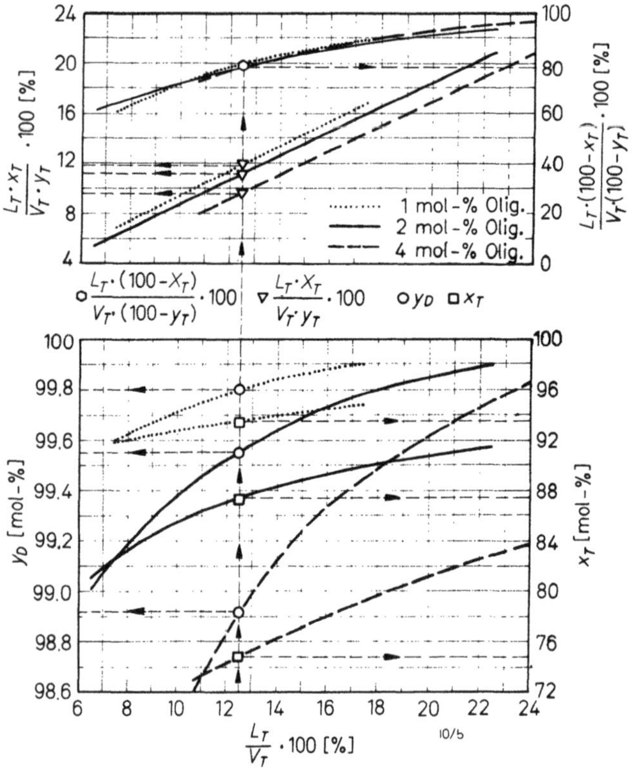

Fig. 10.5 Example 2. Typical data in oligomer drawoff from solvent vapor by partial condensation.

This is plotted in Fig. 10.5. The monomer y_D (mol-%), monomers x_T (mol-%) in the residual vapor and in the condensate of the partially liquefied vapor are calculated from Eq. (10-4).

The oligomers recovered, expressed as a percentage of the oligomers in the entering vapor

$$\frac{L_T}{V_T} \frac{100 - x_T}{100 - y_T} 100$$

and the monomers recovered by partial condensation, expressed as a percentage of the monomers in the entering vapor

$$\frac{L_T}{V_T} \frac{x_T}{y_T} 100$$

are presented. The light dotted lines indicate the relative quantities. Thus, if the oligomers in the entering vapor amount to 2 mol-%, of which 12.5% are to be partially condensed, theoretically 78.5% of the oligomers are removed, together with 11.2% of the monomers. This gives a monomer content of 87.4 mol-% in the condensate; the residual vapor is enriched to 99.55 mol-% of the monomer.

Through use of an optimally operated dephlegmator, theoretically a relatively high percentage of oligomer can be recovered.

It is seen that partial condensation is often a suitable means for removing certain undesirable materials perpetually formed in recirculating solvents and preventing their accumulation to any unacceptable extent.

2. PARTIAL COUNTERCURRENT DIRECT CONDENSATION IN COLUMNS

Evaporative cooling with partial condensation and the removal of a light fraction is illustrated in Fig. 10.3. Another method that can be used is partial countercurrent direct condensation as is shown in Fig. 10.6. Here the feed is preheated by passing it through a column countercurrent to the ascending hot vapors from the reactor. As a result of this direct heat exchange, only a comparatively small proportion of the vapor is collected in the condenser. Whether all or part of this condensate is removed depends on the quantity concerned and on the amount of heat available. If only a part is removed, the remainder is returned to the column together with the feed. This method is particularly suitable for cold feedstocks. If operating conditions demand, part of the feed may also be led direct into the reactor.[14,15]

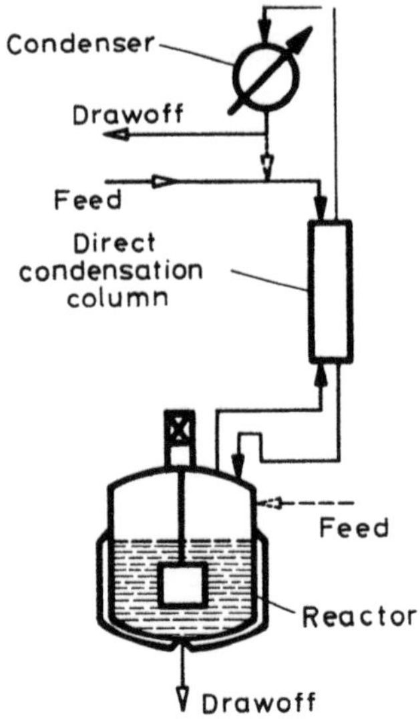

Fig. 10.6 Low-boiler drawoff from vapors of an agitated reactor by partial countercurrent direct condensation.

A. Separation Efficiency and Throughputs

The analysis consists of considering an infinitely small element of volume (Fig. 10.7) in the partial countercurrent direct condensation process; the following model is made

a) the vapor and the liquid pass through the column in ideal countercurrent flow

b) the liquid flows into the element of volume at a rate L (kmol/h) and contains x (mol-%) of the more volatile substance

c) the vapor flows into the element of volume at a rate V (kmol/h) and contains y (mol-%) of the more volatile component

d) any infinitely small amount dV (kmol/h) condensing from the vapor phase contains x_E (mol-%) of the more volatile substance.

Since x_E is less than y, the concentration of the more volatile substance in the vapor phase is increased by an amount dy (mol-%). These changes in quantities and concentrations at any given instant of time can be expressed in terms of the more volatile substance by means of the differential equation[14,15]

$$dVx_E = Vy - (V - dV)(y + dy) \tag{10-19}$$

Since the product $dVdy$ is infinitely small and can thus be neglected, Eq. (10-19) can be rearranged to give

$$\frac{dV}{V} = \frac{dy}{y - x_E} \tag{10-20}$$

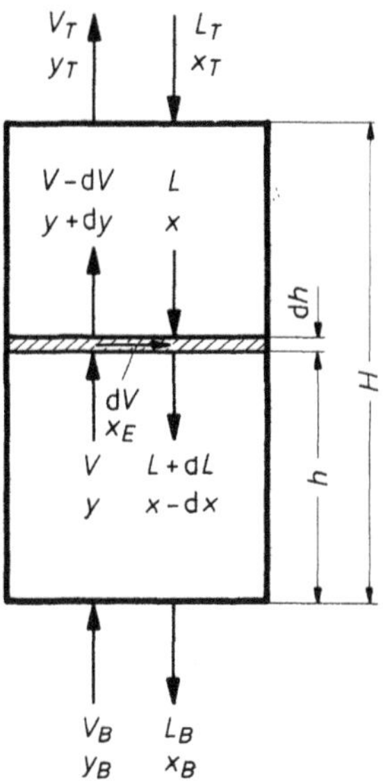

Fig. 10.7 Column with partial countercurrent direct condensation. Masses and concentrations involved.

In passing through the column, the amount of vapor is reduced from V_B (kmol/h) at the bottom to V_T (kmol/h) at the top. The corresponding proportions of volatiles are y_B and y_T (mol-%). Thus if the height of the column is H (m), the total amount of vapor that has been condensed V_C in heating up the feed, L_T (kmol/h), is given by

$$V_C = V_B - V_T \tag{10-21}$$

Similarly, the amount condensed, V_c, when the vapor passes through any given length, h (m), of the column is

$$V_c = V_B - V \tag{10-22}$$

If the variable V in Eq. (10-20) is replaced by the expression

$$V = V_B - V_c \tag{10-23}$$

and both sides of the equation are integrated between the limits $V_c = 0$ to $V_c = V_C$ and $y = y_B$ to $y = y_T$, Eq. (10-24) is obtained

$$\ln \frac{V_B}{V_T} = \int_{y_B}^{y_T} \frac{dy}{y - x_E} \tag{10-24}$$

This can be solved by expressing x_E as a function of y, that is if

$$x_E = f(y) \tag{10-25}$$

For this purpose, it is assumed that the concentration of condensate at the liquid/vapor interface is roughly the same as the phase equilibrium value at the associated vapor concentration. Thus a graphical solution of Eq. (10-24) can be found for binary systems; in this case the value to be taken for $y - x_E$ at any given concentration y, is the vertical distance between the equilibrium curve and the $y = x$ diagonal. Alternatively, Eq. (10-24) may be solved mathematically by considering the relative volatility α in the range of concentration concerned, to be constant; then

$$x_E = \frac{y}{\alpha\ 100 - y\ (\alpha - 1)}\,(100) \tag{10-26}$$

Hence, Eq. (10-24) for the total amount of vapor condensed in the column, V_C, becomes

$$V_C = V_B - V_T = V_T\left[\exp \int_{y_B}^{y_T} \frac{dy}{y - x_E} - 1\right] \tag{10-27}$$

Substituting x_E from Eq. (10-26) and integrating gives

$$V_C = V_B - V_T = V_T\left[\left(\frac{y_T}{y_B}\right)^{\frac{\alpha}{\alpha - 1}} \left(\frac{100 - y_B}{100 - y_T}\right)^{\frac{1}{\alpha - 1}} - 1\right] \tag{10-28}$$

Now if λ_C (kcal/kmol) is the molar latent heat of condensation of the vapor, the amount of heat required to raise the liquid from the initial enthalpy, $(h_L)_T$ (kcal/kmol), to the liquid enthalpy at the boiling point, $(h'_L)_B$ (kcal/kmol) is given by

$$V_C\lambda_C = (V_B - V_T)\lambda_C = L_T\,[(h'_L)_B - (h_L)_T] \tag{10-29}$$

Then

$$V_C = V_B - V_T = L_T\,\frac{(h'_L)_B - (h_L)_T}{\lambda_C} \tag{10-30}$$

Combining Eqs. (10-28) and (10-30)

$$\frac{(h'_L)_B - (h_L)_T}{\lambda_C} \frac{L_T}{V_T} = \left(\frac{y_T}{y_B}\right)^{\frac{\alpha}{\alpha - 1}} \left(\frac{100 - y_B}{100 - y_T}\right)^{\frac{1}{\alpha - 1}} - 1 \tag{10-31}$$

Eq. (10-31) expresses the relationship between the inlet vapor concentration, y_B, and the outlet vapor concentration, y_T, for a given liquid-to-vapor ratio, L_T/V_T.

Now

$$V_B - V_T = L_B - L_T \tag{10-32}$$

and if the amount of liquid at the bottom of the column, L_B, is expressed in terms of the amount at the top, L_T, by means of factor f_i, that is if

$$L_B = L_T f_i \tag{10-33}$$

then

$$V_C = V_B - V_T = L_T f_i - L_T = L_T(f_i - 1) \tag{10-34}$$

Equating Eq. (10-30) with Eq. (10-34) finally gives

$$f_i = 1 + \frac{(h'_L)_B - (h_L)_T}{\lambda_C} \tag{10-35}$$

In many systems, the molar latent heat of condensation, λ_C is practically independent of the concentration. If this is not the case, a mean value of λ_C for a binary system can be obtained from Eq. (10-36)

$$\lambda_C = \lambda_1 \frac{(x_E)_m}{100} + \lambda_2 \frac{100 - (x_E)_m}{100} \tag{10-36}$$

where $(x_E)_m$ (mol-%) is the mean proportion of low-boiling component in the amount of vapor V_C (kmol/h), that has condensed in the column and λ_1, and λ_2 (kcal/kmol) are the molar latent heats of condensation for the low-boiling and high-boiling vapors respectively.

From the mass balance in the column $(x_E)_m$ is given by

$$(x_E)_m = \frac{L_B x_B - L_T x_T}{V_C} \tag{10-37}$$

Substituting the values from Eqs. (10-21), (10-32), (10-33), and (10-34) gives

$$(x_E)_m = \frac{f_i x_B - x_T}{f_i - 1} \tag{10-38}$$

Thus $(x_E)_m$, and thus λ_C and f, can be determined from Eqs. (10-35), (10-36), and (10-38).

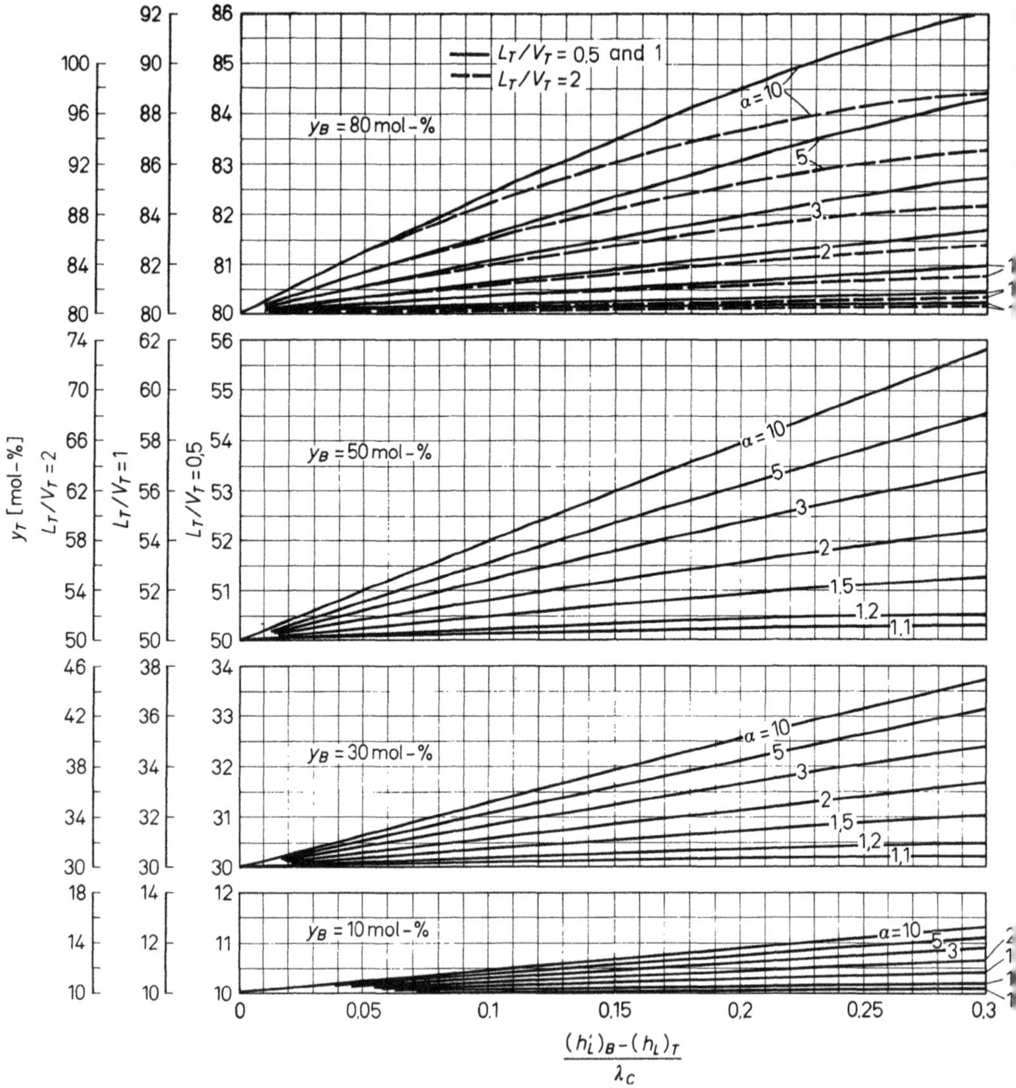

Fig. 10.8 Partial countercurrent direct condensation. Enrichment of low-boiler in vapor phase.

Eq. (10-31) allows the extent to which low boilers in the vapor phase have been enriched by countercurrent direct condensation to be calculated for given vapor-to-liquid ratios. For instance, in Fig. 10.8 the concentration of the vapor in the outlet, y_T, is shown as a function of the expression $[(h'_L)_B - (h_L)_T]/\lambda_C$ defining the extent to which the liquid has been heated up. The concentration of the vapor in the inlet, y_B, and the liquid-to-vapor ratio, L_T/V_T, are the parameters, and a few systems of different relative volatility, α, have been selected as examples.

The higher the values of α and L_T/V_T and the larger the expression $[(h'_L)_B - (h_L)_T]/\lambda_C$, the greater the difference between y_T and y_B, that is, the greater the degree of enrichment of low-boilings in the vapor phase.

Theoretically it is also possible to apply the above analysis of partial countercurrent direct condensation to the feed zone of continuously operating rectification columns when the feed is extremely cold or to the head of the enrichment section of a column if the reflux is fairly strongly cooled. By this means the scope of the analysis presented in Chapters 3 and 7 can be extended by dividing the column into heating and rectification zones.

B. Height of the Column

The height of the column can be calculated by considering what takes place in an infinitely small section of height dh (m). From Eqs. (10-20) and (10-23), the amount of vapor that condenses over the length dh is given by

$$dV = \frac{dy}{y - x_E}(V_B - V_c) \tag{10-39}$$

The amount of heat transferred at the point, h, on the infinitely small area of contact, dA (m^2), between the vapor and the liquid is obtained from Eq. (10-40)

$$dV\lambda = dfk\vartheta \tag{10-40}$$

where k (kcal/m^2hC) is the overall coefficient, ϑ(C) is the temperature difference between vapor and liquid, and λ (kcal/kmol) is the molar latent heat of condensation.

Now, if d_S (m) is the diameter of the column and a (m^2/m^3) is the area of vapor liquid contact per unit of volume, dA can be expressed as

$$df = dh\, d_S^2 \frac{\pi}{4} a \tag{10-41}$$

Combining Eqs. (10-39), (10-40), and (10-41) gives the following differential equation

$$\frac{k\, dh\, d_S^2 \frac{\pi}{4} a\vartheta}{\lambda} = \frac{dy}{y - x_E}(V_B - V_c) \tag{10-42}$$

An exact solution to this is difficult to find. The relationship between y, x_E, V_c and φ can, no doubt, be determined for a binary mixture from the equilibrium curve and the boiling and dew point curves if it is assumed that the liquid was heated up linearly. Furthermore, no great error would be incurred by inserting a mean value for λ. The difficulty, however, is that k for the case of direct heat transfer cannot be calculated exactly from the equations given in the literature, particularly since the area on which transfer takes place, a, also cannot be exactly stated. In addition the function $V_c = f(h)$ must be known.

A simplifying assumption that could be made is that the following linear relationship is valid for the range of concentration considered

$$V_c = V_C - V_C \frac{h}{H} = V_C \left(1 - \frac{h}{H}\right) \tag{10-43}$$

In this case, Eq. (10-42) becomes

$$\frac{dh}{V_B - V_C \left(1 - \dfrac{h}{H}\right)} = \frac{\lambda}{k d_S^2 \dfrac{\pi}{4} a\vartheta} \frac{dy}{y - x_F} \tag{10-44}$$

which transforms into

$$\int_{h=0}^{h=H} \frac{dh}{1 + \dfrac{V_C}{V_T} \dfrac{h}{H}} = \frac{V_T}{d_S^2 \dfrac{\pi}{4} a} \int_{y=y_B}^{y=y_T} \frac{\lambda}{k\vartheta} \frac{dy}{y - x_E} \tag{10-45}$$

This assumes that a is independent of h. After integration, the left side of this equation assumes the form

$$\int_{h=0}^{h=H} \frac{dh}{1 + \dfrac{V_C}{V_T} \dfrac{h}{H}} = H \frac{V_T}{V_C} \ln \left(\frac{V_C}{V_T} + 1\right) \tag{10-46}$$

The assumption that $V_c = f(h)$ entails that $k\vartheta/\lambda$ equals a constant which equals $k_C \vartheta_C/\lambda_C$ [Eq. (10-40) and (10-41)], in which case k_C, ϑ_C, and λ_C represent mean values over the height H. Thus[14,15]

$$H = \frac{V_C}{\ln \left(\dfrac{V_C}{V_T} + 1\right) d_S^2 \dfrac{\pi}{4} a} \frac{\lambda_C}{k_C \vartheta_C} \int_{y=y_B}^{y=y_T} \frac{dy}{y - x_E} \tag{10-47}$$

As a rule, the integral in Eq. (10-47) can be solved graphically if the equilibrium curve is known. However, if it is again assumed that Eq. (10-26) is valid with α equal to a constant, the expression can be integrated, i.e.

$$\int_{y=y_B}^{y=y_T} \frac{dy}{y - x_E} = \ln \left[\left(\frac{y_T}{y_B} \right)^{\frac{\alpha}{\alpha - 1}} \left(\frac{100 - y_B}{100 - y_T} \right)^{\frac{1}{\alpha - 1}} \right] \tag{10-48}$$

Inserting this solution and substituting for V_C Eq. (10-28) converts Eq. (10-47) into the following expression for H

$$H = V_T \left[\left(\frac{y_T}{y_B} \right)^{\frac{\alpha}{\alpha - 1}} \left(\frac{100 - y_B}{100 - y_T} \right)^{\frac{1}{\alpha - 1}} - 1 \right] \frac{\lambda_C}{k_C \, \vartheta_C a d_S^2 \, \frac{\pi}{4}} \tag{10-49}$$

This, in turn, can be combined with Eq. (10-31) to form

$$H = L_T \frac{(h'_L)_B - (h_L)_T}{k_C \vartheta_C a d_S^2 \, \frac{\pi}{4}} \tag{10-50}$$

Eq. (10-49) contains a term for the condensation, whereas the equivalent term in Eq. (10-50) is for the preheating. Another factor affecting the main dimensions of a separation column with countercurrent direct condensation is the nature of the internals that have been installed in the column to increase the effective area of contact between the liquid and the vapor. The equations known in the literature for the various types of column internals may be taken for the determination of the column diameter, d_S. However, the relationships affecting heat and mass transfer in partial countercurrent direct condensation must be known for the exact determination of the column height, H.

For instance, for given equipment and given operating conditions, the relationship may be known between the average heat-transfer coefficient, k_C, and the associated mean temperature difference, ϑ_C, between the vapor on condensing and the liquid on heating up. In this case, the height of the column may be calculated from Eqs. (10-49) and (10-50). Thus if a mixture of methanol and ethanol is separated in a bed of 50 × 50 mm Pall rings under the conditions shown in Table 10.1, the desired relationship may be read off under the column headings "ϑ_C" and "k_C".

Table 10.1. Typical data of a partial countercurrent direct condensation. Metallic Pall rings 50 × 50 mm. Methanol/ethanol, 760 mm Hg.

l_T (kg/m² h)	ϑ_c (°C)	k_c (kcal/m² h °C)	n_c (1/m)	$\frac{k_c \vartheta_c}{n_c} \cdot 10^{-3}$	$\frac{k_c \vartheta_c}{n_c l_T}$
3240	8.65	804	0.895	7.76	2.40
3300	13.15	420	0.696	7.92	2.40
3300	19.65	372	0.913	7.99	2.42
3280	22.15	288	0.841	7.58	2.31
4210	9.99	742	0.756	9.78	2.32
4210	19.79	312	0.630	9.79	2.32

It is an obvious step to take the expression enclosed in brackets on the right of Eq. (10-27) as a measure of the column efficiency, denoted by n, that is,

$$n = \exp \int_{y_B}^{y_T} \frac{dy}{y - x_E} - 1 \tag{10-51}$$

If the relative volatility, α, in the binary system is constant, Eqs. (10-26) and (10-51) give

$$n = \left(\frac{y_T}{y_B}\right)^{\frac{\alpha}{\alpha - 1}} \left(\frac{100 - y_B}{100 - y_T}\right)^{\frac{\alpha}{\alpha - 1}} - 1 \tag{10-52}$$

The efficiency per unit height of column

$$n_C = \frac{n}{H} \tag{10-53}$$

is a magnitude defining the relative efficiency of the column internals. It can generally be determined experimentally. Then the effective height of column H is the ratio of n desired from Eqs. (10-51) and (10-52) to the empirical value n_C, that is,

$$H = \frac{n}{n_C} \tag{10-54}$$

Table 10.1, which has been drawn up for a methanol/ethanol system at atmospheric pressure, lists the values of efficiency per unit height, n_C, that can be expected in a 50–mm Pall ring column for given values of liquid load per unit area of column cross section, l_T (kg/m^{2}h), and for given values of the mean temperature difference, ϑ_C, between vapor temperature, t_V, and liquid temperature, t_L. l_T and ϑ_C are defined by Eqs. (10-55) and (10-56)

$$l_T = \frac{L_T \mu_L}{d_S^2 \frac{\pi}{4}} \tag{10-55}$$

and

$$\vartheta_C = t_V - t_L \tag{10-56}$$

According to Table 10.1 the efficiency per unit height, n_C, can also be related to the rate of heat transfer per unit area, q_C (kcal/m^{2}h), in which case q_C is the product of the heat transfer coefficient, k_C, and the temperature difference ϑ_C, that is,

$$q_C = k_C \vartheta_C \tag{10-57}$$

Then the ratio q_C/n_C can be regarded as a characteristic expression for the relationship between the mass transfer and the heat transfer. Thus if the liquid load per unit area, l_T, is constant at the head of the column, the diameter of the packing is d, and the liquid-vapor ratio, L_T/V_T, is a constant equal to one

$$K_1 = \left(\frac{k_C \vartheta_C}{n_C}\right)_{l_T,\, d,\, L/V = 1} = \text{constant} \tag{10-58}$$

and with varying liquid load, but the same d and $L_T/V_T = 1$

$$K_2 = \left(\frac{k_C \vartheta_C}{n_C\, l_T}\right)_{d,\, L/V = 1} = \text{constant} \tag{10-59}$$

C. Application

From Fig. 10.8 it can clearly be seen under which conditions the partial countercurrent direct condensation process can be considered for removing the one component from a stream of vapors. However, in most cases, a detailed calculation is essential to determine the degree of removal and an example of this is given below.[14,15]

> **Example 3** The degree to which a light fraction of the feed can be removed with the layout shown in Fig. 10.6 has to be determined. Four cases will be considered, viz., for volatility ratios $\alpha = 10$ and $\alpha = 30$ between the light fraction and the carrier and a liquid-to-vapor ratio at the head of the column $L_T/V_T = 2$ and $L_T/V_T = 4$.

The liquid-to-vapor ratio depends on the amount of vapor produced in the reboiler, the amount of feed, F, and the enthalpy corresponding to the expression $[(h'_L)_B - (h_L)_T]/\lambda_C = 0.3$. The proportion of light fraction in the vapor tops, which are removed without reflux, is y_D.

The problem is to determine the proportions of light fraction l_D and of carrier s_D in the tops $D = V_T$, for various proportions of light fraction in the feed x_F and to determine the proportion of residual light fraction x_W run off at the bottom of the boiler. The amounts present in the tops l_D and s_D are given by

$$l_D = \frac{Dy_D}{Fx_F} 100 \qquad (10\text{-}60)$$

and

$$s_D = \frac{D(100 - y_D)}{F(100 - x_F)} 100 \qquad (10\text{-}61)$$

Answer The amounts and concentrations concerned are represented schematically in Fig. 10.9. Simplifying assumptions made are that molar changes in the boiler are negligible and that the amount run off from the boiler is equal to the difference between the feed and the tops. Thus the layout in

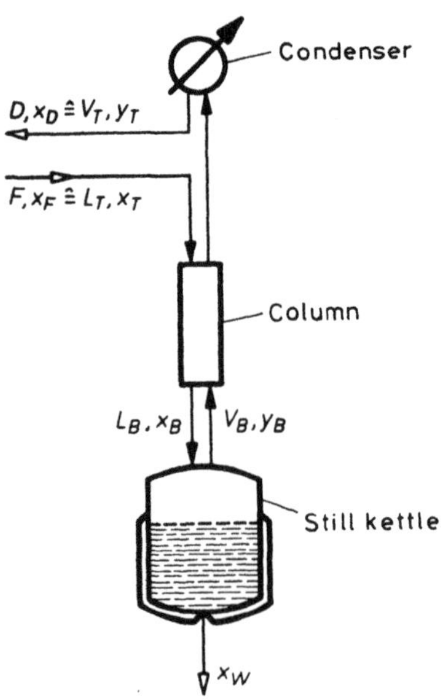

Fig. 10.9 Low-boiler drawoff of Fig. 10.6. Masses and concentrations involved.

Fig. 10.9 can be considered as continuous distillation in which the cold feed is preheated directly in a column by a countercurrent of vapor generated in the boiler.

Since all the condensate formed in the condenser is withdrawn

$$\frac{L_T}{V_T} = \frac{F}{D} \tag{10-62}$$

and

$$y_T = x_D \tag{10-63}$$

from Eq. (10-31), the relationship between x_D and y_B for a given F/D is obtained, that is,

$$\frac{(h'_L)_B - (h_L)_T}{\lambda_C} \frac{F}{D} = \left(\frac{x_D}{y_B}\right)^{\frac{\alpha}{\alpha - 1}} \left(\frac{100 - y_B}{100 - x_D}\right)^{\frac{1}{\alpha - 1}} - 1 \tag{10-64}$$

If the proportion of light fraction, y_B, in the vapor flowing into the column is in phase equilibrium with the proportion of light fraction in the boiler runoffs, x_W, then the relationship between x_W and y_B according to Eq. (10-26) is

$$x_W = \frac{y_B}{\alpha 100 - y_B(\alpha - 1)} 100 \tag{10-65}$$

The equation of mass balance in this case is

$$x_F = x_W + \frac{x_D - x_W}{\frac{F}{D}} \tag{10-66}$$

Numerical solution of Eqs. (10-64), (10-65), and (10-66) gives rise to the curves in Fig. 10.10. The effect of the relative volatility, α, and the liquid-vapor ratio, L_T/V_T, on the degree of enrichment and the degree of removal can be clearly seen in this figure; x_D increases with α and with L_T/V_T, whereas $D\,x_D/F\,x_F$ increases with α but not with L_T/V_T.

According to Fig. 10.10, a given set of conditions, denoted by L_T/V_T, α, and x_F, might arise in which a practically pure light fraction, i.e., $x_D \rightarrow 100$ mol-%, could be removed for a given value of $[(h'_L)_B - (h_L)_T]/\lambda_C$. However, in this case, a finite

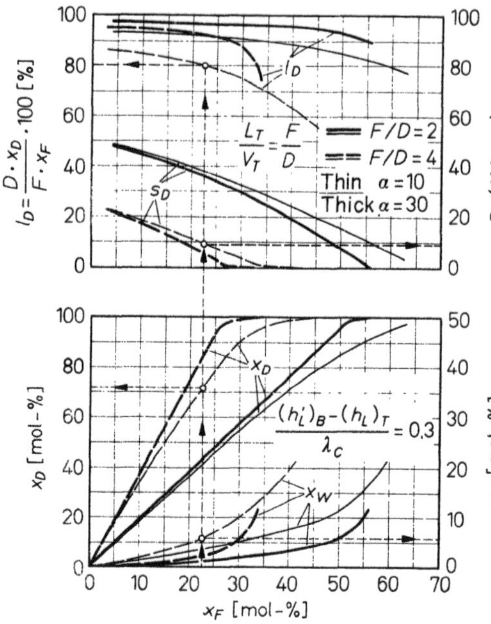

Fig. 10.10 Results of partial countercurrent direct condensation. Layout of Fig. 10.9.

amount of light fraction would still be contained in the bottoms, that is, $x_W > 0$.

In principle it is also possible to combine partial countercurrent direct condensation with stripping if x_W were too high. Thus it would appear that an enrichment column might be required if the proportion of volatiles in the tops, x_D, were well below 100%, that is, if comparatively large amounts of heavy fraction were also removed. But, in practice, this depends on the process and operating conditions of the entire plant, of which the unit represented in Fig. 10.9 is only a part. Another means of increasing the proportion of light fraction in the tops is deep cooling and recycling part of the condensate. Still another is to separate the vapors at the head of the column by partial condensation. Whenever the question arises, each case must be considered on its own merits.

A rough figure for the mean temperature difference, ϑ_C, can be obtained from the inlet and outlet temperatures of the vapor and the liquid. If an exact figure is required, the changes that take place throughout the length of the column must be known, that is, the ratio of the concentrations of vapor, y, and liquid, x, at any given cross section. This ratio can be determined by making a mass balance at the given point, h, in the column (Fig. 10.7). Thus

$$V_B y_B + Lx = Vy + L_B x_B \tag{10-67}$$

which converts into the straight-line form

$$y = \frac{L}{V} x + \frac{V_B y_B - L_B x_B}{V} \tag{10-68}$$

Since

$$L = L_B - V_c \tag{10-69}$$

and since V_c is a function of y, L must vary with y. From Eq. (10-23) the same reasoning applies to V. Then integrating both sides of Eq. (10-20) between the limits 0 and V_C and y_B and y respectively and substituting for Eq. (10-23) gives

$$V_c = V_B \left[1 - \frac{1}{\exp \displaystyle\int_{y_B}^{y} \frac{dy}{y - x_E}} \right] \tag{10-70}$$

If the relative volatility α is constant, Eq. (10-70) can be converted to

$$V_c = V_B \left[1 - \frac{1}{\left(\dfrac{y}{y_B}\right)^{\frac{\alpha}{\alpha - 1}} \left(\dfrac{100 - y_B}{100 - y}\right)^{\frac{1}{\alpha - 1}}} \right] \tag{10-71}$$

The amount of vapor at the bottom of the column is given by

$$V_B = V_T + V_C \tag{10-72}$$

V_C can be expressed either as function of V_T by means of Eq. (10-28) or as a function of L_T by means of Eq. (10-30).

The relationship between x_B, y_B, and x_W must be known to enable the linear equation for the mass balance to be drawn up for the layout in Fig. 10.9. A mass balance gives

$$x_B = \frac{V_B}{L_B} y_B + \frac{L_B - V_B}{L_B} x_W \tag{10-73}$$

Here y_B is a function of x_W and can be calculated from Eq. (10-65). The total balance with $x_F = x_T$ and $x_D = y_T$ is given by Eq. (10-66).

If the numerical example leading to Fig. 10.10 is again taken, Fig. 10.11 represents the mass balance curve for $\alpha = 10$, $L_T/V_T = F/D = 2$, and $x_F = 30$ mol-%. The points T, B, and W represent the concentrations at the head of the column, the base of the column, and the foot of the boiler.

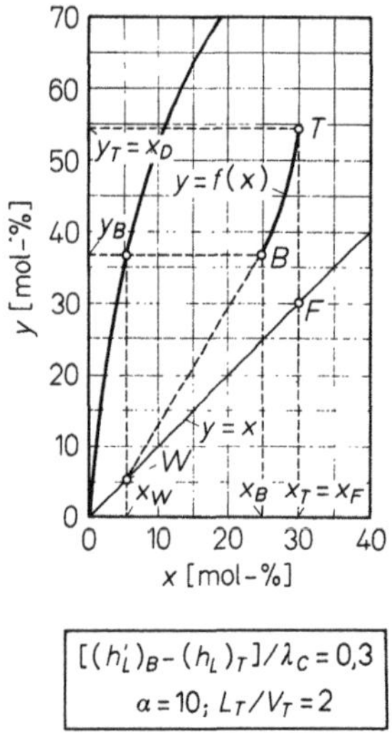

Fig. 10.11 Operation of Fig. 10.10. Material balance in y/x plot.

Fig. 10.12 shows the qualitative relationship between y and x and the associated vapor to liquid temperature difference ϑ, in any given horizontal section of the column. The $y = f(x)$ curve and the temperature-concentration curves ($y = f(t)$ and $x = f(t)$) are presented.

In principle, it is possible to improve the efficiency by lengthening the column, because separation would take place by stripping in the lower part of the column. The qualitative relationship between vapor and liquid concentrations in this case is shown in Fig. 10.13. The $\overline{T'B'}$ portion of the mass balance curve applies for the partial countercurrent-direct condensation section of the column, whereas the portion $\overline{B'W'}$ is the characteristic for the separation that takes place in the extended lower section. The dotted-line curve $\overline{BT}$ represents the conditions for pure countercurrent direct condensation for the same value of $x_F = x_T$ and the same value of $[(h_L')_B - (h_L)_T]/\lambda_C$; it has been included for comparison purposes. At B', the slopes of the mass balance and stripping curves are equal. The projection of the tangent to the stripping curve at B' intersects the line $y = y_T$ at i. The following relationship can be obtained by a mass and heat balance for the abscissa, x_i, of this point

$$x_i = \frac{x_F}{f_i} + \left(1 - \frac{1}{f_i}\right) y_T \qquad (10\text{-}74)$$

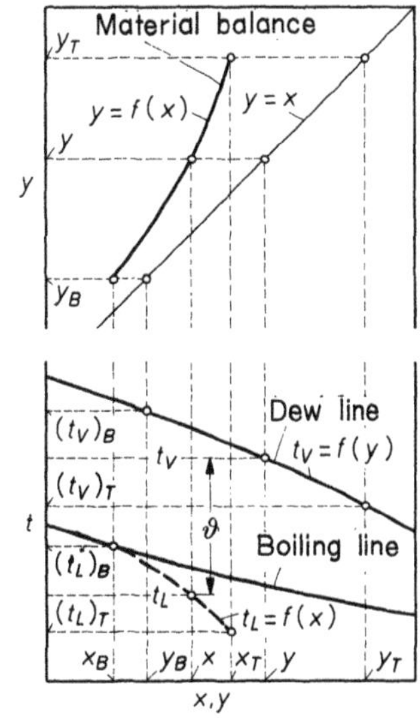

Fig. 10.12 Qualitative representation of relationship between vapor and liquid concentration and temperature differences.
$y = f(x)$–material balance
$t = t_V = f(y)$–dew line
$t = f(x = x_E)$–boiling line
$t = t_L = f(x)$–liquid heating
ϑ = temperature difference, vapor to liquid
Subscripts: V Vapor, L Liquid, T Column top, B Column bottom.

For a qualitative representation of Fig. 10-13, the direct condensation column would have to be lengthened by an amount equivalent in efficiency to two theoretical stages of the stripping section.

Fig. 10.14 shows the degree of enrichment that can be obtained by adding a stripper separation stage to the partial countercurrent direct condensation.

The above discussion on separation by partial countercurrent direct condensation has been based on a phase equilibrium condensation model. It allows the stages to be calculated fairly rapidly and lucidly. Any deviations from the theory are embraced in the term for the efficiency, η, which can be obtained from Eqs. (10-31), (10-51), and

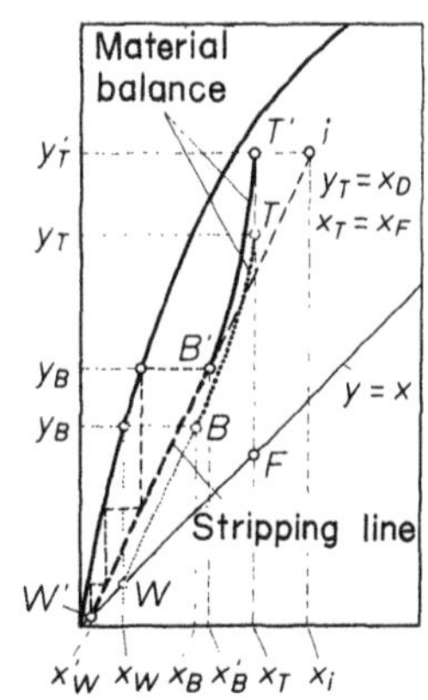

Fig. 10.13 Qualitative representation of relationship with combined partial countercurrent direct condensation and stripping rectification in y/x plot.

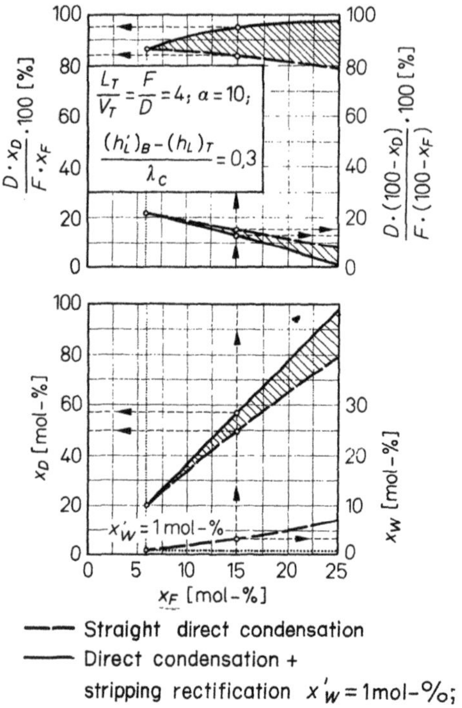

Fig. 10.14 Partial countercurrent direct condensation, combined with stripping rectification. Improvement in separation.

(10-52), that is,

$$\eta = \frac{n_W}{(f_i - 1)\,\dfrac{L_T}{V_T}} \tag{10-75}$$

in which n_W is the actual degree of separation, which can be derived from the measured values of vapor inlet and outlet concentrations with the aid of Eqs. (10-51) and (10-52). Substituting for V_T (see Eq. [10-34]) and rearranging Eq. (10-75) gives

$$\eta = n_W\left(\frac{V_B}{L_T}\,\frac{1}{f_i - 1} - 1\right) \tag{10-76}$$

In retrospect the annotations made regarding partial countercurrent direct condensation will also apply to the supply zone of rectifiers by having extremely low-temperature feed or to the column head by making the reflux extremely cold. In this way, columns may be calculated in detail by means of the concept of preheat and rectification zones, which is beyond the scope traced in Chapters 3 and 7.

As regards the mass- transfer coefficients mentioned in those passages, the reader is referred to the current bibliography for closer definitions.

XI Laboratory Columns and Pilot Plants

Design specifications for industrial-scale equipment are frequently based on test results obtained in laboratories or reduced-scale installations.

If the phase equilibria for a nonideal multiple mixture are unknown, a test is run on a laboratory rectifier. First, one determines the number of theoretical steps of that laboratory column for an ideal test mixture of known boiling performance. From the resulting number of steps for the test mixture and the result of the trial rectification of the actual multicomponent mixture, conclusions are drawn for the equivalent number of theoretical steps to be specified.

If, however, separation tests of an actual multicomponent mixture of unknown phase equilibrium are made on a reduced-scale pilot installation, results are transferred directly to a detailed specification for the industrial-size plant.

The layout, operation, and performance of such test installations, and the use of results obtained in full size plant construction, is discussed in the following section.

1. DISTILLATION COLUMNS WITH MINIATURE SIZE PACKING

Packing of regular size, as discussed in Chapter 6, yields the characteristic saddle-shaped curves giving theoretical steps versus vapor load, or vapor velocity versus capacity. As opposed to this, diminutive packing elements give differently shaped load curves. This fact points to a difference in the flow profile for the packing in the main velocity range and in the vicinity of the load range. This is confirmed by the circumstance that in packed columns a correlation exists between the separation and flow condition[1,2,3] that depends on vapor velocity, irrigation intensity, liquid proportioning, the column diameter, and the physical properties of the mixture to be separated, and eventually the size, shape, and surface condition of the packing.

A. Theory of Flow

Observations on the flow of liquid along packing surfaces leads to a body of information that permits the derivation of equations and correlations for finding the capacities of large industrial packed columns. Analogously, the maximum vapor velocity is established for columns with diminutive packing elements. In this connnection, the liquid volume per unit of packing surface is much smaller than formerly. Thus when Raschig rings of 25 mm size are compared with wire net rings of 3 mm diameter braced internally, the surfaces per cubic meter of column volume represent roughly a 1 to 10 ratio. For a dumped volume of small packing elements, the volume of reflux liquid flowing across one square centimeter of packing surface is only one-tenth of that in an industrial column for equal mass velocities. Observations made on glass-shell columns reveal that due to adhesion forces active between liquid and packing surface the liquid is spread into a thin film. Hence, the inertia of the liquid is disregarded. Strands of liquid that form drops at the lowest points of large packing elements are absent with small bodies. Obviously, the liquid stream is regarded as a film flowing downward along a wall. The wall is regarded as vertical. An inclination from the vertical is expressed by a coefficient that is included in the final relation. If inertia is neglected, inside the film the forces of shear and gravity balance, while at the film surface the vapor promotes friction. If that friction is increased by a growing vapor velocity the fluid film will be entrained upward and the limit capacity will be achieved.

Fig. 11.1 illustrates diagrammatically a height element ΔH. Let the liquid film flow downward along the surface of the packing element, counter to the vapor.[4]

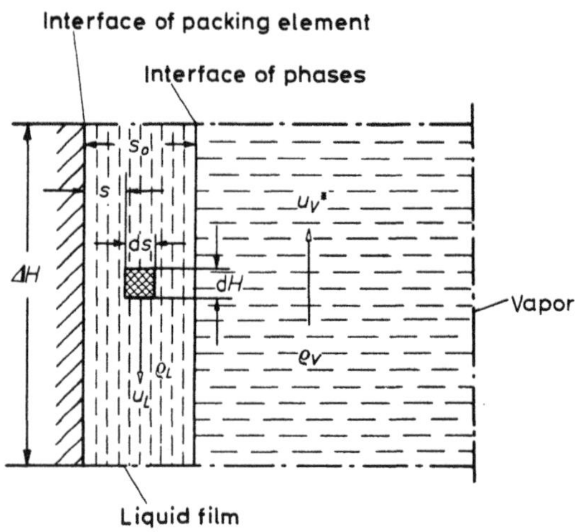

Fig. 11.1 Film type flow, model and characteristic variables.

Because of the change in velocity du_L/ds at point s in the liquid film of thickness s_o, there is produced along the infinitesimal height of dH, at the infinitesimal width ds of the infinitesimal volume element, a shear force $dS_L dH$, if the depth of the element is considered unity in a general way.

According to Nusselt's film theory[5] the shear force $dS_L dH$ balances the weight $\rho_L ds dH$ of the particle, disregarding acceleration[4]

$$dS_L dH + \rho_L ds dH = 0 \tag{11-1}$$

or

$$dS_L + \rho_L ds = 0 \tag{11-2}$$

where ρ_L = the specific weight of the liquid.

Shear stress S_L equals, with the velocity of the liquid particle u_L and η_L, its kinematic viscosity

$$S_L = \eta_L \frac{du_L}{ds} \tag{11-3}$$

and

$$dS_L = \eta_L \frac{d^2 u_L}{ds^2} ds \tag{11-4}$$

From Eqs. (11-2) and (11-4)

$$\frac{d^2 u_L}{ds^2} = -\frac{\rho_L}{\eta_L} \tag{11-5}$$

and from Eq. (11-5) by integration

$$u_L = -\frac{\rho_L}{2\eta_L} s^2 + C_1 s + C_2 \tag{11-6}$$

For $s = 0$, $u_L = 0$ because there the liquid adheres to the packing. From that boundary condition, integration constant $C_2 = 0$.

Let U = the circumference of a flow channel and let A = its cross-sectional area. Column packing may be visualized as composed of many such flow channels, and an infinitesimal height element dH shares in pressure drop dp of the vapor moving through the channels. In the differential equation

$$\frac{dp}{dH} = \xi \frac{u_V^{2*}}{2g} \rho_V \frac{U}{A} \tag{11-7}$$

ξ is the resistance between the vapor and the wetted packing, ρ_V the specific weight of the vapor in the flow channel, and u_V^* the vapor velocity. With S_V = shear stress at circumference U of a channel

$$S_V = \frac{A}{U} \frac{dp}{dH} \tag{11-8}$$

By combining Eqs. (11-7) and (11-8), consequently

$$S_V = \xi \frac{u_V^{2*}}{2g} \rho_V \qquad (11\text{-}9)$$

By differentiation at point $s = s_o$, Eq. (11-6) will yield

$$\left(\frac{du_L}{ds}\right)_{s_o} = -\frac{\rho_L}{\eta_L} S_o + C_1 \qquad (11\text{-}10)$$

By inserting Eq. (11-10) into Eq. (11-3), the shear stress of the liquid at the surface of the packing becomes

$$(S_L)_{s_o} = -\rho_L s_o + \eta_L C_1 \qquad (11\text{-}11)$$

Since $S_V = -(S_L)_{s_o}$, Eqs. (11-9) and (11-11) will give

$$C_1 = -\frac{\xi}{\eta_L}\frac{u_V^{2*}}{2g}\rho_V + \frac{\rho_L}{\eta_L} s_o \qquad (11\text{-}12)$$

Constant C_1 is therefore controlled by the state of flow of the vapor.

First, the likelihood of flooding is investigated. The correlated velocity is labeled "admissible".[1,2,4] It is presumably reached when velocity u_L of the liquid surface adjoining the vapor becomes 0. For a total thickness of the liquid film s_o and boundary condition of $s = s_o$, u_L becomes zero. By Eq. (11-6) the integration constant becomes

$$C_1 = \frac{\rho_L}{2\eta_L} s_o \qquad (11\text{-}13)$$

By equalizing relations (11-12) and (11-13), and solving for u_V^* the limiting vapor velocity becomes

$$u_V^* = u_{Vz}^* = \left(\frac{g}{\xi}\frac{\rho_L}{\rho_V} s_o\right)^{1/2} \qquad (11\text{-}14)$$

By Eq. (11-14) flooding will occur first in that column section where the term $\rho_L/\xi\rho_V$ becomes a minimum.

When the admissible vapor velocity u^*_{Vz} is exceeded through increased column load, the vapor may begin to entrain liquid upward. In the liquid film there must be a point $0 < s_{th} < s_o$ where the velocity u_L becomes nil. Liquid in the range $0 < s < s_{th}$ flows downward, liquid in the range $s_{th} < s < s_o$ moves upward due to the friction of the moving steam and backs up. That point s_{th} depends on the vapor velocity, and knowing it allows for the estimation of the liquid quantity entrained by the vapor.

By Eq. (11-6) velocity u_L in the range $0 < s < s_o$ becomes, by inserting Eq. (11-12)

$$u_L = \left(- \frac{\xi}{\eta_L} \frac{u_V^{2*}}{2g} \rho_V + \frac{\rho_L}{\eta_L} s_o \right) s - \frac{\rho_L}{2.\eta_L} s^2 \qquad (11\text{-}15)$$

It will supply the location of point s_{th} in the liquid film where $u_L = 0$

$$s_{th} = 2\, s_o - \frac{\xi}{g} \frac{\rho_V}{\rho_L} u_V^{2*} \qquad (11\text{-}16)$$

Conceivably at limit condition $s_{th} = 0$ vapor velocity u_V^* can be increased to entrain all of the liquid upward, thus emptying the column. Then Eq. (11-16) supplies the relation[4]

$$u_V^* = (u_V^*)_{fl} = \left(\frac{2g}{\xi} \frac{\rho_L}{\rho_V} s_o \right)^{1/2} \qquad (11\text{-}17)$$

Relation (11-14) is also obtained from Eq. (11-16) by replacing s_{th} with s_o.

Eqs. (11-14) and (11-17) are arrived at by means of an idealized picture of the flow phenomena involved. At the points of contact of the randomly distributed packing elements the liquid collects by preference because of the interfacial tension, and film thickness s_o will assume different magnitudes at different levels of the packing. Furthermore the vapor phase is discontinuous if the liquid backs up in the column. Also, liquid bridges between adjacent points of packing elements may vibrate.[3] If liquid has accumulated at one point, turbulence may develop. Such disturbances affect capacity but are hard to define analytically.

Packing elements are mostly dumped into place and may arrange themselves in various ways, which affects column holdup; for this allowance must be made.[6] The properties of irrigated packing are best described in terms of clearance ϵ per unit volume

$$\epsilon = \frac{V_S - V_F - J}{V_S} \qquad (11\text{-}18)$$

where V_S = volume of column, V_F = compact volume of packing, and J = volume of liquid content, consisting of a static and dynamic component, the former being essentially the liquid at contact points.[3]

Introducing the hydraulic intermediate radius

$$e = \frac{\epsilon V_S}{A_o} \tag{11-19}$$

which is the hydraulic radius of a channel of constant section, whose volume is $(V_S - V_F - J)$ and whose surface is that of the interstice A_o, Eq. (11-7) will yield

$$\Delta p = \xi \frac{u_V^{2*}}{2g} \rho_V \frac{H}{e} \tag{11-20}$$

where u_V^* = vapor velocity in the interstitional volume, H = height of packed column, and Δp = total pressure loss of the packed column.

The vapor velocity u_V is the surficial velocity for that column, and

$$u_V = \epsilon u_V^* \tag{11-21}$$

The resistance factor follows from Eq. (11-20)[4]

$$\xi = \frac{\Delta p 2 g e \epsilon^2}{u_V^2 \rho_V H} \tag{11-22}$$

Given the proper resistance or flow coefficient for different packing types, and the mean film thicknesses s_o, a combination of Eqs. (11-21) and (11-14) yields a relation for the admissible vapor velocity u_{Vz}[4]

$$u_{Vz} = \epsilon \left(\frac{g}{\xi} \frac{\rho_L}{\rho_V} s_o \right)^{1/2} \tag{11-23}$$

The effective load limit is reached when the vapor velocity is increased beyond u_{Vz} to such an extent that the mean integral value becomes zero

$$u_{Lm} = \frac{1}{s_o} \int_0^{s_o} u_L ds = 0 \tag{11-24}$$

Integration of relation (11-24), noting Eqs. (11-15) and (11-21) supplies for the corresponding vapor velocity $(u_V)_g$ the term

$$(u_V)_g = \epsilon \left(\frac{4}{3} \right)^{1/2} \left(\frac{g}{\xi} \frac{\rho_L}{\rho_V} s_o \right)^{1/2} \tag{11-25}$$

where film thickness s_o is given, either by liquid content J and surface A_o

$$s_o = \frac{J}{A_o} \tag{11-26}$$

or it is the difference of the hydraulic intermediate radii with and without liquid content

$$s_O = (e)_{J=0} - e \tag{11-27}$$

which are found by Eqs. (11-18) and (11-19).

B. Results of Laboratory Tests

Fig. 11.2 presents a test assembly suitable for operation at normal pressures and in vacuo. To minimize heat losses the rectifying column is fitted with a reflecting double jacket, silvered inside, with an annular evacuated space.

This setup gives the results presented in Fig. 11.3 when rectifying the mixture of *n*-heptane and methyl cyclohexane, at column head pressures of 760, 500, 250, and 150 mm Hg. The packing was 2 x 2 mm full helices and 3 x 3 mm full helices of stainless type V4A.

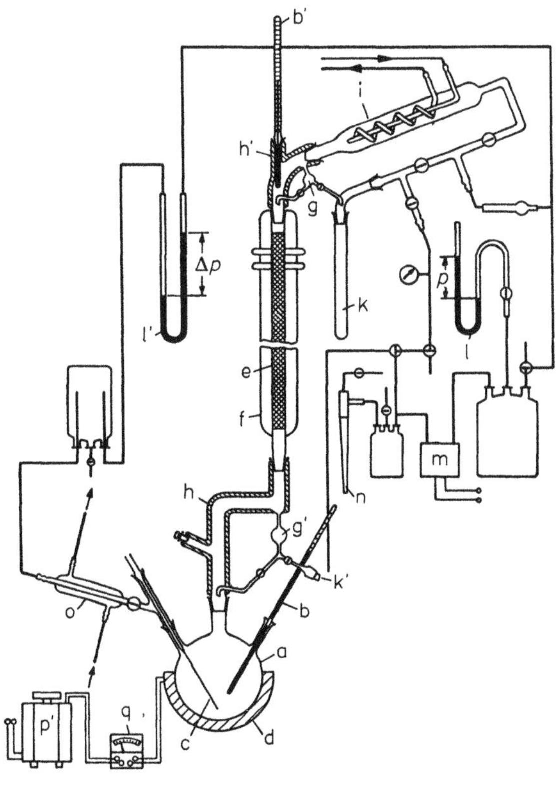

Fig. 11.2 Layout for laboratory distillation. a) still, b) thermometers, c) capillary, d) heating jacket, e) column 0.5 m packed height, 0.02 m diameter, f) vacuum jacket, g) interceptors, h) lagging, i) reflux condenser, k) sampler, l) pressure gages, m) pressure controller, n) ejector, o) condenser, p) variable transformer, q) wattmeter.

In a log–log coordinate system the theoretical steps n_t equivalent to 0.5 m packing depth are plotted as a measure of separation and the corresponding pressure loss, versus the vapor velocity u_V based on the full column area.

The separation effect is high, compared to that of industrial columns discussed in Chapter 8. However, the loads are low, considering that the maximum velocities at normal pressures, indicated by dash-dotted lines, have coordinated load factors $u_V \rho_V^{1/2}$ of 0.24 $m^{-1/2}$ s^{-1} $kg^{1/2}$ for the 2 × 2 mm full helices and 0.29 $m^{-1/2}$ s^{-1} $kg^{1/2}$ for the 3 × 3 mm full helices, which means a throughput of about one-tenth of the capacity of the industrial columns referred to a unit area.

Compared with load curves for industrial packed columns, those in Fig. 11.3 show practically no maximum at heavy loads near the admissible vapor velocity. The configuration of the curve deviates from the saddle shape customary with industrial columns. This is presumably due to the variation in flow pattern in packed laboratory columns filled with small elements across which the liquid flows as a film, as has been demonstrated.

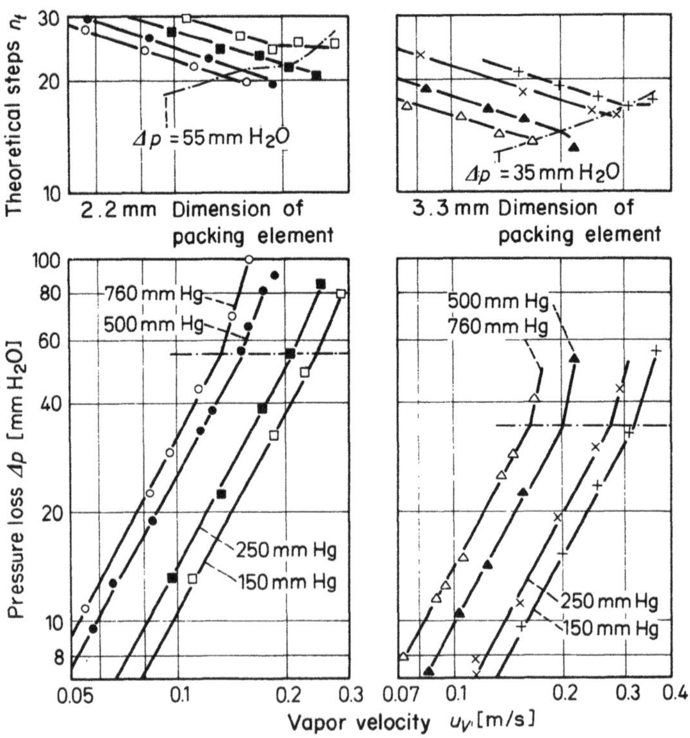

Fig. 11.3 Separation and pressure loss versus vapor velocity at column top. Packing depth 0.5 m, column diameter 20 mm. Mixture *n*-heptane/methyl cyclohexane. Still concentration 15 mol-% of *n*-heptane. Total reflux. Full helices.

Furthermore, industrial packed columns exhibit a greater dependence of the liquid content on the load than do laboratory columns; this applies especially to the velocity range below u_{Vz} where separation mostly increased with increasing velocity.

Lines similar to those of Fig. 11.3 are also found for other systems.[7,8] A checkup on Fig. 11.3 shows that[4] the liquid content of the packing in the main range did not depend excessively on the load. Furthermore it is estimated that the content of operating liquid is almost the same for methyl cyclohexane as for *n*-heptane; differences are within tolerance limits. It also seems evident that the values of Fig. 11.3 are not much affected by the pressure, as far as the liquid holdup at given velocities is concerned. For the velocities given for $p = 760$ mm Hg and 250 mm Hg, 2 × 2 mm full helices give $J = 17.3$ and 17.7g of *n*-heptane liquid volume at limit load. These correspond to about $J = 2.58 \times 10^{-5}/m^3$ of mixture. For 3 × 3 mm full helices the average liquid content at the admissible vapor velocity is $J = 1.9 \times 10^{-5}/m^3$ of mixture.

To evaluate Eq. (11-22) giving resistance coefficient ξ the admissible vapor velocities u_{Vz} may be taken from Fig. 11.3; the horizontal in the figure cuts these off. In the case of the 2 × 2 mm full helices these correspond to a pressure loss of $\Delta p = 55$ mm water, and in the case of the 3 × 3 mm full helices to $\Delta p = 35$ mm water. All of these are below the velocity at which the bulk of the liquid is entrained by the vapor and the separation drops to zero.

Table 11.1

Data on full helices. Values of ϵ (with J), e (with J), s_o and ξ_m refer to maximum vapor velocity

Mixture of *n*-heptane/methyl cyclohexane Column 20-mm diameter, 500-mm high

Packing Data			Packing: 2 × 2 mm Full Helices	Packing: 3 × 3 mm Full Helices
Packing Weight		(kg)	0.3122	0.2645
Surface Area		(m^2)	0.511	0.360
Total Volume of Packing		(m^3)	$3.98 \cdot 10^{-5}$	$3.38 \cdot 10^{-5}$
ϵ	without J		0.747	0.785
	with J		0.582	0.664
e (m)	without J		$22.95 \cdot 10^{-5}$	$34.20 \cdot 10^{-5}$
	with J		$17.90 \cdot 10^{-5}$	$28.92 \cdot 10^{-5}$
s_o (m)			$5.05 \cdot 10^{-5}$	$5.28 \cdot 10^{-5}$
ξ_m			2.62	2.09

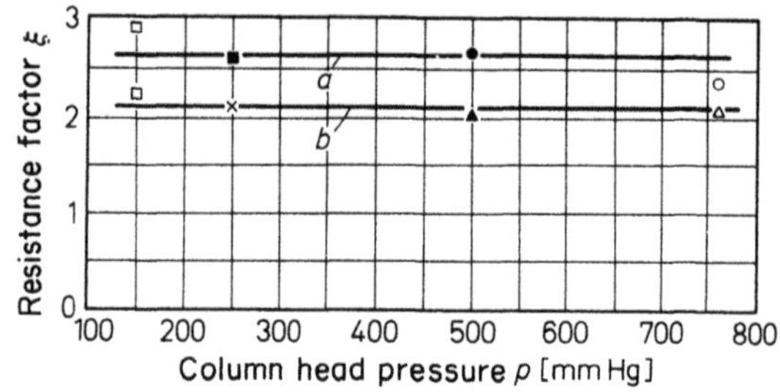

Fig. 11.4 Resistance or flow coefficient between vapor and liquid versus column head pressure. Linear 2 × 2 mm full helices. Line b, 3 × 3 mm full helices. Other data in Fig. 11.3 and Table 11.1.

In Fig. 11.4, the resistance coefficient calculated from Eq. (11-22) for the admissible vapor velocity is almost independent of the pressure. It is based on Fig. 11.3 and Table 11.1, ϵ, e and s_o. For 2 × 2 mm full helices it is an average of $\xi_m = 2.62$, and for 3 × 3 mm full helices $\xi_m = 2.09$.

Using the resistance coefficients ξ of Fig. 11.4 in evaluating Eqs. (11-23) gives the juxtaposition of measured and calculated velocities in Fig. 11.5. For the test mixture applied, there is good agreement of theory and test results.

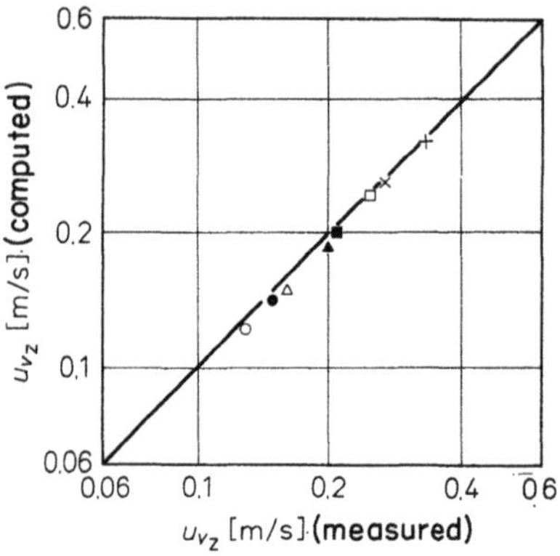

Fig. 11.5 Computed versus measured vapor velocities. Other data in Fig. 11.3 and Table 11.1

Another investigation[4] found that the blowby frequently observed in full-scale packed columns are largely due to an orderly arrangement of packing inside the shell is immaterial for laboratory units and diminutive packing.

The tests used an undulated shell of 22 mm minimum and 24 mm maximum diameter, with all other dimensions of the assembly unchanged. At any event, that undulation eliminated the blowby and results were as before.

C. Results of Pilot Plant Tests

By the use of highly effective packings as described, even difficult separations requiring many theoretical steps, such as isomers and isotopes, are achieved industrially. In general the specific separation effect becomes less as the column diameter increases. Therefore to counteract unequal phase distribution, parallel columns of relatively small diameter are combined and served by the same vaporizing and condensing equipment.

A semiindustrial pilot installation (Fig. 11.6) proved the suitability of column shells of 100 mm diameter, with 1 m of packing depth. The packing consisted of metal wire net rings 3 × 3 mm. To prevent the loss of packing, grating was used as in the small column of Fig. 11.2. The test mixture was *n*-heptane/methyl cyclohexane, p = 760 mm Hg.

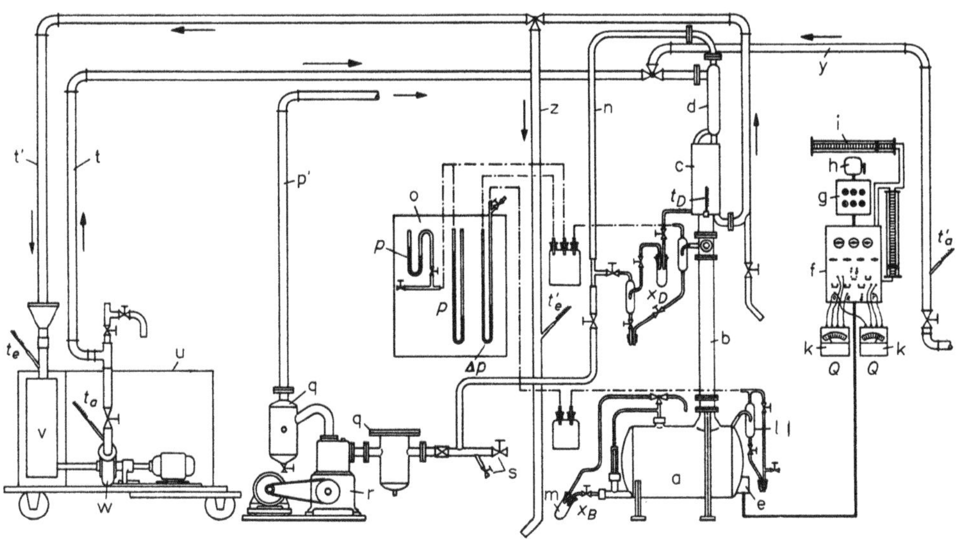

Fig. 11.6 Layout of a semi-pilot plant. Diameter of rectifying column 100 mm, packing depth 1 m. Packing 3 × 3 mm wire net rings.
t_D = temperature of distillate, t_a = initial temperature of coolant, t_e = final temperature of coolant, t'_a = initial temperature of cooling water, t'_e = final temperature of cooling water, Q = heating energy, P = absolute pressure, Δp = pressure loss, x_B = still concentration, x_D = distillate concentration.
a) still, b) packed column, c) reflux condenser, d) aftercooler, e) heater , f) control board, g) fuses, h) main cutout, i) variable resistor, k) wattmeter, l) separator, m) sampler, n) suction line, o) pressure gage, p) pressure line, q) separator, r) vacuum pump, s) pressure regulating valve, t) coolant supply, t') coolant return, u) coolant regenerator, v) countercurrent evaporator, w) coolant circulator.

Results are presented in Fig. 11.7, where separation and pressure loss are plotted versus the vapor velocity. For evaluating Eq. (11-22) that vapor velocity employed is correlated to Δp = 95 mm water pressure loss. Fig. 11.7 shows this to equal u_V = 0.24 m/s.

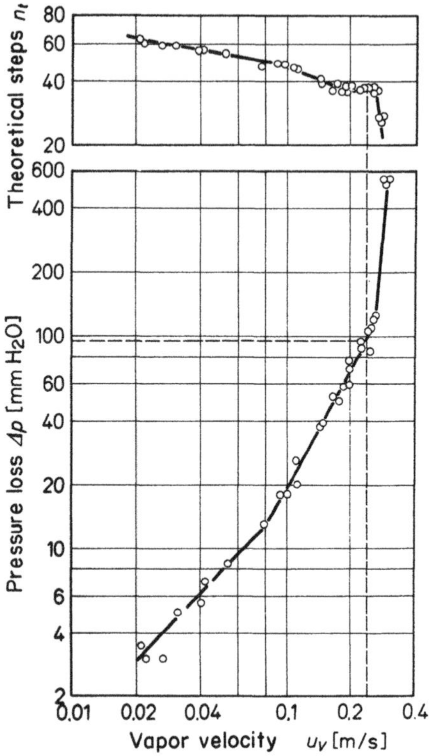

Fig. 11.7 Separation and pressure loss versus velocity at column head. 3 × 3 mm wire net packing. 1-m packing depth. Column diameter 0.1 m. Mixture of *n*-heptane/methyl cyclohexane. Still concentration 18 mol-% of *n*-heptane. Total reflux. 760 mm Hg.

The total weight of the wire web rings is $G = 4.734$ kg, their compact volume $V_F = 0.603 \cdot 10^{-3}$ m^3, and surface $O = 22.05$ m^2. At the vapor velocity of 0.24 m/s the liquid content of the column is $J = 1.535 \cdot 10^{-3}$ m^3. With this and a column volume of $V_s = 7.85 \cdot 10^{-3}$ m^3, Eq. (11-18) yields $\epsilon = 0.922$, disregarding J, and $\epsilon = 0.728$, considering J, and from Eq. (11-19) $e = 32.8 \cdot 10^{-5}$ m, disregarding J, and $e = 25.9 \cdot 10^{-5}$ m, considering liquid volume J.

From the difference of these e-values–note Eq. (11-27), or from the liquid volume and the surface–note Eq. (11-26), a mean film thickness $s_o = 6.9 \cdot 10^{-5}$ m may be derived.

Finally, Eq. (11-22) gives $\xi = 1.34$, and Eq. (11-23) gives

$$u_{Vz} = \left(0.728\right)\left[\left(\frac{9.81}{1.34}\right)\left(\frac{651}{3.31}\right) 6.9 \cdot 10^{-5}\right]^{1/2} = 0.23 \text{ m/s}$$

which is slightly less than the value found by testing.

It should be noted that s_o is actually a mean fictive variable required only to evaluate Eq. (11-23). The true average film thickness is greater, since the effective packing area is less than the assumed one. Other than that, the idealized flow picture presented especially for the contact points of the packing elements cannot be considered in defining s_o.

The separations obtained in Fig. 11.7 for the 100–mm column, per unit packing height are not inferior to those in Fig. 11.3 for the 20–mm column. Use of such elements in columns of semiindustrial size therefore offers no problems and offers possibilities in special separations. The construction heights are moderate, and the pressure losses encountered at maximum loads are below 3 mm water. Drawbacks are the low specific throughput and a relatively high investment.

2. TRANSFERRING DATA GAINED FROM SEMIINDUSTRIAL UNITS TO FULL-SCALE

Having systems of unknown phase equilibrium to investigate, optimum separation conditions are often found by experiment. Analytical methods would also demand much equipment to uncover physical data or equilibria. Then, transferring of answers gotten in pilot plants to production-size installations means a fresh challenge. Ratiocination then may still be straightforward.

A. Conversion of the Packing Height of Columns from Pilot Plant Scale to Production Scale

To gather useful information for the preconstruction planning of full-scale equipment, in many instances 100–mm size columns suffice. Such small-diameter units are mostly packed with standard Raschig rings of 10–mm size. When it comes to laying out an industrial packed column from pilot plant data, the column diameter is conveniently determined from the information presented in Chapter 8. To determine packing height H_T for a full-scale column giving the same separation as the pilot column with H_F (m) of packing, Eq. (8-17) offers a start, and applies to the main velocity range immediately below the maximum.

Application of that equation to both the pilot and the full-scale column (subscripts P and T, respectively), gives a separation

$$n_t = (n_t)_P = (n_t)_T \tag{11-28}$$

and the correlation

$$H_T = 1 + (H_P - 1)\frac{10 + d_T}{10 + d_P} + \xi_{1T}\frac{1 - \xi_{1T}/\xi_{1P}}{\xi_{1T}/\xi_{1P}}\,\frac{10 + d_T}{923} \tag{11-29}$$

The quotient ξ_{1T}/ξ_{1P} of the flow coefficients is plotted in Fig. 11.8 against the ring size $d = d_T$ for the industrial column; the ring diameter of the pilot column being around $d_P = 10$ mm. This curve was found from experiments[9] with Raschig ring column

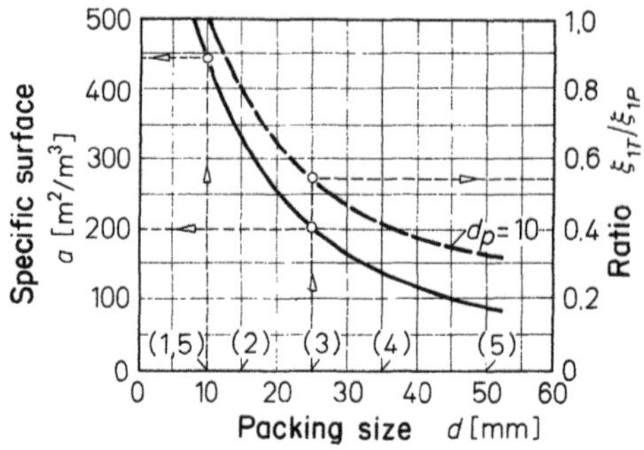

Fig. 11.8 Flow coefficients and specific surfaces for ceramic Raschig rings.

at optimum load. The flow coefficient of the full-scale column packed with Raschig rings of size $d = d_T$ mm) is, by Eq. (8-16)

$$\xi_{1T} = \frac{2g}{u_V^2 \rho_V} \frac{\Delta p}{H} d_T \, 10^{-3} \tag{11-30}$$

where $\Delta p/H$ (mm water) = pressure loss of the vapor flowing at velocity u_V (m/s) of a specific weight ρ_V (kg/m^3) along a packed layer 1-m deep.

The conversion of height H_P of the pilot column to height H_T of the equivalent industrial column as regards separation is thus possible. However, the packing depth thus obtained must be corrected for diameter ratio and for maldistribution according to Fig. 8.2.

Following another source,[10] the formula

$$H_T = H_P \left(\frac{a_P}{a_T}\right)^{1.3} \left(\frac{\epsilon_T}{\epsilon_P}\right)^{1.3} \tag{11-31}$$

gives the correlation of packing depth H_P of a pilot column and packing depth H_T of an industrial column of equivalent separation, both functioning in an optimum load

Table 11.2

Specifications of ceramic Raschig rings

Ring Dimensions		Specific Surface	Free Volume
d (mm)	Thickness (mm)	a (m²/m³)	ϵ (m³/m³)
10	1.5	440	0.70
15	2	330	0.70
25	3	200	0.73
35	4	140	0.78
50	5	90	0.785

range, where a (m^2/m^3) is the specific packing surface, given in Fig. 11.8 for packing size d (mm), and ϵ is the free volume for which the values of Table 11.2 may be used (m^3/m^3). Subscripts P and T again refer to the pilot and the actual column.

Example 1 Tests on a pilot column of 100–mm diameter established that optimum overhead and bottom compositions are achieved by separation of a mixture of unknown phase equilibrium in a stripper packed with ceramic Raschig rings of 10–mm size at optimum load range and for a liquid/vapor ratio of two.
How tall should a full-scale column be if packed with 25–mm ceramic Raschig rings? The specific weight of the liquid is 800 kg/m^3 and that of the vapor is 3 kg/m^3. Packing depth of the pilot column equals 1.2 m.

Answer From Eq. (8-12) and Fig. (8.6)

$$(u_V)_{o\,L/V=1} = 4.55\cdot 10^{-2}\left(\frac{800}{3}\right)^{1/2} = 0.73 \text{ m/s}$$

and with this, by Eq. (8-13) the vapor velocity at top load

$$u_V = (u_V)_{o,\ L/V=2} = \frac{0.73}{2^{0.3}} = 0.58 \text{ m/s}$$

which, by Table 8.3, indicates a pressure loss around $\Delta p/H = (\Delta p/H)_{o1}$ = 140 mm water. With this, from Eq. (11-30)

$$\xi_{1T} = \frac{2g}{(0.58)^2\ (3)}\ 140\cdot 25\cdot 10^{-3} = 68$$

and from Eq. (11–19), with $\xi_{1T}/\xi_{1P} = 0.54$, from Eq. (11–31) gives,

$$H_T = 1 + (1.2 - 1)\frac{10 + 25}{10 + 10} + 68\ \frac{1 - 0.54}{0.54}\cdot\frac{10 + 25}{923} = 3.54 \text{ m}$$

with the values picked from Table 11.2 and Fig. 11.8

$$H_T = 1.2\left(\frac{440}{200}\right)^{1.3}\left(\frac{0.73}{0.7}\right)^{1.3} = 3.53 \text{ m}$$

The results obtained from two different equations are surprising close, which is not necessarily typical, however.

Maintaining this packing depth in the full-scale tower should achieve the same separation as that of the pilot plant, provided that in both cases the liquid is distributed uniformly, that vapor/liquid ratios are identical, and that the relative column loads are the same. Both methods of calculation are based on optimum operation, which is immediately below the maximum admissible vapor velocity. For lower vapor velocities the packing depths must be suitably increased.

B. Theoretical Number of Steps, and Operating Conditions of a Column to Decompose a Multiple Mixture of Unknown Phase Equilibrium

Suppose that a pilot rectifier has provided data on the optimum operating conditions and the necessary packing depth to obtain separation of the mixture of unknown boiling performance into overhead and bottom products of proper composition. Then the preconstruction planning of an industrial-scale plate-type column demands knowledge of the theoretical number of steps $n_t = (n_t)_P$ equivalent to the pilot column of packing depth H_P, for the rectifying portion and for the stripping portion, in view of Eq. (11-28).

Since, however, the phase equilibrium is still unknown, the theoretical steps n_t cannot be found from any data on feed or products; such figures as flow conditions (pressure loss and velocities) of vapor of a given specific weight must be put to use. That kind of correlation can be made for a pilot column packed with ceramic Raschig rings when, in Eq. (8-17) the flow factor ξ_1 is replaced with the term of Eq. (8-16), and the relation solved for n_1. Then, with $H = H_P$ and $d = d_P$

$$n_t = (n_t)_P = 25.5 \cdot 10^{-4} \frac{\Delta p}{H_P} \frac{d_P}{u_V^* \rho_V} + 120 \frac{H_P - 1}{10 + d_P} \tag{11-32}$$

In the event that the pilot column is not packed but filled with small transfer plates of suitable design, one must find the equivalent theoretical step numbers for their optimum operating conditions and actual number of plates $(n_p)_P$

$$n_t = (n_t)_P = (E_c)_P (n_p)_P \tag{11-33}$$

for the rectifying and the stripping function, from the overall separation effects $(E_C)_P$ of the experimental plates. To this end, one can start from the equation[11]

$$(E_C)_P = \frac{\log \left(\frac{\Delta x}{100 - C_2 \Delta x}\right)_o + \log \left(\frac{100 - C_2 \Delta x}{\Delta x}\right)_u}{(n_p)_P \log \left[\frac{\alpha}{\left(\frac{L}{V}\right)\left(\frac{C_1}{100}\right)^2}\right]} \tag{11-34}$$

which must be evaluated for a binary of comparable physical properties, of constant relative volatility, and constant ratio L/V of the molar liquid and vapor quantities for the conditions of the production material. In Eq. (11-34)

$$\Delta x = x - (x_i)_E \tag{11-35}$$

and the difference between liquid concentration x (mol-%) and composition $(x_i)_E$ (mol-%) at the intersection of interchange line and equilibrium curve

$$C_1 = 100 + (\alpha - 1)(x_i)_E \tag{11-36}$$

a constant, and finally

$$C_2 = \frac{C_1\left(\frac{L}{V}\right)(\alpha - 1)}{100\left[\alpha - \left(\frac{L}{V}\right)\left(\frac{C_1}{100}\right)^2\right]} \tag{11-37}$$

a constant limited as noted earlier. Subscripts "o" and "u" refer to the topside or underside of the column section in question.

Under certain conditions, the separations thus obtained are identical for the binary and for the multiple mixture. The circumstances for which deviations must be expected have been mentioned for a ternary.[12]

When the tests are run at total reflux, in agreement with full-scale operation, Eq. (11-34) becomes

$$(E_C)_P = \frac{\log\left(\frac{100 - x}{x}\right)_u + \log\left(\frac{x}{100 - x}\right)_o}{(n_p)_P \log\alpha} \tag{11.38}$$

and the overall separation of the pilot column is calculated from the measured bottom composition and overhead composition at constant relative volatility α and known actual total number of plates $(n_p)_P$: $(E_C)_P$.

Eq. (11-34) and (11-38) are also useful when samples are taken from different plates, since a linear interrelation exists between the plate number in question and the logarithm of the concentration terms of these equations; the experimental results are correlated in a plot of the line, which is two-dimensional

$$\log\left(\frac{\Delta x}{100 - C_2\Delta x}\right) = f(n_p) \tag{11-39}$$

Eventually, a modified mean relative volatility is determined for a system of unknown phase equilibrium by means of Eq. (11-38). The system is rectified at total reflux in a test column of known theoretical number of stages $n_t = (E_C)_P\,(n_p)_P$. By the measured mol-% x of the component at the top and bottom of the column, α follows from Eq. (11-38). If the concentration outline along the column height is traced, $\alpha = f(x)$ is determined by Eq. (11-38). Thus, for example, the boiling characteristic of a key component of a multiple mixture can be predicted within certain limits, offering a ready means for calculating a rectifying column.

Example 2 Find the optimum conditions for separation in a projected plant for stripping cyclohexanone/cyclohexanol from an acid mixture.

The mean composition of the mixture is approximately: 2.0% by weight of cyclohexanone/cyclohexanol, 68.07% by weight of water, 21.78% by weight of residual acid, and 8.25% by weight of adipic acid.

Since cyclohexanone/cyclohexanol forms a partially insoluble mixture with acidic water, the reflux formed in the reflux condenser is recovered in two layers by a separator. The top layer is mostly cyclohexanone/cyclohexanol obtained as the overhead, and the bottom layer, mostly water, is recycled to the column as reflux.

Column dimensions and operation conditions should be specified for the least heat consumption and thus the minimum heat exchanging area in the reboiler and the reflux condenser, and the minimum column diameter. There is no information on the distillation equilibrium of this multiple mixture; therefore an analytical solution of this problem is not feasible, and experimentation must be resorted to.

Answer Fig. 11.9 is a diagram of the test equipment. There is the Robert-type reboiler a, a packed column assembled from five sections b_1 through b_5 of 100-mm diameter. Sections b_1 to b_4 have packing depths of 1.3m, and section b_5 has 0.8-m packing depth. The packing consists of 8-mm Raschig rings of V2A stainless. The total packing depth is 6 m. The lower end of each section has a reflux collector that recycles the liquid centrally, eliminating blowby to some extent. External heating elements c_1 to c_5 can be adjusted according to changes in boiling point. Top column section c_6 is empty but likewise heated, and leads to reflux condenser d.

From feed tank e, stripping acid first flows to a steam heated preheater f, passes meter g, and enters the column at one of the three feedpoints h_1, h_2, h_3. In that way, the column may operate, either as a combined stripper and rectifier with feedpoints h_1 or h_2, or as a straight stripper with feedpoint h_3. The mixed vapors issuing from the steam heated reboiler enter the column where the material transfer with the reflux liquid takes place, and condense in the reflux condenser. The condensate enters the separator i, from which the lower phase rich in water passes orifice j and is recycled as reflux to the column head, while the upper phase rich in cyclohexanone/cyclohexanol is removed by pump k_1. The top-of-column vacuum measured at I is maintained by pump m. Pump k_2 moves the runoff acid from the reboiler to cooler o_1.

The calibrated tanks p_1 and p_2 receive the overhead and the bottom product, and p_3 receives the steam condensate cooled in cooler o_2. Gages q, q_1 to q_6 indicate heating steam pressures. Sightglasses r_1 and r_2 also permit taking liquid samples of mixed vapor condensate ahead of or after the separator. Mercury gage s indicates the total pressure loss through packing. Thermometer t_1 indicates the inlet temperature, and t_2 the exit temperature, t_3 the head temperature.

To find the optimum separation of cyclohexane/cyclohexanol from the stripping acid, the reboiler steam load must be varied, with constant feed rate,

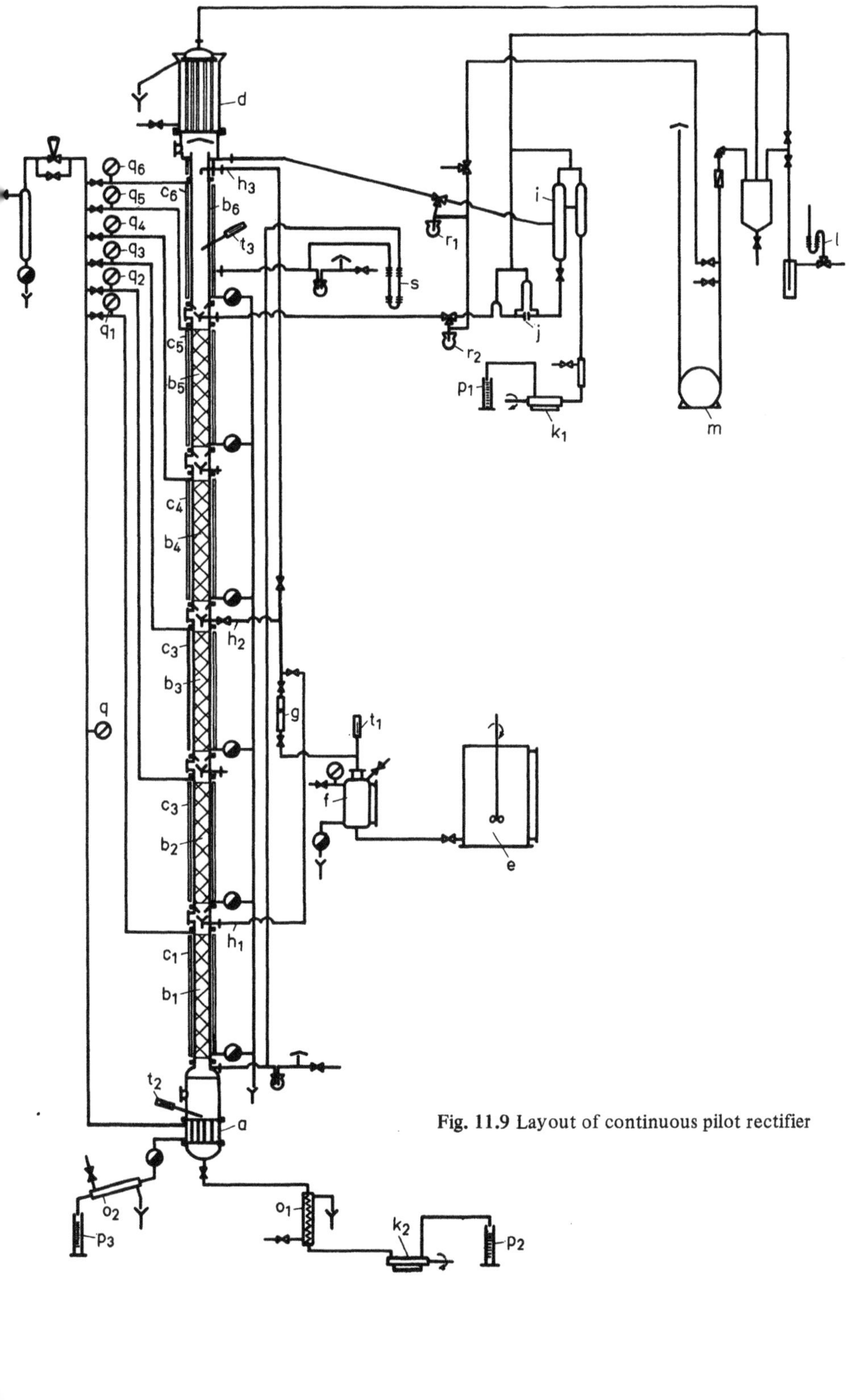

Fig. 11.9 Layout of continuous pilot rectifier

h_1, h_2, or h_3 in Fig. 11.9, and the ensuing variables: the feed quantities and temperatures, the heating steam condensate quantities, the heating steam pressures, the quantities of overhead product and reflux, the overhead and bottom temperatures, and the pressure loss of mixed vapors across the column, are then noted.

Reflux ratio r as the ratio of the reflux quantity $L = R$ (kmol/h) and overhead product quantity D (kmol/h) may be measured via quantities or via concentrations. If y_D equals the concentration of cyclohexanone/cyclohexanol in mixed vapor liquefied in the condenser V, x_V equals the concentration of the total mixed vapor condensate, x_R equals the cyclohexanone/cyclohexanol concentration in the lower liquid layer in the separator or reflux R, and x_D = the cyclohexanone/cyclohexanol concentration of the top layer or overhead product (kmol/h), the mass balance becomes

$$r = \frac{x_D - x_V}{x_V - x_R} \tag{11-40}$$

from which the reflux ratio can be calculated. If a sample of the mixed vapor condensate entering the separator is analyzed volumetrically, at steady conditions the result of Eq. (11-40) must also be identical with the ratio of mol quantity G_R/μ_R of the volume of the lower layer and mol quantity G_D/μ_D of the top layer of a sample.

$$r = \frac{\dfrac{G_R}{\mu_R}}{\dfrac{G_D}{\mu_D}} \tag{11-41}$$

where G = weights (kg/h) and μ = molecular weights (kg/kmol).

The overhead product quantity is by Eq. (3-16)

$$D = \frac{R}{r} \tag{11-42}$$

The runoff concentration of cyclohexanone/cyclohexanol for a feed quantity F (kmol/h) is by Eq. (3-23)

$$x_W = \frac{x_F - \dfrac{D}{F} x_D}{1 - \dfrac{D}{F}} \tag{11-43}$$

The quantities may be checked by these correlations. When using weight units, the molar quantities D (kmol/h) and F (kmol/h) must be multiplied by the proper molecular weights, μ_D and μ_F (kg/kmol).

Tests run at reduced pressure do not furnish the expected enrichment of cyclohexanone/cyclohexanol in the overhead product. This is even remote from the result calculated for total insolubility of the cyclohexanone/cyclohexanol in water, figuring from vapor pressure curves. In theory, distilling vapor with roughly equal parts of water and cyclohexanone/cyclohexanol should be generated if total insolubility is assumed. In fact, the tests supply overhead vapor concentrations around ten, up to 15% of cyclohexanone/cyclohexanol in the mixed vapor. It should be remembered that cyclohexanone/cyclohexanol is much more soluble in the stripping acid than in pure water. Partial solubility of mixtures having a minimum boiling point increases the share of the more volatile component in the vapor of the azeotropic point.

On theoretical grounds, the azeotropic point for a supposedly insoluble mixture must shift slightly toward the side of the pair of components cyclohexanone/cyclohexanol. Therefore, investigations were extended over the range of ordinary pressures. Of the three feedpoints available, the one with a packing depth of 2.1 m in the enriching section was most favorable, and the one with a 3.9 m depth was favored in the stripping section.

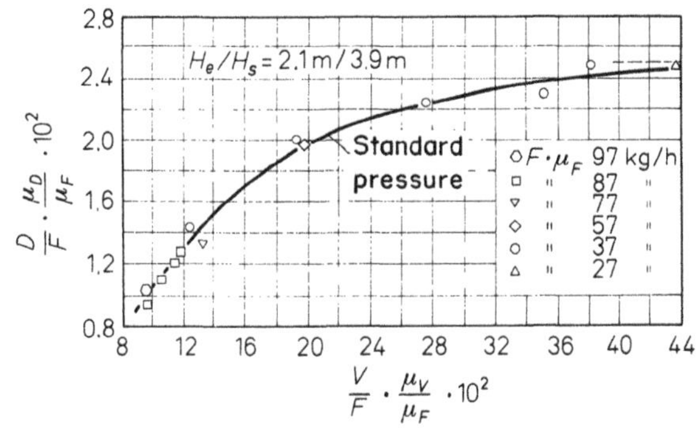

Fig. 11.10 Example 2. Overhead product versus vapor quantity (both referred to feed quantity). Data from pilot installation in Fig. 11.9.

Fig. 11.10 presents the measured values $D\mu_D/F\mu_F$ as a yardstick for the recovery of cyclohexanone/cyclohexanol versus the values $V\mu_V/F\mu_F$ as a yardstick for the vapor load of the rectifying column. The boundary value of the maximum obtainable cyclohexanone/cyclohexanol recovery is situated at ordinate $2.5 \cdot 10^{-2}$. That is confirmed by the relation resulting from Eq. (11-43) for $x_W = 0$

$$\frac{D_{max}}{F} = \frac{x_F}{x_D} \tag{11-44}$$

Evaluating this in terms of weight units, knowing the share of 80% of cyclohexanone/cyclohexanol in the top layer and therefore in the overhead product

$$\frac{D_{max}}{F} \frac{\mu_D}{\mu_F} = \frac{2}{80} = 2.5 \cdot 10^{-2}$$

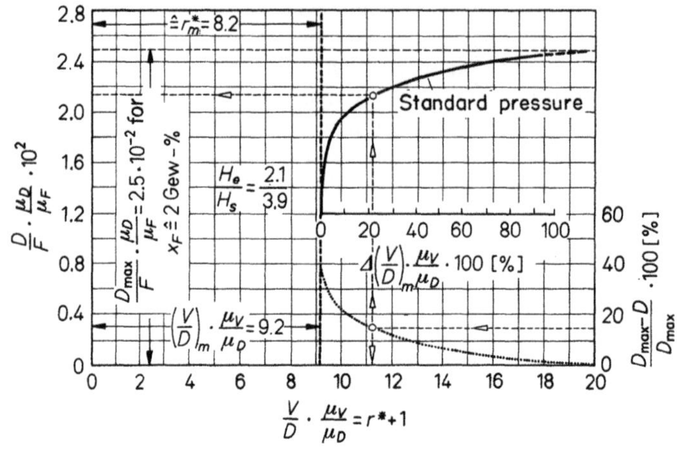

Fig. 11.11 Quantity of overhead product related to feed quantity and percentual loss of overhead product versus vapor load at column top, referred to overhead product. Data from Example 2 and pilot plant Fig. 11.9.

In connection with the test results of Fig. 11.10 it should be noted that the shape of the curve is also influenced by the load characteristic of the test column; the relative import of the number of theoretical steps as an index of the separation by a packed column is strongly affected by the prevailing load.

Starting from the curve in Fig. 11.10, the curves in Fig. 11.11 have been drawn, where coordinated abscissa and ordinate values of the former plot have been entered versus the V/D ratio derived from Eq. (3-40)

$$\frac{V}{D} = r + 1 = \frac{\mu_D}{\mu_V}(r^* + 1) \tag{11-45}$$

where r^* equals the reflux ratio at the column head expressed in weight units.

Over this, again the mass of top product is plotted, note the full line, as well as the loss percentage l_D (%)

$$l_D = \frac{D_{max} - D}{D_{max}} 100 \tag{11-46}$$

of cyclohexanone/cyclohexanol; note the stippled curve. The lines tend to show that a certain minimum ratio of mixed vapor at the top of the column and top product quantity must be maintained, or too much cyclohexanone/cyclohexanol is lost. That ratio approximately equals

$$\left(\frac{V}{D}\right)_m \frac{\mu_V}{\mu_D} = 9.2 = r_m^* + 1, \text{ or } r_m^* = 8.2$$

The abscissa scale inserted into the body of the diagram indicates the percentage by which the $V\mu_V/D\mu_D$ exceeds the $(V/D)_m$ (μ_V/μ_D). Knowing the cost of the operation and the investment, and the cost of the overhead product, the optimum operating point for rectification at ordinary pressures can be located.

Now the theoretical steps are determined for the column that gives the separation of the pilot column for the operating point given in Fig. 11.11. The percentage loss of the top product is

$$\frac{D_{max} - D}{D_{max}} 100\% = 15\%$$

The corresponding top product quantity per kilogram feed is

$$\frac{D}{F} \frac{\mu_D}{\mu_F} = 2.14 \cdot 10^{-2}$$

which is about 22% more than the minimum of 9.2. Thus for a feed quantity in the pilot plant of

$$F\mu_F = 37 \text{ kg/h}$$

the result will be

$$D\mu_D = 2.14 \cdot 10^{-2} \ (37) = 0.792 \text{ kg/h}$$

and

$$V\mu_V = (11.25) \ (0.792) = 8.92 \text{ kg/h}$$

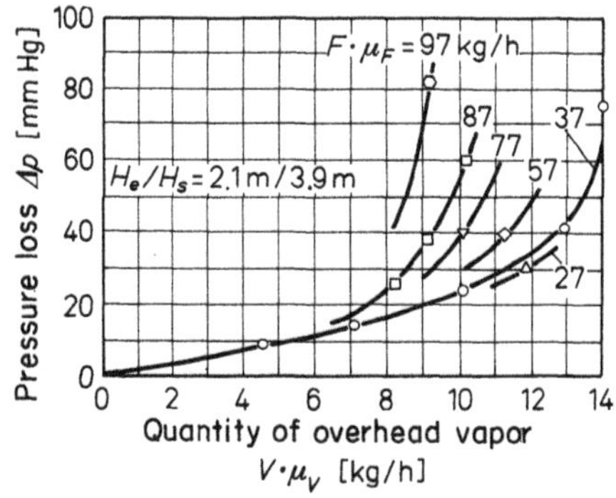

Fig. 11.12 Example 2 and pilot plant data of Fig. 11.9. Pressure loss versus vapor quantity overhead.

In Fig. 11.12 the pressure loss found in tests at normal pressure is plotted versus the vapor quantity at the column top for various feed quantities. It gives

$$\Delta p = 19 \text{ mm Hg}$$

converted to 1 m of packing depth

$$\frac{\Delta p}{H} = \frac{19}{6}(13.6) = 43 \text{ mm water}$$

For improved precision, the pressure losses in the stripping and rectifying parts would have to be measured separately, and evaluated.

Assuming a specific weight of the vapor mixture of about

$$\rho_V = 0.6 \text{ kg/m}^3$$

the equation gives

$$u_V = \frac{1}{3600} \frac{V\mu_V}{\rho_V d_S^2 \dfrac{\pi}{4}} \tag{11-47}$$

or a vapor velocity of

$$u_V = 0.523 \text{ m/s}$$

Eq. (11-32) will supply the equivalent theoretical steps, for $d_P = 8$ mm rings, and a packing depth $H_P = H_e = 0.8 \text{ m} + 1.3 \text{ m}$ in the enriching part

$$(n_t)_P = n_{te} = 11.4$$

As the liquid is fed centrally, a maldistribution factor of around $u = 0.4$ is expected, and from Fig. 8.2, for $\phi' = 12.5$ a column efficiency around $\eta = 70\%$ is assumed. Then the equivalent number of steps for separation in the enriching part of the column becomes

$$(n_t)_P = n_{te} = (0.7)(11.4) = 8$$

and per meter of packing height

$$\frac{n_{te}}{H_e} = \frac{8}{2.1} = 3.8$$

The vapor quantity in the top of the stripper becomes, from Eq. (3-44), for a feed temperature of 80 C

$$V\mu_V = 9.92 \text{ kg/h}$$

and the corresponding vapor velocity is, by Eq. (11-47)

$$u_V = 0.58 \text{ m/s}$$

For this, Eq. (11-32) gives, as the number of theoretical steps required and equivalent to the packing height $H_P = H_s = (3)\,(1.3)$m

$$(n_t)_P = n_{ts} = 19$$

Because of the high specific liquid load, this figure must be corrected, regardless of the central liquid delivery in the three layers, by a factor of approximately 0.9

$$(n_t)_P = n_{ts} = (0.9)\,(19) = 17$$

With this, the theoretical steps per 1 m of packing depth become

$$\frac{n_{ts}}{H_s} = \frac{17}{3.9} = 4.4$$

The calculated theoretical number of steps of the pilot column becomes

$$(n_t)_{PT} = n_{te} + n_{ts} = 25$$

From this, the actual number of plates for an industrial-scale column for identical separation is obtained.

Example 3 A pilot column equipped with $(n_p)_P = 38$ rectifying plates has treated a binary test system having a mean relative volatility α equal to 1.26. The liquid/vapor ratio L/V was 0.99, the share of low boiler in the overhead product $x_D = 93.6$ mol-%, and in the runoff of the lowermost enriching plate $x_i = 4.66$ mol-%. The composition at the intersection of the enriching line and equilibrium curve is $(x_i)_E = 2.5$ mol-% of low boiler. What is the mean overall separation efficiency $(E_C)_P$?

Answer Eq. (11-35) yields for $x = x_D$

$$\Delta x = (\Delta x)_o = 93.6 - 2.5 = 91.1 \text{ mol-\%, and for } x = x_i$$

$$\Delta x = (\Delta x)_u = 4.66 - 2.5 = 2.16 \text{ mol-\%}$$

From Eq. (11-36)

$$C_1 = 100 + (1.26 - 1)\,2.5 = 100.65$$

From Eq. (11-37)

$$C_2 = \frac{(100.65)(0.99)(1.26-1)}{100\left[1.26 - 0.99\left(\frac{100.65}{100}\right)^2\right]} = 1.0077$$

From Eq. (11-34)

$$(E_C)_P = \frac{\log \dfrac{91.1}{100-(1.0077)(91.1)} + \log \dfrac{100-(1.0077)(2.16)}{2.16}}{38 \log \left[\dfrac{1.26}{0.99\left(\dfrac{100.65}{100}\right)^2}\right]} = 0.72$$

By Eq. (11-33) the separation of the plates in the pilot column is identical with that of

$$n_t = (n_t)_P = (0.72)(38) = 27$$

theoretical steps.

Distillation in laboratory and pilot type columns is not limited to experimentation in search of suitable separation criteria for systems that are hard to compute for lack of data. Rather, it will frequently serve to verify results based on computation from previously unconfirmed equilibrium and vapor pressure information.

Other than the devices discussed in this chapter, there are numerous other pieces of equipment for performing reduced-scale distillations.[13,14] Special mention should be made of apparatus for measuring the phase equilibria of vapors and liquids.[13,15,16]

XII Distillation in Fine and High Vacuum

1. MOLECULAR DISTILLATION

The advantages offered by the thin-layer distillation technique are mentioned in Chapter 9, regarding the recovery and refining of thermally unstable materials. Now molecular distillation is understood as a special case of thin-layer treatment, involving the least disruptive of thermal separations, and hence favored in the pharmaceutical industries.

The basic principle in a molecular distillation layout is that the evaporating and the condensing surface are located in a single equipment unit, at a distance of a few centimeters only, the mean free path length l_m (cm) of the molecules, which is computed from

$$l_m = 2.3 \cdot 10^{-20} \frac{T}{p} \frac{1}{r_M^2} \qquad (12\text{-}1)$$

assuming ideal behavior of the molecules of vapor whose radius is r_m (cm). The distillate quantity D' (kg/m^2 h) recoverable in theory per unit of vaporizing surface follows from[2]

$$D' = 2100\, p \left(\frac{\mu_D}{T} \right)^{1/2} \qquad (12\text{-}2)$$

where: p = pressure (mm Hg), T = absolute temperature (K), and μ_D = molecular weight of distillate (kmol/kg).

Earlier types of molecular stills had a tapering vaporizer surface on which the distillate introduced from below was spread out by centrifugal force, moving up in a thin layer.[3] Recent designs embody the principle discussed in Chapter 9 which deals with a falling-film vaporizer with rotating elements. Both types have internal condensing surfaces (Fig. 12.1).

Laboratory-scale molecular stills are made for 0.2 to 1 kg/h throughput. The viscosity of the liquid at boiling temperature may be up to 60,000 cP.

The brevity of the holdover time is a special advantage of this equipment. It may be calculated from the mean film thickness and the vaporizing surface for the capacity intended. Although diameters are limited, as is the construction height, molecular

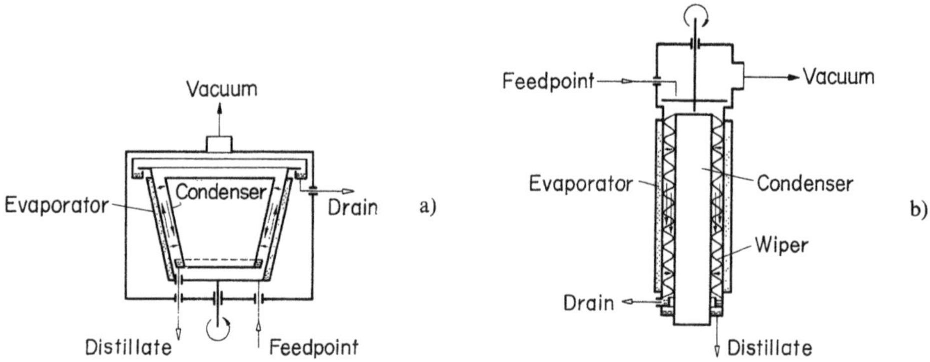

Fig. 12.1 Molecular stills, schematic. a) spinning taper and ascending film. b) free-falling film mechanically conveyed.

stills using the falling-film principle equipped with rotating elements may have throughputs of 1 ton/h and better.[4]

Along with thermal influences, catalytic effects may modify sensitive molecules; therefore, molecular stills should be built using catalytically neutral materials only.[5]

The format of distillation calculations in molecular distillation is identical with that in falling-film equipment with revolving elements as outlined in Chapter 9 for thin-layer distillation.

2. THIN-FILM DISTILLATION

One very gentle distillation method is characterized by extreme thinness of the liquid film.[8,9] An industrial version of the Liprotherm evaporator using centrifugal action is suitable for satisfactory flow rates (Fig. 12.2).

As in the molecular unit type a) in Fig. 12.1, the rotor is a reversed truncated cone heated externally.[10] The feed enters from below and is spread uniformly by the centrifugal forces, producing a thin film that partly evaporates, and partly overflows at the top end where it collects. The film thickness, a function of rotor speed and feed rate, does not exceed an average of 0.1 mm under usual conditions.

Now in the case of pseudo-laminar flow of boiling liquids, minimum thickness is essential to good heat transfer. On the heated side of the rotor, rotation promotes good heat transfer. Heating steam when condensing will form droplets whose size will diminish as the rotor speed increases. Since the condensate droplets are thrown off right after being formed, the heat transfer resistance is much less than in conventional systems. Thus good transfer conditions exist on both sides of the transmitting wall, and the transfer rates achieved are two to four times those of conventional falling-film evaporators.

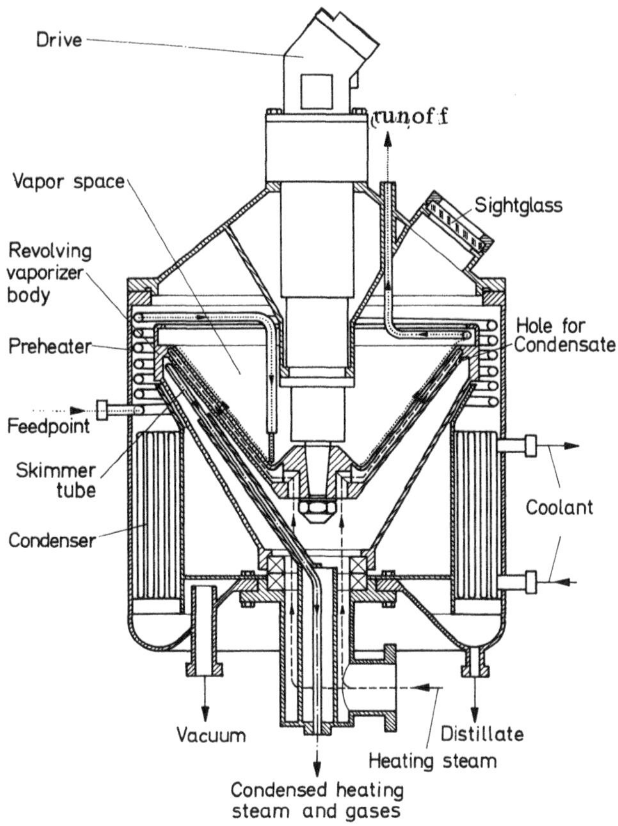

Fig. 12.2 Thin-film evaporator, Liprotherm system, made by Sibtec of Stockholm, Sweden.

Main design characteristics of these evaporators are thus a relatively small exchange area, minimum film thickness, and low holdover time. This allows handling of thermally unstable mixtures and cuts down the processing time. Also, startup and resetting for new variables are made easier, and high viscosity stock can be worked. The Liprotherm handles stocks of a viscosity that would preclude their processing in conventional falling-film equipment. With optimum heat transfer, the evaporator becomes relatively compact. The setup of Fig. 12.2 accommodates a preheater and a vapor condenser. The flow path of the vapor between the vaporizer and condenser is more extended than that in the molecular distillation equipment of Fig. 12.1, but less so than for separate vaporizers and condensers as in Fig. 9.16. Thus the Liprotherm fills the gap between conventional falling-film evaporators and molecular stills, and it is useful in cases where presently a separation could be obtained only through molecular stills.

3. MECHANISM OF SEPARATION

The mechanism of the separation by partial distillation in equipment of this category is understood from the general theory of thin-layer distillation.[9] Referring to the mass-concentration plot in Fig. 12.3, and the general scheme of equipment featuring tapered vaporizers (Figs. 12.1 a) and 12.2) let: F = the mass of feedstock, x_F = the concentration of the low boiler in feedstock, D = the mass of distillate, and x_D = the concentration of the low boiler in the distillate.

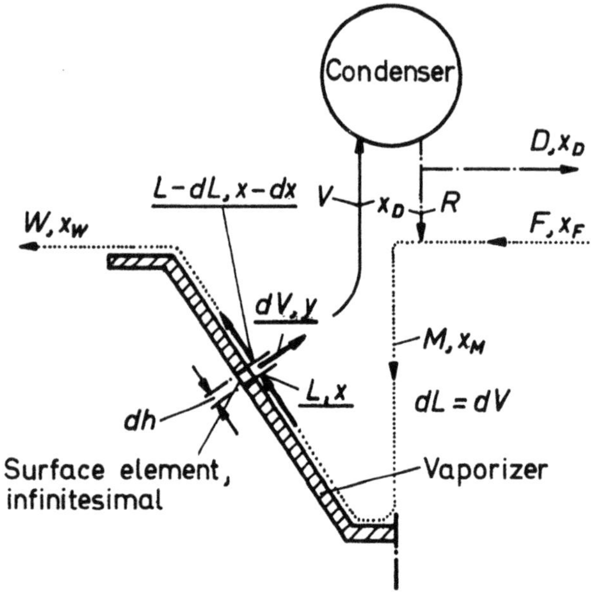

Fig. 12.3 Masses and concentrations in thin-film distillation.

Usually the procedure is one of continuous distillation and once-through operation. Occasionally a portion R of the distillate is recycled and added to the feed in order to increase the output and improve the separation, provided that the thermal stability of the processed material is adequate.

Consider an infinitesimal path element of the radial flow component on the vaporizer area of length dh. Then the mass of vapor escaping along dh from liquid quantity L having concentration x is $dV = dL$, of concentration y, yielding

$$\frac{dL}{L} = \frac{dx}{y - x} \tag{12-3}$$

Integration between limits of $L = M$ to $L = W$ and of $x = x_M$ to $x = x_W$ gives

$$M = W \exp \int_{x_W}^{x_M} \frac{dx}{y - x} \tag{12-4}$$

The ratio of runoff quantity and feedstock quantity

$$w = \frac{W}{M} = \frac{W}{F + R} \tag{12-5}$$

and the vaporization quotient

$$v = \frac{V}{M} = \frac{D + R}{F + R} \tag{12-6}$$

are interrelated by

$$v = 1 - w \tag{12-7}$$

Thus the sum of the distillate and reflux quantities corresponding to the stream flowing from the vaporizer to the condenser equals

$$V = D + R = (1 - w)M = (1 - w)(F + R) \tag{12-8}$$

It can be evaluated when the term for the quantity ratio is known. This will follow from Eqs. (12-4) and (12-5)

$$w = \frac{1}{\exp \displaystyle\int_{x_W}^{x_M} \frac{dx}{y - x}} \tag{12-9}$$

For the mixture

$$M = F + R \tag{12-10}$$

where F equals the feedstock quantity having concentration x_F of the low boiler and R = the reflux quantity having the concentration $x_R = x_D$ of the low boiler. From this

$$x_M = \frac{Fx_F + Rx_D}{F + R} \tag{12-11}$$

By introducing the reflux ratio

$$r = \frac{R}{D} \tag{12-12}$$

and the distillate ratio

$$d = \frac{D}{F} \tag{12-13}$$

there results the "distillation characteristic"

$$x_M = \frac{x_F + rdx_D}{1 + rd} \tag{12-14}$$

When total condensation occurs in the condenser (as assumed), distillate concentration x_D is identical with the vapor concentration y_D at the condenser inlet. Since

$$\frac{V}{D} = r + 1 \tag{12-15}$$

the reflux ratio may also be regarded as the multiple vaporized in order to obtain the distillate quantity D, over and above that required in simple once-through distillation.

For a binary system of virtually constant relative volatility α within the concentration range in question, Eq. (12-9) may be integrated, assuming that the vapor formed on an infinitesimal surface element of a concentric periphery of the vaporizing area has the phase equilibrium composition of the liquid of the same element. Then there exists the relationship

$$x = \frac{y}{y + \alpha\,(100 - y)}\,(100) \tag{12-16}$$

between the liquid of composition x and the liberated vapor of concentration y. When this is inserted into Eq. (12-9), integration will supply a working equation for

$$w = \frac{1}{\left(\dfrac{x_M}{x_W}\right)^{\frac{1}{\alpha - 1}} \left(\dfrac{100 - x_W}{100 - x_M}\right)^{\frac{\alpha}{\alpha - 1}}} \tag{12-17}$$

With no refluxing ($R = 0, r = 0$), $x_M = x_F$, and Eq. (12-5) gives the surficial mass ratio

$$(w)_{R=0} = \frac{W}{F} = m \tag{12-18}$$

of runoff quantity and feed quantity. However

$$\frac{m}{W} = \frac{F + R}{F} \tag{12-19}$$

and

$$D = (1 - m)F \tag{12-20}$$

or

$$d = 1 - m \tag{12-21}$$

The surficial mass ratio is given for the problem in hand. The quantity ratio concerning reflux operating is

$$w = m \frac{1}{(1 + rd)} = m \frac{1}{1 + r(1 - m)} \tag{12-22}$$

Then Eqs. (12-12) and (12-13) in combination with Eq. (12-22) and the reflux ratio as the distillation parameter gives the relation

$$r = \frac{1}{1 - m}\left(\frac{m}{w} - 1\right) \tag{12-23}$$

Finally mass and concentration balances yield the distillation concentration $x_D = y_D$, both as a function of the quantity ratio, w

$$x_D = \frac{x_m - wx_w}{1 - w} \tag{12-24}$$

or as a function of the surficial mass ratio m

$$x_D = \frac{x_F - mx_w}{1 - m} \tag{12-25}$$

and for m there follows

$$m = \frac{x_D - x_F}{x_D - x_w} \tag{12-26}$$

The significant data for a separation by distillation can now be found in the following sequence: for an assumed $x_M = x_F$, find w by Eq. (12-9) or (12-17). Find x_D by Eq. (12-24) and v by Eq. (12-7), then m and d by Eqs. (12-26) or (12-21), and finally r by Eq. (12-23). In this manner, any number of operating conditions characterized by w, x_D, v, m, d, and r may be determined for any pair of variables comprising a given x_F and specified x_W. The characteristic concentrations for distillations with and without refluxing have been presented in Fig. 12.4 for a binary mixture, in a qualitative way.

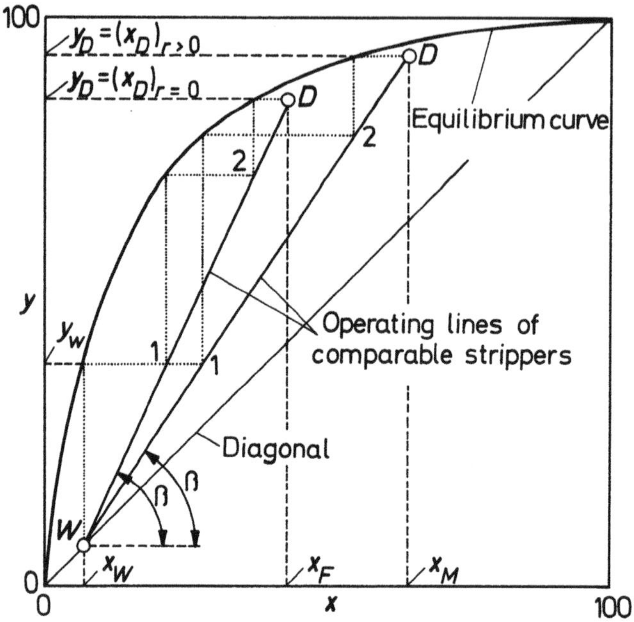

Fig. 12.4 Thin-film distillation of a binary mixture. Characteristic concentrations in y/x plot for thin-film distillation and theoretical steps in stripper provided with a recirculating vaporizer type reboiler of equivalent separation.

The final distillate concentration $y_D = x_D$ always exceeds the equilibrium vapor concentration y_W for the runoff concentration x_W such as it can be obtained, theoretically, in the revolving evaporator of a stripper column; therefore one would be tempted to compare the separative effect of the Liprotherm design with that of a circulating evaporator plus a number of stripping stages. In a juxtaposition to the operation of the Liprotherm without refluxing that rectifier would have to be charged with the equivalent liquid/vapor ratio, corresponding to a slope of the operating line of

$$\tan\beta = \frac{L}{V} = \frac{F}{D} = \frac{1}{1-m} = \frac{(x_D)_{r=0} - x_W}{x_F - x_D} \tag{12-27}$$

In a juxtaposition to operation with reflux with an equivalent liquid/vapor ratio the corresponding slope is

$$\tan\beta = \frac{L}{V} = \frac{F+R}{D+R} = \frac{\dfrac{1}{1-m} + r}{1-r} = \frac{(x_D)_{r>0} - x_W}{x_F - x_D} \tag{12-28}$$

The number of theoretical steps for a given separation, then, is identical with the number of corners, 1, 2, etc., between points W and D, of the sequence between the operating line and the equilibrium line.

To what extent reflux operation is economical must be determined in each individual instance. Part of the load on the evaporating surface is due to the reflux; for a given specific liquid load this reduces the absolute output. Also the energy cost will increase for increasing reflux ratios.

Figs. 12.5–12.8 render the quantitative results[11] of an evaluation of the above system of equations. It was found expedient to represent the separation by the difference

$$\Delta x = x_D - x_W \qquad (12\text{-}29)$$

between the concentrations of low boiler, x_D in the distillate, and x_W in the runoff.

As an example, Fig. 12.5 gives the relation

$$(\Delta x)_{\alpha,\, x_F = \text{constant}} = f(r, \text{parameter } m) \qquad (12\text{-}30)$$

where the influence of the reflux ratio r is obvious. The outlines at different mass ratios m reveal that for each system optimum separation is obtained for different operating conditions.

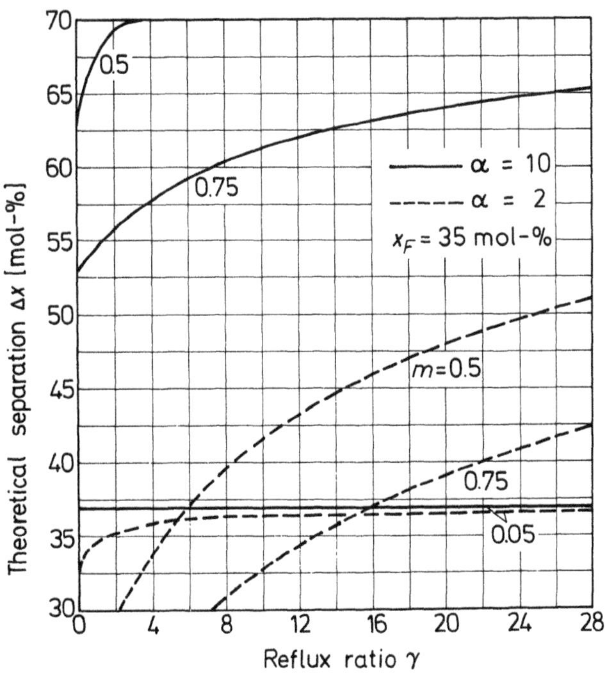

Fig. 12.5 Theoretical separation versus the reflux ratio for variable surficial mass ratios. Thin-film distillation.

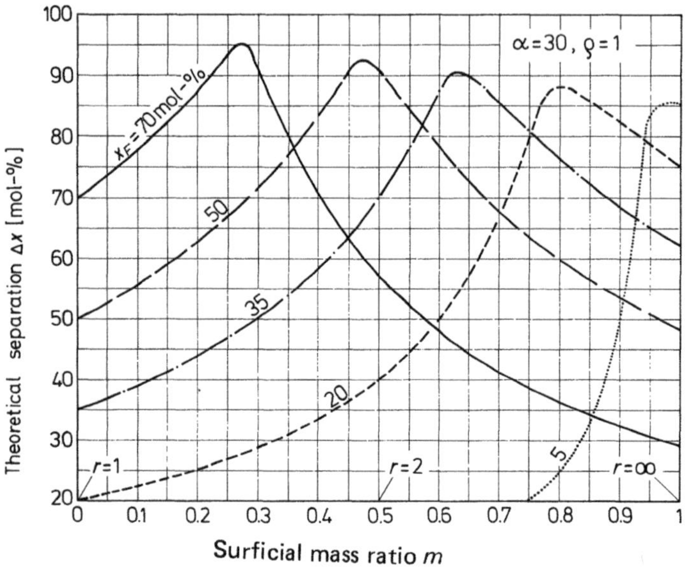

Fig. 12.6 Thin-film distillation. Theoretical separation versus surficial quantities, for various feed concentrations.

Fig. 12.6 makes this plain. For a given ratio

$$\rho = \frac{R}{F} \tag{12-31}$$

of reflux and feed it illustrates the relation

$$(\Delta x)_{\alpha,\, \rho\, =\, \text{constant}} = f\,(m, \text{parameter } x_F) \tag{12-32}$$

The higher the concentration x_F of low boiler in the feed, the more the maximum separation moves up, and the lower the corresponding surficial mass ratio m and reflux ratio r because of the relation

$$r = \frac{\rho}{1 - m} = \frac{\rho}{d} \tag{12-33}$$

Distillation ratio d will increase because

$$d = \frac{\rho}{r} \tag{12-34}$$

and so will evaporation ratio v which follows from a combination of Eqs. (12-7), (12-22), and (12-33)

$$v = 1 - \frac{m}{1 + \rho} \tag{12-35}$$

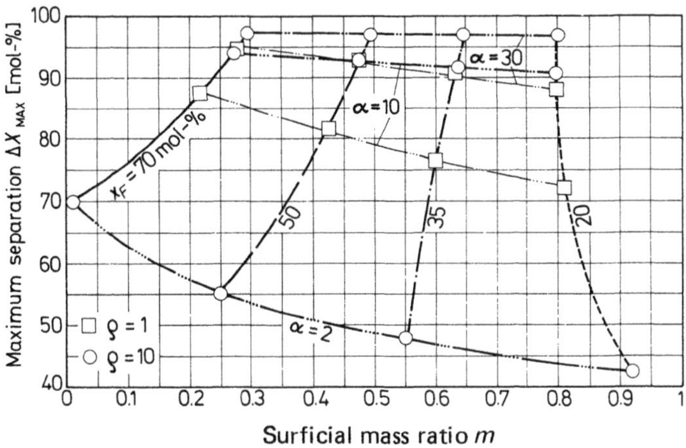

Fig. 12.7 Thin-film distillation. Maximum separation versus surficial quantities.

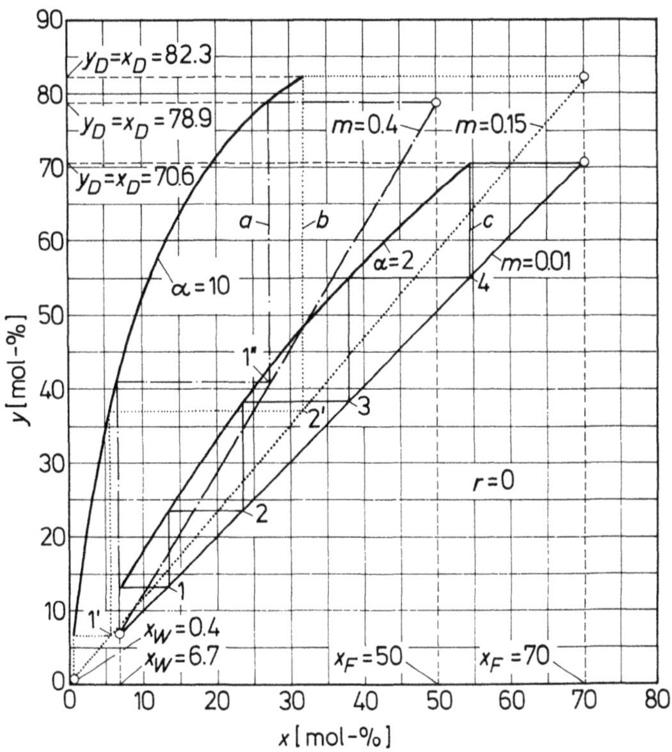

Fig. 12.8 Thin-film distillation. Finding the theoretical steps equivalent for a given separation. $a \triangleq 1$, $b \triangleq 2$, $c \triangleq 4$ theoretical stripping steps.

The curves in Fig. 12.7 render the relationship

$$(\Delta x_{max})_{\rho\,=\,\text{constant}} = f\,(m, \text{parameters } x_F \text{ and } \alpha) \qquad (12\text{-}36)$$

Note the changes in maximum separation for a certain ratio ρ of reflux and feed for systems with different relative volatilities α for various contents x_F of the low boiler. Clearly good separation or Δx_{max} values over 95 mol-% are theoretically feasible only in distilling mixtures of relative volatilities α at operational pressures exceeding 10, $(\Delta x)_{max}$ being the larger, r being less by Eq. (12-33), and v being less by Eq. (12-35) the more x_F is elevated.

Finally there is Fig. 12.8, which was plotted in analogy to Fig. 12.3, with zero reflux. Depending on the relative volatility of the system, the composition of the feed, and the surficial quantity ratio, the theoretical effectiveness of the Liprotherm is equal to that of a rectifier comprising a conventional bottoms type evaporator and one to four stripping steps. This demonstrates the advantages of the Liprotherm as a reboiler in the separation of thermally unstable mixtures; the absence of stripping steps reduces the holdover time in the column and the corresponding investment

To what extent, and under what conditions the actual separation can deviate from the theory, and what factors would be responsible has not been investigated to date. The answer in any event depends on the effective loads in continuous operation, and the distillation equilibrium variations caused by changes in load and mixing effects intervening during operation.

XIII Components of a Separation Plant

A complete separation plant requires various components, the proper selection of which, with judicious design and choice of materials, warrants reliable operation. Thus dependable function of a packed column depends as much on top-of-column liquid distribution and inside flow control in other places as on proper base support. The selection of condensers and cooling media such as air or water is often important in plant optimization. In vacuum work the proper pressures must be held by suitable means. Finally maintenance of product quality and production rate are achieved by proper measurement and controls.

1. INTERNAL COMPONENTS

In the discussion of column internals in Chapters 6 and 8, attention was drawn to the need for good liquid flow management for effective separation. The distributor details in Figs. 6.14, 6.21, 6.22, 6.70, and 8.26 are some examples among many.

Other components include top plates, base screens, retainers, drawoff fittings, and bafflings (Fig. 13.1).

Fig. 13.2 shows liquid redistributors as used at the top and at feed levels of packed towers. The box type, 13.2a), is employed for average to high liquid loads in columns exceeding 0.6 m diameter. For specific liquid loads below 5 m^3/m^2h, an even liquid distribution is hard to maintain. Individual box sections, Fig. 13.3, may be leveled by adjusting bolts; still use of sieve plates (Fig. 13.2.b), nozzle plates (Fig. 6.21), or nozzle sets (Fig. 13.4) is preferable. Sieve plates and spiders as well as nozzle plates are suitable for any diameter. They are still useful at liquid loads of around 0.75 m^3/m^2h. For columns of median and large size they are supported by a ring. With small diameters, brackets may be used (Fig. 13.4).

Liquid distributors as presented in Fig. 13.2c, d, and e, and Fig. 6.22 will have a spray effect ensuring even provision of sprays on the parts covered. They may be successful for small specific loads such as 0.25 m^3/m^2h. With proper design the column diameter is relatively immaterial. For small liquid quantities the capillary distributor (Fig. 6.70) has been successful.

In some cases, as in sporadic flashing or overloads when starting or shutting down a packed column, use of a bottom plate as shown in Fig. 13.5 is recommended. They are fastened between packing and liquid distributors on support rings held inside the shell.

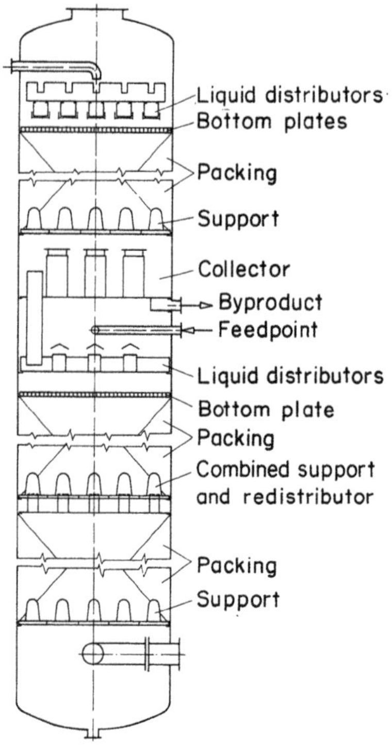

Fig. 13.1 Components of packed columns.

Load fluctuations or vibrations may cause packing of carbon or ceramics to fracture. To prevent this, use of holddown plates as shown in Fig. 13.5 b) and c) is a precautionary measure.

In the design of bottom and retaining plates, the clearances they allow must be more ample than that of the packing. Liquid distribution, pressure loss, and throughput all impose certain boundaries. The same holds for the supports if no risk of flooding can be entertained. Depending on packing type and weight the supports may be of the lightweight or the substantial type; note Fig. 13.6. Lightweight type a) made of expanded metal does not support high loads, and may serve for plastic or thin-walled metal packing of ample clearances. Grid plates as in b) are fairly substantial. Designs c) and d) restrain the passage of liquid along vapor paths; pressure losses stay low and the load on the packing remains favorable.

Fig. 13.7 gives the principal dimensions of a vapor injection plate made of metal of 2-m size and built-up construction. This particular design imparts great stiffness; an 8-*m* packing depth for 3.5-m columns is admissible. Such plates may be made of the standard construction metals, also of ceramics and plastics (Fig. 13.8).

In packed columns of larger diameters, low loads may cause nonuniform distribution more readily the greater this distance is from the feedpoint.[1] This impairs

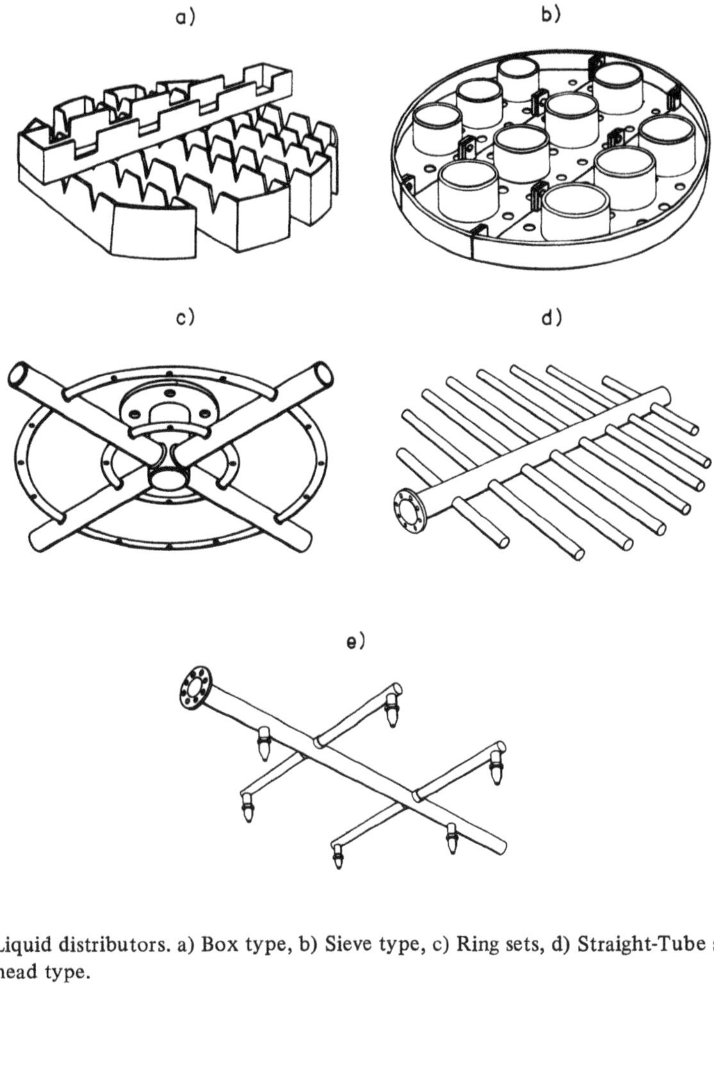

Fig. 13.2 Liquid distributors. a) Box type, b) Sieve type, c) Ring sets, d) Straight-Tube sparger, e) Showerhead type.

separation. As a countermeasure, redistributors are provided at various levels of tall towers. The metal construction in Fig. 13.9 can be described as a combination of a support plate and a distributor plate in which the vapor outlets of the latter extend slightly into the raised vapor nozzles of the former.

The redistributor made of ceramic or plastic material illustrated in Fig. 13.10 is, however, applicable only to towers below 2 m in diameter (Fig. 13.10). With *d* (mm) packing size, the spacing for redistributors should be up to 120 *d* (mm).

Nonuniform liquid distribution may cause, especially in large-diameter columns, appreciable decrease in specific separation along the packing depth. Any device

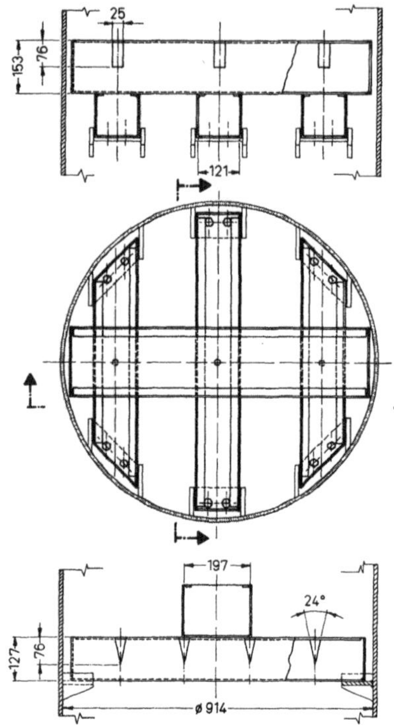

Fig. 13.3 Box-Type distributor. Dimensions in mm. Made by Koch International of Bergamo, Italy.

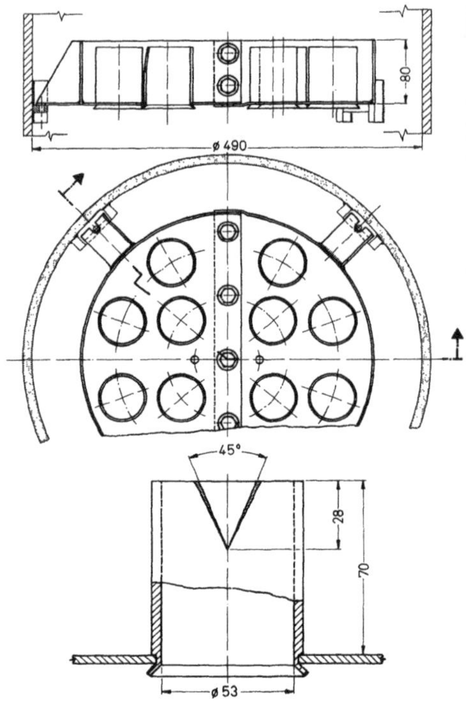

Fig. 13.4 Nozzle-Type distributor. Dimensions in mm. Made by Koch International of Bergamo, Italy.

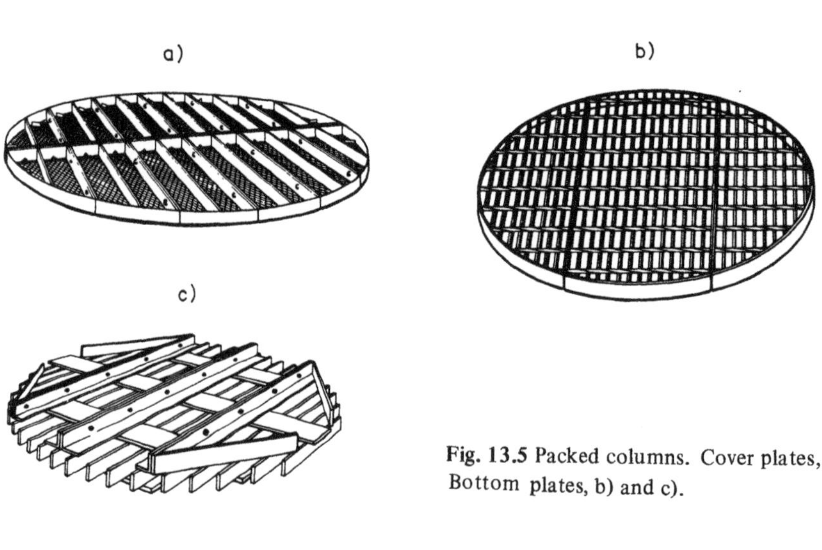

Fig. 13.5 Packed columns. Cover plates, a). Bottom plates, b) and c).

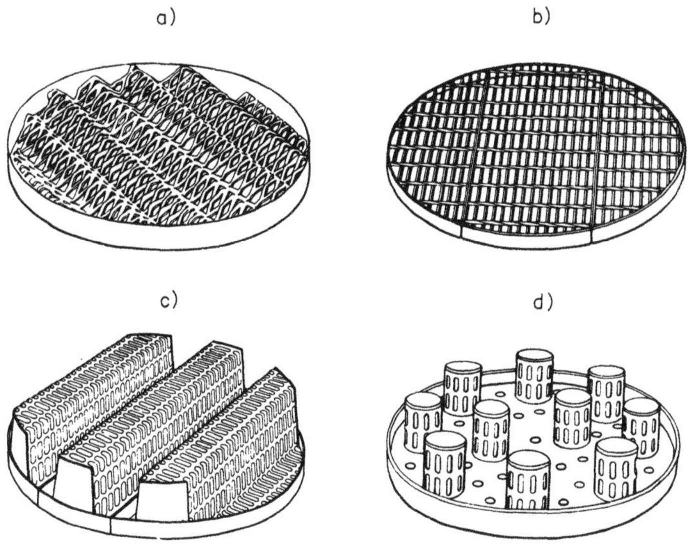

Fig. 13.6 Bottom plates of packed columns. a) Expanded metal, b) Grid type, c) Vapor injection plate, d) Perforated vapor risers.

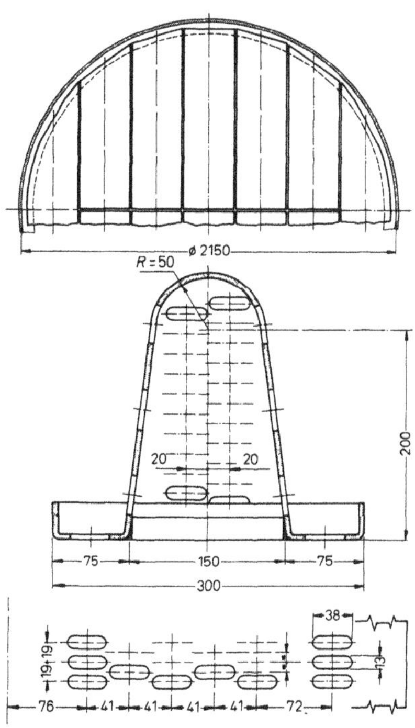

Fig. 13.7 Vapor injection plate. Dimensions in mm. Made by Koch International of Bergamo, Italy.

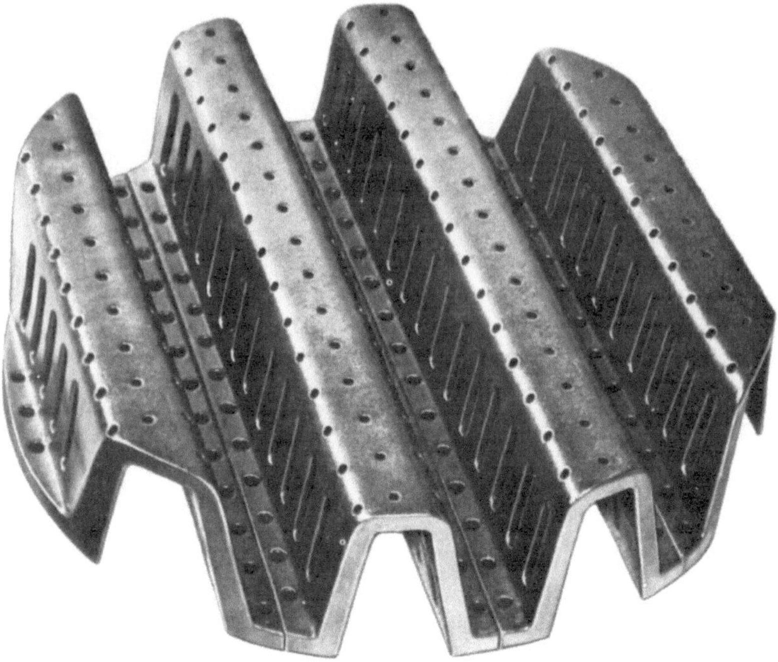

Fig. 13.8 Vapor injection plate made of ceramics or plastics, by Vereinigte Fuelkoerper Fabriken, Baumbach, Germany.

providing a collection of the liquid and simultaneous mixing and redistribution would appear opportune, such as the design in Fig. 13.11. Wires attached to a distributor plate may extend downward to the trickling centers, preventing sudden discharges of blobs; note comments in Chapter 6. This device is expensive but yields a load flexibility ten times the ordinary.

When a side product is to be withdrawn in the liquid state, a drain plate as in Fig. 13.12 is installed.

If an insoluble component accumulates inside a column, and must be removed, a device according to Fig. 13.13 is employed. It collects the reflux and allows separation and removal of the insoluble material. Frequently it will be found convenient to provide the shell with a number of observation windows or sightglasses in order to inspect the function of column internals. Fig. 13.14 presents a good design. Only if hermetic tightness is essential should such conveniences be waived.

2. HEAT EXCHANGERS

The reboilers dealt with in Chapter 9 may serve in rectifying schemes to provide the vapor mixture, in which case the design guides laid down remain unchanged.

As a rule, external heater units are preferred for large evaporator duties (Fig. 13.15).

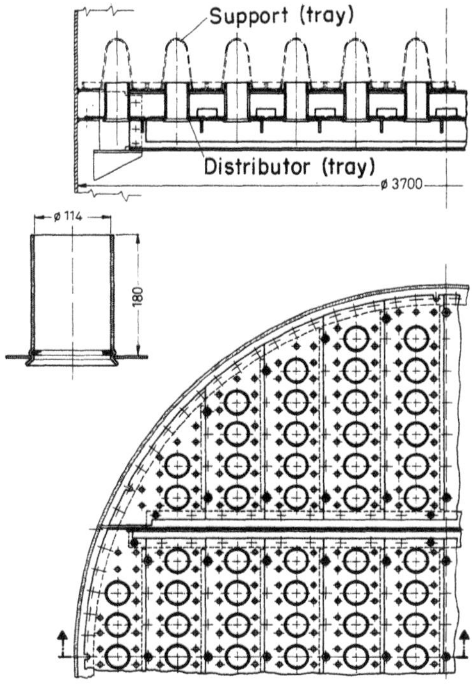

Fig. 13.9 Liquid redistributor for packed columns, Dimensions in mm. Made of metal by Koch International of Bergamo, Italy.

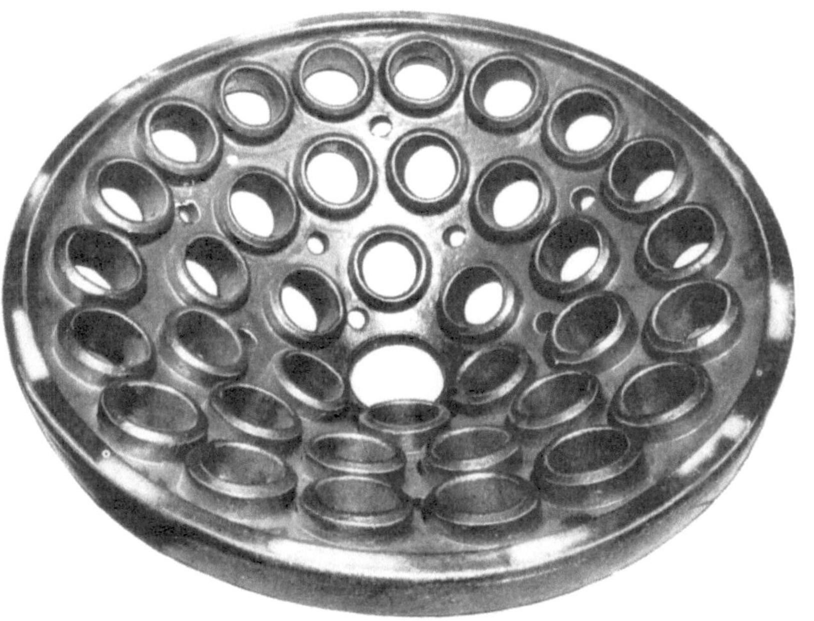

Fig. 13.10 Redistributor made of ceramic or plastics by Vereinigte Fuellkoerper Fabriken, of Baumbach, Germany.

Fig. 13.11 Packed columns. Redistributor featuring a) Collector and mixer, b) Predistributor, c) Nozzle plate with run-off wires. Made by ACV Stage-Schmidding-Heckmann of Koeln, Germany.

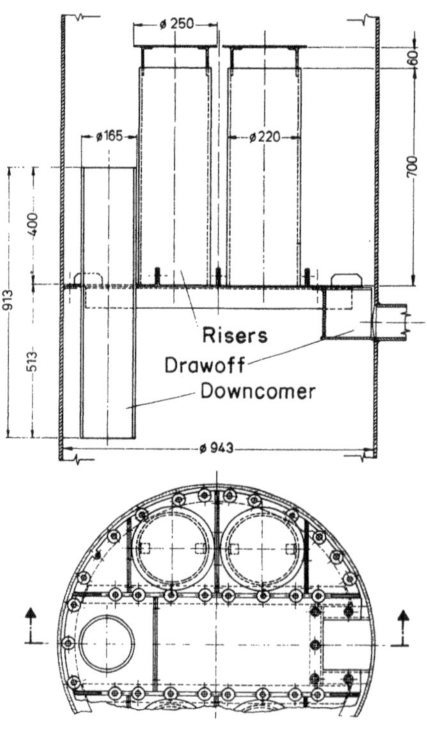

Fig. 13.12 Removal of liquid side-products. Dimensions in mm.

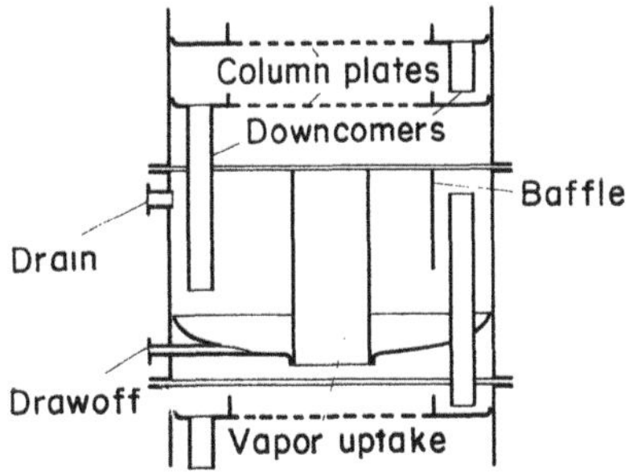

Fig. 13.13 Recovery of insolubles from column.

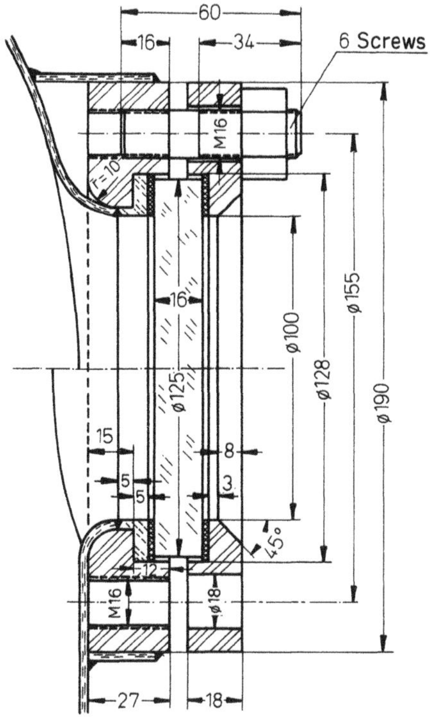

Fig. 13.14 Sightglass in stainless steel column. Dimensions in mm.

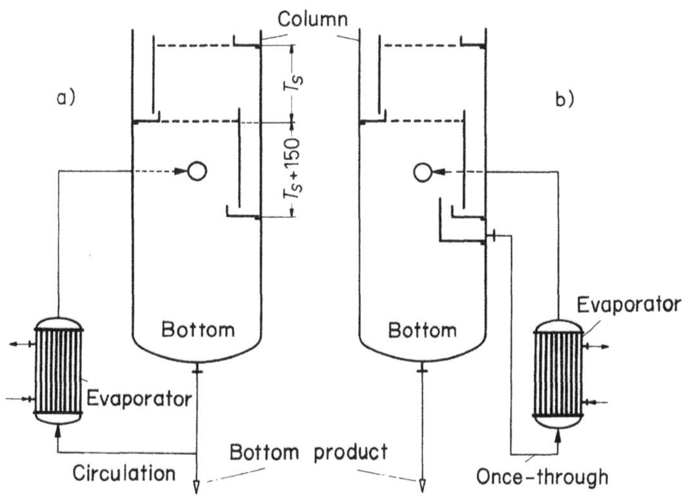

Fig. 13.15 External heater hookup. a) Thermo syphon, b) Once-through. Dimensions in mm.

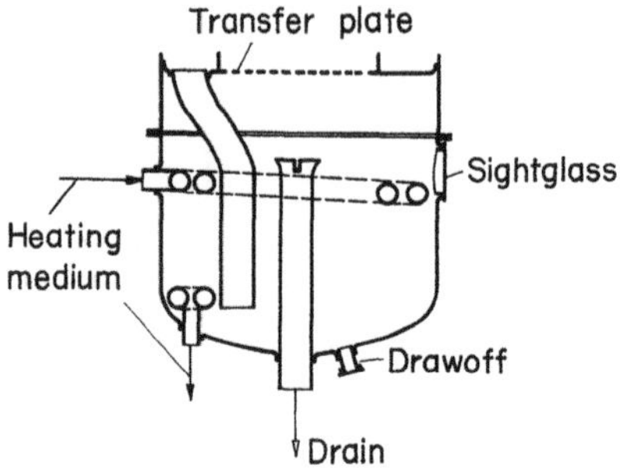

Fig. 13.16 Reboiler with coils.

Depending upon specifications for operation and separation, the hookup may be of a) the thermo-syphon type, or of b) the once-through type. The exchanger may be arranged vertically or horizontally, for the thermo-syphon principle. The calculation of the heating surface has been discussed in Chapter 9.

If the construction volume is limited, use of a Robert-type evaporator as the reboiler should be considered, as such a unit can be installed immediately below the column. Its buildup is shown in Fig. 9.2.

If the mixture to be rectified is of the nonfouling type, small to average throughputs can be handled by a reboiler equipped with a heating coil (Fig. 13.16).

If the runoff from a rectifying coil is water, direct steam heat by injection is the most economical and hence preferable answer. In the rectifier of Fig. 13.17, the aqueous bottoms of an atmospheric column are ejected to the ambience. If the mass of water is large, a direct heating system may be combined with an ejector (Fig. 13.18) to good thermal advantage as explained in Chapter 7. Such steam operated ejectors are relatively inexpensive, and they are dependable inasmuch as they have no

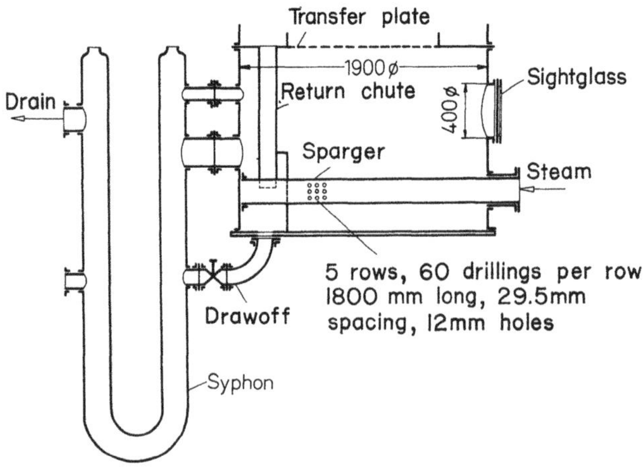

Fig. 13.17 Direct steam heat for column with water as the bottom product. Dimensions in mm.

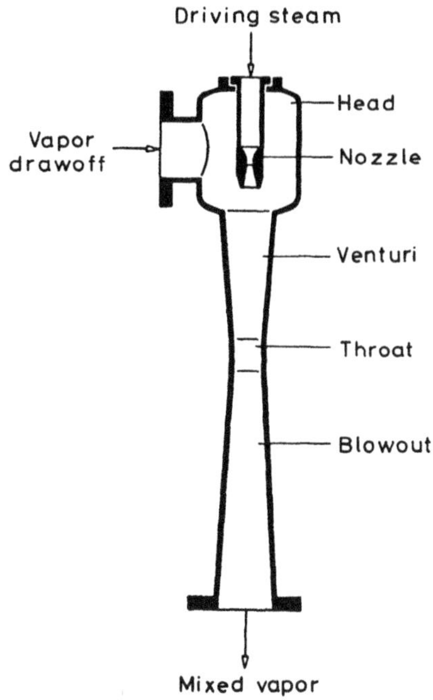

Fig. 13.18 Steam ejector for direct heating of columns. Made by Wiegend, of Ettlingen, Germany.

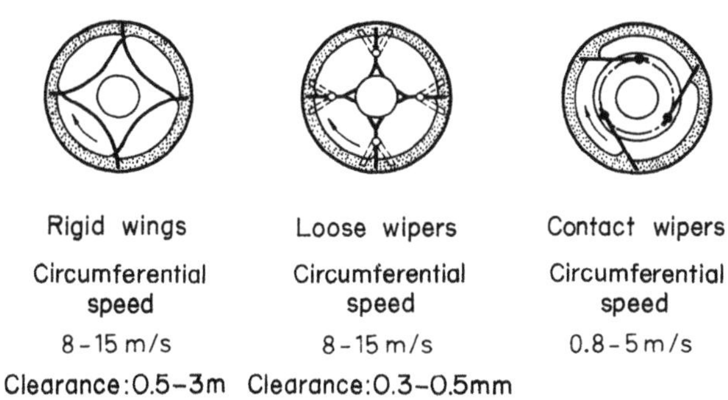

Fig. 13.19 Rotor types in Luwa snd Sambay vaporizers.

moving parts. When made of corrosion-resisting materials such equipment is long-lived. In Fig. 7.64 it is shown how it is installed on a column base.

To rectify temperature-sensitive mixtures, columns are fitted with the falling-film vaporizers of Chapter 9 (Fig. 9.4) or with thin-layer vaporizers (Figs. 9.5 and 9.6). Depending on consistency and performance, the bottom product is processed in vaporizers equipped with rotors of the kinds shown in Fig. 13.19.

The use in distilling schemes of thin-layer evaporators as separators is shown in Figs. 6.65, 7.54, and 9.18, for instance.

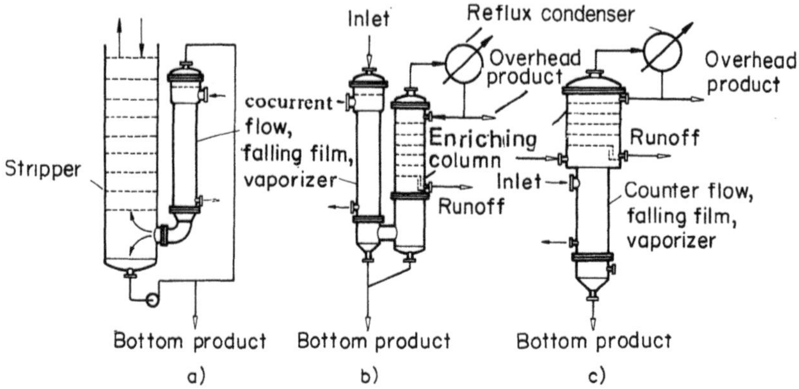

Fig. 13.20 Use of falling-film evaporators as reboilers, a) as bottom heaters with recirculation, b) combined with straight enriching column and cocurrent vaporization, c) combined with enriching column and countercurrent vaporization.
Made by Wiegand of Ettlingen and by SMS of Butzback, Germany.

Fig. 13.20 reviews some instances of falling-film reboilers on columns. Hookup a) was also employed in the installations of Figs. 6.61 and 6.69.

Condensers that provide reflux are water-cooled or air-cooled. Both systems are equivalent in terms of engineering[2] so that factors of a localized or processing nature may dominate in the choice. Air as a coolant may be economically attractive for temperature differences exceeding 20 C between the product and air. For elevated product temperatures, the air cooler is advantageous, as water cooling may entail fouling phenomena on transfer surfaces.

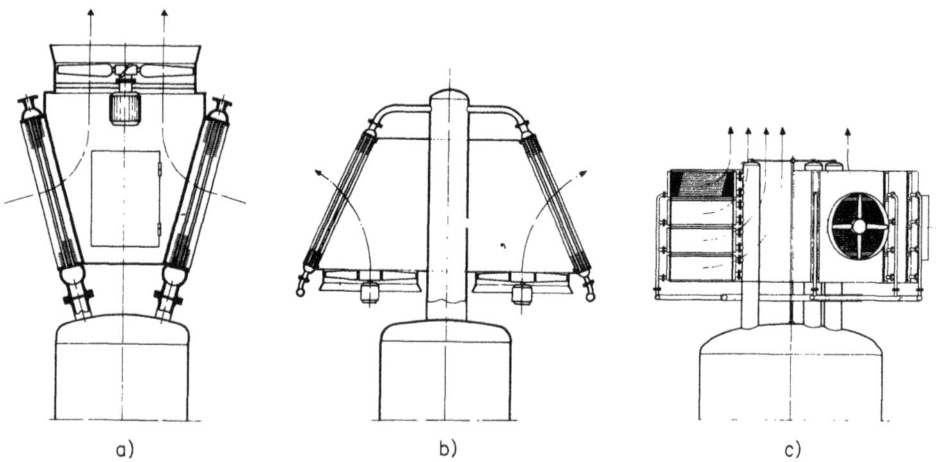

Fig. 13.21 Air-Cooled reflux condensers, overhead mounted, a) countercurrent of vapor and condensate, b) fan in stream of cold air, c) flexible condensing systems. GEA Luftkuehler, of Bochum, Germany.

Fig. 13.22 Multistage rectifier installation with air-cooled reflux condensers mounted overhead. Made by GEA Luftkuehler, of Bochum, Germany.

If water is the coolant, expenses for water supply loom large, while in the case of air the investment for the equipment is the main item. The latter amounts to a multiple of the cost of a water-cooled condenser. The total of investment and operating expense may still be less, even allowing an amortization period of five years.

Depending on column dimensions and local factors, reflux condensers may be mounted either directly on top of the column head, or at a suitable elevation alongside the tower, even at the top base level.

Fig. 13.21 illustrates the top mounting of air-cooled reflux condensers. What is typical of the three arrangements is the difference in the extended-surface exchangers and air fans, and in the flow of vapor and condensates. In example a) for average throughputs, the vapor and condensate flow countercurrently, and the condensate will not become undercooled. The advantage in example b) is that the fans are located in the stream of cool air. For flexibility, example c) excels, as the vapor is divided up into several condenser paths.

Fig. 13.22 is a view of a multistage rectifying plant featuring air-cooled reflux condensers at column heads.

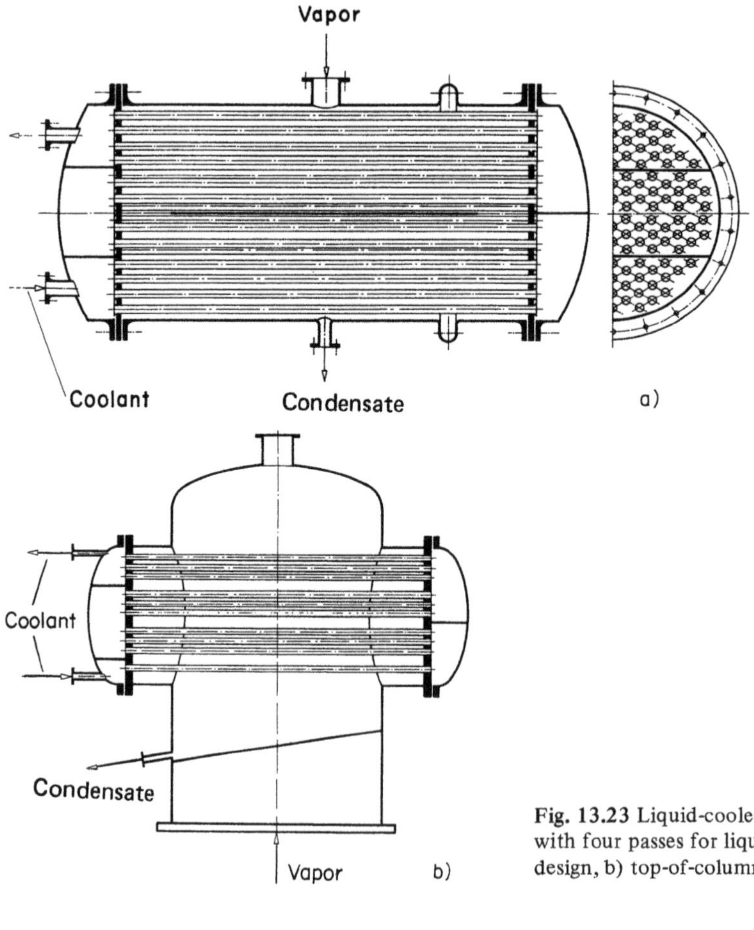

Fig. 13.23 Liquid-cooled reflux condensers with four passes for liquid, a) standard design, b) top-of-column type.

When air as a coolant is unavailable, and cooling water must be used, or an adequate quantity of feed at a sufficiently low temperature is available, or other liquids can serve as coolants, having to be warmed up for processing reasons, then condensing is best done in tubular exchangers. Depending on local conditions, the vertical arrangement (Fig. 7.62) or a horizontal design (Fig. 6.13) may be chosen. Depending on heat transfer conditions, one or more passes of the coolant may be indicated (Figs. 6.61 and 13.23).

Exchangers serving merely to cool or heat liquids are generally tubular equipment. Special designs are rarely warranted. Thus in Fig. 13.24 several coils serve to cool several products in a common cooling water enclosure fitted with baffling. This design is economical, and suitable for cooling small quantities of liquid products that are nonfouling in nature.

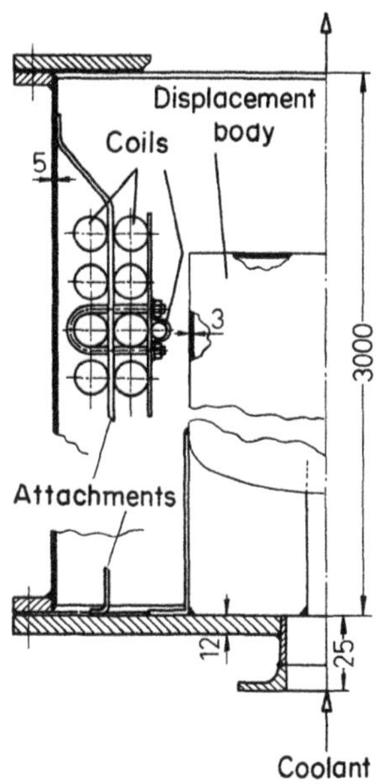

Fig. 13.24 Product cooler with external cooling coils. **Dimensions in mm.**

3. PUMPS

Among refinery pumps those for generating a vacuum stand out. Choice of the optimum type may be difficult since installation, maintenance, and service life all must be considered. Water quality and temperature are important as are pressure considerations.

In Fig. 13.25 various pumping devices are presented. A comparison of these, within the range of practical utility, as to investment and operation only[3,4] yields, based

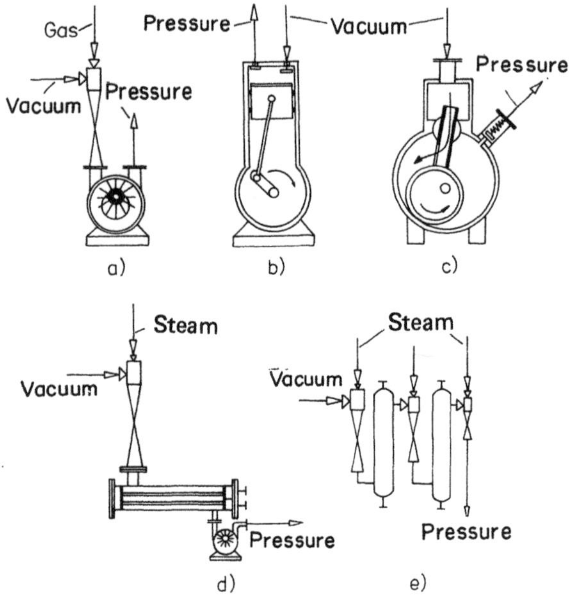

Fig. 13.25 Vacuum pumps. a) Water-ring pump with ejector, 8–60 mm Hg, b) Piston pump 760–1 mm Hg, c) Rotary piston pump, 760–0.1 mm Hg, d) Ejector plus mechanical pump (water-ring), 60–0.1 mm Hg, e) Ejector set, 760–0.1 mm Hg.

on three years amortization, 8000 operating hours yearly, and a vacuum of 5 mm Hg, the result shown in Fig. 13.26. Obviously, the advantage is on the side of the combined steam jet-water ring vacuum pump and of the straight steam jet.

A comparison worked out for pressures over 50 mm Hg, involving a steam ejector and a water ring pump would indicate the cost advantage tending in favor of the water ring pump.[4]

Since a single-stage steam ejector has a limited compression ratio, its maximum suction being around 100 mm Hg, lower pressures require staging of ejectors as illustrated in Fig. 13.27 (five-stage arrangement). Fig. 3.18 shows the buildup and function of a single unit. In between stages, condensers are useful to liquefy the steam of the previous stage and any condensable vapors, and thus partly unload the successive stage or stages. The temperature of the cooling water must always be below the condensation temperature. In this way, vacuums down to $5 \cdot 10^{-3}$ mm Hg can be produced.

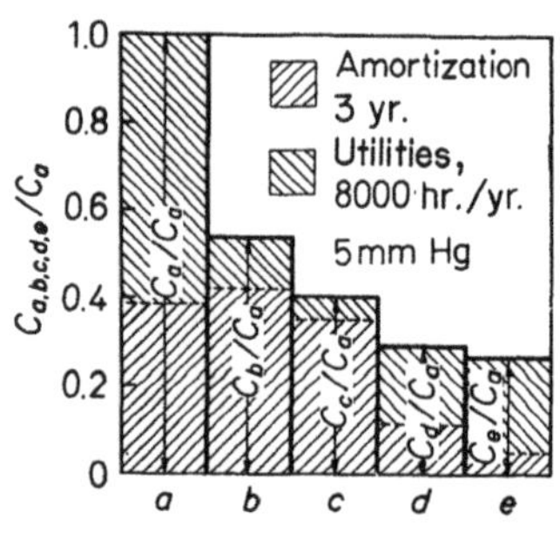

Fig. 13.26 Cost of various vacuum pumps. Subscripts of Fig. 13.25. Amortization, 3 yr. Utilities, 8000 h/yr.

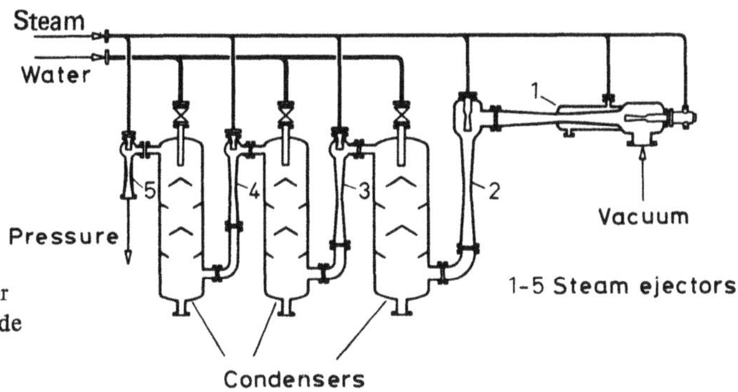

Fig. 13.27 Five-Stage ejector with mixing condensers, made by Wiegand of Ettlingen, Germany.

4. MEASURING AND CONTROLS

The mechanism of separation in a distilling and rectifying plant involves a large number of independent and interdependent variables; therefore trouble-free operations occur only with the help of measuring and control equipment. The scope and selection of that equipment is governed by process requirements and conditions as well as by the desired product quality. The volume of the column in operation is a major factor; its performance in time within the control system and that of the controller itself determine the effectiveness of the control program. If allowance is made for the inevitable deviations in the program, a maximum of control equipment is not necessarily the optimum. Rather, the scope of the control equipment as to measurement and as to regulation depends on the limitations imposed by any deviations.

Most rectification columns are not self-sufficient but they are part of a more or less circumscribed process. From the controls viewpoint, however, they mean a separate cycle that must be coupled with other departments in order to automate the whole installation. The purpose of rectification is, in any event, to split off some component or components from a mixture, singly or as a group; the goal of regulation must be to make such separation as close as desired. To that end, the concentrations of the various components along the column must be surveyed to meet product specifications. Generally, rectification takes place at constant pressure, and therefore the duties of control include that of maintaining the proper pressure which may vary depending on whether the column works at reduced or at increased pressure, whether gases are liberated, etc. Instead of the pressure regulation, frequently a temperature control is preferred; in accordance with the laws of mixtures, temperature, pressure, and concentrations of the components are all interrelated, and both the pressure and the concentration must be under control. Assuming pressure as a constant, for binaries there is a direct interrelation of concentrations and temperatures along a column. Concentration by itself (actually of primary interest)–rectification attempts to provide a definite product or a spectrum of products–is rarely followed up in a continuous fashion, since no suitable instruments exist. Therefore generally the temperatures along a column are controlled or simply held constant across a given zone, and this indirectly takes care of the concentrations.

Regarding the general theory of process control and its terminology note the bibliography.[5,6] Control schemes mostly use certain standardized abbreviations, such as:

CRC	Concentrations recorder and controller
FI	Quantity indicator
FR	Quantity recorder
FRC	Quantity recorder and control
LA	Level alarm
LIC	Level indicator and control
LRC	Level recorder and control
PI	Pressure indicator
PIC	Pressure indicator and control
PRC	Pressure recorder and control
TI	Temperature indicator
TR	Temperature recorder
TRC	Temperature recorder and control

A control cycle generally consists of a converter that transforms the physical or chemical variable to be controlled into a standard variable, a comparator that compares the standard variable against a set value and processes the difference of the two, the deviation in accordance with set parameters, to come up with a resetting element that corrects the deviation and restores the desired condition.

The controller components are symbolized by letters and numbers. The first letter indicates the kind of variable controlled, as *P* for pressure, *T* for temperature, *F* for flow, *L* for level, and *C* for concentration. Another letter *I* or *R* indicates whether the variable is to be indicated or recorded. A third letter, *C*, means control. A serial number is added for sequence. The measuring lines are full, the impulse lines dotted or light.

Fig. 13.28 outlines the control of the first runnings tower of a refinery that is to remove gases dissolved in crudes along with the lowest boiling cut. The bottom product passes to an atmospheric tower. The low boilers pass overhead and collect in a condenser; part of them stay in gaseous form, the balance is liquefied. In that way, a gas cushion forms in the condenser which helps in maintaining an even pressure. In controller PRC–1, a valve will open when the pressure rises, letting gas escape so the pressure goes down. The reverse happens when the pressure drops. The temperature is controlled by TRC–2 of which the steam valve is part. The temperature sensor is located at that point of the column that has the largest temperature gradient, thus reducing the inertia in the controller to a minimum. To prevent load fluctuations in the following refinery departments, the bottom product must be withdrawn via a flow controller, FRC–3.

A rectifying tower will require reflux, taken from the overhead condensate, of which the quantity is normally held constant. Controller FRC–4 works so that a constant amount is returned to the column from the condenser. Controller LRC–5 maintains the

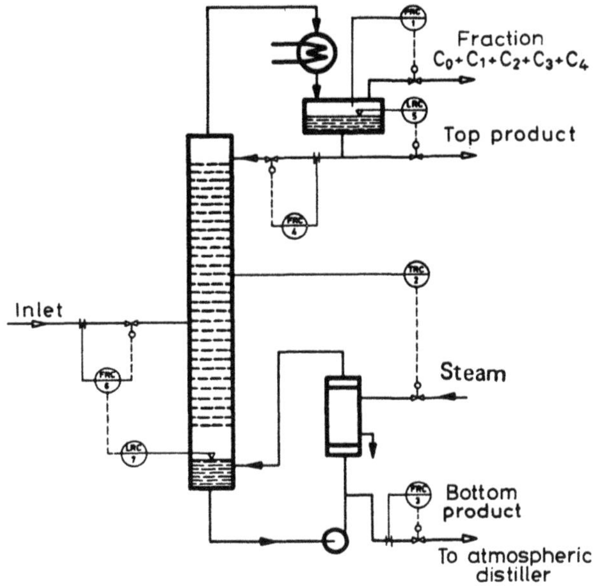

Fig. 13.28 Controls in topping column of a refinery. Made by Siemens & Halske of Karlsruhe, Germany.

condensate level, letting the proper quantity of overhead product run off. Some level fluctuation must be admitted, always retaining enough condensate for the reflux demand. FRC–6 steadies the column feed. The bottom level is maintained in two ways, either by way of FRC–3 or by way of FRC–6. A variation in FRC–3 may disturb the later atmospheric still. Therefore a correction via FRC–6 is preferred. LRC–7 controls the set point of FRC–6 by the proper level.

This assembly of controllers, together with column temperature control, constitutes a primary automation. In the event that the quantity, composition, or temperature of the feed material fluctuate, the heat supply and reflux quantity must likewise be adjusted. In other words, the next step in automation is a superimposed control that ties together the miscellaneous control cycles described above. Actually, these are not mutually independent but interdependent.

In Fig. 13.29 a rectifying plant is presented for decomposing a feedstock into its two components, the overhead and the bottoms. Also the feedstock contains inert gases that must be absent from the top product. Throughput, temperature, and composition of the feedstock are known. The composition cannot be held constant, but column load and temperature remain steady. This is achieved through controls FRC–1 and TRC–2. To prevent a change of temperature inside the column, feed control FRC–1 and heat control FRC–3 are interconnected; when the feed rate changes, the heat is increased or reduced in proportion automatically. To control gas outflow from the head, a cooler unit provides total liquefaction of a low boiler. Again the temperature stays above the dewpoints of the gases which pass through the cold zone and are drawn off by a steam jet. Since the negative pressure generated is below the operating

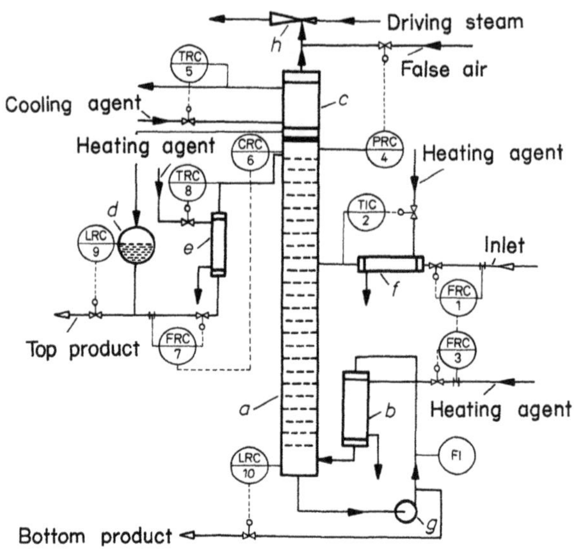

Fig. 13.29 Controls of a plate-type column. a) Column, b) Reboiler, c) Reflux condenser, d) Storage, e) and f) Preheaters, g) Circulator, h) Ejector.

pressure of the column, control PRC–4 permits outside air to enter and build up pressure. When large quantities of gas ascend through the column, the air valve is closed in proportion, and vice versa, maintaining constant pressure.

The aforementioned cold zone at the column top is maintained by control TRC–5; the rate of coolant is such that the temperature in the cooler stays below the condensation point of the overhead product and the gases.

The concentration profile along the column adjusts so that at the top the low boiler concentration is a maximum, and the concentration of high boiler is a minimum; the reverse holds for the base. In this case the significant composition is at the top where best product purity is expected. For this reason, the concentration is measured and adjusted by device CRC–6. If the concentration drops off, the reflux is augmented in order to have the enrichment in high-boilers farther downward. Hence controls CRC–6 and FRC–7 are cascade mounted; the output of CRC–6 shifts the set point of control FRC–7 which handles the reflux flow. As, however, the reflux can return to the column at variable temperature, which could cause disturbances in the transfer of materials and heat, the temperature is regulated immediately prior to column inlet by TRC–8, which avoids trouble from that quarter. The overhead product condenses in the cooler and collects in the receiver equipped with level control LRC–9 that acts upon the valve for the top product, and prevents both overflowing and drainage. The high boilers build up in the lower part of the column and collect in the base section. Control LRC–10 maintains the level within specified limits.

The mass of liquid in the column establishes the sequence of measurements and reactions of instantaneous conditions; therefore the controls must be selected to react to impulses inside time allowances. Therefore, the dynamics of a rectifying column with plates (Fig. 13.29) differ from those of a thin-layer rectifier (Fig. 13.30) or those of

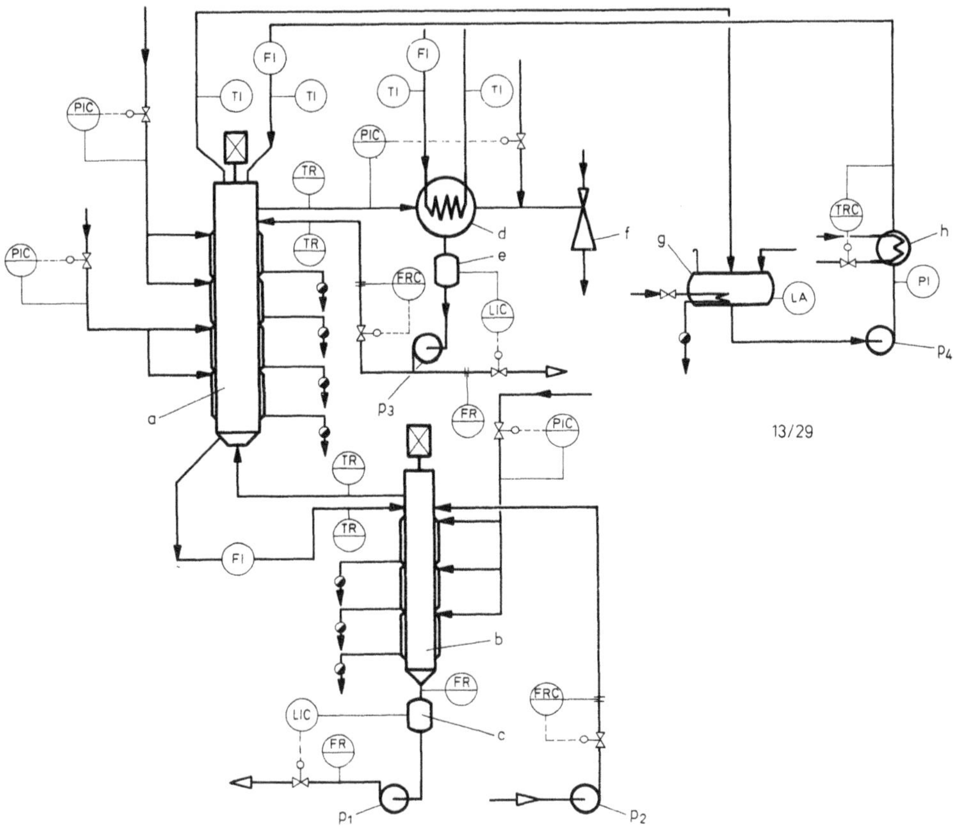

Fig. 13.30 Controls in thin-layer rectifying plant. a) Rectifier, b) Evaporator, c) Receiver, d) Reflux condenser, e) Surge tank, f) Steam ejector, g) Coolant supply, h) Preheater, p_1 to p_4) Pumps. Luwa Systems of Zurich, Switzerland.

a discontinuous packed column (Fig. 13.31). Obviously, a disturbance of the operating pattern, affecting an external variable such as the temperature, feed rate and composition, separation and therefore compositions of top and bottom products, is more marked in a plate column of large liquid volume than in a thin-layer rectifier designed for limited holdup. Control system inertia depends basically on the extent of liquid volume comprised between the measuring point and drawoff, as compared with that between the measuring point and disturbance location. For columns of small liquid holdup, as rectifiers or film type columns, the system inertia is less than for packed columns, and much less than for plate type columns. Therefore deviations must be detected promptly, and correcting action must occur readily. However controls meeting this requirement are relatively costly.

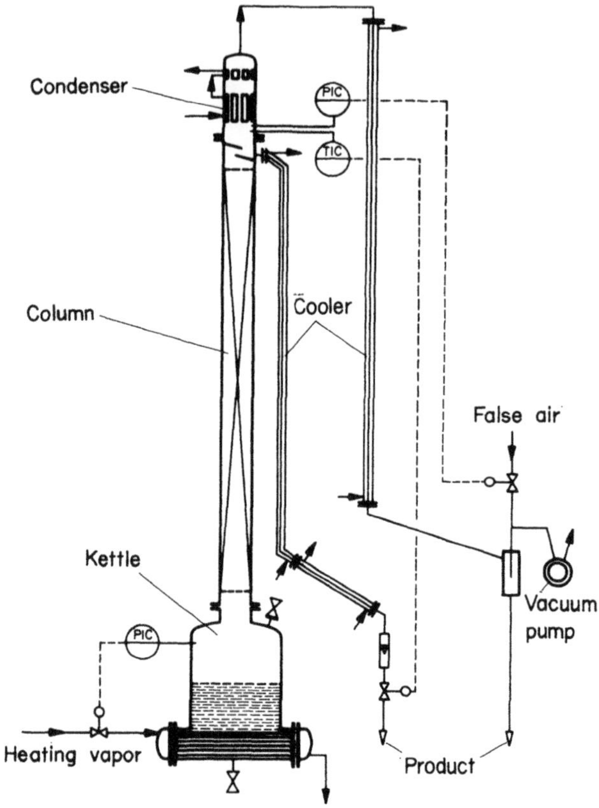

Fig. 13.31 Controls of batch rectifier for constant overhead composition.

For columns of little liquid reserve, the inertia can, within limits, be enhanced by means of accumulators. Because of the sensitivity of the separation to load fluctuations in columns of small liquid content it is advantageous to hold the feed rate, the reflux quantity, the heating steam pressure and the column head pressure constant (Fig. 13.30). Given that type of control, the reflux ratio is ample enough to compensate for fluctuations in the composition of the feedstock.

Those rectifiers whose liquid volumes are relatively limited, and which therefore must meet the above recommendations, include the thin-layer rectifier, the various film type columns, and the wire web columns of Chapter 6.

If capacities are large, or the materials recovered are of high value, optimum control of rectifying columns can be assisted by electronic computers.[7] In that event, beginning with the input data, the probable perturbances are prefigured and immediately compensated for by introducing a counteracting variable. The time constant involved is thus neutralized for all practical purposes.

For the control of a discontinuous rectifying plant, the dynamic performance of the column is significant if the overhead concentration must be held. In that event, the reflux ratio requires steady adjustment; refer to Chapter 4. Then, depending on the kind of column internals, even at constant vapor load the liquid content and therefore the dynamic condition of the plant changes, and reflux control becomes sensitive. A plant of the type in Fig. 13.31, having to work discontinuously a binary as defined by the y/x and $t/x, y$ plots of Fig. 13.32, at constant overhead concentration x_D, requires an increase in the reflux ratio ($r = r_1 = x_D/y_{o1} - 1$) to ($r = r_2 = x_D/y_{o2} - 1$), in order to compensate for the drop in still concentration from $x_B = x_{B1}$ to $x_B = x_{B2}$.

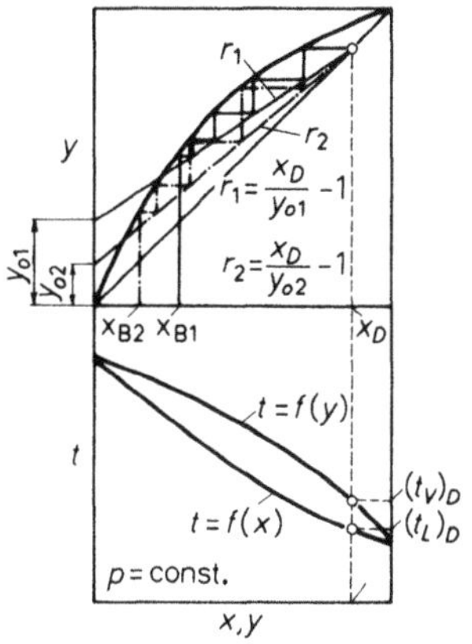

Fig. 13.32 Temperature versus overhead concentration in separation of binary, as in Fig. 13.30.

If the temperature gradient at the column head is adequate, the impulse for the product control is the temperature at that point; otherwise, the results obtained by a suitable continuous analyzer of sufficient sensitivity can serve as control pulses.

In the case of a continuous setup, working a binary, temperature measurement for concentration control is made at that point of the column section where the largest temperature gradient is expected from the boiling characteristic.

In discontinuous operation, and for a ternary, the distillation triangle in Fig. 13.33 reveals that reflux control for maintaining the overhead concentration via the vapor product temperature $(t_V)_D$ at the column head is feasible within limits only, since a ternary mixture of a given temperature has an infinite number of possible compositions; note Chapter 2.

Maintaining a constant temperature $(t_V)_D$, and referring to Fig. 13.33 and the caption, for the composition of product D, many compositions are possible for points a and b. The greater the share of low boiler in the top product (generally desirable), and the flatter the isothermal, the smaller the eventual differences of low boiler in the top product.

To maintain an even load in a batch plant of the type in Fig. 13.31, the controls must provide constant pressures at both the top and the bottom of the column. By an absolute pressure transmitter the pressure in the vapor space of the still is signalized to the final element, a pneumatic vapor flow valve. Likewise, the head pressure is

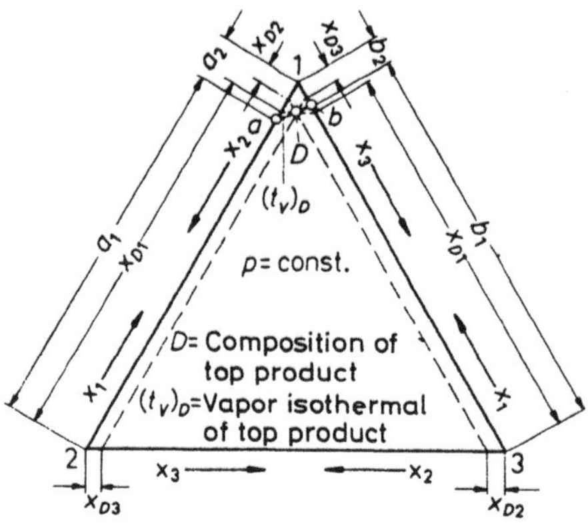

Fig. 13.33 Temperature versus overhead concentration in the splitting of ternary as in Fig. 13.31
Point a: Reduction of component 1, $(x_{D1} - a_1)$,
Increase in component 2, $(a_2 - x_{D2})$,
Component 3 = nil.
Point b: Increase in component 1, $(b_1 - x_{D1})$,
Increase in component 3, $(b_2 - x_{D3})$,
Component 2 = nil.

controlled by pulses sent via a transmitter to a control actuating an air valve in the line.

To sum up, the scope of the control equipment should respond to the separation requirements stemming from the physical properties of the system and from the dynamic loads of the column and the capacity of the plant. In retrospect, the control systems presented in Figs. 12.38, 13.29, 13.30, and 13.31 are only a few examples. To handle specific problems, the bibliography[8–16] offers much advice. On the whole, the goal is to obtain a maximum yield of high-quality products by means of a minimum expenditure of energy, and to economize in manpower. Moreover, it may be necessary to adapt a plant to a given market situation. To answer the request measuring and control instruments must be arrayed as well as computers in order to give a clear insight into the process mechanism, and the required information for immediate action in emergencies, or for automatized correction when errors begin to build up.

No standardized control of rectifying columns appears feasible; rather, an optimized control pattern must be developed in every instance, and there are boundaries always set by available financial resources.

XIV Use of Computers

The preceding sections describe the essential premises in the preconstruction engineering efforts required to specify equipment and plants for the thermal separation of materials by distillation. The numerical relations and graphs presented should lend themselves to evaluation by conventional methods. Their application should suffice in many instances of daily practice, and the values obtained be accurate enough for use in project work, costing, and operation of a distilling plant.

The mechanism of decomposition has been analyzed, and the details of construction have also been dissected. The facts on flow and separation of materials were gained from numerous tests, and the correlations derived from much research are such that they can be developed further by means of computer techniques. This subject is likely to gain in importance and should be reviewed at greater length.

1. COMPUTERS AS AIDS IN PROCESS CALCULATIONS AND BASIC DESIGN

The technical and economical benefits of computerization become evident especially at the optimization stage of distillation engineering, and accordingly the use of electronic aid has become widespread, as in resolving complex equations for columns dealing with actual multicomponent mixtures, which in earlier days could be done only by laborious experimentation.

Programs existing in computerization for representing phase equilibrium relations include operating data for various numbers of steps, the sizing of packing, and establishing diameters and heights of towers.

Recently, programming is based on the works of Wilson[1] and of Renon and Prausnitz[2] on real equilibria of multicomponent systems. The latter even discloses phase equilibria of partially insoluble multicomponent systems from experimental data on the mixing gap at boiling, and the composition of the heteroazeotropic point.[3]

The basics and logistics of recent methods are dealt with in other publications.[4–12]

Virtually all programs intended to yield the theoretical steps as a function of the reflux ratio are founded on the same basic concept proceeding by gradual sequence.[13–16] The concentration profiles thus generated are especially meaningful when byproducts are branched off, or compositions change continuously as in batch runs.[17,18] In this connection, numerical treatment[19] of the process becomes meaningful.

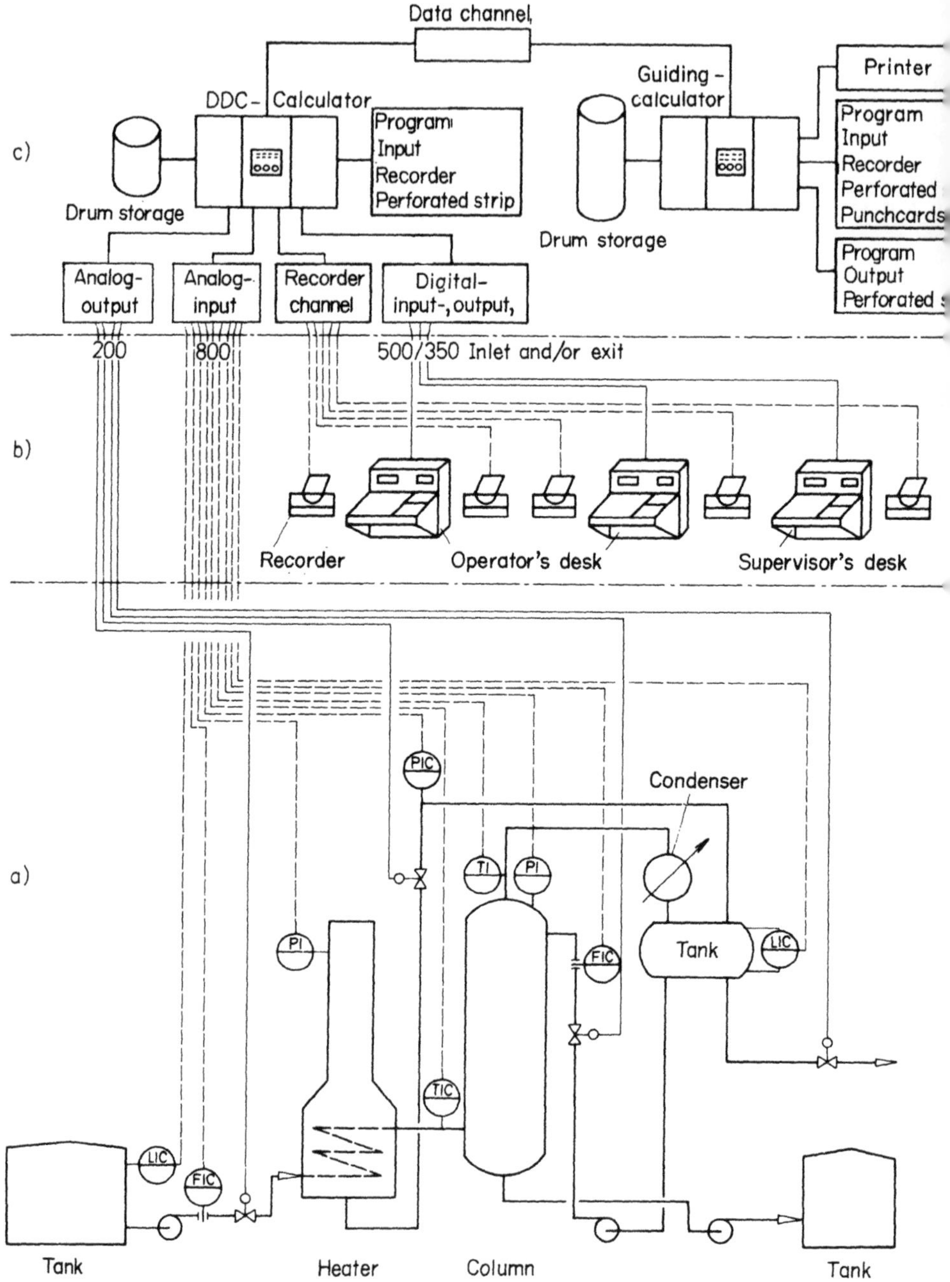

Fig. 14.1 Mobil Oil Refinery at Woerth, Germany. Computer control scheme.

Existing computerization schemes for column intenals may serve to recheck available equipment or to decide on new items; in either case the process may be subject to improvement.[20] If packed columns are involved, the entire load range should be investigated to review the factual effectiveness.[21]

2. APPLICATION OF COMPUTERS IN PLANT OPERATION

Whether or not the use of computers, as reviewed in Chapter 13, Section 4, in order to optimize plant control[22–31] is justified must be decided on the merits of the case. The economic aspects are especially important. They in turn depend on continuous operation, based on stable control of process variables, and this can be achieved only by means of suitable computerization.

Fig. 14.1 presents an example of computerized process control installed in a modern refinery (Europe's first[32]). In this somewhat condensed picture, a) is the production plant, b) the supervisory center, and c) the computer department. Of the production plant, only the storage facilities, the crude oil heater, a rectifying column with a reflux condenser, a condensate tank, and various pumps are shown; some typical measuring and control points have been included.

The nucleus of computer control comprises two process calculators, one guiding calculator, and one DDC calculator (direct digital control). The former sets the standards for the controls from the interrelations of variables revealed by mathematical calculations, in order to safeguard uninterrupted production. The latter defines the set points by direct digital action and thus holds the operational controls on the line, and assures stability. As a safety measure in the event of computer failure, analog controls are installed parallel with the direct digital controls. The standards developed in the guiding calculator are fed into the digital program by the data channel. The calculators are served by the input and output units, the drum storage and the extensions to plant and to personnel.

The supervisory desk and two operating desks permit the supervisors in the control room to immediatelyintervene at any point of the installation, which comprises around 800 measuring inputs and 200 control outputs. To these are added 500 digital inputs and 350 digital outputs. The use of semiconductor switches allows calls at the rate of 10,000/s.

Core storage capacity of computers is 16,000 words, that of drum storage in the guide computer is 256,000 words, and that of drum storage in the DDC computer, 64,000 words.

The almost perfect flexibility of direct digital control, based on the calculations of the required set point corrections, is essential to overall control, which is a set-point decision. Improved plant operation is a principal argument for computerization, in the face of increasing demands on production yield and quality combined with minimum investment and energy expenses attached to the particular product. Secondary applications are in the establishment of plant criteria from material and heat balances, efficiency calculations, and operational statistics.

Fig. 14.2 Partial view of the Mobil Oil Refinery at Woerth, Germany.

Fig. 14.2 illustrates the refinery for which Fig. 14.1 rendered the control scheme by computer; columns and auxiliaries are in the foreground. The use of computers is obviously justified in view of the many interrelated components.

3 USES OF COMPUTERS FOR ENGINEERING

Heat Transfer: Coefficients for convection, *F*–factors for radiation can be determined from nomograms or by special slide rules, charts, tables, or graphs. For conduction, both steady-state and three-dimensional, relaxation techniques are available. Results are more accurate and more quickly obtained when programs for heat transfer "packages" are written. When mass and heat are transferred simultaneously, the design involves iterative methods that require many computer loops to converge.

Fluid flow also involves iterative calculations. Network problems, surge and resonance phenomena are solved by computer analysis.[33,34]

Flow Sheet Design: General packages, such as one developed at Cambridge University Chemical Engineering Laboratory in 1972, provide several inputs ranging from question-and-answer routine, to teletype use, and to employment of a light pen and graphics facility to produce a flowsheet by computer display. The flowsheet is analyzed to find loops in the system, and calculation follows. The user has the option for various types of outputs of stream variables and convergence information, and eventually a hard copy of a flowsheet.

An important advantage is achieved in the optimization of construction of any chemical process through the choice of the best flow sheet, and this is made possible by the use of computers. The automatic manufacture and evaluation of process and instrumentation diagrams is of particular significance for the early stages of process engineering decisions, in order, for example, that the best separation method may be found for the mixture to be separated by using distillation or extraction technology. This is particularly so when little information is available about a process.[35]

The application of various processes such as BALTAZAR leads to the determination of the optimum flow sheet involving the minimum of investment costs.[36,37]

Equipment Design: Computers are applied in the design of pressure vessels, heat exchangers, separators, and so on, for thermal and mechanical design, and on material listings; standard codes have been issued by ASME, TEMA, and others. For stress analyses, computer packages exist likewise.

Piping and Plant Layout: In the siting of plant equipment, models can be constructed of blocks representing vessels, and plastic tubing representing piping. They help in detailing pipe runs and for materials takeoff. It may be less costly to describe the layout in computer language, and display it on the screen of an interactive graphics satellite computer. The layout can be studied from every angle and moved at will. Piping can be routed to the best advantage. The results of the study and graph plotters can yield various sections, and pipe isometrics can be drawn for material take-off. This detailing may save expense.

Integrated Process Plant Design: This is based on a design chain for several contributing offices, with physical transfer of information and drawings, and a central computer system. Each office has interactive design programs which read as "input" the output of the previous office. Software must permit rapid feedback when an impasse is reached. Integrated process plant is ultimately a time and cost saver.

Dupont has already successfully applied a system for the quantitative preparation of piping and instrumentation diagrams. It was extended to electrical and logic diagrams which were controlled by means of computers. Because of the fact that all computer systems are connected and are dependent on one another, the total plant design, including cost controlling, is possible through the use of the so-called Integrated Project Engineering System (I.P.E.S.).[38]

XV Distillation and Environmental Protection

Environmental problems are an incidental corollary of process planning and development. The demand for maximum exploitation of basic raw materials has imposed tasks of a peculiar nature on chemical engineers. This explains why so many highly efficient items of equipment and processes have been developed to protect the environment; particularly great impetus has been imparted to distillation processes.

Fig. 15.1 shows the product streams in a production plant. Means of operating to reduce pollution are shown on it schematically. It can be seen from this that the

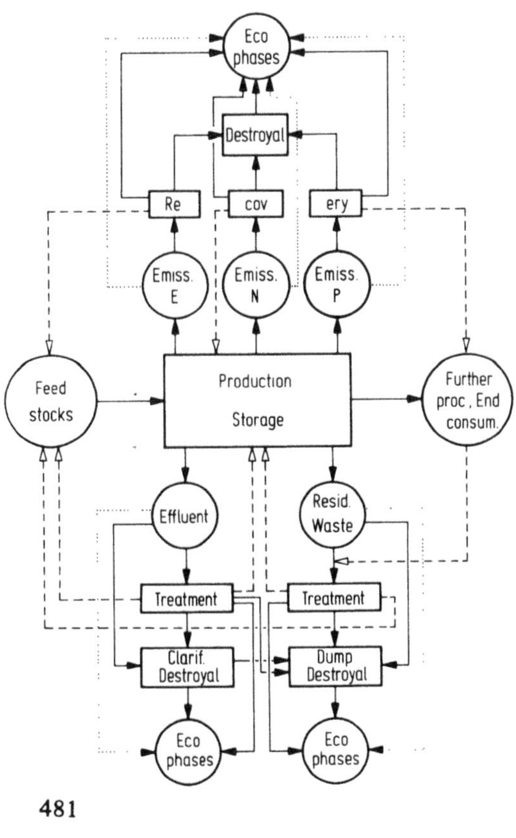

Fig. 15.1 Schematic diagram showing product streams in a plant. The measures taken to avoid pollution are marked. Symbols: E = feedstocks, N = by-products, P = product.

entire "Production and Environment" complex must be considered as an integrated whole.

Pollution from the residues of thermally unstable substances is not uncommon, e.g., from polymerization reactions or thermal cracking processes. An example is provided by liquid phases whose residence times at fairly high temperatures at the foot of rectification columns are quite long. In order to restrict the formation of these undesirable byproducts to a minimum, special separation equipment must be used with the minimum possible residence times for the product concerned. They must be designed to operate with the minimum pressure drop. Optimization methods have been developed for this purpose.[1] These give rise to the optimum conditions and design of separation equipment that allows a specific product to be obtained with the minimum capital and running costs.

The economics of a process are based on the application of process and equipment optimization with due allowance for ecological demands. A feature of the expenditure on equipment for environmental protection is that unit operations can be resorted to in almost all production techniques concerned. For example, distillation techniques are adopted for processing residues. Fig. 15.2, for purifying spent air or recovering valuable substances from spent air, Fig. 15.3, and for disposing of process water

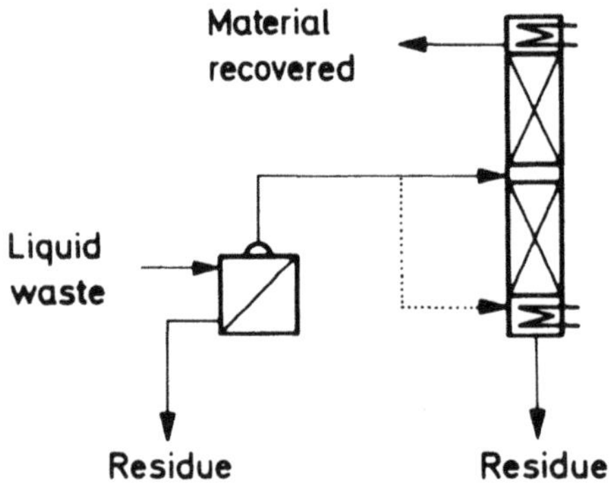

Fig. 15.2 Recovery of valuable materials from liquid residues or waste by evaporation followed by rectification.

and reducing its pollutant load, Fig. 15.4. Thus these chemical engineering measures do more than prevent or reduce pollution. In many cases, they also reduce the losses of valuable substances and thereby lead to favorable amortization of the expenditure incurred.

Thus some characteristic, basic, chemical-engineering principles are usually applied to nonpolluting production plants. The expenditure depends on the number of process stages, the number and amounts of feedstocks, and their physical properties; it also depends on the number, amounts, and physical properties of the byproducts or waste products.

A typical example is the separation of a raw material by vacuum distillation from the process effluent representing the low-volatility component of a mixture from medium-boiling by products and from high-boiling impurities destined for incineration. A feasible solution would be a combined evaporation–rectification plant, Fig. 15.5.

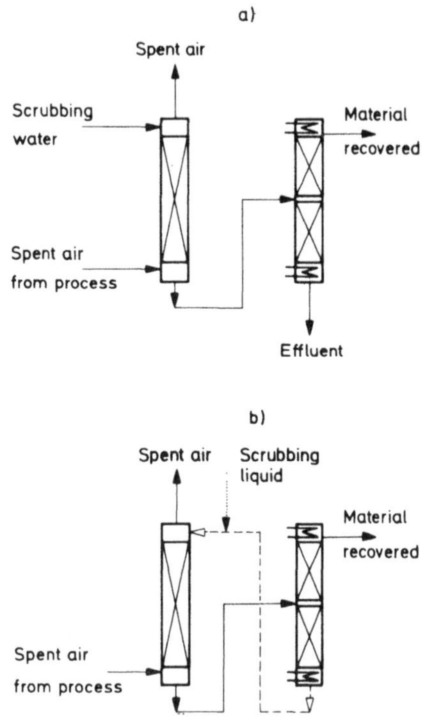

Fig. 15.3 Purifying spent air or recovering valuable materials from the spent air in production processes:
a) by absorption in water and rectification of the mixture thus obtained.
b) by absorption in a suitable scrubbing fluid and rectifying the ensuing mixture with recycling of the scrubbing fluid.

In order to ensure that the amount of effluent requiring treatment is kept at a minimum, the principle of surface condensation and not of mixed condensation must be realized in the vacuum steam injectors. The off-gases from all the vacuum stages in this plant must be scrubbed with water in a common absorber.

Hence in the planning of new installations, unit operations are affected to a considerable extent by ecological demands. This is particularly evident from the fact that the most efficient means of solving ecological problems in production processes are those that ensure the minimum of waste products and emissions. An attendant requirement is that the substances that favor the formation of waste products be removed from the product stream at some suitable point in the process. Or, alternatively, the unavoidable byproducts or waste products must be removed before they have time to accumulate. Despite these measures, the capital and running costs must be kept at a minimum. This demand is achieved by optimization of the process and judicious selection and design of the equipment and machinery.

Nowadays production effluents that have to be cleaned are usually run into a clarification plant with a biological treatment basin. The cost of biological treatment in the clarification plant depends on the pollutant load and the biological degradability. The pollutant load can be reduced in two separate stages. Firstly physical means adapted to

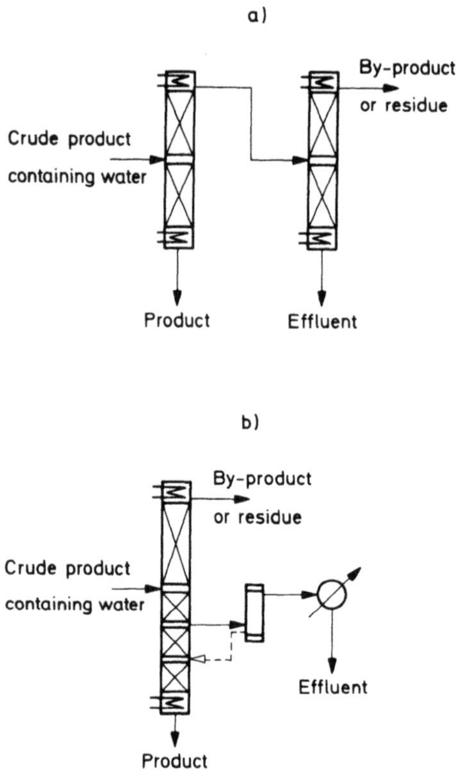

Fig. 15.4 Separation of process effluent and reduction of pollutant load:
a) by rectification
b) by rectification combined with dephlegmation.

suit local conditions are resorted to on the production site and are followed by biological treatment (Fig. 15.6, a). This is fundamentally more economical than it is to allow all the water purification stages to be carried out exclusively in a clarification plant (Fig. 15.6, b).

In certain cases, optimal results may be achieved by local application of a purely physical method (Fig. 15.6, c). Physical treatment of effluents is not merely a question of economics. It would become a necessity if the pollutants in the effluent could not be degraded by biological techniques. In such cases, chemical engineering offers the fundamentals for modern effluent treatment.

One of the most expensive physical methods of water treatment is rectification, which is so frequently resorted to in chemical engineering. It has therefore been selected for the optimization method developed below for the physical pretreatment of effluents before they enter the biological clarification plant.[2–5]

The total costs C_T (DM/t) for integrated water treatment are given by

$$C_T = C_I + C_O + C_W \tag{15-1}$$

where:
C_I (DM/t) are the capital investment costs for the physical treatment,
C_O (DM/t) are the running costs for the physical treatment, and
C_W (DM/t) are the costs of treatment in the clarification plant.

The capital investment C_I (DM/t) is defined as the cost of the equipment required for physically separating pollutants in a rectification plant from one ton of effluent flow-

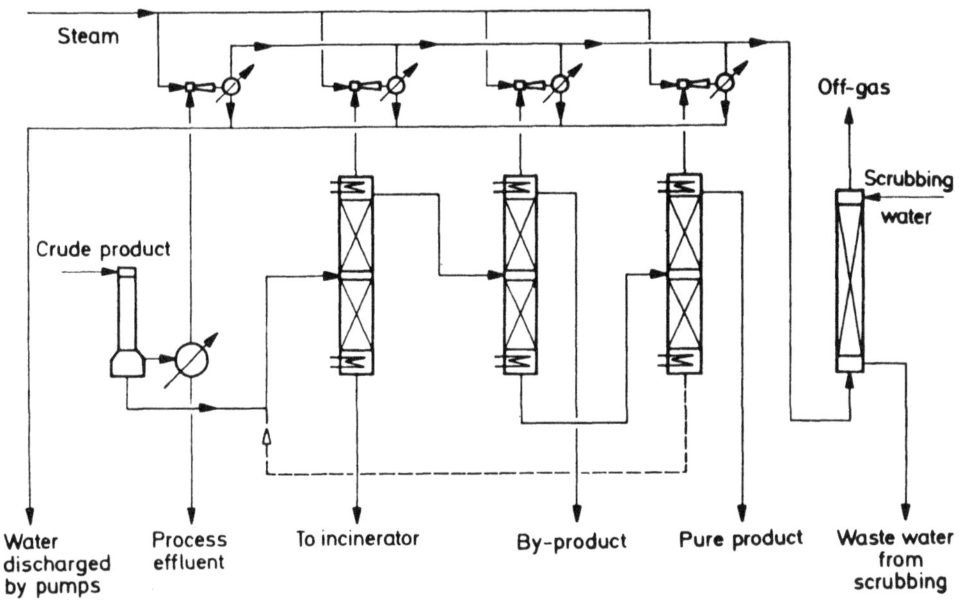

Fig. 15.5 Plant for removing the following from a crude product by distillation: the process effluent as the low-boiling ingredient, the byproduct as the medium-boiling ingredient, and the waste product as the high-boiling ingredient with local off-gas scrubbing.

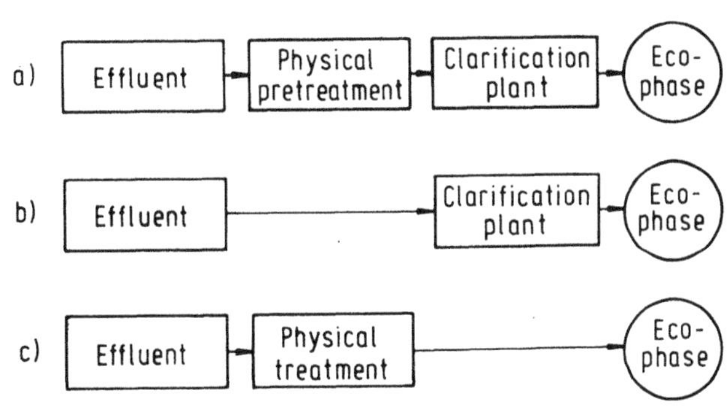

Fig. 15.6 Schematic diagram showing means of effluent treatment.

ing at a molar rate of F (kmol/h) and having a molecular weight of μ_F (kg/kmol). It can be broken down into the costs for the vaporizer C_V (DM), the costs for the column C_D (DM), and the costs for the reflux condenser C_R (DM). If the capital is to be paid back in n_A years or t_A hours when

$$n_A = \frac{t_A}{8640} \tag{15-2}$$

and if z is the annual interest rate, the total interest with linear amortization is given by

$$Z = z \frac{n_A + (n_A - 1) + (n_A - 2) + \ldots + 1}{n_A} \tag{15-3}$$

The following equation is thus valid for the capital investment costs per ton of effluent.

$$C_I = (C_V + C_S + C_R)(1 + Z) \frac{1000}{t_A F \mu_F} \tag{15-4}$$

The vaporizer costs C_V (DM) can be determined from

$$C_V = f_V A_V^{0.6} \tag{15-5}$$

where f_V is a factor depending on the material, design, and heat transfer area A_V (m^2) of the vaporizer. A_V is given by

$$A_V = \frac{1}{3.6} \frac{Q}{k_V \vartheta_V} \tag{15-6}$$

where k_V ($W/m^2 K$) is the heat transfer coefficient, ϑ_V (K) is the logarithm of the associated difference in temperature between the heat carrier and the boiling liquid, and Q (kJ/h) is the amount of heat transferred in the vaporizer. The following relationship can be derived for Q_V, Chapter 3, Eq. (3-62):

$$Q = F \left\{ \frac{x_F - x_W}{x_D - x_W} [(r_D + 1) \lambda_0 + h'_D - h'_W] + h'_W - h_F \right\} \tag{15-7}$$

where:

x_F, x_D, and x_W (mol-%) are the pollutant concentrations in the effluent, the distillate, and the outlet of the rectification plant;
λ_D (kJ/kmol) is the latent heat of condensation at the top
h'_D (kJ/kmol) is the enthalpy of the liquid at the boiling point at the top
h'_W (kJ/kmol) is the enthalpy of the liquid at the boiling point at the bottom of the column,
h_f (kJ/kmol) is the enthalpy of the effluent fed to the plant, and

$$r_D = f\, r_m \tag{15-8}$$

is the reflux ratio at the top of the column.

For the rectification of a binary system with a relative volatility α, r_m can be calculated from the minimum reflux ratio[6] i.e.,

$$r_m = \frac{1}{\alpha - 1}\left(\frac{x_D}{x_F} - \alpha\frac{100 - x_D}{100 - x_F}\right) \tag{15-9}$$

A value of between 1.05 and 2 is usually taken for the factor f in Eq. (15-8).

Assume that the spacing of the plates in the column is T_S (m), that the column efficiency is E_C, and that a total of n_t theoretical stages is required for the desired degree of pollutant removal by distillation. Then, if the price per plate is c_T (DM) and the price per unit length of column shell is c_S (DM/m), the prime costs for the column are given by

$$C_S = \frac{n_t}{E_C}\left(c_T + c_S T_S\right) \tag{15-10}$$

The numerical value for the number of theoretical plates n_t can be obtained by known methods, and that for the column efficiency E_C can be derived from the literature for the more common types of plate. The price per plate for a column of diameter d_S (m) can be estimated with the aid of Eq. (15-11).

$$c_T = f_T\,(0.7 + d_s^2) \tag{15-11}$$

The price per unit length of column shell is given by

$$c_S = f_S\, d_S^{1.45} \tag{15-12}$$

f_T and f_S are factors that depend on the material, design, and diameter. Thus the following equation can be obtained for the column diameter, (Chapter 3):

$$d_S = 0.057\left[\frac{F}{F_V}\,\frac{x_F - x_W}{x_D - x_W}\,(r_D + 1)\left(\mu_V\,\frac{T}{P}\right)^{1/2}\right]^{1/2} \tag{15-13}$$

where F_V ($m^{-1/2}\ s^{-1}\ kg^{1/2}$) is the vapor capacity factor, which is a measure of the dynamic load on the column fittings selected, μ_V (kg/kmol) is the molecular weight of the vapor, T (K) is the absolute temperature, and p (mbar) is the pressure in the column.

Let us now consider a packed column whose efficiency per meter of bed of packing is given by [Chapter 6, Eq (6-2)]:

$$n_s = \frac{n_t}{H} \tag{15-14}$$

In other words, the efficiency is the same as that of a column with n theoretical stages. If the cost per unit of packed volume is c_F (DM/m^3), then the column costs are given by Eq. (15-15) instead of Eq. (15-10):

$$C_S = \frac{n_t}{n_s}\left(c_F\, d_S^2\,\frac{\pi}{4} + c_S\right) + c_E \tag{15-15}$$

c_E (DM) allows for the costs of accessories such as supporting grids, distributors, etc. The n_s values for conventional packings are known from previous investigations (Chapter 8).

In analogy to Eq. (15-5), the function in Eq. (15-16) is valid for the costs of the reflux condenser C_R (DM)

$$C_R = f_R\, A_R^{0.6} \tag{15-16}$$

Again f_R is a factor depending on the material, design, and area A_R (m^2) of the reflux condenser. The area is derived from k_R (W/m^2 h) the heat transfer coefficient, ϑ_R (K) the logarithm of the associated difference in temperature between the condensing steam and the cooling medium, and Q_V (kJ/h) the amount of heat removed, that is,

$$A_R = \frac{1}{3.6}\;\frac{Q_V}{k_R\, \vartheta_R} \tag{15-17}$$

If all the vapor condenses, Q_V is given by (Chapter 3, Eq. 13–56):

$$Q_V = F\,\frac{x_F - x_W}{x_D - x_W}\,(r_D + 1)\,\lambda_D \tag{15-18}$$

The capital investment costs C_I (DM/t) for physical effluent treatment by rectification can now be determined from Eqs. (15-4)–(15-18). The associated running costs per ton of effluent C_O (DM/t) are simply the sum of the costs for the superheated steam with a latent heat of evaporation r_H (kJ/kg) and those for the cooling medium, which has a specific heat q_K (kJ/kg K) and has been heated through a temperature range Δt_K (K), that is

$$C_O = \left(\frac{Q}{r_H}\,c_H + \frac{Q_V}{q_K\,\Delta t_K}\,c_K + \mu_A\,A\,c_A\right)\frac{t_O}{t_A}\;\frac{1}{\mu_F}\;\frac{1}{F} \tag{15-19}$$

where c_H, c_K and c_A (DM/t) are the specific costs per ton of steam, ton of cooling medium, and ton of pollutant concentrate respectively, incurred in an operating period t_O (h) in the amortization period t_A (h). If the pollutant concentrate can be used as fuel or made into a useful product, c_A becomes negative. However, if the concentrate is burned without recovering utilizable energy, c_A will be positive. The amount of pollutant concentrate A, having a molecular weight of μ_A (kg/kmol), is smaller by a factor a than the amount of effluent treated, that is,

$$A = a\,F \tag{15-20}$$

If the pollutants boil at a lower temperature than water, that is, if the depolluted water is run out at the bottom of the column, then (Chapter 3)

$$a = \frac{x_F - x_W}{x_D - x_W} \tag{15-21}$$

If, however, the water is the more volatile, that is, if the pollutants are run off at the bottom of the column, then

$$a = 1 - \frac{x_F - x_W}{x_D - x_W} \tag{15-22}$$

For the purposes of this study, it can be assumed that the estimated costs for the effluent treatment in the clarification plant are mainly governed by the pollutant load and the amount of effluent. If the pollutants are predominately organic, the costs can be estimated from the amount of organic carbon. Thus the costs of treating effluents in clarification plant are given by

$$C_W = (1 - a)\left(\mu_W c_W + \frac{x_W}{100}\,\mu_C i_C c_C\right)\frac{t_O}{t_A}\,\frac{1}{\mu_F} \tag{15-23}$$

where c_W (DM/t) is the cost of treating unit weight of effluent with a molecular weight of μ_W (kg/kmol) and c_C is the cost of removing unit weight of organic carbon with a molecular weight of μ_C (kg/kmol), and i_C is the number of carbon atoms in the pollutant molecule.

If the effluent is not subjected to any physical treatment, that is, if $x_W = x_F$ and thus $a = 0$, Eq. (15-23) for the cost of treating one ton of effluent in the clarification becomes

$$(C_W)_{x_W = x_F} = \left(\mu_W c_W + \frac{x_F}{100}\,\mu_C i_C c_C\right)\frac{t_O}{t_A}\,\frac{1}{\mu_F} \tag{15-24}$$

Since c_I and c_O in Eq. (15-1) are zero,

$$(C_T)_{x_W = x_F} = (C_W)_{x_W = x_F} \tag{15-25}$$

With the aid of the above equations, the total costs for integrated effluent treatment can be presented as a function of pollutant load in the form

$$C_T = f(x_W, \text{parameter } \alpha, x_F, x_D, i_C) \tag{15-26}$$

This is shown qualitatively in Fig. 15.7. According to this, the total costs of integrated effluent treatment C_T may be less than the costs of treating the effluent exclusively in the clarification plant $(C_W)x_W = x_F$, if the relative volatility is large. Now the limiting case is $x_W = x_F$, when the condition $C_T = C_W$ must apply. On the other hand, if x_W is low enough, that is, less than $x_W = x_u$, then $C_T > C_W$, owing to the attendant high costs of physical treatment. Therefore, at a given pollutant load, $x_W = (x_W)_{opt}$, the cost curves must pass through a minimum $(C_T)_{min}$. Again, if the relative volatility α is comparatively low, the integrated treatment will always be more expensive than treating the effluent exclusively in the clarification plant. Therefore, a definite set of α, x_F values must exist at which the costs for the two methods of treatment are practically the same. In this case, the slope of the costs curve must be zero at $x_W = x_F$, that is,

$$\alpha = f(x_F) \text{ for } \left(\frac{dC_T}{dx_F}\right)_{x_W = x_F, x_D = \text{const.}} = 0 \quad\text{and}\quad C_T = (C_W)x_W = x_F \tag{15-27}$$

This relationship is shown qualitatively in the lower part of Fig. 15.7. Pairs of α, x_F

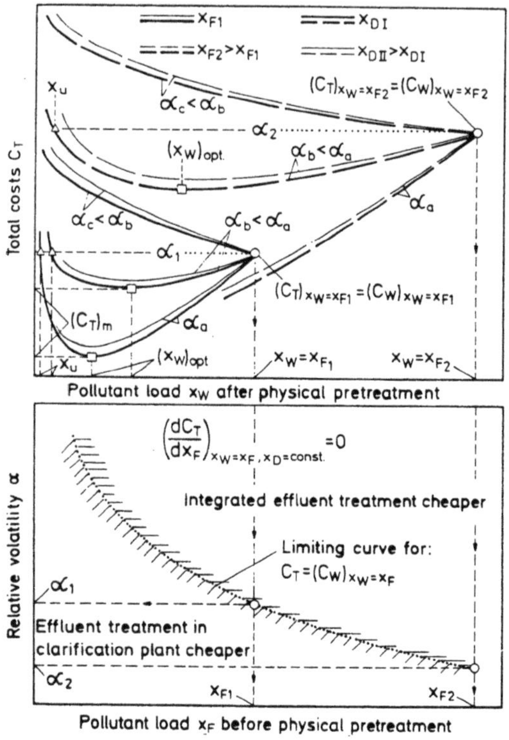

Fig. 15.7 Qualitative representation of total costs for integrated effluent treatment as a function of the pollutant load and separation conditions.

values above the limiting curve represent the conditions under which integrated effluent treatment in the $x_U < x_W < x_F$ range of pollutant concentration is the more economical. The sets of α, x_F values below the limiting curve represent conditions under which it is always cheaper to treat the effluent exclusively in the clarification plant.

Fundamentally, the total cost of integrated effluent treatment increases as α and x_W become smaller and as x_F and x_D become larger.

The figures quoted in the following example can be used as a basis in evaluating the equations numerically.

Example 1 The data are listed in the seauence in which they appear in Eq. (15-1)–(15-27), which were solved with the aid of a computer (7).

t_A = 43,200 h
z = 0.08
F = 1000 kmol/h
μ_F = 18 kg/kmol
f_V = 1250 for $A_V < 100$ m^2, carbon steel, based on 1975 price

k_V = 580 W/m^{2}h
ϑ_V = 40 K
x_F = 0 – 2.4 mol-%
x_W = 0 – 2.4 mol-%
x_D = 60 mol-%
λ_D = 38,000 kJ/kmol
h'_D = 8,400 kJ/kmol
h'_W = 8,000 kJ/kmol
h'_F = 7,500 kJ/kmol
f = 1.2
α = 2–10, pollutants assumed to have a lower boiling point than that of water
E_C = 0.75
T_S = 0.5
c_T = 420 for $d_S < 4$m, carbon steel, based on 1975 prices
c_S = 2000 for $d_S < 4$ m, carbon steel, based on 1975 prices
F_V = 1.5 m$^{-1/2}$ s^{-1} kg$^{1/2}$, average capacity
μ_V = 18 kg/kmol
T = 373 K
p = 1000 mbar
f_R = 1250 for $A_R < 100$ m^2, carbon steel, based on 1975 prices
k_R = 460 W/m^{2}h
ϑ_R = 40 K
r_H = 2100 kJ/kg for saturated steam at 5 bar,
c_H = 21.10 DM/t for saturated steam at 5 bar, based on 1975 prices
q_K = 4.19 kJ/kg K
Δt_K = 7 K
c_K = 0.05 DM/t for cooling water, based on 1975 prices
c_A = 0, assumed as limiting case
t_O = 40,000 h
μ_W = 18 kg/kmol
c_W = 0.10 DM/t, figure assumed for large chemical companies
μ_C = 12 kg/kmol
i_C = 1, assumed as limiting case
c_C = 1000 DM/t, figure assumed for large chemical companies

The concentrations x_F and x_W of organic pollutants in the example are on the high side. If they exceed the maximum permissible figures for the clarification plant, the effluent concerned must be mixed with others that are less concentrated. Thus it is not the concentration in the individual effluents that counts but the average concentration in all the effluents mixed together. The general rule is that pretreatment must be considered if the pollutant load is greater than 1% by weight and that burning the effluent is the most economic method if the pollutant concentration is higher than 10%.

The figures submitted for the parameters c_W and c_C are averages for large chemical works and cannot be applied to all cases. They depend largely on the design, size, and exploitation factor of the clarification plant and on local conditions.

It is assumed that in the range of concentrations considered the relative volatility is practically independent of concentration.

Fig. 15.8 shows a few characteristic total cost curves. They were compiled from the numerical evaluation along the lines indicated by Eq. (15-26). The main features are: the minima obtained for certain pairs of α, x_F values, and the limiting conditions (see curve shown in dotted lines) under which there is no difference in costs between integrated treatment and treatment of the effluent exclusively in the clarification plant. In other words, the condition imposed in Eq. (15-25) is satisfied. The resultant

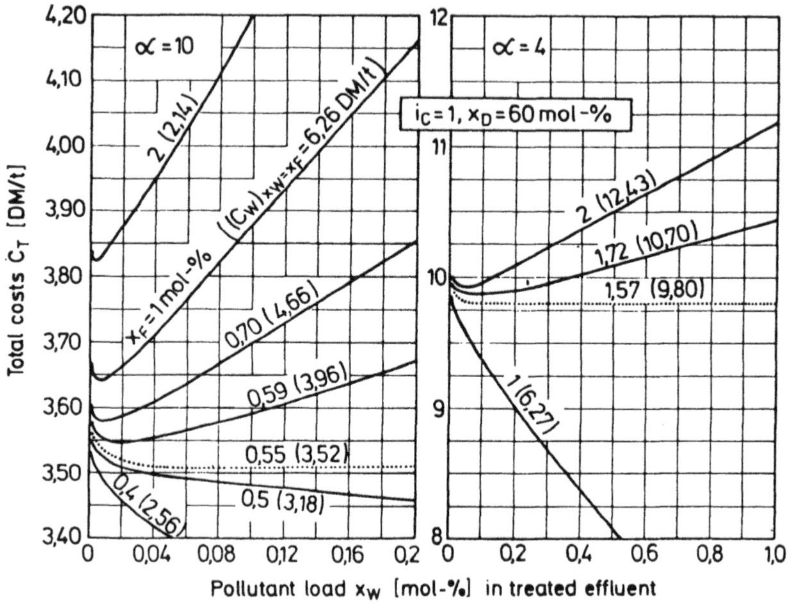

Fig. 15.8 Characteristic total costs curves for integrated effluent treatment (Example No. 1).

limiting curve is shown, again in dotted lines, in Fig. 15.9. It is analogous to that shown in the lower part of Fig. 15.7. By definition, it satisfies the conditions implied in Eq. (15-27). The area above this curve embraces pairs of α, x_F values for effluents that can be more economically purified by integrated treatment than by treatment in the clarification plant alone. The maximum reduction in costs can be determined with the aid of the full line curves in Fig. 15.9. These represent the following function

$$\left(1 - \frac{(C_T)_m}{(C_W)_{x_W = x_F}}\right)_{x_D = \text{const.}} 100 = f(x_F, \text{ parameter } \alpha) \qquad (15\text{-}28)$$

linking the minimum total costs $(C_T)_m$ for the integrated effluent treatment, the costs $(C_W)_{x_W = x_F}$ for the treatment in the clarification plant, the pollutant load x_F, and

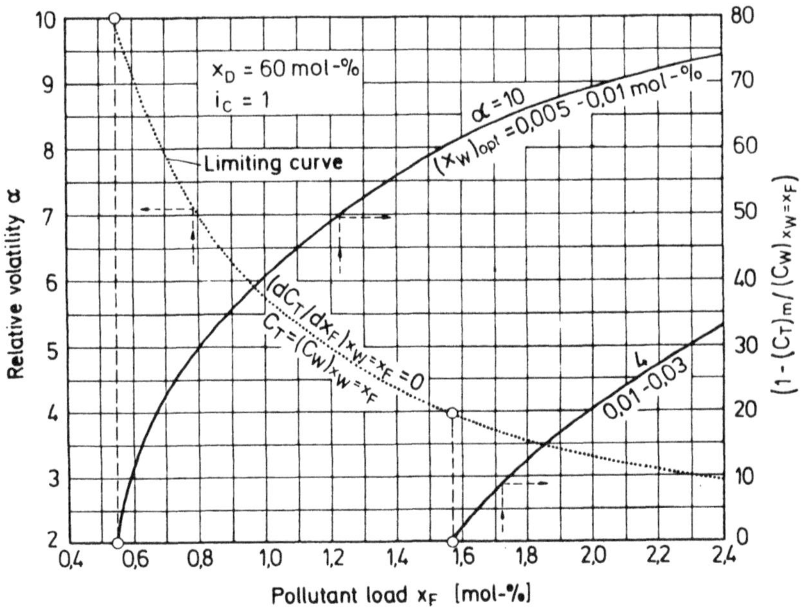

Fig. 15.9 Limiting curve for effluent treatment showing the zones in which the costs are in favor of integrated treatment (Example No. 1).

the relative volatility α. In the concentrate obtained from the thermal pretreatment, there are pollutant loads $(x_W)_{opt}$ [mol-%] which are optima from the economic aspect and at which the total costs minima are attained. These are also shown to complete the picture. The left-hand terminal points of these curves are marked by a circle. These points have an ordinate value of 0 and thus conform to the conditions that must pertain if C_T and $(C_W)_{x_W=x_F}$ are to be equal. The associated pair of α, x_F values must necessarily lie on the limiting curve; they are also marked with a circle.

It is obvious from the shape of the full-line curves in Fig. 15.9 that integrated effluent treatment may be much cheaper if the pollutant load is high and the volatility of the pollutants is high relative to that of water. However, Figs. 15.8 and 15.9 are valid only for i_C = 1, that is, for effluents whose pollutant molecules contain only the one carbon atom. If i_C is greater than unity, the costs for treatment in the clarification plant increase considerably, with the result that the integrated effluent treatment is favored even more.

Some limiting curves for various i_C values are presented in Fig. 15.10, which unequivocally shows the economic limits for thermal pretreatment of effluents. The curves have been based on the assumption that the boiling point of the pollutants is lower than that of water. However, it can be easily demonstrated that the shape of the curves would not be apprecialy changed if the situation were reversed, that is, if the water were the ingredient with the lower boiling point.

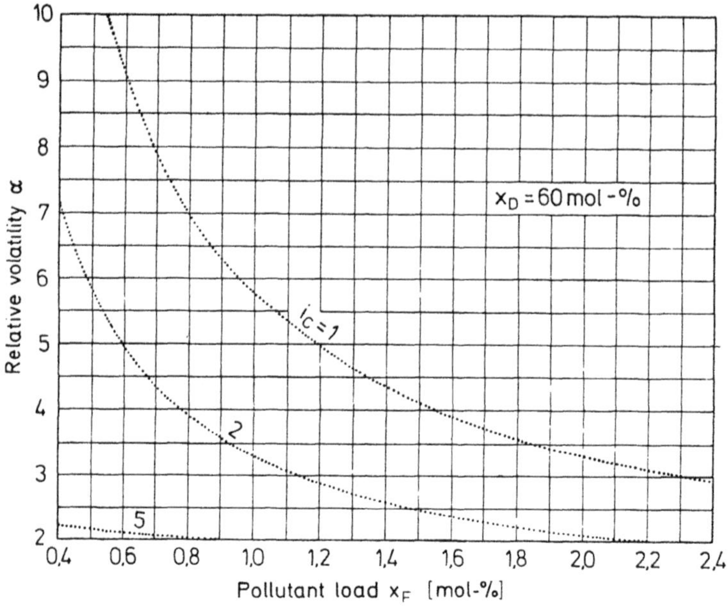

Fig. 15.10 Limiting curves for factory effluents containing organic pollutants (Example No. 1).

If certain factory effluents are discharged into the main effluent stream, the maximum permissible organic pollutant load in the feed to the clarification plant may be exceeded. Such effluents always demand treatment before clarification unless it would be more economical to burn them.

Let us now take the limiting case of very high relative volatility α. It can be seen from the shape of the limiting curves that here thermal pretreatment followed by biological purification offers the greatest advantage, even if the pollutant concentration in the effluent is comparatively low. In this case, the thermal treatment may consist simply of evaporation, which can give rise to economical installations if it is carried out in stages.[5,8]

The manifold tasks in technological protection of the environment thus inevitably increase the scope for applying distillation and rectification technology and the equipment required for it. In other words, ecological protection has exerted a great influence on equipment trends in chemical engineering. Plant design must be governed by a number of aspects that depend on the properties of the feedstocks, intermediates, and final products, and on their behavior under process conditions.

An example is a plant for distillation and removal of process water and residues from a crude product. The flow chart is shown in Fig. 15.11. First the crude product is freed from process water in a rectification column "a" operating at atmospheric pressure. The water is collected together with byproducts with a lower boiling point at the head of the column. Any water remaining at the foot of this column is removed in

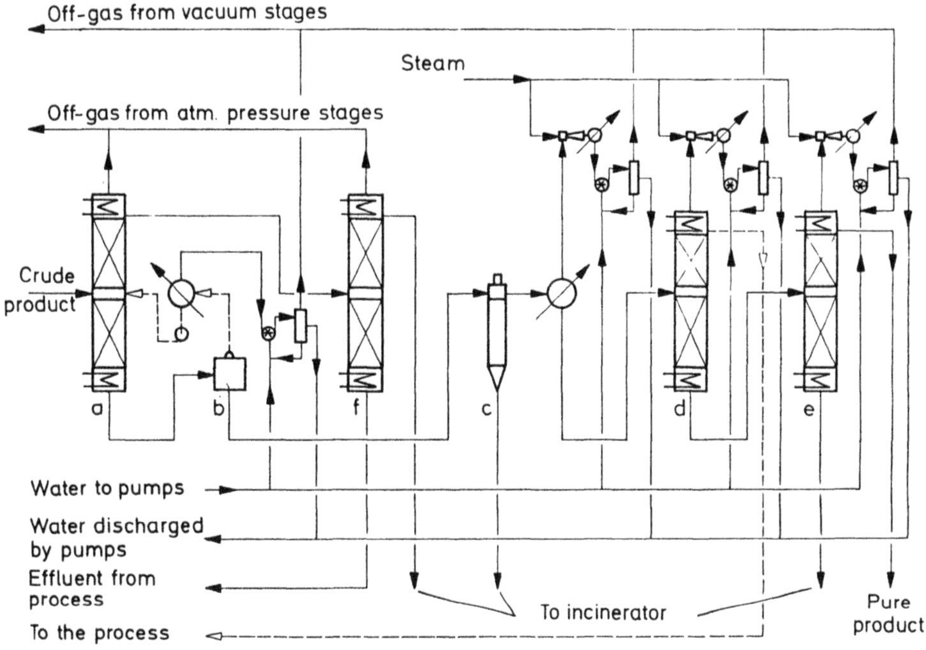

Fig. 15.11 Flow chart of a plant for purifying a crude product by removal of process effluent and residues by distillation.

a flash evaporator "b" operating under vacuum. Thus allowance is made for the low thermal stability of the product and the tendency of residues to form crusts on the heated surfaces of tube vaporizers. For the same reason, the residue is separated with a short residence time under vacuum in a thin-film evaporator "c" with rotating fittings. The low-boiling and high-boiling impurities are removed under mild conditions for the product in two rectification columns "d" and "e", which are arranged in series and which are equipped with contact fittings that have a low pressure drop and a low liquid hold-up and which operate under vacuum. Another rectification column "f" operates at atmospheric pressure and removes organic impurities from the process water withdrawn from the top of column "a" before it is run into the biological treatment plant. The residue from evaporator "c" and the impurities from columns "e" and "f" are incinerated.

This is a particularly characteristic example of including environmental protection aspects in process planning. It must be first realized that the main causes of damage to the product are high temperatures and long residence times. The formation of thermally cracked products and thus the emission of pollutants in the gases leaving these four vacuum stages are reduced to a minimum by the flash vaporizer "a", the thin-film evaporator "c" and the columns "d" and "e" with the low pressure drop and low liquid content, and by operating these under vacuum. In addition, column "f" allows economic effluent treatment along the lines indicated in Example No.. 1.

If the boiling points of the components of the mixture to be separated do not differ very much that is, if the relative volatility of the binary system is comparatively small, another advantage of using column fittings with a low pressure drop is a reduction in costs. Owing to the successive pressure drops, the pressure in the column increases in the direction of top to bottom. As a result of this, the relative volatility may become much less along the length of the column. Thus the minimum for the total costs that is, the sum of the prime and the running costs, which could be expected to occur at the optimal reflux ratio, is displaced in the direction of lower values if the pressure drop per theoretical stage is small. In Fig. 15.12 the number of theoretical

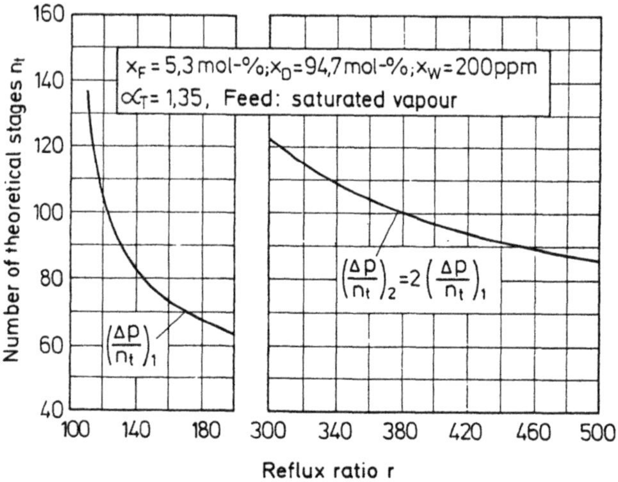

Fig. 15.12 Number of theoretical stages as a function of the reflux ratio for different column fittings with different pressure drops per theoretical stage. The relative volatility of the mixture depends on pressure.

stages required for separating a binary system with a low relative volatility is plotted against the reflux ratio for two different types of column fittings. The specific pressure drops for these two fittings differ by a factor of two. It can be seen from these curves that the difference in costs could be very large. Another advantage is that cheaper steam at a lower pressure may be used or the vaporizer at the foot may have a smaller heat transfer area if the pressure at the foot of the column is low as a result of the small pressure drop per theoretical stage.

The example demonstrates that it is possible to satisfy ecological demands ideally by treating the product under mild conditions in a plant designed to run at optimal costs.

XVI Outlook

From the viewpoint of industrial application and problem solving the present status of the distillation technology can best be described by the question, "to what extent can fractionation equipment be designed and operated for highest economy?"

New plants are designed with the prime target being the determination of the optimum process requirement as well as the selection of the most cost-effective apparatus combined with the optimization of their sizing. For a conceptual design of a fractionation plant an intimate knowledge of the relationship between the separation effect and the fluidity behavior, and thus of the design variables is needed.

As is shown in Chapter 15 the distillation and rectification technology has received a strong impulse through environmental protection requirements. This is true for the process design of production units of chemical and related industries in that, for example, the separation equipment of some processes have now to be designed for higher product purity and other process requirements as well as for specific measures for the protection of environment, that is, measures for the treatment of process water and air. Because of the resulting substantial increase in the investment and energy costs, the planning engineer is more than ever pressed to design the separation equipment for optimal economy. Generally this calls for the knowledge of characteristic data about the process assessment of the apparatus on one hand and the thermodynamics of the separation of the components on the other hand, whereby the extent of the calculation must allow for the production and operating requirements—resulting in complex evaluations.

When designing a plant, for example, for the separation of a binary system of which the relative volatility α at the pressure p_T at the top of the column is relatively low and which as a consequence of the expected increase in pressure, decreases further towards the bottom of the rectification column, then one is confronted with an economical limit[1] of the pressure drop $\Delta p/n_t$ of the selected column internals which is relative to a theoretical stage. This is a consequence of a shift in the minimum total cost C_T, which is the sum of equipment and operating costs, towards a greater reflux ratio r as well as towards a greater number of theoretical stages $n_t = f(r, \alpha)$ and consequently towards high absolute values $(C_T)_m$, when $\Delta p/n_t$ is large (Fig. 16.1).

In sample studies it can readily be shown that a deviation from the optimum in the design of a separation unit can bring about a large increase in cost.

It is thus obvious that it is an indispensable prerequisite for a comprehensive optimization to have exact knowledge about the phase equilibrium of the system, of the separation effect and the pressure drop of the selected column internals, and the auxiliary equipment.

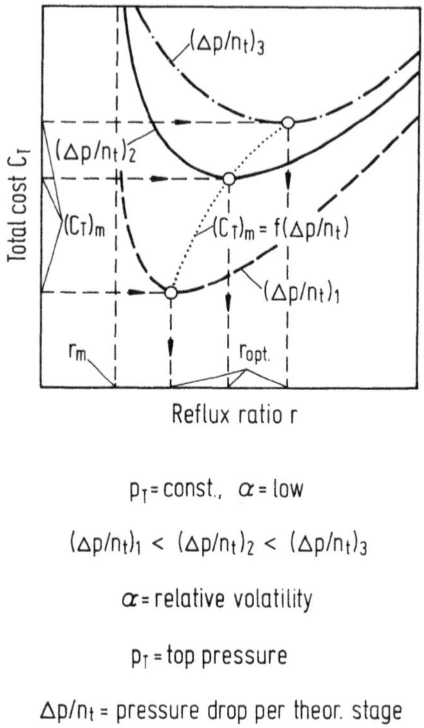

Fig. 16.1 Influence of the pressure drop per theoretical stage upon the result of the cost optimization of separation column.

When scrutinizing the pertinent literature one can only conclude that it is rather deficient in presenting data for the evaluation of common and conventional column internals. While the pressure drop-loading dependence can be predetermined for nearly all common designs within the regime of technical significance, the calculation methods for the mass transfer efficiency (resulting from the predetermination of the pressure drop per theoretical stage) even for conventional equipment is rather inadequate for some systems. Experimental investigations which serve as the basis for planning are thus at this time unavoidable in many cases. The importance of these interrelations for the selection of the optimal column internals can be derived clearly from Chapter 7.

Finally, reference should be made to the progress in the development of the industrial application of the solid–liquid rectification (not covered here) for difficult separations with high purity requirements, where distillation or extraction is economically or technically not feasible. The apparatus shown schematically in Fig. 16.2 developed by UC Australia can be employed advantageously for eutectic systems;[2] pilot plants and also industrial applications have shown good results. Detailed basic research is still needed to develop exact methods for the determination of optimal design parameters through the knowledge of the best mass transfer and fluid dynamic requirements. Activity in this area is on a worldwide scale.[3]

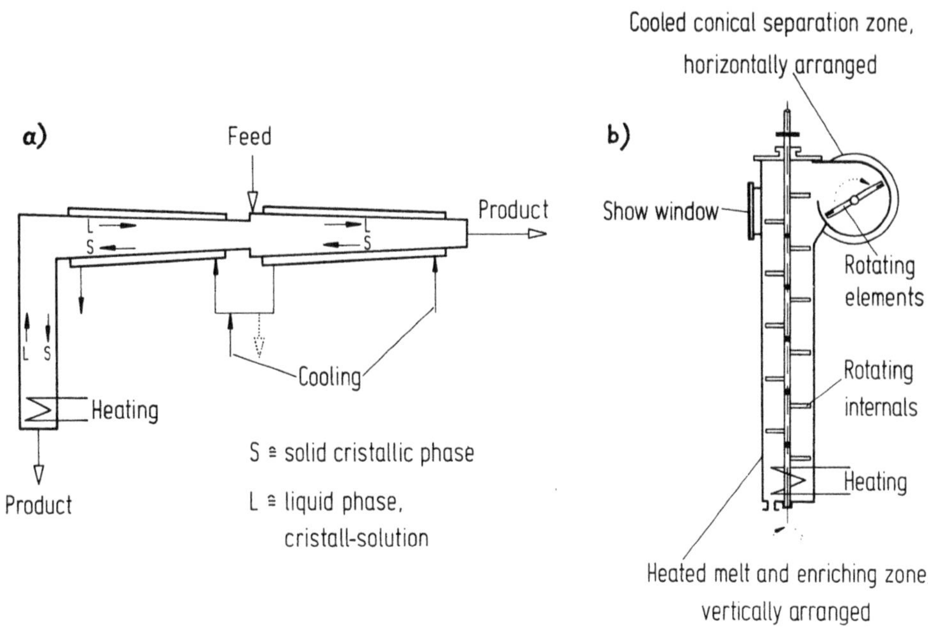

Fig. 16.2 Apparatus for solid-liquid rectification according to Union Carbide, Australia: a) Working principle, b) schematic of the cross section.

Bibliography

Chapter 1

1) E. Kirschbaum: Distillation and Rectification. Springer, Berlin-Heidelberg-New York 1969.
2) E. Krell: Manual of Laboratory Distillation. VEB-Deutscher Verlag der Wissenschaften, Berlin, 1958.
3) R. Billet: Fundamentals of Thermal Separation of Liquids. Verlag Bibliographisches Institut, Mannheim 1962.

Chapter 2

1) H. Ulich: Physical Chemistry, 5th edition. Th. Steinkopf, Dresden and Leipzig 1948.
2) E. Kirschbaum: Distillation and Rectification. Springer, Berlin-Göttingen-New York 1969.
3) K. Nesselmann: Applied Thermodynamics, 1st edition. Springer, Berlin-Göttingen-Heidelberg 1950.
4) P. Duhem, Trav. et Mem. Fac. Lille, Ne. 13, 75 (1894).
5) M. Margules, S. ber. Akad. Wiss. Wien *104*, 1243 (1895).
6) J. J. van Laar, Phys. Chem. *72*, 723 (1910); *83*, 599 (1913).
7) D. F. Othmer, L. Ricciardi u. M. Thakar, Ind. Engng. Chem. *45*, No. 8, 1815 (1953); see also R. F. Baumann, Chem.-Ing.-Techn. *28*, 396 (1956).
8) A. J. Teller, Chemical Engineering *61*, 9, 168 (1954).
9) F. Neumann, Erdöl u. Kohle *4*, 561 (1951).
10) J. H. Perry: Chemical Engineers Handbook, 3rd Edition, McGraw-Hill Book Co. Inc., New York, Toronto, London 1950.
11) K. Thormann: Distillation and Rectification. O. Spamer, Leipzig 1928.
12) L. H. Horsley, Ind. Engng. Chem. Anal. *19*, No. 8, 508 (1947).
13) C. V. Rechenberg: Simple and Fractional Distillation. Selverlag von Schimmel u. Co., Miltiz bei Leipzig 1923.
14) A. & E. Rose & A. Glasebrook, Distillation, Vol. 4 of Tech. of Organ. Chem., Interscience Publishers, Inc., New York 1951.
15) R. Billet: Principles of Thermal Separation of Liquids. Verlag Bibliographisches Institut, Mannheim 1962.
16) Carlson, A. P. Colburn, Ind. Engng. Chem. *34*, 581 (1942).
17) E. G. Scheibel, Friedland, Ind. Engng. Chem. *39*, 1329 (1947).
18) W. Jost, Naturforschung *1*, 576 (1946).
19) A. J. V. Underwood, Chem. Engng. Progr. *44*, 603 (1948).
20) M. R. Fenske, Ind. Engng. Chem. *24*, 482 (1932).
21) E. R. Gilliland, Ind. Engng. Chem. *32*, 1220 (1940).
22) O. Nagel & R. Sinn, Chem.-Ing.-Techn. *38*, 265 (1966).
23) O. Nagel, H. Smits, D. Wolf & H. Günzler, Chem.-Ing.-Techn. *41*, 983 (1969).
24) O. Redlich & A. T. Kister, Ind. Engng. Chem. *40*, 345 (1948).
25) J. M. Prausnitz & F. H. Shair, A. I. Ch. E. Journal 7, 682 (1961).
26) R. V. Orye & J. M. Prausnitz, Ind. Engng. Chem. *57*, 19 (1965).
27) G. M. Wilson, J. Am. Chem. Soc. *86*, 127 (1964).

Chapter 3

1) R. Billet, Chem.-Ing.-Techn. *30*, 513 (1958).
2) E. Kirschbaum, VDI Beih. Verf., No. 1, 15 (1943).
3) E. Kirschbaum, Chem.-Ing.-Techn. *21*, 293 (1949).
4) H. Ulich: Physical Chemistry, 5th edition. Th. Steinkopf, Dresden and Leipzig 1948.
5) R. Billet: Principles of Thermal Separation of Liquids. Verlag Bibliographisches Institut, Mannheim 1962.
6) E. Kirschbaum: Distillation and Rectification, Springer, Berlin-Heidelberg-New York 1969.
7) J. H. Perry: Chemical Engineers Handbook, 3rd Edition. McGraw-Hill Book Co. Inc., New York, Toronto, London 1950.
8) W. L. McCabe & E. W. Thiele, Ind. Engng. Chem. *17*, 605 (1925).
9) M. R. Fenske, Ind. Engng. Chem. *24*, 482 (1932).
10) A. J. V. Underwood, Chem. Engng. Progr. *44*, 603 (1948).
11) A. P. Colburn, Trans. Am. Inst. Chem. Engrs. *37*, 805 (1941).
12) W. D. Harbert, Ind. Engng. Chem. *37*, 1162 (1945).
13) H. U. Kohrt, Chem.-Ing.-Techn. *20*, 117 (1948).
14) H. U. Kohrt, Chem.-Ing.-Techn. *21*, 384 (1949).
15) E. G. Scheibel & C. F. Montros, Ind. Engng. Chem. *38*, 268 (1946).
16) E. J. Jenny, Trans. Am. Inst. Chem. Engrs. *35*, 635 (1939).
17) R. N. Schiras, D. N. Hanson & C. H. Gibson, Ind. Engng. Chem. *42*, 871 (1950).
18) E. R. Gilliland, Ind. Engng. Chem. *32*, 918 & 1220 (1940).
19) R. L. Smith, Trans. Am. Inst. Chem. Engrs. *37*, 333 (1941).
20) G. G. Brown & D. E. Holcomb, Petrol. Engr. *11*, 27 (1940).
21) E. Kirschbaum & G. Lipphardt, Chem.-Ing.-Techn. *29*, 393 (1957).
22) G. Lipphardt: Diss. T. H. Karlsruhe, 1957.
23) E. Tröster: Diss. T. H. Karlsruhe, 1960.
24) E. Kirschbaum & F. Gerstner, VDI-Beih. Verfahrenstechnik No. 1, 10 (1939).

Chapter 4

1) R. Billet, Chem.-Ing.-Techn. *30*, 407 (1958).
2) R. Billet, Der Maschinenmarkt *65*, No. 30/31, 76 (1959).
3) E. H. Smoker & A. Rose, Trans. Am. Inst. Chem. Engrs. *36*, 285 (1940).
4) R. Billet: Principles of Thermal. Separation of Liquids. Verlag Bibliographisches Institut, Mannheim 1962.
5) D. Mohnhaupt, Erdöl & Kohle *11*, 858 (1958).
6) C. D. Hodgman, R. C. Weast, S. M. Selby: Handbook of Chemistry and Physics, 37th Edition Cleveland, Ohio, Chemical Rubber Publishing 1955.
7) K. Höffer, Chem.-Ing.-Techn. *42*, 933 (1970).

Chapter 6

1) E. Kirschbaum: Chem.-Ing.-Techn. *28*, 639 (1956).
2) E. Kirschbaum, VDI *75*, 1212 (1931).
3) T. H. Chilton & A. P. Colburn, Ind.-Eng.-Chem. *27*, 255 (1935).
4) H. Klingenspor, Chem.-Ing.-Techn. *31*, 598 (1959).
5) F. H. Garnn & Coworkers., J. Inst. Petroleum *43*, 399 (1957).
6) T. Hobler, Chemia Stosowana *2*, 105 (1957).
7) H. S. Bell: American Petroleum Refining. Van Nostrand, New York 1945.
8) R. Billet. J. Chem. Engng. Japan *2*, 107 (1969).
9) R. Billet: Optimization in Rectification. Verlag Bibliographisches Institut, Mannheim 1967.
10) R. Billet & L. Raichle, Chem. Eng. Feb. 13, 145 (1967); Feb. 27, 149 (1967).

11) R. Billet & L. Raichle, Chem.-Ing.-Techn. *38*, 825 (1966).
12) EFChE: Daily Report, Chem.-Ing.-Techn. *39*, (1967).
13) R. Billet, Chem.-Ing.-Techn. *40*, 43 (1968).
14) R. Billet, Chem. Engng. Progr. *63*, 53 (1967).
15) Patent owned by Sulzer of Winterthur, Switzerland. U.S. License to Koch Engineering, Wichita, Kansas.
16) R. Billet: Chem. Eng. Feb. *21*, 68 (1972).
17) A. Sperandio, M. Richard & M. Huber, Chem.-Ing.Techn. *37*, 322 (1965).
18) Design by F. Glitsch of Dallas, Texas. German license to GHH Sterkrade.
19) R. Billet & O. Nagel, Chem.-Ing.Techn. *39*, 1311 (1967).
20) German manufacturer: GHH Sterkrade-Oberhausen.
21) German manufacturer: Vereinigte Füllkörperfabriken, Baumbach, Westerwald, Dr. F. Raschig GmbH, Ludwigshafen; in England: Hydronyl Ltd., London; in America: U.S. Stoneware Co., Akron/Ohio.
22) L. Raichle & R. Billet, Chem.-Ing.-Techn. *35*, 831 (1963).
23) L. Raichle & R. Billet, Erdöl & Kohle, Erdgas-Petrochemie *17*, 715 (1964).
24) R. Billet & O. Nagel, Chem.Ing.-Techn. *36*, 1257 (1964).
25) Stanslaus and Smith: Paper read at International Symposium on Distillation, Brighton, The Institute of Chem. Eng., London. 1960.
26) L. Raichle & R. Billet, Chem.-Ing.-Techn. *37*, 669 (1965).
27) Based on research done by Fractionation Research for F. Glitsch of Dallas.
28) Koch Flexitray Design Manual, Bulletin 960, Wichita (1960).
29) Design by Koch Engineering, Wichita, DBP 1014078, 1958.
30) R. Billet, Chem.-Ing.-Techn. *40*, 377 (1968).
31) R. Billet, Ing. Chim. Ital. *4*, 106 (1968).
32) R. Billet: Paper read at A.I.Ch.E.–64th National Meeting, Preprint 9G, New Orleans (1969).
33) R. Billet, Brit. Chem. Eng. *14* Nr. 4, 491 (1969).
34) R. Billet: Paper read at International Symposium on Distillation, Brighton, England (1969), The Institution of Chemical Engineers, London (1970).
35) R. Billet, Chem.-Ing.-Techn. *42*, 910 (1970).
36) R. Billet: Note for: Primera Convención Mundial de Ingenería Química 1965, 558, México, D. F., Instituto Mexicano de Ingenieros Químicos, A. C., Congress Publication. (1966).
37) F. Neumann, Chem.-Ing.-Techn. *33*, 439 (1961).
38) R. Billet: Columns for Separation in Process Work. Verlag Bibliographisches Institut, Mannheim 1971.
39) R. Billet: Paper read at A.I.Ch.E.–68th National Meeting, Houston, Texas (1971).
40) R. Billet: Lectures at the Meeting of the European Federation of Chemical Engineers at Amsterdam 1965 and Moscow 1966.
41) R. Billet: Verfahrenstechnik *6*, Nr. 3 (1972) sowie W. Meier: Discussion contribution.
42) E. Kirschbaum: Distillation and Rectification, 3rd Edition, Springer, Berlin-Göttingen-Heidelberg 1960.
43) R. Billet: Lecture at Japan Annual Meeting, Chemie-Ingenieure in Kyoto (1968).
44) R. Billet, S. Conrad & C. M. Grubb: Paper read at International Symposium on Distillation Brighton/England (1969).
45) Kirschbaum-Stora, system made by Müller-Schuss KG, Weidenau, Sieg, DBP 1024924, 1958.
46) Büche System, design by Luwa AG, Zürich, DBP 957119, 1957.
47) Jantzen System, design by Schmidding, Köln.
48) R. Billet: Vortrag, Paper read at Conference 'Recent Advances in Mass Transfer' in Bradford, England (1964).
49) L. Raichle & R. Billet, Chem.-Ing.-Techn. *37*, 367 (1965).
50) L. Raichle & R. Billet, Ind. Eng. Chem. *57*, 52 (1965).

52) F. Neumann: Paper read at 83rd. DECHEMA-Colloquium in Frankfurt/Main (1960).
53) P. M. Heertjes: Paper read at International Symposium on Distillation, Brighton/England (1960).
54) F. Neumann, Chem.-Ing.-Techn. *33*, 485 (1961).
55) V. Stevanovic: Diss. E. T. H. Zürich, 1959.
56) V. Stevanovic: Paper read at International Symposium on Distillation, Brighton/England (1960).
57) V. A. Malyusov & N. M. Zhavoronkov: Paper read at International Symposium on Distillation, Brighton/England (1960).
58) G. Adolphi & R. Fleischer, Chem. Technik *7*, 638 (1955).
59) G. Adolphi, Chem. Technik *7*, 324 (1955).
60) G. Adolphi, Chem. Technik *13*, 25 (1961).
61) K. Bevers: Publication "f2" of the firm Julius Montz, Hilden (1963).
62) W. Kuhn, Chem.-Ing.-Techn. *25*, 12 (1953).
63) E. Jantzen & O. Wieckhorst, Chem.-Ing.-Techn. *26*, 392 (1954).
64) J. W. Westhaver, Ind. Eng. Chem. *34*, 126 (1942).
65) Kloss System, made by Montz, Hilden, DBP 1084699, 1960.
66) G. Adolphi, Chem. Technik *13*, 25 (1961).
67) Kloss System, made by Montz, Hilden.
68) FRI-Fractionation Design Handbook, South Pasadena, Cal., USA.

Chapter 7

1) R. Billet, Contribution to: Primera Convención Mundial De Ingenieria Quimica 1965, México, D. F. Instituto Mexicano de Ingenieros Quimicos, A. C., Congress Publication 1966.
2) R. Billet: Lecture at Nitrogen Research Center, USSR in Moscow (1966).
3) R. Billet & L. Raichle, Chem.-Ing.-Techn. *39*, 167 (1966).
4) R. Billet & L. Raichle, Chem. Eng. Feb. 27, 149 (1967).
5) R. Billet: Optimization in Rectification, Verlag Bibliographisches Institut, Mannheim 1967.
6) R. Billet, Der Maschinenmarkt *67*, No. 43, 15 (1961).
7) R. Billet: Fundamentals of Thermal Separation of Liquids, Verlag Bibliographisches Institut, Mannheim 1962.
8) R. Billet, Notes of the BASF Test Group, 14 (1965).
9) R. W. Merriman, Journal of the Chemical Society of London *103*, 1790 u. 1801 (1913).
10) F. Neumann, Erdöl u. Kohle *4*, 56 (1951).
11) O. v. Keussler: Thesis, Darmstadt, 1926.
12) H. Guinot & F. W. Clark, Trans. Inst. Chem. Eng., London, *16*, 181 (1938).
13) R. Fritzweiler, VDI *48*, 1373 (1938).
14) W. Jost, Naturforschung *1*, 576 (1946); Chem.-Ing.-Techn. *3*, 64 (1951).
15) C. L. Dunn, R. W. Millar, G. J. Pierotti, R. W. Shiras & M. Souders, Trans. Am. Inst. Chem. Engrs. *41*, 631 (1945).
16) H. Ratte, Erdöl u. Kohle *1*, 25 (1961).
17) P. Faldix & H. Stage, Chem.-Ing.-Techn. *41*, 1265 (1969).
18) W. D. Harbert, Petroleum Refiner *29*, No. 10, 117 (1950).
19) P. Duhen, Chim. Ind., Gén. chim. *83*, No. 5, 125 (1960).
20) H. Ramer, Chem. Technik *22*, 217 (1970).
21) F. B. Petlyuk, V. M. Platonov & V. S. Aretyan, Chem. Ind. 865 (1966).
22) F. B. Petlyuk, V. M. Platonov & D. M. Slavinskii, Chem. Ind. 206 (1965).
23) B. G. Bergo, V. M. Platonov & Ya. D. Monko, Internat. Chem. Engng. *3*, 269 (1963).
24) H. Stage, Chem. Zeitung *94*, 955 (1970).
25) R. Billet u. K. Höffer: Evaporators and Calculations, Vogel-Verlag, Würzburg 1959.

Chapter 8

1) R. Billet, S. Conrad & C. M. Grubb, Verfahrenstechnik *4*, No. 7 (1970).
2) Koch Bulletin KF3, Koch International Sp. A, Bergamo (1969).
3) E. Kirschbaum, Chem.-Ing.-Techn. *28*, 639 (1956).
4) E. Kirschbaum: Distillation and Rectification. Springer, Berlin-Göttingen-Heidelberg 1950.
5) M. Huber & R. Hiltbrunner, Chem. Engineering Science *21*, 829 (1966).
6) T. K. Sherwood, C. H. Shipley & F. A. L. Holloway, Ind. Eng. Chem. *30*, 765 (1938).
7) M. Leva: Tower Packings and Packed Tower Design. U. S. Stoneware Comp., Akron/Ohio 1951.
8) J. S. Eckert, Chem. Eng. Progr. *59*, 5, 76 (1963).
9) R. Billet: Thesis, Albeit T. H. Karlsruhe, 1953.
10) E. Kirschbaum & A. David, Chem.-Ing.-Techn. *25*, 592 (1953).
11) E. Kirschbaum, W. Busch u. R. Billet, Chem.-Ing.-Techn. *28*, 475 (1956).
12) R. Billet, Chem.-Ing.-Techn. *32*, 517 (1960).
13) A. David: Diss. T. H. Karlsruhe, 1953.
14) C. H. G. Hands & F. R. Whitt, J. appl. Chem., London, *1*, 135 (1951).
15) R. Billet: Fundamentals of Thermal Separation of Liquids. Verlag Bibliographisches Institut, Mannheim 1962.
16) BASF Utility Patent 1891 441, 1964.
17) R. Billet: VDI paper "Material Transfer and Rectification" Haus der Technik, Essen (1967).
18) R. Billet: Paper read at 61st A.I.Ch.E. National Meeting, Houston, Texas (1967).
19) R. Billet, Notes of BASF Test Group, Vol. 18 (1967).
20) R. Billet: Lecture at Discussion Meeting of European Federation of Chemical Engineers, Ludwigshafen (1967).
21) R. Billet, Chem. Eng. Progr. *63*, 53 (1967).
22) R. Billet, Chem.-Ing.-Techn. *40*, 43 (1968).
23) Recently published data for pressure loss calculations give values notably different from those in Fig. 8.19.
24) E. Kirschbaum, Chem.-Ing.-Techn. *28*, 639 (1956).
25) From measurements by author and staff made for BASF at Fractionation Research Inc., Alhambra, Cal.
26) A. G. Kassatkin: Chemische Verfahrenstechnik II. VEB Verlag, Leipzig 1961.
27) A. Meersmann, Chem.-Ing.-Techn. *37*, 218 (1965).
28) R. Billet: Paper read at the 68th A.I.Ch.E. National Meeting, Houston, Texas (1971).
29) R. Billet: Contribution to the International Symposium on Distillation, Brighton, England (1969).
30) R. Billet, Chem. Eng. Progr. *66*, 1, 41 (1970).
31) R. Billet, Chem.-Ing.-Techn. *42*, 910 (1970).
32) R. Billet, S. Conrad u. C. M. Grubb: Beitrag zum International Symposium on Distillation, Brighton/England (1969).
33) K. Hoppe & M. Mittelstrass: Fundamentals of Dimensioning Column Heads. Th. Steinkopf, Dresden 1967.
34) R. Billet: Columns for Process Separations. Verlag Bibliographisches Institut, Mannheim 1971.
35) R. Billet & O. Nagel, Chem.-Ing.-Techn. *36*, 1267 (1964).
36) R. Billet & O. Nagel, Chem.-Ing.-Techn. *39*, 1311 (1967).
37) E. V. Murphree, Ind. Eng. Chem. *17*, 747 (1925).
38) H. Hausen, Chem.-Ing.-Techn. *25*, 595 (1953).
39) G. Standart, Chem. Eng. Sci. *20*, 611 (1965).
40) A. P. Colburn, Ind. Eng. Chem. *28*, 526 (1936).
41) A.I.Ch.E.: Bubble-Tray Design Manual. American Institut of Chemical Engineers, New York 1958.

42) E. Kirschbaum, Chem.-Ing.-Techn. *26*, 461 (1954).
43) W. Hübner & E. U. Schlünder: Contributions to International Symposium on Distillation, Brighton/England (1969).
44) DECHEMA: DECHEMA-Erfahrungsaustausch, Destillieren, Frankfurt/Main (1952).
45) W. M. Newton, J. W. Mason, T. B. Metcalfe & C. O. Summers, Petroleum Refiner *31*, No. 10, 141 (1952).
46) R. R. Hughes & E. R. Gilliland, Chem. Eng. Progr. *48*, 497 (1952).
47) G. G. Brown: Unit Operations. John Wiley & Sons, Inc., New York 1960.
48) F. J. Lockhart & C. W. Leggett: Advances in Petroleum Refining and Chemistry, I. Interscience Publishers Inc., New York 1958.
49) W. J. Thomas & A. N. Shah, Trans Instn. Chem. Engrs. *42*, 71 (1964).
50) W. J. Thomas & M. Campbell, Trans Instn. Chem. Engrs. *45*, 53 (1967).
51) S. P. S. Andrew: Beitrag zum International Symposium on Distillation, Brighton/England (1960)
52) E. Kirschbaum, Chem.-Ing.-Techn. *28*, 713 (1956).
53) E. Kirschbaum, VDI *98*, 1797 (1956).
54) J. R. Fair & R. L. Matthews, Petrol. Refiner *37*, 153 (1958).
55) W. L. Bolles, Petroleum Processing *11*, No. 2, 3, 4 and 5 (1956).
56) E. Kirschbaum, Chem.-Ing.-Techn. *31*, 779 (1959).
57) P. Grassmann, Chem.-Ing.-Techn. *28*, 270 (1956).
58) P. Grassmann, Chem.-Ing.-Techn. *29*, 497 (1957).
59) H. Stage, Chem.-Ing.-Techn. *42*, 669 (1970).
60) K. Sigwart, VDI-Beih. Verf. No. 3, 79 (1943).
61) R. Billet, Verfahrenstechnik *5*, No. 8 (1971) and Brit. Chem. Eng. 17, No. 9, 705 (1972).
62) F. D. Mayfield, W. I. Church, A. C. Green & R. W. Rasmussen, Ind. Engng. Chem. *44*, 2238 (1952).
63) R. G. H. Prince: Notes from International Symposium on Distillation, Brighton/England (1960).
64) K. Blechschmidt, Chem. Technik *13*, 567 (1961).
65) J. D. Chase, Chem Engng. 105 (1967).
66) C. d'A. Hunt, D. N. & C. R. Wilke, A.I.Ch.E. Journal *1*, 441 (1955).
67) L. Gyökhegyi: Diss. Hochschule für Schwermaschinenbau Magdeburg, 1960.
68) L. Sum-Shik, Khim, promysl., 66 (1968).
69) Fig. 8.81 (A, B) is based upon data by Koch Engineering Wichita, Kansas/USA and Koch International, Bergamo/Italy.
70) K. Nitschke, Chem. Technik *20*, 411 (1968).
71) W. Bruckert & D. Wang: US-Patent 3 282 576 (1966) and British Patent 1 013 547 (1965), Union Carbide Corp.
72) J. C. Frank, G. R. Geyer & H. Kehde, Chem. Eng. Progr. *65*, 79 (1969).
73) W. K. Lewis, Ind. Eng. Chem. *28*, 399 (1936).
74) From data by Union Carbide Deutschland, Düsseldorf.
75) W. Bruckert: US-Patent 3 410 540, Design by Union Carbide Corp.
76) Registered patent of Stahl of Viernheim.
77) Fig. 8.83 is based upon data from Stahl, Viernheim, Germany.
78) H. Stage & L. Gemmeker, Chemical Plant and Processes, *69* (1970).
79) J. Majeweski, Brit. Chem. Engng. *4*, 336 (1959).
80) D. H. Stanislas & A. J. Smith: Notes from International Symposium on Distillation, Brighton, England (1960).
81) Shell Development Comp., Chem. Eng. Progr. *50*, 57 (1954).
82) K. Hoppe & G. Krüger: Lecture at ACHEMA (1970).
83) K. Hoppe & G. Krüger: DBP Invention disclosure 1 719 451, Registered by VEB-Germania, Karl-Marx-Stadt.

84) M. Leva, Chem. Eng. Progr. *67*, 65 (1971).
85) P. G. Nygren & G. K. S. Connolly, Chem. Eng. Progr. *67*, 49 (1971).
86) H. Kutzer, Chem.-Ing.-Techn. *42*, 128 (1970).

Chapter 9

1) R. Billet: Evaporator Engineering. Verlag Bibliographisches Institut, Mannheim 1964.
2) W. Nusselt, VDI *60*, 541 (1916).
3) E. Gnam, VDI-Forschungsheft 382 (1937).
4) Dechema Monographie, Vol. 20, No. 244. Verlag Chemie, Weinheim 1952.
5) Z. Rant: Theory and Practice of Evaporation. Th. Steinkopf, Dresden and Leipzig 1959.
6) M. Jacob: Heat Transfer, vol. 1., John Wiley & Sons, New York 1950.
7) H. Kraussold, VDI-Forschungsheft B351 (1931) und VDI-Forschungsheft A, Vol. 4 (1933).
8) R. Billet & K. Höffer, Der Maschinenmarkt *65*, No. 12, 5 (1959).
9) R. Billet, Chem.-Ing.-Techn. *29*, 733 (1957).
10) E. Kirschbaum & K. Dieter, Chem.-Ing.-Techn. *30*, 715 (1958).
11) R. Bressler, VDI *100*, 630 (1958).
12) W. Linke, Kältetechnik *5*, 275 (1953).
13) G. D. Kerns, Petroleum Refiner *39*, (7) 168 (1960).
14) Lord Rayleigh, Phil. Mag. *4* (6), 521 (1902).
15) R. Billet: Fundamentals of Thermal Separation of Liquids. Verlag Bibliographisches Institut, Mannheim 1962.
16) H. Vorkauf, Forschung a. d. Gebiet des Ingenieurwesens, B 341 (1931).

Chapter 10

1) E. Kirschbaum: 2. Dechema-Monographie. Verlag Chemie, Berlin 1930.
2) E. Kirschbaum & G. Lipphardt, Chem.-Ing.-Techn. *29*, 393 (1957).
3) G. Lipphardt: Diss. T. H. Karlsruhe 1957.
4) E. Tröster: Diss. T. H. Karlsruhe 1960.
5) E. Kirschbaum, Chem.-Ing.-Techn. *29*, 595 (1957).
6) E. Kirschbaum, Ann. Phys. *20*, H. 3 (1934).
7) E. Herrmann: Diss. T. H. Karlsruhe 1954.
8) E. Kirschbaum & E. Tröster, Chem.-Ing.-Techn. *32*, 395 (1960).
9) D. J. Ward, Petro Chem. Engr. *32*, 11, C42 (1960).
10) R. Billet: Fundamentals of Thermal Separation of Liquids, Verlag Bibliographisches Institut, Mannheim 1962.
11) G. Hochgesand: Diss. T. H. Hannover 1961.
12) E. Bergemann: Lecture at Meeting on Thermal Separation of Gas and Liquid Mixtures, 1970.
13) E. Rose, Ind. Engng. Chem. *33*, 596 (1941).
14) R. Billet, Paper read at 4th International Congress of Chemical Engineering CHISA, Prag, CSSR (1972).
15) R. Billet: CZ-Chemie-Technik 1, No. 9, 427 (1972).

Chapter 11

1) E. Kirschbaum & A. David, Chem.-Ing.-Techn. *25*, 592 (1953).
2) E. Kirschbaum, W. Busch & R. Billet, Chem.-Ing.-Techn. *28*, 475 (1956).
3) H. Brauer, Chem.-Ing.-Techn. *29*, 520 (1957).
4) R. Billet, Chem.-Ing.-Techn. *32*, 544 (1960).
5) W. Nusselt, Ver. dtsch. Ing. *60*, 541, 569 (1916).
6) W. Barth, Chem.-Ing.-Techn. *23*, 289 (1951).
7) R. T. Struck & C. R. Kinney, Ind. Engng. Chem. *42*, 77 (1950).
8) H. Kolling, Chem.-Ing.-Techn. *24*, 405 (1952).

9) R. Billet, Chem.-Ing.-Techn. *32*, 517 (1960).
10) V. V. Kafarov: Contribution to International Symposium on Distillation, Brighton, England (1960).
11) E. H. Smoker, Am. Inst. Chem. Engrs. *34*, 165 (1938).
12) A. Vogelpohl: Lecture at VDI meeting on Thermal Separation of Gas and Liquid Mixtures (1971).
13) E. Krell: Manual of Laboratory Distillation. VEB Verlag, Berlin 1958.
14) S. R. M. Ellis, Laboratory Practice *14*, 815 (1965).
15) E. Müller & H. Stage: Experiments on Vapor-Liquid Phase Equilibria. Springer, Berlin-Heidelberg 1961.
16) E. Hála, J. Pick, V. Fried & O. Vilím : Vapour-Liquid Equilibrium. Pergamon Press, Oxford 1967.

Chapter 12

1) G. Burrows, Trans. Instn. Chem. Engrs. *32*, 24 (1954).
2) S. Langmir, Phys. Rev. *8*, 149 (1916).
3) K. C. D. Hickmann, Ind. Engng. Chem. *29*, 968 (1937).
4) A. Schiffers, Chemiker-Ztg. *86*, 189 (1962).
5) W. Frank & D. Kutsche. Glastechn. Ber. *39*, 167 (1966).
6) W. Frank & D. Kutsche: Manual of Process Control. Krauskopf-Verlag. Mainz 1969.
7) R. Billet, Chem.-Ing.-Techn. *29*, 733 (1957).
8) R. Billet: CZ-Chemie-Techn. 2 (1973) 461.
9) R. Billet: CZ-Chemie-Techn. 3 (1974) 15.
10) R. Billet: Nachr. Chem. Techn. 20 (1972) 475.
11) A program for the system of equations evaluating separations was processed at the BASF Computer Center, rendered in Fortran IV, and evaluated numerically by a Honeywell-Bull Computer Series 6.000 (HB 605).

Chapter 13

1) R. Billet: Trennkolonnen für die Verfahrenstechnik. Verlag Bibliographisches Institut, Mannheim 1969.
2) H. Kassat, Chem.-Ing.-Techn. *38*, 987 (1966).
3) J. Wiegand, in: Dechema-Monographie, Vol. 40. Verlag Chemie, Weinheim 1962.
4) W. Hummel, VDI-Nachrichten *17*, 6 (1971), and CZ-Chemie-Tech. *1*, 63 (1972).
5) J. Hengstenberg, B. Sturm & O. Winkler: Measurements and Control in Chemical Technology. Springer, Berlin-Heidelberg 1960.
6) J. Bleisteiner u. W. v. Mangoldt: Manual of Process Control. Springer, Berlin-Heidelberg 1961.
7) J. G. Dobson, Regelungstechnik *8*, 393 (1960).
8) H. H. Rosenbrock, Trans. Instn. Chem. Engrs. *38*, 280 (1960).
9) H. W. Haines, Ind. Eng. Chem. *52*, 663 (1960).
10) Conference report in Brit. Chem. Engng. 7, 359 (1962).
11) D. E. Lupfer & J. R. Parsons, Chem. Eng. Progr. *58*, 37 (1962).
12) H. Stage: Paper read at the DGF Conference, Heidelberg (1969).
13) G. Bogenstätter u. K. Hengst. Chem. Ing.-Techn. *31*, 425 (1959).
14) J. S. Anderson & J. McMillan: Contribution to International Symposium on Distillation, Brighton (1969).
15) W. L. Luyben: Notes from International Symposium on Distillation, Brighton (1969).
16) W. Y. Sorcek u. H. W. Wilson, Chem. Eng. Progr. *67*, 45 (1971).

Chapter 14

1) G. M. Wilson, Journal of Am. Chem. Soc. *86*, 127 (1964).
2) H. Renon & G. M. Prausnitz, A. I. Ch. E. Journal *14*, 135 (1968).

3) D. Hesse & D. Wolf: Paper read at VDI Meeting on Thermal Separation of Gas and Liquid (1971).
4) V. A. Kuznecov, Chem. Industrie (russ.) *4*, 62 (1965).
5) B. A. Suckov, Chem. Technol. Brennst. (russ.) *2*, 42 (1965).
6) B. Aristovich, Khim. Prom. (russ.) *5*, 62 (1967).
7) V. Aristovich, Int. Chem. Engng. *8*, 1 (1968).
8) H. Ramer, Chem. Techn. *22*, 217 (1970).
9) D. Gelbin, Brit. Chem. Engng. *10*, 301 (1965).
10) O. Nagel, Chemie-Ing.-Techn. *39*, 671 (1967).
11) W. W. Watermann, Hydrocarbon Process. *47*, 155 (1968).
12) G. W. Boynton, Hydrocarbon Process. *49*, 153 (1970).
13) H. W. Brandt & K. H. Reissinger: Paper on EFCE-Conference in Venice (1969).
14) J. C. Mecklenburgh & E. Hartland, Chem. Eng. Sci. *24*, 1259 (1969).
15) O. Nagel, H. Smits, D. Wolf & H. Günzler: Notes from International Symposium on Distillation, Brighton, England (1969).
16) S. O. Megantz, Petroleum Refiner *41*, 169 (1962).
17) L. G. Mitten, Chem. Engng. Progr. Symp. Sci. 60/64/50, 53/59.
18) B. S. Sverschinsky, Zh. Prikl. Khim. (russ.) 1524 (1966).
19) J. W. Gentry, Canad. Journ. of Chem. Engng. *48*, 451 (1970).
20) J. Marec, Chem. Techn. *14*, 405 (1962).
21) R. Schmidt: Dissertation Univ. Karlsruhe (1971).
22) J. W. Lane, Instrumentation Technol. *15*,46 (1968).
23) R. A. Heidemann, Theoret, Chem. Engng. Abstr. *4*, 1399 (1967).
24) V. Gervasiv, Petro Chem. Engr. *41*, 18 (1969).
25) O. Winkler, Verfahrenstechnik *3*, 523 (1969).
26) R. A. JR. Baxley, Instrumentation Technol. *16*, 75 (1969).
27) J. E. Lee, Chem. Anlagen Verf. *3*, 63 (1969).
28) J. E. Lee, Rubber Plastics AGE *49*, 419 (1968).
29) B. Inskip, Measurement and Control 2, 9 (1969).
30) M. Wagner, Chem. Techn. *18*, 348 (1966).
31) M. L. Johnson, Chem. Engng. Progr. *62*, 75 (1966).
32) H. Hentschel: Oel-Zeitschrift f. d. Mineralölwirtschaft, 342 (1971).
33) B. Gay, P. E. Preece,: Matrix Method for Solution of Fluid Network Problems. Fourth Annual Research Meeting of the Institute of Chemical Engineers–April 4–6 (1977).
34) W. R. Patterson, J. W. Ponton, R. A. W. Donaldson: The Estimation of Uncertainty in Complex Networks Using a Modular Process Simulator. Fourth Annual Research Meeting of the Institute of Chemical Engineers–April 4–6 (1977).
35) W. R. Johns: Automatic Generation and Evaluation of Process Flowsheets. Fourth Annual Research Meeting of the Institute of Chemical Engineers–April 4–6 (1977).
36) V. Mahalec, R. L. Motard: Comput. Chem. Enging. Vol 1 p. 149 (1977).
37) V. Mahalec, R. L. Motard: Comput. Chem. Enging. Vol. 1 p. 57 (1977).
38) P. P. O'Neill: Interactive Graphics in Du Pont Plant Design. Fourth Annual Research Meeting of the Institute of Chemical Engineers–April 4–6 (1977).

Chapter 15

1) R. Billet: Verfahrenstechnik *5*, 3 (1974), 65.
2) R. Billet: Paper read at 75th Int. Chemical Plant Engineering Congress, Tokyo 9.10.1975, Session A-III, 193 (1975).
3) R. Billet: 75-12 Distillation and Separation Process Engineering (Japan) *3*, 23 (1975).
4) R. Billet: Paper read at Congress on Chemical Engineering, Amsterdam *1*, 7.1976.
5) R. Billet: Chemie-Technik 5, 397 (1976).
6) A. G. V. Underwood: Chem. Engng. Progr. 44 (1948). 603
7) Programmed in Fortran by J. Weiss in BASF Computer Center.
8) R. Billet: Evaporation Technique, Bibliographisches Institut Mannheim (1965).

Chapter 16

1) R. Billet: Presentation at the 2nd Austrian Session on Process Technology in Graz on March 3 (1973).
2) G. A. Brodie: Mech. and Chem. Eng. Transactions of the Inst. of Engineers, Australia, MC7, 37 (1971).
3) C. Gerber: Chem. Techn. 25, 6 324 (1971).

SYMBOLS AND UNITS

a (mol-%)	activity
a (m^2/m^3)	specific surface of a packing
A (m^2)	area
B (kmol)	charge of still
c (kcal/kg C)	specific heat
C (cost units/m^3)	column cost per unit of volume
C_A (cost units/m^3/sec)	specific column cost referred to unit separation and unit throughput of comparable atmospheric air
C_p (%)	percentual cost
C_S	specific cost factor
C_V (cost units/m^3/sec)	specific column cost referred to unit separation and unit vapor throughput
d (mm)	size of packing
d (kg/m^2/h)	specific distillate quantity
d_c (mm)	diameter of bubble cap
d_h (m)	size of drillings in rectifier plates for passing vapor
d_S (m)	column diameter
D (kmol/h, or kmol/kmol)	mass of overhead product or distillate
D' (kg/m^2 h)	in molecular distillation, mass of distillate referred to evaporating surface
e (m)	intermediate hydraulic radius for dumped packing
E	efficiency
f (kmol/.n^2/h)	specific feed load of a column
F (kmol/h)	feed load
f_i	coefficient depending on thermal condition of feed for rectifying column
F_V (m$^{-1/2}$ s^{-1} kg$^{-1/2}$)	vapor capacity factor of a rectifying column
h (kcal/kmol)	enthalpy of liquid
h (m)	height
h_w (mm)	height of weir
H (m)	height
J (m^3)	liquid content of column
k (kcal/m^2 h C)	heat transfer coefficient
K	constant
l' (1/s m)	specific liquid load of weir
l_W (m)	length of weir
L (kmol/h)	quantity of liquid
L^+ (m^3/h)	quantity of liquid
L^* (m^3/m^2 h)	specific liquid load of column section
L_R (m^3/h)	reduced liquid load
m	ratio of quantities (in thin layer vaporizer)

m^*	coefficient (of stripping line)
n	(coefficient of separation (in partial condensation)
n_c	number of caps on tray
n_h	number of drillings in sieve plate
n_p	number of effective plates
n_{RK}	coefficient of effectiveness of thin-layer rectifier
n_t	number of theoretical stages
n_t	number of theoretical stages for separation
N (kW or HP)	power
Nu	Nusselt number
NTU	number of transfer units
P (mm Hg)	total pressure of vapor mixture
p_c (mm Hg)	partial pressure of component c of vapor mixture
Δp (mm H_2O)	pressure loss of vapor flowing through column
P (mm Hg)	boiling pressure of pure component
Pe	Peclet number
Pr	Prandtl number
q (kcal/m^2 h)	specific flux of heat, or load on heating surface
q_h (kcal/kmol)	heat liberated in rectification of 1 kmol of high-boiling component
q_l (kcal/kmol)	heat absorbed in rectification of 1 kmol of low-boiling component
Q (kcal/h)	heat requirement of continous still
r	reflux ratio of column (ratio of molar reflux quantity and molar overhead quantity)
r_e	ratio of liquid entrained by vapor on to next upper tray, and reflux quantity
R (kmol/h)	reflux quantity at column top
Re	Reynolds number
t (C)	temperature
t_R (s)	duration
s (m)	thickness of wall or layer
S (kmol/h)	sidestream, byproduct
T (K)	absolute temperature
T_S (m)	spacing of trays in column
u	nonuniformity of liquid in packed columns
u_V (m/s)	vapor velocity in the free cross section of a rectifying column
v_A (m^3/m^3/s)	specific volume of column, referred to unit separation and comparable throughput of atmospheric air
v_V (m^3/m^3/s)	specific volume of column, referred to unit separation and unit vapor throughput
V^+, V (kmol/h)	vapor quantity
V_R (m^3/s)	reduced vapor load
w	ratio of liquid quantity at column base and quantity of bottom product
W (kmol/h)	quantity of residue or bottom product
We	Weber number
x (mol-%)	concentration of low boiler in a liquid mixture
y (mol-%)	concentration of low boiler in a vapor mixture
Z (h)	time of rectification in discontinuous work
α	relative volatility
α (kcal/m^2 h C)	individual coefficient of heat transfer
β	slope of operating line

α_{cn}	relative volatility between components c and n of a mixture of *n* components
γ	activity coefficient
ϵ	relative free interstitial volume in packed column
η	efficiency
η (kg h/m)	dynamic viscosity
ϑ (C)	temperature difference
λ (kcal/kmol)	molar heat of evaporation (or condensation)
λ (kcal/m h C)	conductivity of heat
μ (kg/kmol)	molecular weight
ν (cP)	viscosity
χ	flow coefficient
ρ (kg/m^3)	density
σ (dyn/cm)	surface tension of liquid versus its vapor
ϕ (%)	percent openings ratio for rectifying tray
ϕ'	ratio of column diameter versus packing size
ϕ_a	ratio of column section versus operating area outside bubble caps
ϕ_c	ratio of bubble cap area versus column surficial area
ϕ_h	open area of rectifier tray referred to column surficial area
ϕ'_h	open area of sieve plate, referred to perforated area

INTERNATIONAL GLOSSARY

(English–French–Spanish–German–Dutch)

E Separation of liquids
F Séparation des liquides
S Separación de liquidos
G Trennung von Fluessigkeiten
D Scheiding van vloeistoffen

E Pilot installation
F Atelier pilote
S Planta piloto
G Versuchsanlage
D Proefproces

E Production equipment
F Establissement industriel
S Planta industrial
G Industrie-Anlage
D Industrieel apparatuur

E Flowsheet
F Diagramme d'écoulement
S Esquema funcional
G Flussbild
D Stromingsdiagram

E Throughput
F Débit
S Caudal
G Durchsatz
D Doorstroming

E Stage
F Etage
S Piso
G Stufe
D Stoep

E Operation
F Conditions de marche
S Condiciones operatorias
G Betriebsweise
D Werkwijze

E Boiling point
F Point d'ébullinion
S Punto de ebullición
G Spiedepunkt
D Kookpunt

E Boiling range
F Intervalle d'ébullition
S Ambito de ebullidión
G Siedebereich
D Omvang van kookpunt

E Boiling process
F Evaporation
S Evaporación
G Verdamfung
D Verdamping

E Kettle
F Chaudière
S Caldera
G Destillierblase
D Ketel

E Reboiler
F Bouilleur
S Evaporador
G Umlauf-Verdampfer
D Verdamper

E Plate column
F Colonne à plateaux
S Columna de bandejas
G Bodenkolonne
D Plaatkolom

E Bubble cap
F Cloche
S Campana de borbeteo
G Austauschglocke
D Bel

E Weir
F Déversoir
S Rebasadero
G Ueberlaufwehr
D Stuwdam

E Packed column
F Colonne à garnissage
S Columna de relleno
G Fuellkoerper-Kolonne
D Pakkingkolom

E Reflux ratio
F Taux de reflux
S Relación de reflujo
G Ruecklauf-Verhaetnis
D Refluxverhouding

E Efficiency
F Efficacité
S Eficacia
G Wirkungsgrad
D Rendement

E Cut
F Coupe de distillation
S Corte de destilación
G Fraktion
D Fractie

E Sidestream
F Soutirage latéral
S Toma lateral
G Nebenprodukt
D Zijproduct

E Rectifying column
F Colonne de rectification
S Torre de rectificatión
G Rektifizierkolonne
D Rectificatiekolom

E Operating line
F Ligne opératoire
S Linea de operación
G Austauschgerade
D Bedrijfslijn

E Sharpness of separation
F Finesse de séparation
S Eficacia de separación
G Trennschaerfe
D Fijnheid van de scheiding

E Partial pressure
F Pression partielle
S Presión parcial
G Teildruck
D Gedeeltelijk druk

E Mixtures: Binary, ternary, multicomponent
F Mélanges: Binaire, ternaire, multicomposant
S Mezclas: Binaria, Ternaria, de varios componentes
G Genische: Zweistoff, Dreistoff, Mehrstoff
D Mengsels: Binaire, Ternaire, multicomposities

E Instrumentation
F Instrumentation
S Instrumentación
G Messinstrumente
D Meetinstrumenten

Index

www.ingramcontent.com/pod-product-compliance
Ingram Content Group UK Ltd.
Pitfield, Milton Keynes, MK11 3LW, UK
UKHW041132071026
15587UKWH00009B/444

9 780820 602158